Advances in

ORGANOMETALLIC CHEMISTRY

VOLUME 24

CONTRIBUTORS TO THIS VOLUME

Neil G. Connelly

Gerhard Erker

William J. Evans

William E. Geiger

Gregory L. Geoffroy

Wayne L. Gladfelter

Carl Krüger

Gerhard Müller

Daniel B. Pourreau

William N. Setzer

Paul von Ragué Schleyer

Nils Wiberg

Advances in Organometallic Chemistry

EDITED BY

F. G. A. STONE
DEPARTMENT OF INORGANIC CHEMISTRY
THE UNIVERSITY
BRISTOL, ENGLAND

ROBERT WEST
DEPARTMENT OF CHEMISTRY
UNIVERSITY OF WISCONSIN
MADISON, WISCONSIN

VOLUME 24

1985

ACADEMIC PRESS, INC.

(Harcourt Brace Jovanovich, Publishers)

Orlando   San Diego   New York   London
Toronto   Montreal   Sydney   Tokyo

ACADEMIC PRESS, INC.
Orlando, Florida 32887

United Kingdom Edition published by
ACADEMIC PRESS INC. (LONDON) LTD.
24–28 Oval Road, London NW1 7DX

LIBRARY OF CONGRESS CATALOG CARD NUMBER: 64-16030

ISBN: 0-12-031124-0

PRINTED IN THE UNITED STATES OF AMERICA

85 86 87 88   9 8 7 6 5 4 3 2 1

Contents

The Remarkable Features of (η^4-Conjugated Diene)zirconocene and -hafnocene Complexes

GERHARD ERKER, CARL KRÜGER, and GERHARD MÜLLER

Organometallic Metal Clusters Containing Nitrosyl and Nitrido Ligands

WAYNE L. GLADFELTER

The Electron-Transfer Reactions of Polynuclear Organotransition Metal Complexes

WILLIAM E. GEIGER and NEIL G. CONNELLY

Organometallic Lanthanide Chemistry

WILLIAM J. EVANS

Silyl, Germyl, and Stannyl Derivatives of Azenes, N_nH_n: Part II. Derivatives of Triazene, N_3H_3, Tetrazene, N_4H_4, and Pentazene, N_5H_5

NILS WIBERG

Photochemistry of Alkyl, Alkylidene, and Alkylidyne Complexes of the Transition Metals

DANIEL B. POURREAU and GREGORY L. GEOFFROY

X-Ray Structural Analyses of Organolithium Compounds

WILLIAM N. SETZER and PAUL VON RAGUÉ SCHLEYER

Contents

Contributors

Numbers in parentheses indicate the pages on which the authors' contributions begin.

NEIL G. CONNELLY (87), *Department of Inorganic Chemistry, University of Bristol, Bristol BS8 1TS, England*

GERHARD ERKER[1] (1), *Abteilung für Chemie der Ruhr-Universität, Bochum, Federal Republic of Germany*

WILLIAM J. EVANS (131), *Department of Chemistry, University of California, Irvine, Irvine, California 92717*

WILLIAM E. GEIGER (87), *Department of Chemistry, University of Vermont, Burlington, Vermont 05405*

GREGORY L. GEOFFROY (249), *Department of Chemistry, The Pennsylvania State University, University Park, Pennsylvania 16802*

WAYNE L. GLADFELTER (41), *Department of Chemistry, University of Minnesota, Minneapolis, Minnesota 55455*

CARL KRÜGER (1), *Max-Planck-Institut für Kohlenforschung, Mülheim a. d. Ruhr, Federal Republic of Germany*

GERHARD MÜLLER[2] (1), *Max-Planck-Institut für Kohlenforschung, Mülheim a. d. Ruhr, Federal Republic of Germany*

DANIEL B. POURREAU (249), *Department of Chemistry, The Pennsylvania State University, University Park, Pennsylvania 16802*

WILLIAM N. SETZER[3] (353), *Institut für Organische Chemie, Friedrich-Alexander-Universität Erlangen-Nürnberg, D-8520 Erlangen, Federal Republic of Germany*

PAUL VON RAGUÉ SCHLEYER (353), *Institut für Organische Chemie, Friedrich-Alexander-Universität Erlangen-Nürnberg, D-8520 Erlangen, Federal Republic of Germany*

NILS WIBERG (179), *Institut für Anorganische Chemie der Universität München, 8000 Munich 2, Federal Republic of Germany*

[1]Present address: Max-Planck-Institut für Kohlenforschung, Mülheim a.d. Ruhr, Federal Republic of Germany.

[2]Present address: Anorganisch-chemisches Institut der Technischen Universität München, Garching, Federal Republic of Germany.

[3]Present address: Department of Chemistry, University of Utah, Salt Lake City, Utah 84112.

The Remarkable Features of (η^4-Conjugated Diene)zirconocene and -hafnocene Complexes

GERHARD ERKER[1]

Abteilung für Chemie der Ruhr-Universität
Bochum, Federal Republic of Germany

CARL KRÜGER and GERHARD MÜLLER[2]

Max-Planck-Institut für Kohlenforschung
Mülheim a. d. Ruhr, Federal Republic of Germany

I

INTRODUCTION

(η^4-Conjugated diene)transition metal complexes are important as starting materials or reactive intermediates in many catalytic processes (*1*). In addition, they play an increasingly important role as stoichiometric substrates or reagents in organic synthesis (*2*). Metal complexes of general formula (η^4-diene)ML_n undergo a variety of reaction types depending

---

[1] Present address: Max-Planck-Institut für Kohlenforschung, Mulheim a. d. Ruhr, Federal Republic of Germany.
[2] Permanent address: Anorganisch-chemisches Institut der Technischen Universität, München, Garching, Federal Republic of Germany.

upon the nature of the metal fragment ML_n; for example, thermally induced coupling with olefinic substrates leads to allyltransition metal compounds (1). Depending on constitution and electric charge, η^4-diene complexes can also be transformed into η^3-allylmetal species by addition of either nucleophilic or electrophilic reagents (3). Moreover, (conjugated diene)metal complexes in their turn are often easily accessible from η^3-allyl transition metal species by proton or hydride abstraction (4, 5).

This pronounced ability to undergo facile mutual transformations with both allyl (6) and pentadienyl systems (7) is a major reason for the synthetic potential of (η^4-conjugated-diene)transition metal complexes. This diversity in chemical reactivity of η^4-diene complexes of different mononuclear transition metal fragments is contrasted, however, by a rather uniform pattern with regard to a number of other characteristics. Thus, many spectroscopic properties (8), as well as X-ray crystallographic data (9), show a surprising similarity for 1,3-diene complexes of different transition metals. Classification of complexes of conjugated dienes simply as belonging to an olefin-metal complex "subgroup" could be inferred from many of these data. However, notable exceptions to this include the rather small number of metal complexes formed by very reactive conjugated dienes [e.g., cyclobutadiene (10), o-quinodimethane (11, 12), or fulvene systems (13)] and the existence of diene complexes of both main group elements (14, 15) and very electropositive early transition metals (16, 17). (η^4-Conjugated diene) complexes of the "bent" metallocenes bis(η-cyclopentadienyl)zirconium and -hafnium occupy an exceptional position among the latter. Some of their structural, spectroscopic, and chemical properties are so different from those found for the majority of known (conjugated diene)transition metal complexes that a comprehensive review of the synthesis and properties of (η^4-conjugated diene)zirconocene and -hafnocene complexes is justified.

II

FORMATION OF (η^4-1,3-DIENE)ZIRCONOCENES AND -HAFNOCENES

A. *General Considerations*

The potential surface of the parent conjugated diene system, 1,3-butadiene, exhibits two local minima. These correspond to the two possible planar conformations of the C_4H_6 molecule, distinguished by the arrangement of the vinyl moieties relative to their connecting σ-bond, accordingly referred to as *s-trans-* (**1**) and *s-cis*-butadiene (**2**). These

1   **2**

s-trans – butadiene – s-cis

$\underline{1} \rightarrow \underline{2}$ (25°C): $\Delta G^{*} = 7$; $\Delta\Delta G^{°} = 3$ kcal/mol

conformers are rapidly interconverted at ambient temperature. *s-cis*-Butadiene is separated from the *s*-trans form by an activation barrier of only about 4 kcal/mol (*18*). Under equilibrium conditions *s-trans*-butadiene is the markedly favored conformer. According to Mui and Grunwald (*19*), *s-cis*-butadiene is about 3 kcal/mol higher in energy, i.e., free butadiene at ambient temperature adopts the more stable *s*-trans conformation in excess of 99%.

The situation is changed when the butadiene molecule binds to a metal center. Most mononuclear transition metal systems which are sufficiently unsaturated are characterized by a spatial arrangement of valence orbitals that clearly favors η^4-complexation of the less stable, but readily available, *s-cis*-butadiene conformer (*20*). This geometric preference of most transition metal fragments ML_n for binding **2** over **1** is so pronounced that until recently only (*s-cis-η^4*-diene)transition metal complexes were known (*21*). The development of novel routes to prepare conjugated diene complexes of the bent metallocenes Cp_2Zr ($Cp = \eta^5$-C_5H_5) and Cp_2Hf opened the way for the experimental realization of an apparent exception to this rule.

B. *The Unique (s-trans-η^4-Diene)metallocene System*

Some years ago we observed that Group IV metallocenes, i.e., the elusive monomeric units bis(η-cyclopentadienyl)zirconium and -hafnium, readily form 1:1 complexes with conjugated dienes in their *s*-trans conformation (*22*). To our knowledge, these complexes (**3**) are the first and, at the time, only known examples of mononuclear (*s-trans-η^4*-diene)metal complexes, i.e., compounds in which both double bonds of the *s*-trans conformer of a conjugated diene are coordinated to the same transition metal center (*21*). Some examples of this class of transition metal complexes are stable, isolable compounds even at room temperature (*23*).

Two major factors are likely to be responsible for the relative stability of (*s-trans-η^4*-diene)zirconocene and -hafnocene complexes. It seems that the characteristic bonding features of the bent metallocene unit *thermodynamically* favor coordination of the *s*-trans over the *s*-cis conjugated-diene conformer. For bonding to incoming ligands the coordinatively unsaturated 14-electron Cp_2Zr fragment has orbitals available that are arranged only in the x, z molecular plane ($1a_1$, b_2, $2a_1$; see Fig. 1) (24, 25). This coplanar arrangement of valence orbitals allows a much more favorable interaction with the more "linear" π system of the s-trans-butadiene conformer (1) and, as will be discussed in more detail in Section III,B, causes a slightly reduced overlap with 2, as compared with other metal systems ML_n (26).

In addition, the experimental observation of (s-trans-η^4-diene)metallocene complexes is *kinetically* facilitated by the special appearance of the energy profile of the dienemetallocene system. For the parent system there is evidence (27–31) that the stable (butadiene)ZrCp$_2$ isomers [about equal amounts of (s-trans- (3a) and (s-cis-η^4-butadiene)zirconocene (5a) are obtained under equilibrium conditions (22, 23)] are connected through a reactive (η^2-butadiene)zirconocene intermediate (4a) which rapidly equilibrates with s-trans-diene complex 3a, but is separated from the (s-cis-diene)metallocene isomer 5a by a rather substantial activation barrier (Fig. 1).

(Diene)Group IV-metallocene complexes have been prepared by four different routes: treatment of the metallocene dihalide with a conjugated diene dianion (method a) (32, 33), generation of a reactive Cp_2M unit in the presence of a diene as a scavenger (methods b, c) (22, 23, 34), coupling of alkenyl ligands in the coordination sphere of the transition metal

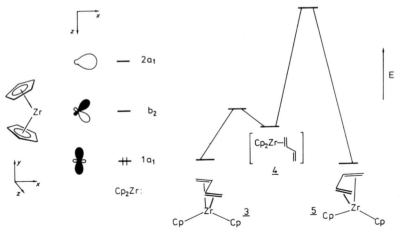

FIG. 1. Valence orbitals of the bent metallocene unit (from ref. 24), and schematic representation of the energy profile of the (butadiene)ZrCp$_2$ system.

(method d) (*35*), and thermally induced exchange of diene ligands (method e) (*22*, *33*).

(*cis*-2-Butene-1,4-diyl)magnesium ("butadiene magnesium") (**7**) is a butadiene dianion (similar to other dienemagnesium adducts) easily available by treating magnesium metal with butadiene in tetrahydrofuran solution (*14*, *15*, *17*) (Scheme 1). Reaction of **7** with metallocene dihalides **6a–6c** yields exclusively (*s-trans*-η^4-butadiene)metallocenes (**3**) if performed under sufficiently mild reaction conditions (*32*). Nucleophilic halide substitution taking place upon treatment of **6** with the bifunctional "butadiene magnesium" reagent could, in principle, give rise to the formation of two different intermediates, i.e., **8** or **9** (*36*). Obviously, **8** forms the (η^2-butadiene)metallocene species **4** in a subsequent step of this reaction sequence, as should **9**. In many instances it has been shown that, under kinetic control, electrophilic attack on a crotyl Grignard reagent never occurs at the α carbon (b) but always takes place at the allylic position (a) (Scheme 1) (*37*). Therefore, the expected product from the intramolecular reaction of the Cp$_2$MCl-substituted crotylmagnesium halide **9** is the

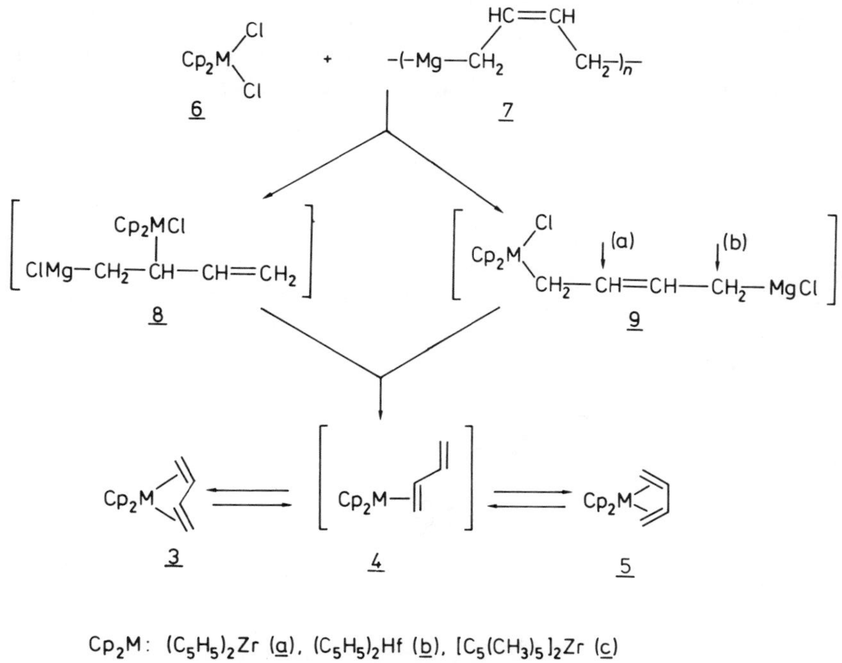

Cp$_2$M: (C$_5$H$_5$)$_2$Zr (**a**), (C$_5$H$_5$)$_2$Hf (**b**), [C$_5$(CH$_3$)$_5$]$_2$Zr (**c**)

SCHEME 1. Mechanistic alternatives for the reaction between metallocene dihalides (**6**) and "butadiene magnesium."

organometallic equivalent of a carbocyclic three-membered ring system
(38), the (η^2-butadiene)metallocene 4.

Subsequent rapid rearrangement then leads to stable (s-trans-η^4-
butadiene)zirconocene and -hafnocene complexes (3). These have
been isolated in high yields from reactions of Cp_2MCl_2 with "butadiene-
magnesium" carried out at sufficiently low temperatures (usually ⩽ 0°C).
They have been obtained free of the (s-cis-butadiene)metallocene isomers
(5) even in cases where the latter, under equilibrium conditions (22, 23),
have turned out to be of equal or higher thermodynamic stability.
Likewise, several (s-trans-diene)zirconocene and -hafnocene complexes
have been obtained as kinetically controlled products upon reductive
dehalogenation of Cp_2MCl_2 with powdered lithium metal in ether or
tetrahydrofuran in the presence of conjugated dienes (23, 34).

Photochemical generation of the bent metallocene (39) has turned out to
be very useful for the preparation of 3. Thus, (s-trans-η^4-butadiene)ZrCp_2
(3a) is obtained in good yield from the photochemically induced reductive
elimination of biphenyl from diphenylzirconocene (10) at low temperatures
in the presence of 1,3-butadiene. This is probably the most versatile and
most widely applicable method to prepare (s-trans-η^4-diene)zirconocenes.
A wide variety of examples of this class of complex [and of (s-cis-
diene)zirconocenes 5, as well] has been prepared by this route using
various substituted conjugated dienes (22, 23) (Scheme 2).

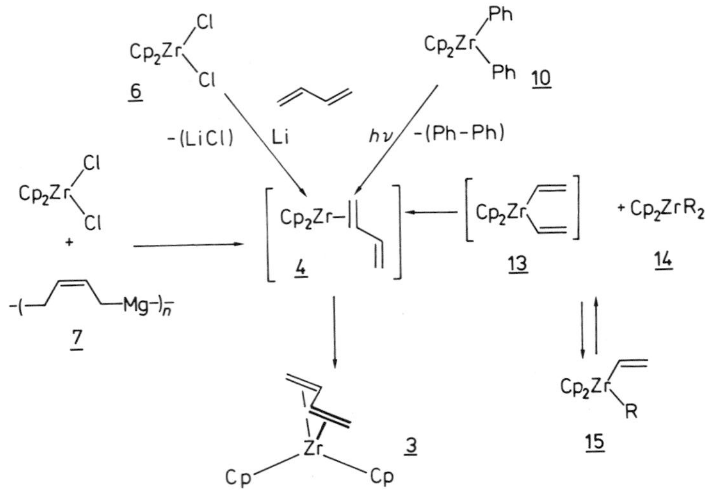

SCHEME 2. Preparative methods for (conjugated diene)MCp_2 complexes, shown for the
example 3a.

The obvious advantage of our photochemical procedure is that it allows the formation of **3** under mild conditions at very low temperature. Thereby, thermally rather unstable derivatives of **3** may be generated and their reactions studied.[3] This enabled us, for example, to show that stereochemical configurations at the diene termini are retained during photolysis, but may be lost by subsequent, rather rapid thermally induced intramolecular rearrangement steps. For example, irradiation of **10** in the presence of pure (Z,Z)-2,4-hexadiene at $-60°C$ yields exclusively (s-trans-η^4-(Z,Z)-2,4-hexadiene)zirconocene [**3h**-(Z,Z)]. At $-40°C$, the stereochemical information at one terminal diene carbon center is lost. A clean

intramolecular rearrangement is observed to give (s-trans-η^4-(Z,E)-2,4-hexadiene)zirconocene [**3h**-(Z,E)]. At higher temperature ($\geq 0°C$), further rearrangement occurs to yield the stable [**3h**-(E,E)]/[**5e**-(E,E) equilibrium mixture (*41*). A similar rearrangement sequence has been observed for (1,4-diphenylbutadiene)zirconocene (**3j**) (*42*).

The formation of an (s-trans-diene)metallocene complex (**3**) by C—C coupling of alkenyl ligands at the metal center has been realized for the most stable member of the (s-trans-diene)zirconocene family, *viz.* (s-trans-η^4-1,4-diphenylbutadiene)ZrCp$_2$ (**3j**). There is evidence for a radical-induced ligand-exchange reaction, starting from readily available alkyl(β-styryl)zirconocenes (*43*). It produces a system of equilibrating σ-complexes from which the bis(alkenyl)zirconocene is rapidly and irreversibly consumed to form the alkenyl coupling product **3j**. There is also evidence that the parent compound, (s-trans-butadiene)zirconocene, can be formed this way, starting from vinyl zirconocene complexes (**15**) (Scheme 2) (*35*).

C. (s-cis-Diene)zirconocenes and -hafnocenes

Many (η^4-diene)metallocenes (**5**) of conjugated dienes, fixed in their s-cis conformation by a rigid carbon framework, have been obtained in high yields by most of the synthetic routes used for the preparation of **3**

---

[3] The first isolated example of a bis(olefin)zirconocene has been prepared by this method. Photolysis of **10** in the presence of 2,3,5,6-tetrakis(methylene)bicyclo[2.2.2]octane yields **11**, which at elevated temperature rearranges to the (conjugated-diene)zirconocene complex **12** (*40*).

TABLE I

s-cis-/s-trans-(Butadiene)metallocene Equilibrium (at 25°C) and
Isomerization Barrier

Cp₂M	(C₅H₅)₂Zr (a)	(C₅H₅)₂Hf (b)	(C₅Me₅)₂Zr (c)	(C₅H₄CMe₃)₂Zr (d)
3/5	45/55	<1/99	40/60	5/95
$\Delta G^{\ddagger}{}_T{}^a$	22.7(10.5)	24.7(60)	$(140)^b$	$(-10)^c$
References	22, 23	27, 32	32	44

^{a} Units: kcal/mol; T (°C) in parentheses.
^{b} $\tau_{1/2}$ ~10 hours.
^{c} $\tau_{1/2}$ ~5 minutes.

TABLE II

3 ⇌ 5 Equilibrium and **3 → 5** Isomerization Barrier for Selected
(Substituted-conjugated-diene)ZrCp₂ and −HfCp₂ Complexes

	Diene	M	Preparation^a	3/5 (25°C)	$\Delta G^{\ddagger}{}_T{}^b$	References
e	Isoprene	Zr	a, b, c	<1/99	19.6(−14)	22, 23
f	Isoprene	Hf	a	<1/99	—	33
g	2,3-Dimethyl-butadiene	Zr	a, b, c	<1/99	18.2(−25)	22, 23
h	2,4-Hexadiene^c	Zr	c	60/40	—	22, 23
i	2,3-Diphenyl-butadiene	Zr	c	<1/99	21.0(+5)	22, 23
j	1,4-Diphenyl-butadiene	Zr	a–e	95/5	—	22, 23
k	1,4-Diphenyl-butadiene	Hf	a, e	24/76	—	33

^{a} Prepared by methods a–e; see text (Section II,B) and Scheme 2.
^{b} In kcal/mol; T (°C) in parentheses.
^{c} Trans,trans isomer.

(see Section II,B). It seems that none of these methods can be used to prepare (*s-cis*-diene)zirconocene and -hafnocene complexes from open-chain conjugated dienes *directly* (*32*). However, the (*s-trans*-η^4-diene)metallocenes (**3**) formed first can be thermally equilibrated readily with their *s-cis*-diene isomers (**5**)[4] (*22, 23, 32*).

The kinetic as well as thermodynamic stability of these dienemetallocene isomers is influenced markedly by variations of the metallocene unit (Table I) (*44*), and the nature and position of substituents on the particular diene ligand involved (Table II). For the butadiene complexes,

---

[4] In all cases looked at, photolysis of **5** at low temperature results in cleanly reforming **3**.

substituents on the η^5-C$_5$H$_5$ rings tend to increase the activation barrier of the **3** → **5** rearrangement, as does substitution of zirconium for haf-

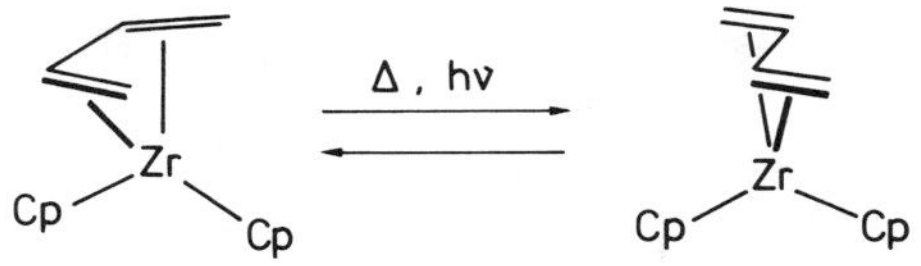

5   **3**

nium. The few sufficiently accurate data presently known also point to a slight thermodynamic preference of the s-cis-diene form for the heavier Group IV metal. At room temperature, (s-trans- and (s-cis-butadiene)metallocene isomers are about equal in energy for zirconium. Substituents at the internal carbon atoms C2 and C3 of the diene chain for presently known examples shift the **3** ⇌ **5** equilibrium completely to the cis form. In contrast, the s-trans-diene isomer is increasingly favored by placing the same substituents (in this study methyl and phenyl groups) at the diene termini C1 and C4.

The observed formation of (s-cis-diene)zirconocene complexes by a high-temperature (60–70°C) exchange reaction of diene ligands can be rationalized easily for the reported examples (**3g, 3j, 3k**) in terms of this pronounced thermodynamic preference for **5** over **3** in (2,3-disubstituted conjugated diene)zirconocene complexes.

D. Dynamic Features of the (s-cis-Diene)metallocenes

In solution, most (s-cis-diene)zirconocene and -hafnocene complexes exhibit dynamic behavior (22, 33, 45). The temperature-dependent appearance of the ^{1}H-NMR spectrum of [s-cis-η^4-1,2,5,6-tetramethyl-3,4-bis(methylene)tricyclo[3.1.0.0^{2,6}]hexane]zirconocene (**5l**) (Fig. 2) illustrates the characteristics of the molecular process responsible for this spectroscopic property of complexes **5**. Inner and outer methylene hydrogens appear as doublets (δ −0.04, 3.58; ^2J = 11 Hz) at sufficiently low temperature. Three methyl singlets appear at δ 1.28 (6 H), 1.43, and 1.59 (3 H each), in addition to two Cp resonances (δ 5.26 and 5.62 in toluene-d_8). Raising the temperature from −55°C results in a rapid broadening of all signals except the singlet at δ 1.28. After pairwise coalescence of methylene, methyl, and cyclopentadienyl resonances, a very simple limiting high-temperature spectrum is obtained with only four singlets at δ 5.44 (Cp), 1.62 (=CH$_2$), 1.49 (—CH$_3$), and 1.28 (—CH$_3$) in an intensity ratio of 10:4:6:6 (45).

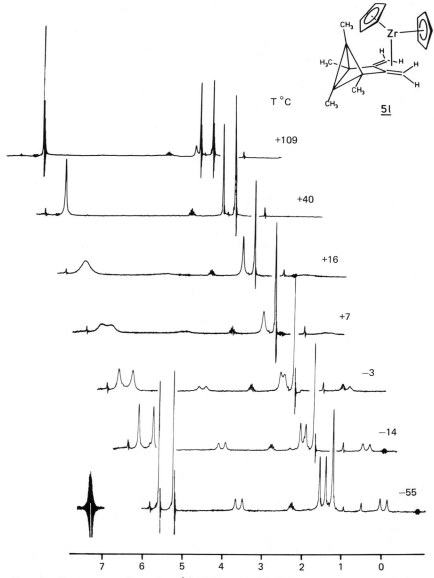

FIG. 2. Temperature dependent ^{1}H-NMR spectrum (60 MHz, toluene-d_8) of [s-cis-η^4-1,2,5,6-tetramethyl-3,4-bis(methylene)-tricyclo[3.1.0.0^{2,6}]hexane]zirconocene (**51**).

This behavior indicates a degenerate rearrangement, rapid on the NMR time scale, by which the characteristic differentiation of diene faces and metallocene sides, created by complex formation, is lost. A process producing this is the migration of the bent metallocene unit from one face of the diene ligand to the other. Decreasing interaction between the metal and the internal carbon atoms C2 and C3 of the diene unit in the course of this reorganization leads to a five-membered metallacyclic structure (not necessarily representing a local energy minimum on the dienemetallocene surface) whose topomerization (*46*), i.e., ring inversion through a probably planar metallacyclopentene transition state (*47*), would readily explain the observed NMR-exchange phenomena (*45, 48, 49*).

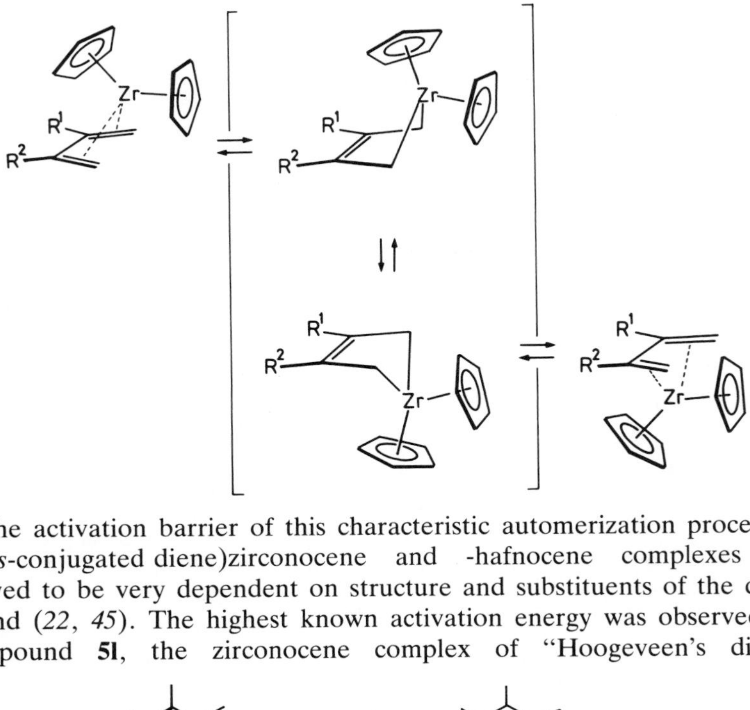

The activation barrier of this characteristic automerization process of (*s-cis*-conjugated diene)zirconocene and -hafnocene complexes has proved to be very dependent on structure and substituents of the diene ligand (*22, 45*). The highest known activation energy was observed for compound **51**, the zirconocene complex of "Hoogeveen's diene"

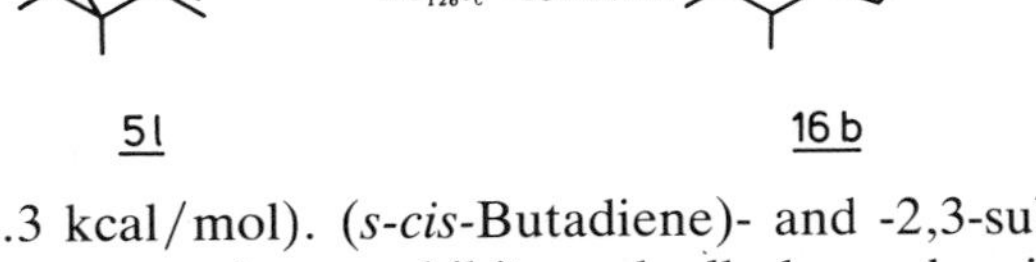

$$\underset{\mathbf{51}}{} \qquad\qquad \underset{\mathbf{16\,b}}{}$$

($\Delta G^{\ddagger}_{11^0C}$ = 14.3 kcal/mol). (*s-cis*-Butadiene)- and -2,3-substituted buta-diene)zirconocene complexes exhibit markedly lower barriers (Table III) (*50–52*). In general, the corresponding bis(η-cyclopentadienyl)hafnium

TABLE III

AUTOMERIZATION ACTIVATION BARRIERS OF SELECTED (s-cis-DIENE)ZIRCONOCENE AND
-HAFNOCENE COMPLEXES (5)

	Diene	M	Preparation[a]	$\Delta G^{\ddagger}_T$[b]	References
l	Tetramethylbis(methylene) tricyclo[3.1.0.0²,⁶]hexane	Zr	b, c	14.3 (+11)	22, 45
m	1,2-Bis(methylene)cyclobutane	Zr	c	12.9 (−19)	34, 50
a	Butadiene	Zr	a, b, c	12.6 (−19.5)	22, 45
b	Butadiene	Hf	a	8.1 (−108)	33, 51
e	Isoprene	Zr	a, b, c	12.1 (−31)	22, 45
g	2,3-Dimethylbutadiene	Zr	a, b, c, e	11.5 (−42)	22, 45
n	2,3-Dimethylbutadiene	Hf	b	8.3 (−105)	51
o	1,2-Bis(methylene)- cyclohexane	Zr	b, c	10.8 (−57)	41
p	1,2-Bis(methylene)- cyclohexane	Hf	b	7.6 (−121)	51
q	2-Phenylbutadiene	Zr	c	12.0 (−30)	42
i	2,3-Diphenylbutadiene	Zr	c	8.0 (−112)	22, 45
r	2,3-Bis(methylene)- bicyclo[2.2.2]octane	Zr	c	7.1 (−131)	52
16b	Tetramethylzirconaindan			6.5 (−140)	45

[a] Prepared by methods a–e; see text (Section II,B) and Scheme 2.
[b] From ¹H-NMR Cp coalescence, kcal/mol; T (°C) is given in parentheses.

compounds seem to undergo this degenerate rearrangement much easier as compared with their zirconium analogues (*33, 51*). Similarly, activation energies almost approaching the limiting value expected for the conformational equilibration of a true metallacylic σ complex, a bis(η-cyclopentadienyl)metallacyclopentene, are observed for the automerization of (s-cis-2,3-diphenylbutadiene)zirconocene (**5i**) ($\Delta G^{\ddagger}_{-112°C} = 8$ kcal/mol) and [2,3-bis(methylene)bicyclo[2.2.2]octane]ZrCp$_2$ (**5r**) ($\Delta G^{\ddagger}_{-131°C} = 7$ kcal/mol). According to an X-ray crystallographic analysis (*12*), a 2-zirconaindan should be a good model for experimentally studying the properties of the hypothetical zirconacyclopent-3-ene. The degenerate ring inversion of 4,5,6,7-tetramethylbis(η-cyclopentadienyl)zirconaindan (**16b**) prepared by thermolysis of **5l** at 120°C, becomes measurably slow on the ¹H-NMR time scale (60 MHz) below −120°C. From the ¹H-NMR Cp coalescence, a value of $\Delta G^{\ddagger}_{-140°C} = 6.5$ kcal/mol is obtained for the activation barrier of the conformational equilibration of this metallacyclopentene model.

In principle, two (possibly cooperative) effects could be responsible for the observed pronounced substituent effect on the automerization rate: different stabilization (or destabilization) of the metallacyclic topomerization transition state and/or rather different ground state structures and relative energies for differently substituted dienemetallocene complexes. A rather extensive X-ray crystallographic analysis (see Section III,B) of (s-cis-η^4-conjugated diene)ZrCp$_2$ (22, 45) and −HfCp$_2$ systems (51) revealed that for most complexes of type 5 the latter effect is by far the most important; a substantial ground state effect is observed.

III

IDENTIFICATION OF THE Cp₂M(DIENE) SPECIES

A. Spectroscopic Methods

IR spectroscopy is of limited practical use for distinguishing between species of types 3 and 5. Both exhibit absorptions typical of a bent metallocene unit (e.g., 3a: ν = 3100, 1440, 1020, 795 cm^{-1}) (53). In the mass spectrum, loss of the diene ligand to give Cp$_2$Zr$^+$ (m/e = 220) seems to represent the dominating fragmentation mode. The mutual rearrangement 3 ⇌ 5 can be followed in many instances by UV/visible spectroscopy. A hypsochromic shift is observed for the long-wave absorption band on going from 5a to 3a [λ_{max} = 430 nm (ε = 875) for 5a; 365 nm (sh) for 3a]. Upon photolysis at low temperature, the deep red (s-cis-η^4-2,3-diphenylbutadiene)zirconocene (5i) [λ_{max} = 452 nm (sh)] rearranges to its (s-trans-diene) isomer (3i), which is dark green in toluene solution (45).

Photoelectron (PE) spectra have been measured for several (s-cis-diene)zirconocenes (5g, 5l, 5s), for one example of the hafnium analogue (5n), and for (s-trans-η^4-butadiene)Zr(C$_5$Me$_5$)$_2$ (3c) (54). For both Cp and Cp* series the highest occupied molecular orbital (HOMO) is separated from subsequent molecular orbitals (MO's) by a substantial energy difference. The observed PE spectra of 5 each exhibit a band around 6–6.5 eV (a_1), well separated from several overlapping bands at 8–9 eV (b_1, b_2, a_2, b_1). The PE spectrum of 3c shows a similar pattern [5.9 eV (a_1); 7.1, 7.5, 7.5, 8.3 eV (a_2, b_2, a_2, b_1)]. The PE bands are in good accord with the calculated MO sequence (26) (given in parentheses), indicating that Koopman's theorem may be used to interpret PE spectra of these complexes.

NMR spectroscopy is a convenient and rather reliable method for identifying dienezirconocene and -hafnocene complexes belonging to the s-cis or s-trans series (23, 33, 45, 48). For most conjugated diene ligands used, isomers 3 and 5 show several characteristically different NMR features. These seem to be quite similar for analogous zirconium and hafnium complexes.

The observed chemical equivalence of the η-C_5H_5 ligands within the Cp_2Zr unit of most complexes of type **3** is determined by the molecular geometry (C_2 symmetry) of the (*s-trans*-diene)metallocene framework [obvious exceptions: Cp_2M(isoprene) (**3e** and **3f**); Cp_2M(2-phenyl-butadiene) (**3q**)]. In contrast, coordination to the *s-cis*-diene conformer always leads to a chemical differentiation of the metallocene Cp ligands. Therefore, all (*s-cis*-diene)zirconocene and -hafnocene complexes exhibit two Cp singlets in the ^1H- or ^{13}C-NMR spectra. However, for the many examples of this class of compounds that show dynamic behavior these two resonances are, of course, only observable in the low temperature limiting spectra.

Characteristic chemical shift differences are observed between configurations **3** and **5**. Relative to **5**, the Cp resonances of (*s-trans*-diene)-metallocene isomers (**3**) are shifted to high field. As shown for butadiene as an example (Table IV), complexation of the *s-cis*-diene conformer to the transition metal causes large chemical shift differences of meso, syn, and

TABLE IV

SELECTED NMR DATA FOR (*s-trans-* AND (*s-cis-*DIENE)ZIRCONOCENE AND -HAFNOCENE COMPLEXES[a,b]

Metal	Conformation	References	H_s	H_m	H_a	$^2J_{H_sH_a}$	$^3J_{H_mH_m'}$
Zr	*s*-trans	23	3.20	2.90	1.20	4.4;[c] 5.0[d]	14.3;[e] 14.0[f]
Hf	*s*-trans	48	3.13	2.83	1.13	4.9	16.5
Zr	*s*-cis	23, 45	3.15	4.85	−0.70	9.7;[c] 10.5[d]	10.6[e]
Hf	*s*-cis	33	3.36	5.04	−0.65	—	11.1
			C1	C2		$^1J_{C1-H}$	$^1J_{C2-H}$
Zr	*s*-trans	23	59.0	96.0		149, 159 (dd)	152 (d)
Hf	*s*-trans	32	57.8	92.8		142, 148 (dd)	147 (d)
Zr	*s*-cis	23, 45	49.0	112.0		138(t)[d]; 144 (t)	165 (d)
Hf	*s*-cis	27, 33	45.0	114.5		140 (t)	156 (d)

[a] For butadiene if not otherwise specified, chemical shift in ppm (δ), J in Hz (multiplicity is given in parentheses).

[b] Designation according to:

$$H_s \diagdown \quad \overset{H_m}{\underset{H_a}{|}} \quad \diagup H_{a'}$$
$$H_a \diagup C=C-C=C \diagdown H_{s'}$$
$$\underset{H_{m'}}{|}$$

[c] 2,3-Dimethylbutadiene.
[d] 2,3-Diphenylbutadiene.
[e] *trans,trans*-2,4-Hexadiene.
[f] *trans,trans*-1,4-Diphenylbutadiene.

anti hydrogens as well as internal (C1) and terminal (C2) carbon centers. The H_{syn} absorption remains practically unaffected on going from **5** to the isomeric (s-trans-η^4-butadiene)zirconocene or -hafnocene (**3**). However, one observes a pronounced high-field shift for H_{meso}, and substantial deshielding for H_{syn}. Similarly, the chemical shift difference between internal and terminal carbon atoms is reduced.

(s-cis- and s-trans-Diene)metallocenes of configurations **5** and **3** show characteristically different J_{HH} coupling constants. Typically, the (s-cis-diene)metallocene isomers exhibit unusually large values for the geminal coupling constant of terminal methylene hydrogens ($^2J \sim$ 10–11 Hz). Corresponding values for **3** are much smaller ($^2J \sim$ 4–5 Hz). The conformational influence is reflected by an increasing numerical value of the vicinal coupling constant between meso hydrogens H_m and $H_{m'}$ on going from **3** to **5**.

B. X-Ray Crystallographic Studies

So far the structures of five (s-cis-η^4-1,3-diene)ZrCp$_2$ complexes and two analogous -HfCp$_2$ compounds of type **5** [1,3-diene = 1,2,5,6-tetramethyl-3,4-bis(methylene)tricyclo[3.1.0.0^{2,6}]hexane (**51**) (*55*); 2,3-bis(methylene)norbornane (**5s**) (*55*); 2,3-dimethylbutadiene (**5g**, **5n**) (*22*, *45*, *56*); 1,2-bis(methylene)cyclohexane (**5o**, **5p**) (*56*); and 2,3-diphenyl-butadiene (**5i**) (*45*)], as well as two (s-trans-η^4-1,3-diene)ZrCp$_2$ complexes of type **3** [1,3-diene = 1,3-butadiene (**3a**) (*22*, *23*) and *trans*, *trans*-1,4-diphenyl-1,3-butadiene (**3j**) (*57*)] have been determined by single-crystal X-ray crystallography.

1. (s-cis-1,3-Diene)zirconocene Complexes

As shown in Fig. 6 for **5g** (see Table II), the coordination of the double bonds of the *cis*-1,3-diene ligands to the bent bis(η^5-cyclopentadienyl)zirconium fragments leads to a pseudotetrahedral geometry of complexes **5**. The Cp—Zr—Cp angle in all the complexes ranges between 123.2 and 126.1°, whereas the angle Cl—Zr—C4 lies between 79.3° in **5i** and 88.7° in **51** (Scheme 3 and Tables V and VI). The diene moiety (plane C1, C2, C3, C4) of the ligands remains essentially planar. All four carbon atoms of the conjugated diene skeleton are clearly within bonding distance of the zirconium center. For all (s-cis-diene)zirconocene complexes the respective bonds between the metal and the diene termini C1/C4 are equal, within standard deviation, as are those to the internal carbon atoms C2/C3. The only slight discrepancy appears to be in the Zr—C1/C4 bond lengths of **5s**, a fact which will be commented on later. There are, however,

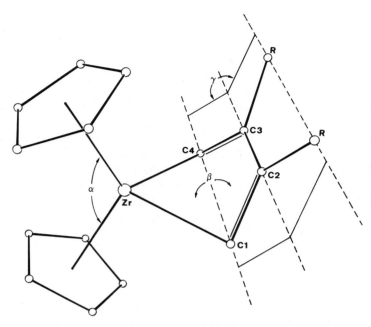

Scheme 3. Definition of some geometric parameters used for the characterization of (s-cis-diene)metallocene complexes (**5**).

TABLE V

Selected Structural Data for (s-cis-Diene)zirconocene Complexes (**5**) and for Bis(cyclopentadienyl)zirconaindan (**16a**)[a,b]

	Compound			
	5l (*55*)[c]	**5s** (*55*)	**5i** (*45*)	**16a** (*12*)
Zr—C1	2.317 (3)	2.307 (2)	2.289 (4)	2.300 (4)
		2.329 (2)		
Zr—C2	2.550 (5)	2.574 (5)	2.714 (5)	2.855 (4)
C1—C2	1.445 (2)	1.454 (3)	1.473 (4)	1.48 (1)
C2—C3	1.398 (3)	1.391 (2)	1.392 (2)	1.42 (2)
C1—Zr—C4	88.7 (2)	87.3 (1)	79.3 (1)	77.4
Zr—C1—C2	81.8 (1)	82.7 (2)	89.7 (4)	95.8
C1—C2—C3	129.6 (3)	128.6 (2)	121.3 (1)	119.4
C3—C2—R	104.4 (3)	106.1 (1)	122.8 (3)	118.0
α[d]	124.4	123.2	123.4	128.4
β	123.4	122.3	119.6	126.9
γ	171.1	178.4	175.3	174.4

[a] See Scheme 3 for definition of dihedral angles and numbering scheme used; bond distances in Å, angles in degrees, estimated standard deviation (esd) values given in units of the last significant figure in parentheses, where available.
[b] Averaged values, where justified.
[c] Mean values from two crystallographically independent molecules.
[d] Angle between normals to planes of Cp rings (Scheme 3).

pronounced differences in bond length between the two types of metal–diene interaction; the Zr—C2/C3 bonds are generally longer than the Zr—C1/C4 bonds. The overall symmetry of (s-cis-η^4-1,3-diene)zirconocene complexes is thus close to m (C_s), the mirror plane bisecting the respective ligands and containing the normals to the Cp rings as indicated in Fig. 6 and Scheme 3. Actually, this molecular symmetry is even imposed crystallographically on **5g**. Deviations arise only from the conformation of the 1,3-butadiene substituents as in **5i** and **5o**, although the overall bonding of the 1,3-dienes remains essentially the same. The structural data collected in Tables V and VI are averaged accordingly.

The above-mentioned unsymmetric bonding of the cis-1,3-diene ligands, with respect to different bond lengths to the diene termini and to the internal C atoms, gives rise to the interesting question of π- versus σ-bonding in the metal-diene linkages. The effect is greatest for **5i** with a difference of 0.43 Å, whereas for **5l** the discrepancy still amounts to 0.23 Å (Fig. 3).

Obviously, there is a decrease in π-bonding compared to that observed in late transition metal conjugated-diene complexes. With the latter, approximately equal metal–carbon interactions are found, with a tendency

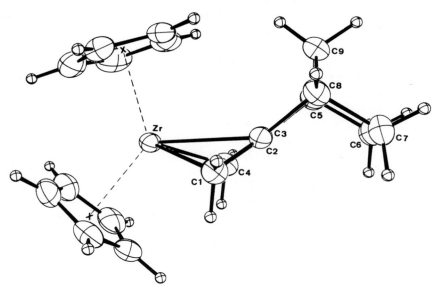

FIG. 3. "Side on" ORTEP representation of [η^4-2,3-bis(methylene)bicyclo[2.2.1]heptane]-zirconocene (**5s**). Zirconium is bonded to the exo side of the bicyclic ligand. Zr—C bonds to C2 and C3 are noticeably longer than those to C1 and C4.

of the metal–C2/C3 bonds to be shorter than those to the diene terminal carbons. The bond lengths are in a range typical for π-bonded systems (58–67). In the (s-cis-diene)zirconocene complexes **5**, however, the bonds between zirconium and the diene termini more closely resemble Zr—C σ—bonds. The observed bond lengths between 2.279(2) Å in **5o** and 2.329(2) Å in **5s** fit well into the overall pattern observed in a variety of Zr—C σ-bonds (12, 68–74). Although slightly longer than the 2.251(6) and 2.277(4) Å separations found in bis(η^5-indenyl)dimethylzirconium (73) and dimethylzirconocene (68), respectively, they are even shorter than those with bulky substituents, as in $Cp_2Zr(CHPh_2)_2$ [2.39(1) Å (69)] or $Cp_2ZrPh[CH(SiMe_3)_2]$ [2.329(6) Å (70)]. Also, in cyclic systems such as **17** or **18**, the Zr—C σ-bonds are appreciably longer [**17**:2.41(1) Å (27); **18**:2.311(7) Å (29)]. There is, however, a striking similarity to the

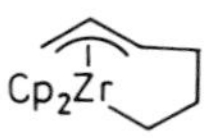

| | 17 | 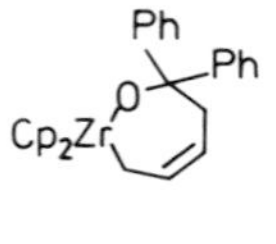 | 18 | 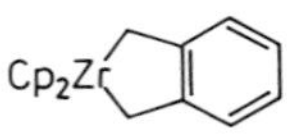 | 16a | 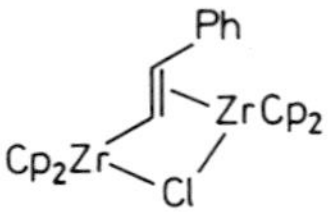 | 19 |

zirconaindan complex **16a** [d(Zr—C1/C4 = 2.300(4) Å (12)], which is a good model for a metallacyclic zirconium compound (see Table V and Fig. 4).

Comparison with **16** however shows a noteworthy difference. Whereas in the latter the Zr—C2/C3 distances are 2.855(4) Å, and have to be regarded as essentially nonbonding, these distances in the compounds **5(l, s, g, o, i)** lie between 2.550(5) and 2.714(5) Å. Although there are no examples of simple $Cp_2Zr(\eta^2$-olefin) moieties that might yield reference data for Zr—C π-bonds, these values might be compared with those found in (s-trans-η^4-1,3-diene)zirconocenes described in Section III,B, 3. Also, the bond lengths to the η^5-bonded cyclopentadienyl rings (~ 2.55 Å) come close to the majority of the Zr—C2/C3 distances in **5** (Tables V and VI), as do those for the π-bonds in **17** (27) or **19** (43). To a first approximation, zirconocene complexes **5** can thus be described as metallacyclic σ^2,π complexes (B). Confirmation for such a bonding model is obtained from

an examination of the C—C bond lengths in the ligand diene moiety.

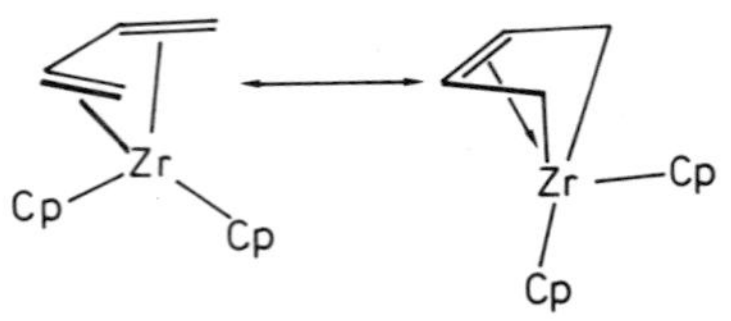

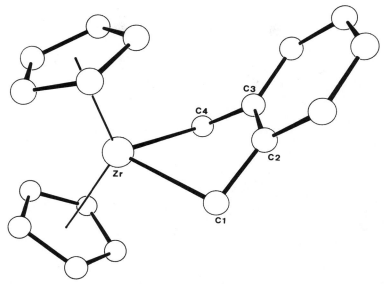

FIG. 4. Perspective view of the molecular geometry of bis(η-cyclopentadienyl)-zirconaindan (**16a**) as determined by X-ray crystallography (*12*).

The observed long–short–long sequence for the coordinated cis ligands (Table V) is the reverse of that found in the free ligands. Thus, for 2,3-dimethylbutadiene, values of $d(C\!\!=\!\!C) = 1.349(6)$ Å and $d(C\!\!-\!\!C) = 1.504(5)$ Å have been determined by electron diffraction (*76*). A comparison with late transition metal conjugated-diene complexes (*58–67*) is again informative. In a systematic study of iron tricarbonyl complexes of cyclic conjugated dienes by Cotton and others (*63*), the central C—C bonds were found to be marginally shorter by about 0.02 Å than the outer bonds. The trend to an equalization of diene bond lengths upon complexation is more pronounced in the zirconocene complexes, with the central C—C bonds being shorter by up to 0.08 Å. The reasoning developed first by Mason and others (*64, 77–80*) for a theoretical understanding of the bonding in transition metal complexes of unsaturated hydrocarbons would indicate a substantial degree of back donation from filled d_π orbitals to the lowest-lying π^*-antibonding level of the 1,3-diene. An analysis of the bonding capabilities of bent bis(cyclopentadienyl)transition metal fragments by Lauher and Hoffmann (*24*) showed that these are determined largely by three orbitals ($1a_1$, b_2, and $2a_1$; see Fig. 1) all lying in the plane bisecting the Cp—M—Cp angle. The primary metal acceptor orbitals are the $2a_1$ and the b_2, while the $1a_1$ is largely nonbonding in

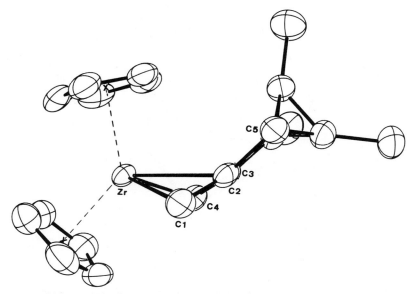

Fig. 5. "Side on" ORTEP representation of the molecular geometry of **5l**. The methyl group at C5 has been omitted for clarity.

Cp_2ML_2 complexes with σ-bonding ligands. It is this orbital, however, that is ideally situated for a back-bonding interaction with the lowest-lying antibonding orbital (ψ_3) of the 1,3-diene (26). Moreover, since the $1a_1$ orbital is directed mainly to the diene termini, rehybridization of C1/C4 from sp^2 to sp^3 would greatly facilitate overlap, thus leading to the observed overall geometry with pseudotetrahedral diene termini.

Another noteworthy feature in the structure of (1,3-diene)zirconocene complexes follows directly from this bonding description. In all complexes, the substituent carbon atoms at C2/C3 do not lie in the plane determined by the diene carbon atoms, but their linkages with C2/C3 are significantly bent *toward* the metal center. This "substituent angle" γ (defined in Scheme 3) may even amount to 171.1° (in **5l**), equivalent to an 8.9° bending of R groups toward zirconium (Fig. 5) (55). The effect is similar to the observed bond bending at the meso carbon atom of η^3-allyl complexes (64, 81–85), and may be observed in late transition metal-butadiene complexes as well.[5] In line with the above reasoning, the antibonding interaction between the π orbitals at C1/C4 and C2/C3, on the diene face opposite to the metal, should increase upon rehybridization of C1/C4 to allow for better overlap with the Zr-$1a_1$ orbital. The observed substituent bending

---

[5] (*o*-Quinodimethane)Fe(CO)$_3$, γ = 172.1° (*11*); [(2,3-dimethylbutadiene)Co(CO)$_2$]$_2$, γ = 176° (*58*).

toward the metal clearly diminishes this repulsive interaction. It is interesting to note that the effect is greatest for **5l** despite the relatively bulky ligand (Fig. 5). In **5s**, however, the effect is almost nonexistent. This might be due to a repulsive interaction between the cyclopentadienyl ring and H_{syn} at the bridging C9 atom of the norbornane ligand, as shown in Fig. 3. This interpretation finds support in the conformation of the Cp ring in question and the slight asymmetry in Zr—C1/C4 bond lengths, the discrepancy being clearly outside the standard deviations. The latter could well be caused by the same steric interaction.

The σ^2,π character of the (1,3-diene)zirconocene complexes of type **5**, as inferred from the results of crystallography, is also evident from the envelope shaped Zr,C1,C2,C3,C4 five-membered metallacycle. Actually, the "angle of fold" β, as listed in Table V (see also Scheme 3), has been used for a comparison of the extent of the σ character in such complexes. However, this value might be influenced by the drastically different geometrical requirements imposed on the angles at C2/C3, depending on whether C2 and C3 are part of rigid bi- and tricyclic systems as in **5l** or **5s**. On going from **5l** to the zirconacyclopentene model **16**, the only significant trend in the structural parameters collected in Tables V and VI appears to be the Zr—C2/C3 distances, the angles Zr—C1—C2/Zr—C4—C3 at the diene terminal carbons, and, to a lesser extent, the C1—C2/C3—C4 bond lengths. Neither the C2—C3 bond lengths, nor the angle β provide definite clues as to the varying extent of metal-to-ligand σ- versus π-bonding. If the first parameters are taken as a sensitive measure of σ^2,π-bonding, a significant increase in metallacyclopentene character is suggested by X-ray crystallographic data upon conversion from **5l** to **5i** and **16**. These pronounced ground state effects appear to exercise a direct influence on reaction rates of these compounds (see Sections II,D and IV,A).

The factors influencing the varying metallacyclopentene character, however, are not as easily apparent from crystallographic data. Clearly, for **16** the aromaticity of the fused benzene ring will effectively prevent the π orbitals at C2/C3 from participating significantly in additional π-bonding to the metal center. Similar arguments might apply for **5i**, where the phenyl groups are tilted by 43.9 and 46.3° against the C1,C2,C3,C4 plane (*45*). A remaining conjugation via the internal C2/C3 atoms could then account for the observed Zr—C2/C3 bond length elongation.

2. Structural Comparison Between Analogous Cp₂Zr- and Cp₂Hf (s-cis-diene) Complexes

The synthesis and structural characterization of the first (*s-cis*-1,3-diene)-HfCp₂ complexes [1,3-diene = 2,3-dimethylbutadiene (**5n**) and 1,2-dimethylenecyclohexane (**5p**)] (*56*) yield valuable reference material, especially on the nature of the metal-to-carbon bonds. This is particularly important, since the differences between Hf—C and Zr—C bonds have been the subject of controversy (68, 86). A detailed comparison with the zirconium analogues should thus allow a deeper insight into the factors determining the bonding mode exhibited in such complexes.

Structural results on **5n** and **5p** reveal both compounds to be closely related to their Zr analogues **5g** and **5o** (56). In particular, the metal–diene portions of the molecules, which are of prime interest for the scope of this article, are truly isostructural (Tables V and VI). As Fig. 6 for **5g** (Zr) and **5n** (Hf) vividly shows, the similarities between both compounds even extend to the thermal behaviour of most atoms. For compounds **5o** and **5p**, some minor differences arise from the rotational and conformational lability of the Cp-rings and the six-membered ring system of the ligand. They leave the overall geometrical parameters of the metal–diene framework essentially unchanged, however, and are further discussed in Kriger et al. (56).

Despite the striking similarities between both compounds and their Zr analogues, there exist subtle differences within the metal-diene frameworks, as may be inferred from Tables V and VI. These are highly significant, however, as they allow (1) an assessment of the differences between Hf—C and Zr—C bonds, and (2) an evaluation of the degree of σ-

TABLE VI

A Comparison of Selected Structural Data for (s-cis-2,3-Dimethylbutadiene)- and [1,2-Bis(methylene)cyclohexane]zirconocene (5g, 5o) and -hafnocene (5n, 5p) Complexes[a,b,f]

	Compound			
	5g (Zr)	**5n** (Hf)	**5o** (Zr)[c]	**5p** (Hf)[c]
M—C1	2.300(3)	2.267(5)	2.279(2)	2.255(8)
M—C2	2.597(3)	2.641(5)	2.635(6)	2.72(1)
C1—C2	1.451(4)	1.472(8)	1.465(3)	1.469(8)
C2—C3	1.398(4)	1.378(8)	1.379(4)	1.373(7)
C1—M—C4	80.2(1)	80.3(2)	82.2(4)	80.8(4)
M—C1—C2	84.4(2)	87.3(3)	86.6(2)	91.2(6)
C1—C2—C3	122.7(3)	121.7(5)	123.5(2)	121.8(4)
C3—C2—R	121.8(3)	122.1(5)	120.5(2)	121.0(4)
α^d	123.9	124.3	125.8	125.3
β^e	112.8	116.5	118.8	124.1
γ^e	172.9	171.1	173.9	174.6

[a-d] See footnotes to Table V.
[e] See Scheme 3.
[f] From refs. 45 (**5g**) and 56 [**5**(**n**, **o**, **p**)].

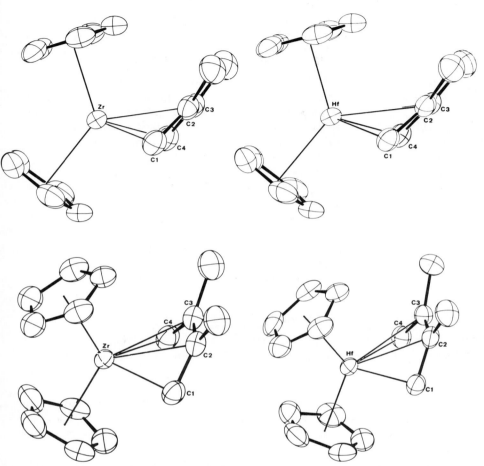

FIG. 6. Perspective representations of the molecular geometries of **5g** (Zr) and **5n** (Hf) in two different orientations showing the close similarities between both compounds. Thermal ellipsoids (ORTEP) are drawn at the 50% probability level.

versus π-bonding among different Hf compounds, as well as a comparison with their Zr analogues.

Differences in atomic radii between Zr and Hf [1.45 *versus* 1.44 Å (*87*)] suggest that covalent bonds of these elements will be of comparable lengths, with a tendency of those to Hf to be shorter. This is what is actually observed in the species **5g, n, o, p**. In the preceding section the Zr—C1/C4 bond lengths were shown to be independent of the varying σ versus π character of the diene complexes. The slightly shorter Hf—C1/C4 bond lengths of 2.267(5) (**5n**) and 2.255(8) Å (**5p**) should thus be reliable comparisons in an assessment of bond length differences. This is a substantiation of observations made recently on Hf—CH_3 bonds (*68*). It is supported by the Hf—$C(\eta^5)$ bond lengths to the Cp rings, which are also slightly shorter (*56*). The Hf—C2/C3 bond lengths of 2.641(5) and 2.72(1) Å are longer than their Zr counterparts in **5g** and **5o**. The observed shortening of M—C bonds on going from Zr to Hf renders these findings especially important. It implies that the Hf complexes examined so far show a distinctly greater σ/π ratio in the metal–diene bonds. Furthermore, the ligand in **5p** should be more σ-bonded than that in **5n**. It should be kept in mind, however, that C2/C3 are still in a range to be π-bonded to the central Hf atoms. These observations may be summarized as follows: (1) Hf—C bonds are slightly shorter than Zr—C bonds; (2) (*s-cis*-η^4-diene)hafnocene complexes show the same bonding characteristics as do their zirconocene analogues; (3) the σ/π ratio in the diene bonding is shifted slightly toward a larger σ character in the case of Hf; (4) the ligand dependency of this σ/π ratio is the same for the hafnocene complexes examined so far as for the respective zirconocene complexes.

3. (s-trans-η^4-Conjugated diene)zirconocene Complexes

The observation that in certain cases the thermodynamic stability of mononuclear zirconocene complexes of *s-trans*-1,3-dienes could be equal to or even higher than that of the corresponding *s*-cis analogues (*21*) offered for the first time the possibility of investigating these complexes by X-ray crystallography (*22, 23, 57*). The most important structural data obtained so far on the complexes **3a** and **3j** are collected in Table VII. Figure 7 shows a model of **3j** (*57*). The zirconium atoms again are pseudotetrahedrally coordinated. The dienes are located in a plane roughly bisecting the Cp—Zr—Cp group. In contrast with the mirror symmetry of most *s-cis*-1,3-diene complexes, the molecular symmetry of **3a** and **3j** is approximately 2 (C_2). The conjugated dienes are bonded through both double bonds to the single zirconium atom. The bond length pattern of the Zr—C and C—C bonds is the reverse of that observed in the *s*-cis complexes. Thus the bonds to the terminal carbon atoms C1/C4 are longer than those to the internal ones C2/C3. The former come close to the values of π-bonded cyclopentadienyl rings. In their short–long–short sequences the diene moieties more closely resemble those observed in free 1,3-dienes, as does their trans geometry (*18, 19*). There is, however, a marked deviation from planarity. The torsion angle (*88*) C1,C2,C3,C4 in **3j** amounts to +126.1°, equivalent to a tilt of the C1—C2 and C3—C4 bonds of ~ 27° *toward* the Zr atom.

TABLE VII

SELECTED STRUCTURAL DATA FOR
(s-trans-η^4-BUTADIENE)ZIRCONOCENE (**3a**) AND
(s-trans-η^4-1,4-DIPHENYLBUTADIENE)-ZIRCONOCENE (**3j**)[a]

	3a	**3j**[b]
Zr—C1	2.477(4)	2.50
Zr—C2	2.331(8)	2.38
C1—C2	c	1.40
C2—C3	c	1.48
α[d]	126.8	126.3
C1—Zr—C4	92.6(2)	92.8
C1—C2—C3—C4[e]	c	126.1

[a] From refs. 22 and 23 (**3a**) and 57 (**3j**); bond distances in Å; angles in degrees, averaged values are given when justified.

[b] Mean values from two crystallographically independent molecules.

[c] High thermal parameters for C2 and C3 result in artificially shortened C—C bond lengths and distorted angles within the ligand; see text for discussion.

[d] Angles between normals to planes of Cp rings.

[e] Angle between directions of C1—C2 and C3—C4 in projection down C2—C3. The C1—C2 and C3—C4 bonds are tilted *toward* the zirconium atom.

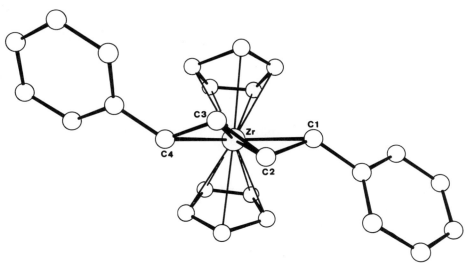

FIG. 7. Perspective view of the molecular geometry of (s-trans-η^4-1,4-diphenyl-butadiene)ZrCp$_2$ (**3j**) as determined by X-ray crystallography (57). Only one of the two crystallographically independent species is shown.

In **3a**, a slight disorder of the C2/C3 atoms that could not be resolved led to abnormally high thermal parameters of C2/C3, thus ill defining the exact geometry of these atoms (21). However, the excellent agreement between **3a** and **3j**, with regard to the geometrical parameters at C1/C4 and Zr, as well as the marked differences when compared to the s-cis complexes, leave little doubt as to the very nature of **3a** as a true s-trans complex.

IV

CHEMISTRY OF THE (s-cis/s-trans-η^4-DIENE)METALLOCENE ISOMER PAIRS

Isomeric (s-cis- and (s-trans-η^4-conjugated diene)zirconocene and -hafnocene complexes exhibit pronounced differences in their characteristic structural data as well as their spectroscopic features. These differences exceed by far the consequences expected to arise simply from the presence of conformational isomers of the 1,3-diene unit. While (s-trans-butadiene)-zirconocene (**3a**) shows a behavior similar to a transition metal olefin π-complex, the (s-cis-diene)ZrCp$_2$ isomer **5a** exhibits a pronounced alkylmetal character (23, 45). Typical features are best represented by a σ^2,π-type structure for **5** (55). However, the distinctly different bonding situation of the butadiene π-system/bent-metallocene linkage is not only reflected in differences in physical data between the dienemetallocene isomers **3** and **5**, but also gives rise to markedly different chemical behavior. Three examples of this are discussed in this section: the reactions of the **3/5** isomeric mixture with carbon monoxide, ethylene, and organic carbonyl compounds.

A. Carbonylation of (Conjugated diene)zirconocene Complexes

By the nature of its molecular mechanism, the carbonyl-insertion reaction represents a typical reaction mode of σ-alkyltransition metal complexes. Formation of the new C—C σ-bond takes place during a 1,2-alkyl-migration step, transforming an alkylmetal carbonyl moiety [cis-M(CO)R] into an acylmetal unit (M—COR) (89). In general, (s-cis-diene)-zirconocene complexes **5** appear to exhibit a substantial alkylmetal character (90). Therefore, it is not too surprising that some members of this class of compounds [in contrast to most other dienetransition metal complexes (91)] react with carbon monoxide with C—C bond formation (45). However, as demonstrated by X-ray structural data for **5** (Tables V

TABLE VIII

REACTION OF (s-cis-CONJUGATED DIENE)ZIRCONOCENE
COMPLEXES WITH CARBON MONOXIDE—RELATIVE
CARBONYLATION RATES AND PRODUCT YIELDS

	Compound				
	5l	5a	5g	5i	16b
Ketone[a]:	—	11[b]	60	69	64
k_{rel}:	1	2.5	5	$\geqslant$250	$\geqslant$5000

[a] Isolated yield in %.
[b] 1:1 Mixture of 2- and 3-cyclopentenone.

and VI), as well as by the activation barriers of the automerization process (Table III), the σ-complex character of 5 is quite different, depending on the substituents on the conjugated-diene ligand. This turns out to be important for the carbonylation reaction.

For these complexes, the rate of carbonylation increases rapidly with increasing σ-alkyl character. Thus, (s-cis-butadiene)ZrCp$_2$ (5a) is carbonylated 2.5 times faster than the zirconocene complex of 1,2,5,6-tetramethyl-3,4-bis(methylene)tricyclo[3.1.0.0^{2,6}] hexane (5l) (ambient temperature and 1 bar CO pressure) taken as a standard. Introduction of methyl groups at the internal carbon centers C2/C3 of the diene chain increases the carbonylation rate by a factor of 5 with phenyl groups at these positions increasing the rate by a factor of 250 (Table VIII). The organometallic reaction products have yet to be isolated or completely identified. Ultimately, zirconium enolate complexes (20) are probably formed in these

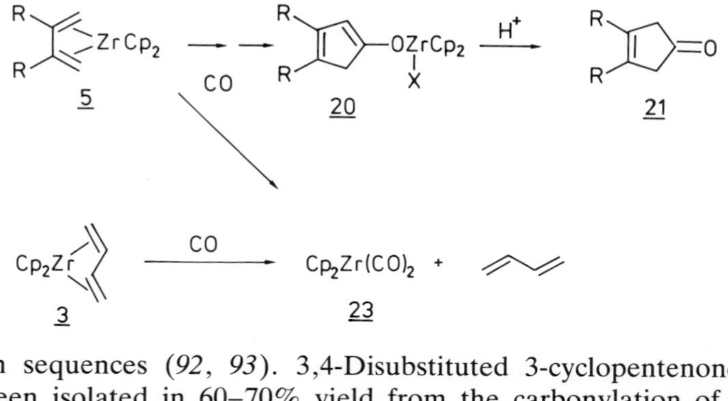

reaction sequences (92, 93). 3,4-Disubstituted 3-cyclopentenones (21) have been isolated in 60–70% yield from the carbonylation of 5g and 5i after hydrolysis (45). Starting from the parent compound, (s-cis-butadiene)zirconocene (5a), a 1:1 mixture of 2- and 3-cyclopentenone

(**21a** and **22a**) is obtained with a combined yield of 11%. The residual
σ-complex character of **5l**, the (*s-cis*-diene)zirconocene used as a reference
in this series, is not sufficient to allow C—C coupling between the diene
ligand and the CO molecule. As the result of a very slow carbonylation
reaction, the diene ligand is set free and the zirconocene dicarbonyl **23** (*94*)
is formed.

In view of these observations, it is not at all surprising that (*s-trans*-η^4-
butadiene)zirconocene (**3a**), which resembles a "normal" olefintransition
metal π complex far more than its *s*-cis isomer **5a**, does not react with
carbon–carbon bond formation when exposed to carbon monoxide. Instead, zirconocene dicarbonyl and butadiene are formed in a very slow
carbonylation reaction.

B. Olefin Insertion versus Metallacyclic Ring Closure— the Reaction of (Butadiene)zirconocene and -hafnocene with Ethylene

The unique antagonistic features of the (butadiene)zirconocene isomers
3a/5a have been used as a probe for the elucidation of organometallic
reaction mechanisms. In some cases it was possible to distinguish between
mechanistic alternatives by simply allowing the isomeric substrates **3** and **5**
to compete for a reagent. An example is as follows. Transition metal-
induced C—C coupling between a conjugated diene and an olefin can
occur by two basically different types of reaction sequence. Either a new
C—C bond can be formed by olefin insertion into a metal–carbon bond of
a (σ-allyl)M-type intermediate (**24**) (*95*), or, alternatively, the alkene may

be taken up by a coordinatively unsaturated (η^2-diene)metal π complex
(**25**). A consecutive electrocyclic ring-closure reaction (*96*) then yields
metallacyclic C—C coupling products (**27** and **28**). A simple competition
experiment, i.e., comparison of the rates of the reaction of the (*s-cis/s-trans*-
butadiene)MCp$_2$ pairs of isomers (**5/3**) with ethylene, provides sufficient
information to decide which of these mechanistic alternatives is favored for

the C—C coupling between butadiene and ethylene at the early transition metal center.

Surprisingly, competition of **3a/5a** for ethylene under various reaction conditions gave no indication at all of C—C coupling via ethylene insertion into a zirconium-to-carbon σ-bond (27). At low temperatures, i.e., under conditions when the (s-trans-/s-cis-butadiene)zirconocene equilibration (**3a ⇆ 5a**) is sufficiently slow (22, 23), it is only the (s-trans-η^4-diene)metallocene isomer (**3a**) that reacts with ethylene. With increasing temperature, increasing amounts of (s-cis-butadiene)ZrCp$_2$ (**5a**) are also consumed, but only at a limited rate that allows restoration of the **3 ⇄ 5** equilibrium, which was disturbed by the rapid reaction of **3a** with the added alkene. Two organometallic reaction products are formed. Incorporation of 2 equivalents of ethylene and liberation of the butadiene lead to bis(cyclopentadienyl)zirconacyclopentane (**30a**). In a competing reac-

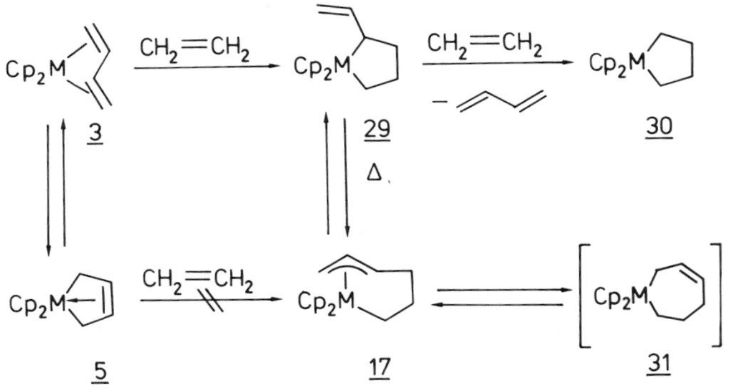

M = Zr (**a**), Hf (**b**)

tion sequence, dominating at high temperature, uptake of 1 equivalent of ethylene and C—C coupling with the butadiene ligand yield the metallacyclic π-allylzirconocene complex **17a**.

The actual reaction path followed becomes more evident if the corresponding hafnium complexes are used as substrates. In this case the crucial intermediate **29**, which can react to form either product **30** or **17**, can be isolated. The thermodynamically favored (s-cis-butadiene)hafnocene (**5b**) (32) turns out to be inert toward ethylene under the conditions applied. Even heating to 120°C in an ethylene atmosphere (1 bar) for several hours does not result in consumption of the metal complex. In contrast, (s-trans-η^4-butadiene)hafnocene (**3b**) rapidly takes up 1 molar equivalent of C$_2$H$_4$ even at -10°C to yield a C—C coupling product, i.e., the five-membered metallacyclic σ-allylhafnocene complex **29b**. Above 0°C, vinylhafnacyclopentane reacts with additional ethylene to form bis(cyclopentadienyl)hafnacyclopentane (**30b**) and free butadiene. In the absence

of ethylene, σ-allylhafnocene (**29b**) rapidly rearranges to the thermodynamically favored isomeric metallacyclic π-allylmetallocene complex **17b**. Rapid
equilibration between **17** and **29**, and also with their seven-membered
metallacyclic σ-allylmetallocene isomer **31**, is implied from the dynamic
NMR behavior of the π-allylzirconocene and -hafnocene complexes **17a** and
17b (27).

In a beautiful series of experiments *Nakamura et al.* showed that
coupling of various organic π systems with isoprene at the Cp_2Zr unit often
occurs regioselectively (28). As exemplified by the product **17e**, identified

| 5e | 4e | 17e |

by X-ray crystallography, C—C coupling at C4 of the isoprene unit
predominates in reactions of (*s-cis*-isoprene)zirconocene with alkenes and
alkynes. The observed regiochemical preference for **17e** has been explained as reflecting basically the steric factors of an olefin insertion
reaction of (*s-cis*-isoprene)zirconocene. However, this is quite unlikely in
view of the chemistry of the related **3a/5a** system (see above). Rather, it
may indicate that of the two possible (η^2-isoprene)MCp$_2$ species, the
intermediate **4e**, leading to coupling of the less substituted isoprene C=C
unit, is formed along the dominating reaction path. This seems to be a very
important factor determining the regiochemical outcome of C—C coupling
reactions in these systems in general (see Section IV,C).

C. *Reactions with Ketones and Aldehydes*

The pronounced alkylmetal character of (*s-cis*-butadiene)zirconocene
should make the diene ligand in **5a** much more amenable to electrophilic
attack than in the isomeric (*s-trans*-η^4-butadiene)ZrCp$_2$ (**3a**). At ambient
temperature, reactions of the (*s-cis*-/*s-trans*-butadiene)zirconocene equilibrium mixture (**5a** $\leftrightarrows$ **3a**) with ketones and aldehydes proceed rapidly
to yield organometallic reaction products **18**. The expected organic
C—C-coupling products **32** and **33** are formed in high yield upon hydrolysis (29). An X-ray structure determination confirms the formation of the
seven-membered metallacyclic σ-allylzirconocene **18a**, which is the expected "direct" reaction product of the carbonyl addition reaction of
(*s-cis*-butadiene)zirconocene (**5a**). However, despite the fact that the
cis-configurated C=C bond (connecting C2 and C3) of the alleged starting
material **5a** is retained in the organometallic reaction product, it is
surprising that **18** is not formed in this apparently obvious way.

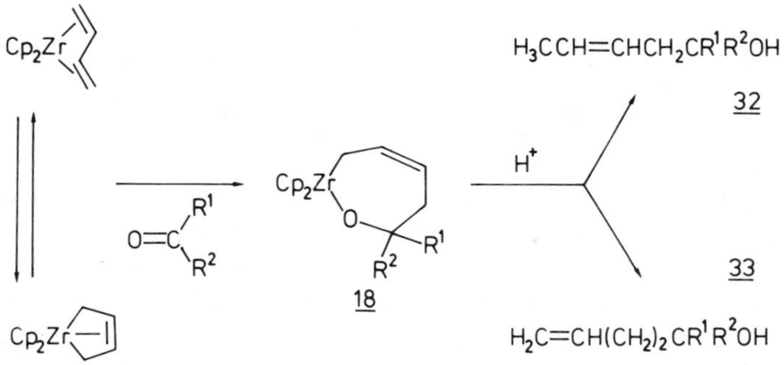

Again it is **3a** that reacts predominantly by far, even with organic carbonyl compounds. For several examples the slower disappearance of **5a** can be monitored at room temperature. This is confirmed by allowing the **3a/5a** mixture to react at low temperature with benzophenone (29). At $-25°C$, i.e., conditions under which the **3a** $\rightleftarrows$ **5a** equilibration is exceedingly slow, it is exclusively (*s-trans*-η^4-butadiene)ZrCp$_2$ (**3a**) that reacts with the ketone to produce the oxazirconacycloheptene **18a**. At $-50°C$ a thermally induced reaction is no longer observed. However, both (butadiene)zirconocene isomers rapidly form the metallacycle **18a** upon photolysis. From these observations, a reaction mechanism is inferred for the C—C coupling of the butadiene ligand with organic carbonyl compounds similar to the reaction of the **3a** $\rightleftarrows$ **5a** system with ethylene (27), i.e., a reaction sequence starting from a coordinatively unsaturated (η^2-butadiene)zirconium species (**4a**) through bis(olefin)- and metallacyclic σ-allylzirconocene-type intermediates (**34** and **35**).

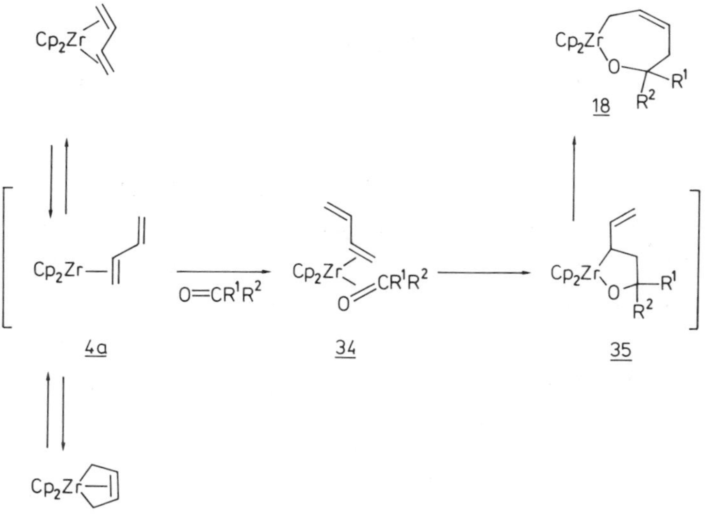

Shifting the equilibrium to the (*s-cis*-diene)metallocene should make it more likely to observe reactions of this isomer. Introduction of a methyl group at C2 of the conjugated-diene chain is sufficient to move the 3 ⇄ 5 equilibrium almost completely to the cisoid side (*22*). (*s-cis*-Isoprene)zirconocene indeed reacts only slowly with several aldehydes and ketones at room temperature. A pronouncedly regioselective C—C coupling is observed to give predominantly the isomer **18B**, which Nakamura *et al.* (*30*) have interpreted as arising from a preferred carbonyl addition at the position of highest electron density (C1) of the diene ligand in the σ^2,π-structured (isoprene)zirconocene **5e**. From the observations made on the (butadiene)zirconocene/olefin system (*27*) (see above), it was obvious that the regioselectivity of the ketone/(isoprene)zirconocene coupling could be controlled by adjusting the reaction conditions to allow participation of the (*s-trans*-η^4-isoprene)zirconocene isomer (**3e**).

This was demonstrated experimentally by treating (isoprene)zirconocene with 3,3-dimethyl-2-butanone (**36**) (*31*). The reaction yielded a mixture of the two expected organometallic regioisomers **18A** and **18B**. C—C coupling at C1 of the isoprene unit (to give **18B**) predominates by

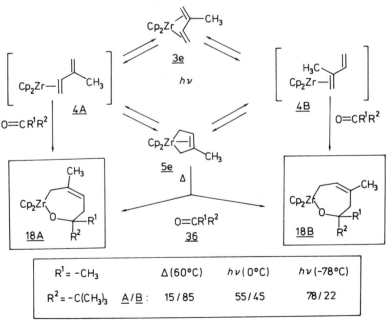

$R^1 = -CH_3$		$\Delta(60°C)$	$h\nu(0°C)$	$h\nu(-78°C)$
$R^2 = -C(CH_3)_3$	$\underline{A}/\underline{B}:$	15/85	55/45	78/22

SCHEME 4. C—C coupling between (isoprene)zirconocene and a ketone—mechanistic control of the regioselectivity.

85:15 if (s-cis-isoprene)zirconocene **5e** is treated with **36** at 60°C. The product ratio is shifted considerably in favor of **18A** if the reaction mixture of **5e** and methyl-t-butylketone is irradiated at 0°C. Under these conditions the reaction path via (η^2-diene)metallocene intermediates **4A** and **4B** can compete successfully with the thermally induced carbonyl addition reaction of **5e** (Scheme 4). Again, this alternative apparently leads to predominant coupling of the reagent, in this case ketone **36**, with the less substituted C=C bond of the isoprene moiety (28, 31).

Photolysis of the **5e/36** mixture at sufficiently low temperature, or a thermally induced reaction of 3,3-dimethyl-3-butanone with the pure (s-trans-η^4-isoprene)zirconocene isomer (**3e**) eventually effects the reversal of the regioselectivity. The regioisomer **18A** resulting from C—C coupling at C4 of the isoprene unit dominates over **18B** by a ratio of 78:22.

V

CONCLUDING REMARKS

The (conjugated-diene)zirconocene and -hafnocene systems reviewed above are unique in that three isomeric organometallic species can be distinguished by physical and/or chemical means. Two of these, **3** and **5**, represent stable isolable molecules, while the third, (η^2-diene)MCp$_2$ (**4**), functions as a chemically detectable intermediate connecting the (η^4-diene)metallocene isomers (see Scheme 1).

The rapid **3** ⇌ **4** equilibrium makes the coordinatively unsaturated mono(olefin)zirconocene (-hafnocene) complex **4** an easily accessible organometallic synthon even at extremely low temperatures, a fact that certainly will make its readily available precursor **3** a good starting material for the preparation of many novel organometallic compounds in the future.

Our observation that almost equally stable isomeric (s-trans- and (s-cis-η^4-butadiene)zirconocenes (**3a** and **5a**), which drastically differ in their chemical behavior, exist at ambient temperature and can even be isolated, should have some impact on the understanding and further development of transition metal-mediated processes of the 1,4-functionalization of conjugated dienes. It is also conceivable that cis/trans product ratios could be controlled through (s-trans-diene)metal complexes being involved in the course of such reactions (97, 98). The advent of the unique mononuclear (s-trans-η^4-conjugated diene)metal unit could turn out to be especially useful for the development of synthetic strategies, as it might be used to introduce an element of chirality into transition metal-induced C—C bond-forming reaction sequences (99, 100). It would, therefore, be very interesting to discover other examples of mononuclear (*s-trans*-η^4-conjugated diene)metal complexes. Good candidates should be other "sufficiently unsaturated" bent metallocene systems, and these may be found in the lanthanide or actinide series. Moreover, we believe that the synthesis of related complexes of late transition metals, e.g., of square-planar Ni, Pd, or Pt(II) systems, should also be possible. Even though it may be difficult to find many more examples of stable and isolable (*s-trans*-η^4-diene)transition metal compounds, their possible role as reactive intermediates equilibrating with the ordinary (*s-cis*-diene)metal complex isomers should always be kept in mind.

Acknowledgments

Our contribution has been made possible by the dedicated work of a small group of very talented co-workers, whose names are given in the references. G. E. thanks Professor W. R. Roth at the Department of Chemistry, Ruhr-Universität Bochum, for continuing encouragement and support. We wish to thank Professor G. Wilke and many scientists at the Max-Planck-Institut für Kohlenforschung (Mülheim a. d. Ruhr) for very helpful discussions and contributions. We are indebted to Professors A. Nakamura, H. Yasuda, and K. Tatsumi of the Osaka group for sending us several manuscripts prior to publication. Generous financial support from the Deutsche Forschungsgemeinschaft, the Minister für Wissenschaft und Forschung des Landes Nordrhein-Westfalen, and the Fonds der Chemischen Industrie is gratefully acknowledged.

References

1. G. Wilke, *Angew. Chem.* **75,** 10 (1963); P. W. Jolly and G. Wilke, "The Organic Chemistry of Nickel," Vol. 1. Academic Press, New York, 1974; Vol. 2 (1975); P. Heimbach and R. Traunmüller, "Chemie der Metall-Olefin-Komplexe," Verlag Chemie, Weinheim, 1970, and references cited therein.
2. H. Alper, "Transition Metal Organometallics in Organic Synthesis," Vol. 1. Academic Press, New York, 1976; Vol. 2 (1978); for the recent examples, see A. J. Pearson, S. L. Kole, and B. Cheu, *J. Am. Chem. Soc.* **105,** 4483 (1983); A. J. Pearson, *Acc. Chem. Res.* **13,** 463 (1980), and literature cited therein.
3. J. P. Collmann and L. S. Hegedus, "Principles and Applications of Organotransition Metal Chemistry." University Science Books, Mill Valley, California, 1980; E. A. Koerner von Gustorf, F. W. Grevels, and I. Fischler, "The Organic Chemistry of Iron." Academic Press, New York, 1978.
4. L. A. Oro, *Inorg. Chim. Acta* **21,** L6 (1977); J. Müller, H. Menig, and P. V. Rinze, *J. Organomet. Chem.* **181,** 387 (1979).
5. J. W. Faller and A. M. Rosan, *J. Am. Chem. Soc.* **99,** 4858 (1977).
6. J. M. Williams, R. K. Brown, A. J. Schultz, G. D. Stucky, and S. D. Ittel, *J. Am. Chem. Soc.* **100,** 7407 (1978).
7. M. Arthurs, M. Sloan, M. G. B. Drew, and S. M. Nelson, *J. Chem. Soc. Dalton Trans.* p. 1794 (1975); M. Arthurs, C. M. Regan, and S. M. Nelson, *ibid.* p. 2053 (1980).
8. S. Ruh and W. von Philipsborn, *J. Organomet. Chem.* **127,** C59 (1977); P. Diehl, A. C. Kunwar, and H. Zimmermann, *ibid.* **135,** 205, (1977).

9. O. S. Mills and G. Robinson, *Proc. Chem. Soc., London* p. 421 (1960); *Acta Crystallogr.* **16**, 758 (1963); M. J. Davis and C. S. Speed, *J. Organomet. Chem.* **21**, 401 (1970); R. L. Harlow, P. J. Krusic, R. J. McKinney, and S. S. Wreford, *Organometallics* **1**, 1506 (1982).

10. G. F. Emerson, L. Watts, and R. Pettit, *J. Am. Chem. Soc.* **87**, 131 (1965); L. Watts, J. D. Fitzpatrick, and R. Pettit, *ibid.* p. 3253; R. H. Grubbs and R. A. Grey, *ibid.* **95**, 5765 (1973).

11. W. R. Roth and J. D. Meier, *Tetrahedron Lett.* p. 2053 (1967); X-ray structure: C. Krüger and T. Mühlenbernd, unpublished (1982); S. D. Chappell and D. J. Cole-Hamilton, *J. Chem. Soc., Chem. Commun.* p. 238 (1980); S. D. Chappell and D. J. Cole-Hamilton, *ibid.* p. 319 (1981); M. F. Lappert, T. R. Martin, C. R. C. Milne, J. L. Atwood, W. E. Hunter, and R. E. Pentilla, *J. Organomet. Chem.* **192**, C35 (1980); M. F. Lappert, C. L. Raston, B. W. Shelton, and A. H. White, *J. Chem. Soc., Chem. Commun.* p. 485 (1981).

12. M. F. Lappert, T. R. Martin, J. L. Atwood, and W. E. Hunter, *J. Chem. Soc., Chem. Commun.* p. 476 (1980).

13. V. G. Andrianov, Y. T. Struchkov, V. N. Setkina, V. I. Zdanovich, A. Zhakaeva, and D. N. Kursanov, *J. Chem. Soc., Chem. Commun.* p. 117 (1975); F. Edelmann and U. Behrens, *J. Organomet. Chem.* **134**, 31 (1977); B. Lubke, F. Edelmann, and U. Behrens, *Chem. Ber.* **116**, 11 (1983).

14. Y. Kai, N. Kanehisa, K. Miki, N. Kasai, K. Mashima, H. Yasuda, and A. Nakamura, *Chem. Lett.* p. 1277 (1982).

15. H. E. Ramsden, U. S. Patent 3, 388, 179 (1968); Y. Nakano, K. Natsukawa, H. Yasuda, and H. Tani, *Tetrahedron Lett.* p. 2833 (1972); K. Fujita, Y. Ohnuma, H. Yasuda, and H. Tani, *J. Organomet. Chem.* **113**, 201 (1976); H. Yasuda, Y. Nakano, K. Natsukawa, and H. Tani, *Macromolecules* **11**, 586 (1978).

16. P. S. Skell and M. J. McGlinchey, *Angew. Chem.* **87**, 215 (1975); W. Gausing and G. Wilke, *ibid.* **93**, 201 (1981); H. Yasuda, K. Mashima, T. Okamoto, and A. Nakamura, *Int. Conf. Organomet. Chem. 11th, 1983* Abstr. No. 79 (1983).

17. S. Datta, S. S. Wreford, R. P. Beatty, and T. J. McNeese, *J. Am. Chem. Soc.* **101**, 1053 (1179); S. S. Wreford and J. F. Whitney, *Inorg Chem.* **20**, 3918 (1981).

18. M. E. Squillacote, R. S. Sheridan, O. L. Chapman, and F. A. Anet, *J. Am. Chem. Soc.* **101**, 3657 (1979).

19. P. W. Mui and E. Grunwald, *J. Am. Chem. Soc.* **104**, 6562 (1982).

20. R. Hoffmann, *Angew. Chem.* **94**, 725 (1982).

21. Several examples of s-trans-butadiene bridging *two* transition metal centers have previously been described: M. J. Grogan and K. Nakamoto, *Inorg. Chim. Acta* **1**, 228 (1967); M. Tachikawa, J. R. Shapley, R. C. Haltiwanger, and C. G. Pierpont, *J. Am. Chem. Soc.* **98**, 4651 (1976); C. G. Pierpont, *Inorg. Chem.* **17**, 1976 (1978); M. Ziegler, *Z. Anorg. Allg. Chem.* **335**, 12 (1967); H. E. Sasse and M. L. Ziegler, *ibid.* **392**, 167 (1972); M. Zöller and M. L. Ziegler, *ibid.* **425**, 265 (1976); K.-H. Franzreb and C. G. Kreiter, *J. Organomet. Chem.* **246**, 189 (1983); Z. *Naturforsch., B: Anorg. Chem. Org. Chem.* **378**, 1058 (1982).

22. G. Erker, J. Wicher, K. Engel, F. Rosenfeldt, W. Dietrich, and C. Krüger, *J. Am. Chem. Soc.* **102**, 6344 (1980).

23. G. Erker, J. Wicher, K. Engel, and C. Krüger, *Chem. Ber.* **115**, 3300 (1982).

24. H. H. Brintzinger and L. S. Bartell, *J. Am. Chem. Soc.* **92**, 1105 (1970); J. W. Lauher and R. Hoffmann, *ibid.* **98**, 1729 (1976); for chemical consequences of the stereoelectronic properties of the bent metallocene unit, see G. Erker and F. Rosenfeldt, *Angew. Chem.* **90**, 640 (1978); *J. Organomet. Chem.* **188**, C1 (1980).

25. C. J. Ballhausen and J. P. Dahl, *Acta Chem. Scand.* **15**, 1333 (1961); N. W. Alcock, *J. Chem. Soc. A.* p. 2001 (1967); J. C. Green, M. L. H. Green, and C. K. Prout, *J. Chem. Soc., Chem. Commun.* p. 421 (1972); M. L. H. Green, *Pure Appl. Chem.* **30**, 373 (1972); J. C. Green, S. E. Jackson, and B. Higginson, *J. Chem. Soc., Dalton Trans.* p. 403 (1975); J. L. Petersen and L. F. Dahl, *J. Am. Chem. Soc.* **97**, 6416, 6422 (1975); D. P. Bakalik and R. G. Hayes, *Inorg. Chem.* **11**, 1734 (1972); C. P. Stewart and A. L. Porte, *J. Chem. Soc. Dalton Trans.* p. 722 (1973); A. G. Evans, J. L. Evans, and E. H. Moon, *ibid.* p. 2390 (1974); A. H. Al-Movali, *ibid.* p. 426 (1980).

26. H. Berke, personal communication (1982); K. Tatsumi, H. Yasuda, and A. Nakamura, *Isr. J. Chem.* **23**, 145 (1983).

27. G. Erker, K. Engel, U. Dorf, J. L. Atwood, and W. E. Hunter, *Angew. Chem.* **94**, 915 (1982); U. Dorf, K. Engel, and G. Erker, *ibid.* **94**, p. 916.

28. H. Yasuda, Y. Kajihara, K. Nagasuna, K. Mashima, and A. Nakamura, *Chem. Lett.* p. 719 (1981); Y. Kai, N. Kanehisa, K. Miki, N. Kasai, K. Mashima, K. Nagasuna, H. Yasuda, and A. Nakamura, *ibid.* p. 1979 (1982); the less substituted isoprene C=C double bond is exclusively complexed by a nickel complex to form (3, 4-η^2-isoprene) Ni (CH$_3$)Cp: H. Lehmkuhl, F. Danowski, R. Benn, A. Rufinska, G. Schroth, and R. Mynott, *J. Organomet. Chem.* **254**, C11 (1983).

29. G. Erker, K. Engel, J. L. Atwood, and W. E. Hunter, *Angew. Chem.* **95**, 506 (1983).

30. H. Yasuda, Y. Kajihara, K. Mashima, K. Nagasuna and A. Nakamura, *Chem. Lett.* p. 671 (1981); M. Akita, H. Yasuda, and A. Nakamura, *ibid.* p. 217 (1983); X-ray diffraction study: N. Kanehisa, Y. Kai, K. Miki, N. Kasai, M. Akita, H. Yasuda, and A. Nakamura, *45th Nat. Meet. Chem. Soc. Jpn., 1982* Abstr. No. 3B31 (1982), cited in Ref. *28*.

31. G. Erker and U. Dorf, *Angew. Chem.* **95**, 800 (1983).

32. U. Dorf, K. Engel, and G. Erker, *Organometallics* **2**, 462 (1983).

33. H. Yasuda, Y. Kajihara, K. Mashima, K. Nagasuna, K. Lee, and A. Nakamura, *Organometallics* **1**, 388 (1982).

34. K. Engel, Dissertation, University of Bochum (1983).

35. G. Erker and P. Czisch, *Int. Conf. Organomet. Chem., 11th, 1983* Abstr. No 63 (1983).

36. M. Yang, K. Yamamoto, N. Otake, M. Ando, and K. Takase, *Tetrahedron Lett.* p. 3843 (1970); M. Yang, M. Ando, and K. Takase, *ibid.* p. 3529 (1971); R. Baker, R. C. Cookson, and A. D. Saunders, *J. Chem. Soc., Perkin Trans. 1* p. 1809 (1976); S. Akutagawa and S. Otsuka, *J. Am. Chem. Soc.* **98**, 7420 (1976).

37. R. A. Benkeser, *Synthesis* p. 347 (1971).

38. J. H. Bahl, R. B. Bates, W. A. Beavers, and N. S. Mills, *J. Org. Chem.* **41**, 1620 (1976); W. J. Richter, *Angew. Chem.* **94**, 298 (1982).

39. G. Erker, *J. Organomet. Chem.* **134**, 189 (1977).

40. G. Erker, K. Engel, and P. Vogel, *Angew. Chem.* **94**, 791 (1982).

41. G. Erker and K. Engel, unpublished observations (1980); cf. K. Engel, Diploma Thesis, University of Bochum (1980).

42. G. Erker and U. Korek, unpublished observations (1983).

43. G. Erker, K. Kropp, J. L. Atwood, and W. E. Hunter, *Organometallics* **2**, 1555 (1983).

44. G. Erker and T. Mühlenbernd, unpublished observations (1983).

45. G. Erker, K. Engel, C. Krüger, and A.-P. Chiang, *Chem. Ber.* **115**, 3311 (1982).

46. G. Binsch, E. L. Eliel, and H. Kessler, *Angew. Chem.* **83**, 618 (1971); for a comparison of activation parameters, see G. Binsch and J. D. Roberts, *J. Am. Chem. Soc.* **87**, 5157 (1965); A. C. Cope, K. Banholzer, H. Keller, B. A. Pawson, J. J. Wang, and H. J. S. Winkler, *ibid.* p. 3644; A. Cope and B. A. Pawson, *ibid.* p. 3649; Y. Inoue,

T. Ueoka, T. Kuroda, and T. Hakushi, *J. Chem. Soc., Chem. Commun.* p. 1031 (1981).

47. G. K. Barker, M. Green, J. A. K. Howard, J. L. Spencer, and F. G. A. Stone, *J. Chem. Soc., Dalton Trans.* p. 1839 (1978); G. Wilke and W. Gausing, unpublished, cf. in W. Gausing, Dissertation, University of Bochum (1979).

48. Previously observed only for a few examples of (conjugated diene)transition metal complexes. For examples, see Faller and Rosan (5); G. Wilke, *Fundam. Res. Homogeneous Catal.* **3**, 1 (1979); R. Benn and G. Schroth, *J. Organomet. Chem.* **228**, 71 (1981); W. H. Hersh, F. J. Hollander and R. G. Bergman, *J. Am. Chem. Soc.* **105**, 5834 (1983).

49. There are other types of dynamic behavior known for (1,3-diene) metal complexes: L. Kruczynski and J. Takats, *J. Am. Chem. Soc.* **96**, 932 (1974); *Inorg. Chem.* **15**, 3140 (1976); C. G. Kreiter, S. Stüber, and L. Wackerle, *J. Organomet. Chem.* **66**, C49 (1974); M. A. Busch and R. J. Clark, *Inorg. Chem.* **14**, 226 (1975); S. D. Ittel, F. A. Van-Catledge, and J. P. Jesson, *J. Am. Chem. Soc.* **101**, 3874 (1979); *J. Organomet. Chem.* **168**, C25 (1979); M. Kotzian, C. G. Kreiter, and S. Özkar, *ibid.* **229**, 29 (1982).

50. G. Erker and W. Frömberg, unpublished.

51. G. Erker, U. Dorf, G. Müller, and C. Krüger, unpublished work (1983).

52. G. Erker, K. Engel, and P. Vogel, unpublished work (1983).

53. H. P. Fritz, *Adv. Organomet. Chem.* **1**, 262 (1964).

54. R. Gleiter, G. Erker, K. Engel, and U. Dorf, unpublished observations (1983).

55. G. Erker, K. Engel, G. Müller, and C. Krüger, *Organometallics* **3**, 128 (1984).

56. C. Krüger, G. Müller, G. Erker, U. Dorf, and K. Engel, *Organometallics* **4**, 215 (1985).

57. Y. Kai, N. Kanehisa, K. Miki, N. Kasai, K. Mashima, K. Nagasuna, H. Yasuda, and A. Nakamura, *J. Chem. Soc., Chem. Commun.* p. 191 (1982).

58. F. S. Stephens, *J. Chem. Soc. A.* p. 2745 (1970).

59. J. Wenger, N. H. Thuy, T. Boschi, R. Roulet, A. Chollet, P. Vogel, A. A. Pinkerton, and D. Schwarzenbach, *J. Organomet. Chem.* **174**, 89 (1979).

60. E. Meier, O. Cherpillod, T. Boschi, R. Roulet, P. Vogel, C. Mahaim, A. A. Pinkerton, D. Schwarzenbach, and G. Chapuis, *J. Organomet. Chem.* **186**, 247 (1980).

61. A. A. Pinkerton, G. Chapuis, P. Vogel, U. Hänisch, P. Narbel, T. Boschi, and R. Roulet, *Inorg. Chim. Acta* **35**, 197 (1979).

62. Ph. Narbel, T. Boschi, R. Roulet, P. Vogel, A. A. Pinkerton, and D. Schwarzenbach, *Inorg. Chim. Acta* **36**, 161 (1979).

63. F. A. Cotton, V. W. Day, B. A. Frenz, K. I. Hardcastle, and J. M. Troup, *J. Am. Chem. Soc.* **95**, 4522 (1973).

64. M. R. Churchill and R. Mason, *Adv. Organomet. Chem.* **5**, 93 (1967).

65. A. Immirzi, *J. Organomet. Chem.* **76**, 65 (1974).

66. C. Barras, R. Roulet, E. Vieira, P. Vogel, and G. Chapuis, *Helv. Chim. Acta* **64**, 2328 (1981).

67. M. G. B. Drew, S. M. Nelson, and M. Sloan, *J. Organomet. Chem.* **39**, C9 (1972).

68. W. E. Hunter, D. C. Hrncir, R. Vann Bynum, R. A. Penttila, and J. L. Atwood, *Organometallics* **2**, 750 (1983).

69. J. L. Atwood, G. K. Barker, J. Holton, W. E. Hunter, M. F. Lappert, and R. Pearce, *J. Am. Chem. Soc.* **99**, 6645 (1977).

70. J. Jeffery, M. F. Lappert, N. T. Luong-Thi, J. L. Atwood, and W. E. Hunter, *J. Chem. Soc., Chem. Commun.* p. 1081 (1978).

71. J. Jeffery, M. F. Lappert, N. T. Luong-Thi, M. Webb, J. L. Atwood, and W. E. Hunter, *J. Chem. Soc., Dalton Trans.* p. 1593 (1981).

72. M. F. Lappert, P. I. Riley, P. I. W. Yarrow, J. L. Atwood, W. E. Hunter, and M. J. Zaworotko, *J. Chem. Soc., Dalton Trans.* p. 814 (1981).

73. J. L. Atwood, W. E. Hunter, D. C. Hrncir, E. Samuel, H. Alt, and M. D. Rausch, *Inorg. Chem.* **14,** 1757 (1975).

74. G. Fachinetti, G. Fochi, and C. Floriani, *J. Chem. Soc., Dalton Trans.* p. 1946 (1977).

75. K. Kropp, Dissertation, University of Bochum (1981).

76. C. F. Aten, L. Hedberg, and K. Hedberg, *J. Am. Chem. Soc.* **90,** 2463 (1968).

77. N. A. Bailey and R. Mason, *Acta Crystallogr.* **21,** 652 (1966).

78. R. McWeeny, R. Mason, and A. D. C. Towl, *Discuss. Faraday Soc.* **47** 20 (1969).

79. R. Mason, *Nature (London)* **217,** 543 (1968).

80. D. M. P. Mingos, *Adv. Organomet. Chem.* **15,** 1 (1977).

81. H. Dietrich and R. Uttech, *Naturwissenschaften* **50,** 613 (1963).

82. R. Mason and D. R. Russell, *J. Chem. Soc., Chem. Commun.* p. 26 (1966).

83. M. R. Churchill and T. A. O'Brien, *J. Chem. Soc., Chem. Commun.* p. 246 (1968).

84. S. F. A. Kettle and R. Mason, *J. Organomet. Chem.* **5,** 573 (1966).

85. R. Goddard, C. Krüger, F. Mark, R. Stansfield, and X. Zhang, *Organometallics* **4,** 285 (1985).

86. W. E. Hunter, J. L. Atwood, G. Fachinetti, and C. Floriani, *J. Organomet. Chem.* **204,** 67 (1981).

87. F. A. Cotton and G. Wilkinson, "Advanced Inorganic Chemistry," p. 824 Wiley, New York, 1980.

88. J. D. Dunitz, "X-Ray Analysis and the Structure of Organic Molecules," p. 240 Cornell Univ. Press, Ithaca, New York, 1979.

89. A. Wojcicki, *Adv. Organomet. Chem.* **11,** 87 (1973); F. Calderazzo, *Angew. Chem.* **89,** 305 (1977); H. Berke and R. Hoffmann, *J. Am. Chem. Soc.* **100,** 7224 (1978).

90. For mechanistic details of the carbonylation of zirconocene complexes, see G. Erker and K. Kropp, *Chem. Ber.* **115,** 2437 (1982); G. Erker, K. Kroop, C. Krüger, and A.-P. Chiang, *ibid.* p. 2447; G. Erker, *Acc. Chem. Res.* **17,** 103 (1984); P. T. Wolczanski and J. E. Bercaw, *ibid.* **13,** 121 (1980).

91. R. F. Heldeweg and H. Hogeveen, *J. Am. Chem. Soc.* **98,** 6040 (1976); B. F. G. Johnson, J. Lewis, and D. J. Thompson, *Tetrahedron Lett.* p. 3789 (1974); C.-W. Yip, P. Au, T.-Y. Luh, and S. W. Tam, *J. Organomet. Chem.* **175,** 221 (1979); W. H. Hersh and R. G. Bergman, *J. Am. Chem. Soc.* **103,** 6992 (1981); thermal or photochemical reactions: M. S. Raasch and C. W. Theobald, British Patent 595, 161 (1947); *Chem. Abstr.* **42,** 2988a (1948); E. J. Corey and S. W. Walinsky, *J. Am. Chem. Soc.* **94,** 8932 (1972); W. R. Dolbier, Jr. and H. M. Frey, *J. Chem. Soc., Perkin Trans. 2* p. 1674; J. E. Baldwin, *Can. J. Chem.* **44,** 2051 (1966); S. Yankelevich and B. Fuchs, *Tetrahedron Lett.* p. 4945 (1967); K. Nakamura, S. Koda, and K. Akita, *Bull. Chem. Soc. Jpn* **51,** 1665 (1978).

92. G. Erker and K. Kropp, *J. Organomet. Chem.* **194,** 45 (1980).

93. J. Blenkers, J. H. de Liefde Meijer, and J. H. Teuben, *Organometallics* **2,** 1483 (1983).

94. J. L. Thomas and K. T. Brown, *J. Organomet. Chem.* **111,** 297 (1976); J. L. Atwood, R. D. Rogers, W. E. Hunter, C. Floriani, G. Fachinetti, and A. Chiesi-Villa, *Inorg. Chem.* **19,** 3812 (1980).

95. A. C. L. Su, *Adv. Organomet. Chem.* **17,** 269 (1979).

96. A. Stockis and R. Hoffmann, *J. Am. Chem. Soc.* **102,** 2952 (1980); regiochemistry: V. Skibbe and G. Erker, *J. Organomet. Chem.* **241,** 15 (1983); stereochemistry:

G. Erker and F. Rosenfeldt, *ibid.* **224,** 29 (1982); K. Kropp and G. Erker, *Organometallics* **1,** 1246 (1982).
97. G. Henrici-Olivé and S. Olivé, "Coordination and Catalysis," pp. 210–232. Verlag Chemie, Weinheim, (1977); R. Warin, M. Julémont, and P. Teyssié, *J. Organomet. Chem.* **185,** 413 (1980); J. Furukawa, *Acc. Chem. Res.* **13,** 1 (1980).
98. J.-E. Bäckvall and R. E. Nordberg, *J. Am. Chem. Soc.* **103,** 4959 (1981); J.-E. Bäckvall, R. E. Nordberg, and S. E. Byström, *Int. Conf. Organomet. Chem., 11th* Abstr. No. 67 (1983).
99. H. Brunner, *Acc. Chem. Res.* **12,** 250 (1979), and literature cited therein.
100. G. A. S. Howell and M. J. Thomas, *J. Chem. Soc., Dalton Trans* p. 1401 (1983).

Organometallic Metal Clusters Containing Nitrosyl and Nitrido Ligands

WAYNE L. GLADFELTER

Department of Chemistry
University of Minnesota
Minneapolis, Minnesota

I

INTRODUCTION

One of the fascinating features to be developed in the area of metal clusters involves the coordination of atomic species to the metal framework. There is substantial interest in developing a basic understanding of the formation, bonding, and reactivity of these unusual organometallic functional groups. Further, some of these efforts are beginning to define a relation between cluster coordinated atomic species and atoms bound to a metal surface. These surface-coordinated atoms are invoked as intermediates in several heterogeneous catalytic reactions. For instance, carbides are proposed intermediates in the reduction of CO in the Fisher–Tropsch synthesis $(1, 2)$. It is in this light that the first study of organometallic nitrido clusters was reported in 1979 (3). Just as some of the surface-catalyzed reactions of CO are believed to proceed through carbides, several important reactions of N_2 and NO are known to proceed

41

through surface coordinated nitrogen atoms (4, 5). Equation (1), the

$$N_2 + 3H_2 \rightarrow 2NH_3 \tag{1}$$

Haber process, has been shown to involve the dissociative chemisorption of molecular nitrogen to give surface adsorbed nitrogen atoms (4). These subsequently form N—H bonds followed by desorption of the ammonia.

The use of air as the oxidant for internal combustion engines also results in the formation of nitrogen oxides causing serious pollution problems. Much of the effort spent on removing these toxic pollutants centers on the reduction of NO using CO (5) or H_2 (6) as the reducing agent and the requirement of a heterogeneous catalyst [Eqs. (2) and (3)]. Surface

$$2NO + 2CO \rightarrow 2CO_2 + N_2 \tag{2}$$

$$2NO + 5H_2 \rightarrow 2NH_3 + 2H_2O \tag{3}$$

adsorbed nitrogen atoms may play an important role in these reactions as well. The connection between the two parts of this article, organometallic nitrosyl and nitrido clusters, is in part derived from the above reactions. It has been discovered that the conversion of NO to N on metal carbonyl clusters is facile. This is true to such an extent that the same methods used to synthesize nitrosyl clusters are also employed to prepare nitrido clusters.

Many excellent review articles exist covering most aspects of metal nitrosyl complexes (5, 7–10). This article covers the methods used to prepare cluster coordinated nitrosyl ligands, their structural and spectroscopic properties with emphasis on the relationship of terminal and bridging nitrosyls, and their unique reactivity. A discussion of nitrido clusters, which are formed using reactions similar to those used for nitrosyl formation, will also cover structural, spectroscopic, and reactivity studies. Although one review article has been published (11) on the nitrido ligand in high-valent inorganic complexes, this is the first review covering organometallic nitrido clusters.

II

NITROSYL CLUSTERS

A. Synthetic Methods

1. Nitric Oxide Gas

The use of NO was successfully applied to $Ru_3(CO)_{12}$ and $Os_3(CO)_{12}$ to make the first carbonyl clusters containing nitric oxide as a ligand (12). The nitrosyl ligands bridge the same edge of the triangle in both

$Ru_3(CO)_{10}(NO)_2$ (1) and $Os_3(CO)_{10}(NO)_2$, which can be isolated in yields

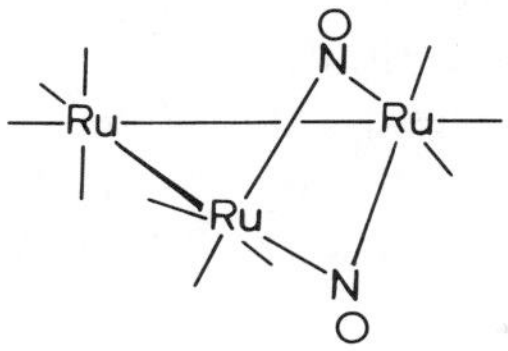

(1)

of 20 and 14%, respectively. The reaction also produces a large amount of insoluble material that has not been characterized. A reexamination of the reaction of NO with $Os_3(CO)_{12}$ under slightly different conditions produced the very interesting species $Os_3(CO)_9(NO)_2$ in low (1%) yield (13). Unlike the decacarbonyl cluster, $Os_3(CO)_9(NO)_2$ has no bridging ligands and both the nitrosyls are coordinated to the same metal. The unique reactivity this imparts to the cluster will be discussed later, and the structure of the $P(OCH_3)_3$ derivative is shown in structure 2 (14).

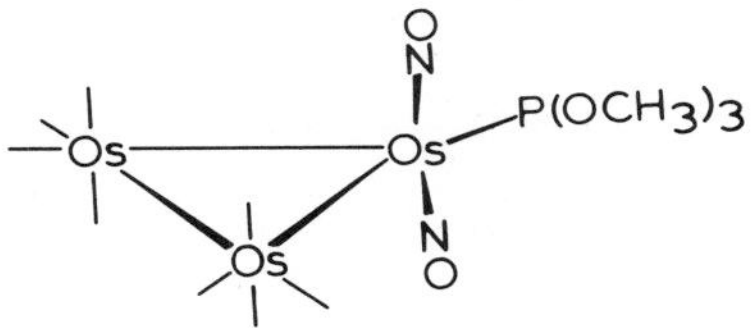

(2)

The basic problem of using NO gas with clusters is that this reagent often cleaves the metal–metal bond as well as substituting for a carbonyl ligand. Since NO donates three electrons, displacement of precisely one CO and one M—M bond is feasible. Equations (4) (15), (5) (16), and (6) (17) are excellent examples of this.

$$Co_2(CO)_8 + 2NO \rightarrow 2Co(CO)_3(NO) + 2CO \qquad (4)$$

$$[(\eta^5\text{-}C_5H_5)Cr(CO)_3]_2 + 2NO \rightarrow 2(\eta^5\text{-}C_5H_5)Cr(CO)_2(NO) \qquad (5)$$

$$Fe_3(CO)_{12} + 6NO \rightarrow 3Fe(CO)_2(NO)_2 + 6CO \qquad (6)$$

The successful isolation of the intact Ru and Os clusters is due in part to the greater strength of the M—M bonds.

2. Nitrosonium Ion

At this point in time NO^+ has been the most successful reagent for the synthesis of nitrosyl carbonyl clusters. It is usually used as the BF_4^- or PF_6^- salt, which should be sublimed prior to use, and in all successful uses of this reagent the starting cluster has been anionic. Equations (7)–(11) show several of the successful uses of NO^+ where the products of the reaction result in CO substitution [Eqs. (7) (18), (9) (20), (10) (21)] or M—M bond cleavage [Eqs. (8) (19) and (11)]. In one case, $[Os_{10}C(CO)_{24}(NO)]^-$, which initially contains a μ_2-NO, reacts over a period of 10 days by CO loss to form $[Os_{10}C(CO)_{23}(NO)]^-$ which has a linear terminal nitrosyl (19).

$$[HM_3(CO)_{11}]^- + NO^+ \rightarrow HM_3(CO)_{10}(NO) + CO \qquad (M = Ru, Os) \qquad (7)$$

$$[Os_{10}C(CO)_{24}]^{2-} + NO^+ \rightarrow [Os_{10}C(CO)_{24}(NO)]^- \qquad (8)$$

(3)

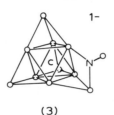

(3)

$$[Fe_4N(CO)_{12}]^- + NO^+ \rightarrow Fe_4N(CO)_{11}(NO) + CO \qquad (9)$$

$$[Re_7C(CO)_{21}]^{3-} + NO^+ \rightarrow [Re_7C(CO)_{20}(NO)]^{2-} + CO \qquad (10)$$

The difficulty often encountered with NO^+ arises from its high oxidation potential which can easily destroy clusters that are not robust. The reaction of NO^+ with $[H_3Os_4(CO)_{12}]^-$ [Eqs. (11) and (12)] exemplifies the different products that can form as well as illustrating the dramatic effect

$$[H_3Os_4(CO)_{12}]^- + NO^+ \xrightarrow[\text{or THF}]{CH_2Cl_2} H_3Os_4(CO)_{12}(NO) \qquad (11)$$

(4)

(30% yield)

$$[H_3Os_4(CO)_{12}]^- + 2NO^+ \xrightarrow{CH_3CN} [H_3Os_4(CO)_{12}(CH_3CN)_2]^+ \qquad (12)$$

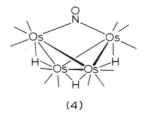

(4)

solvent can exert on these reactions (THF = tetrahydrofuran) (22–25). If $NO(PF_6)$ is used in acetonitrile in which traces of water are present, the reaction with $[H_3Os_4(CO)_{12}]^-$ yields a mixture of $H_4Os_4(CO)_{12}$, $H_3Os_4(OH)(CO)_{12}$, and $H_3Os_4(OPO_3H_2)(CO)_{12}$. The last species arises from hydrolysis of HPF_6 present in solution. It seems likely that the first step in these reactions is a one-electron oxidation of the cluster. In acetonitrile the intermediate must be stabilized by the solvent and subsequently further oxidized by NO^+ before it can react with NO. The greater solubility of NO^+ salts in CH_3CN would also promote the second oxidation. The reaction of neutral clusters with NO^+ commonly results in oxidation and cleavage of one or more of the M—M bonds. With $NOPF_6$ and $H_4Os_4(CO)_{12}$ in CH_3CN the butterfly cluster $[H_4Os_4(\mu\text{-OH})(CO)_{12}]^+$ is formed in 90% yield (26). In CH_3CN the trinuclear clusters $M_3(CO)_9(PPh_3)_3$ (M = Ru, Os; Ph = phenyl) are oxidized by NO^+ without forming nitrosyl complexes (27), while in methanol oxidation of $Ru_3(CO)_9(PPh_3)_3$ still occurs but the product, $[Ru(CO)_2(PPh_3)_2(NO)]^+$, does contain a nitrosyl ligand. Another side reaction sometimes encountered when using NO^+ is protonation of the cluster. Both Eqs. (13) and (14) were conducted under "anhydrous" conditions. The corresponding

$$[Ru_3(CO)_{10}(NO)]^- + NO^+ \rightarrow HRu_3(CO)_{10}(NO) \tag{13}$$

$$[Ru_4N(CO)_{12}]^- + NO^+ \rightarrow HRu_4N(CO)_{12} \tag{14}$$

nitrosylation of $[Fe_4N(CO)_{12}]^-$ [Eq. (9)] goes in 60% yield when using the identical solvent and conditions. The source of the protons has not been established but is likely due to traces of H_2O (28).

3. Nitrite

Although it is not strictly an organometallic cluster, Roussin's Black salt, $[Fe_4(S)_3(NO)_7]^-$, represents the first nitrosyl compound prepared via nitrite. The reaction of $FeSO_4$, $NaNO_2$, and $(NH_4)_2S$ is perhaps the most convenient method of synthesis of this cluster (29). A better defined approach to the use of NO_2^- was first introduced by Hieber with $Fe(CO)_5$ [Eq. (15)] (30). The direct extension of this reaction to $Mo(CO)_6$

$$Fe(CO)_5 + NaNO_2 + NaOCH_3 \rightarrow Na[Fe(CO)_3(NO)] + CO + Na[CH_3OCO_2] \tag{15}$$

produced two unusual polynuclear nitrosyl-containing complexes: $\{Na[Mo_3(CO)_6(NO)_3(OCH_3)_3(O)]_2\}^{3-}$ and $[Mo_3(CO)_6(NO)_3(OCH_3)_4]^-$ (31). Both of these clusters contain a triangular array of $Mo(CO)_2(NO)$ groups bridged by CH_3O^- and O^{2-} ligands, and no direct Mo—Mo bonds are proposed to be present.

PPN(NO_2) [PPN = bis(triphenylphosphine)nitrogen(1+)] has been found to be particularly successful as a mild reagent for converting metal carbonyls into nitrosyl carbonyl complexes (32–37). Equations (16) and

(17) *(32, 33)*, (18) *(33, 34)*, (19) *(35)*, (20) *(36)*, and (21), *(37)* illustrate its use for the nitrosylation of metal clusters. The mechanism of the reaction

$$Ru_3(CO)_{12} + PPN(NO_2) \rightarrow CO + CO_2 + PPN[Ru_3(CO)_{10}(NO)] \qquad (16)$$

$$Os_3(CO)_{12} + PPN(NO_2) \rightarrow CO + CO_2 + PPN[Os_3(CO)_{10}(NO)] \qquad (17)$$

(5)

$$Ru_6C(CO)_{17} + PPN(NO_2) \rightarrow CO + CO_2 + PPN[Ru_6C(CO)_{15}(NO)] \qquad (18)$$

$$Fe_5C(CO)_{15} + PPN(NO_2) \rightarrow CO + CO_2 + PPN[Fe_5C(CO)_{13}(NO)] \qquad (19)$$

$$Os_{10}C(I)_2(CO)_{24} + PPN(NO_2) \rightarrow CO + CO_2 + [Os_{10}C(I)(CO)_{22}(NO)]^{2-} \qquad (20)$$

$$Co_3(CCH_3)(CO)_9 + PPN(NO_2) \rightarrow CO + CO_2 + PPN[Co_3(CCH_3)(CO)_7(NO_2)] \qquad (21)$$

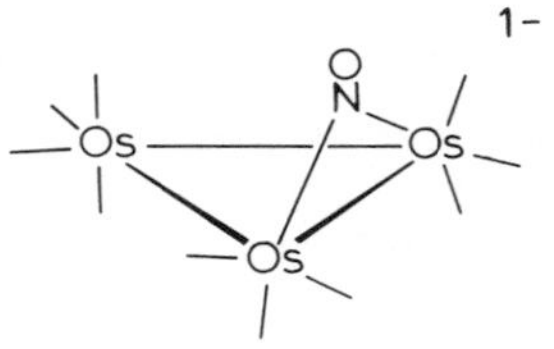

(5)

of PPN(NO$_2$) with Fe(CO)$_5$ has been shown to proceed via nucleophilic attack on a coordinated CO group *(33)*. Since many of the metal carbonyl clusters are readily attacked by base via a similar mechanism, PPN(NO$_2$) should find wide application in this area. Although NO$_2^-$ will react most rapidly with neutral or cationic reagents, there is one example of its reaction with a large anionic cluster [Eq. (22)] *(35)*. Presumably the charge is so effectively delocalized in the larger clusters that nucleophilic attack at any one site is not inhibited. Clusters containing hydride ligands can be deprotonated by PPN(NO$_2$). For instance, HOs$_4$N(CO)$_{12}$ *(38)*, H$_4$Ru$_4$(CO)$_{12}$ *(35)*, and HRu$_3$(CO)$_{10}$(NO) *(35)* react with NO$_2^-$ yielding the corresponding monoanions.

$$[Ru_6C(CO)_{15}(NO)]^- + NO_2^- \rightarrow [Ru_6C(CO)_{13}(NO)_2]^{2-} \qquad (22)$$

Since NO$_2^-$ is best suited for attacking neutral clusters and NO$^+$ is best at nitrosylating anionic clusters, the successive combination of these two reagents offers a feasible method toward polynitrosyl clusters. To date this has been used only once [Eqs. (23) and (24)] *(34)*. It is conceivable (although admittedly unlikely) that such a one-two punch could be

$$Ru_6C(CO)_{17} + PPN(NO_2) \rightarrow PPN[Ru_6C(CO)_{15}(NO)] \qquad (23)$$

$$[Ru_6C(CO)_{15}(NO)]^- + NO^+ \rightarrow Ru_6C(CO)_{14}(NO)_2 \qquad (24)$$

(6)

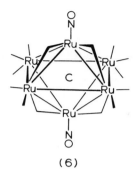

(6)

continued leading to a binary metal nitrosyl cluster. This area offers many possibilities for additional work.

4. Condensation of Nitrosyl Fragments

Few successful reactions are known by which an organometallic nitrosyl product has been formed from the condensation of mononuclear metal nitrosyl complexes. The unique cluster $(\eta^5\text{-}C_5H_5)_3Mn_3(NO)_4$ (39), which contains the only μ_3-NO known (7) (40), is prepared using photochemical or thermal (41) decarbonylation of $(\eta^5\text{-}C_5H_5)_2Mn_2(CO)_2(NO)_2$ [Eq. (25)]. The use of the redox condensation of a nitrosyl carbonylmetallate with a neutral metal carbonyl cluster to produce a nitrosyl cluster has

$$(\eta^5\text{-}C_5H_5)_2Mn_2(CO)_2(NO)_2 \xrightarrow[\Delta]{h\nu} (\eta^5\text{-}C_5H_5)_3Mn_3(NO)_4 \qquad (25)$$

(7)

(~30% yield)

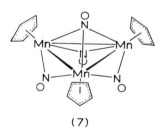

(7)

been reported for $[FeRu_3(CO)_{12}(NO)]^-$ [Eq. (26)] (42, 43). The limitation is clearly the small number of known nitrosyl carbonylmetallates. Prior to 1981 only $[Fe(CO)_3(NO)]^-$ was known (30). However, recent work has generated $[Mn(CO)_2(NO)_2]^-$ (32, 33), $[Mn(CO)_3(NO)]^{2-}$ (44), $[Cr(CO)_4(NO)]^-$ (45), $[V(CO)_4(NO)]^{2-}$ (46), and $[Ru(CO)_3(NO)]^-$ (47).

The ready availability of these anions should open the way to several novel clusters.

$$[Fe(CO)_3(NO)]^- + Ru_3(CO)_{12} \rightarrow [FeRu_3(CO)_{12}(NO)]^- \qquad (26)$$

$$(\mathbf{8})$$

(40% yield)

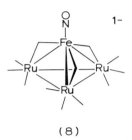

(8)

By far the most pervasive use of the condensation of mononuclear fragments to form nitrosyl-containing cluster has been in the synthesis of cubane-type clusters (48–51). Although most of these are not formally organometallic clusters, the reagents used as starting materials are. Equations (27)–(29) are typical of these condensations.

$$Co(CO)_2(PPh_3)(NO) + (Me_3CN)_2S \xrightarrow[\text{toluene}]{\Delta} Co_4(NCMe_3)_4(NO)_4 \qquad (27)$$

$$Hg[Fe(CO)_3(NO)]_2 + (Me_3CN)_2S \xrightarrow[\text{benzene}]{\text{2 hours, } \Delta} Fe_4(S)_2(NCMe_3)_2(NO)_4 \qquad (28)$$

$$Hg[Fe(CO)_3(NO)]_2 + S \xrightarrow[\text{toluene}]{\text{16 hours, } \Delta} Fe_4(S)_4(NO)_4 \qquad (29)$$

By extensive structural studies of these and closely related cubane clusters a detailed understanding of the cluster bonding has been developed. An interesting series of cubane clusters containing vanadium, iron, and/or cobalt was recently reported using $(\eta^5\text{-}C_5H_4Me)_2V_2S_4$ as the starting framework (52). Equations (30) and (31) show how two Fe—NO

$$(\eta^5\text{-}C_5H_4Me)_2V_2S_4 + Hg[Fe(CO)_3(NO)]_2 \rightarrow (\eta^5\text{-}C_5H_4Me)_2V_2Fe_2(S)_4(NO)_2 \qquad (30)$$

$$(\eta^5\text{-}C_5H_4Me)_2V_2S_4 + Co(CO)_3(NO) \rightarrow (\eta^5\text{-}C_5H_4Me)_2V_2Co_2(S)_4(NO)_2 \qquad (31)$$

and Co—NO fragments can be added, and in the case of iron the product (9) resulting from the addition of one metal was isolated. This species is the first to contain an $M(NO)_2$ group in a closed metal cluster. The interest in this stems from the fact that the $M(NO)_2$ group is isoelectronic with the $M(CO)_3$ group, which is perhaps the most common apical group in carbonyl cluster chemistry. A similar approach to the Fe_2M_2 clusters, where M = Cr and Mo, has been reported using $[Fe(CO)_3(NO)]^-$ (53).

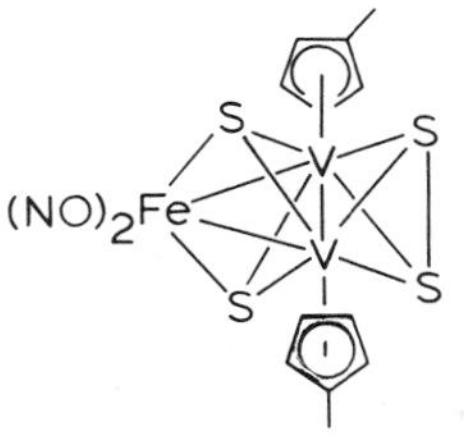

(9)

There has been increasing interest in using Roussin's red salt, $[Fe_2S_2(NO)_4]^{2-}$, as a precursor to polynuclear complexes containing bridging sulfur ligands. The product of the reaction of this anion with cis-Pt $Cl_2(PPh_3)_2$ is $(PPh_3)_2Pt(\mu_3\text{-S})_2Fe_2(NO)_4$ in 90% yield (54). The interesting structural feature is the presence of one linear nitrosyl [172°(ave.)] and one bent nitrosyl [150°(ave.)] on each iron atom.

B. Characterization of Organometallic Nitrosyl Clusters

1. Structural Factors

The bonding of the nitrosyl ligand to metal clusters can occur by several modes. As a terminal ligand, NO can bind as a linear nitrosyl, which is by far the predominant mode, as shown in **2**, **6**, and **8**, or as a bent nitrosyl. Although examples of the latter have been proposed (13), no structurally characterized clusters with a bent nitrosyl exist. As a bridging group there are several examples of a μ_2-NO and these can be further separated into the following subclasses: (1) containing an M—M bond (**10**), and (2) no M—M bond present (**11**). The existence or nonexistence of direct M—M

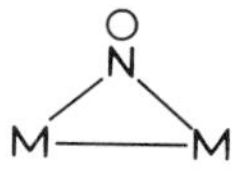

(10) 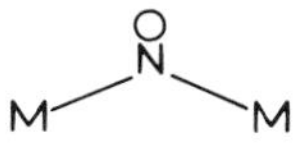   (11)

bonds when one or more bridging ligands are present remains a controversial subject. The criteria used here for deciding whether or not an M—M bond exists is based on electron counting and M—M bond distance arguments. Within the class of compounds discussed there are no ambiguities that arise from applying either of these arguments. A listing of the known nitrosyl clusters is given in Table I.

Table II summarizes the data for the structurally characterized clusters containing the nitrosyl ligand. Some general trends that are distinguishable

TABLE I

Known Nitrosyl-Containing Clusters[a]

Cluster	ν_{NO} (cm^{-1})	δ^{15}N [ppm(NH$_3$)][a]	References
$(\eta^5-C_5H_5)_3Mn_3(NO)_4$	1530, 1480, 1340		39–41
$[(\eta^5-C_5H_5)_3Mn_3(NO)_3(NOH)]^+$			68
$[\eta^5-C_5H_5)_3Mn_3(NO)_3(NH)]^+$			68
$[Ru_3(CO)_{10}(NO)]^-$	1479 (CHCl$_3$)	814 (CH$_2$Cl$_2$)	32, 33
$HRu_3(CO)_{10}(NO)$	1550 (C$_6$H$_{12}$)	807 (CH$_2$Cl$_2$)	18, 33
$HRu_3(CO)_{10-n}(L)_n$			
L = PPH$_3$, n = 1,2,3			18
L = P(OCH$_3$)$_3$[c], n = 3	1460 (KBr)		18
$[Os_3(CO)_{10}(NO)]^-$	1462 (CHCl$_3$)	760 (CH$_2$Cl$_2$)	32, 33, 69
$HOs_3(CO)_{10}(NO)$	1586 (C$_6$H$_{12}$)		18, 33
$HOs_3(CO)_9[P(OCH_3)_3](NO)$	1460 (KBr)		18
$Ru_3(CO)_{10}(NO)_2$	1517, 1500 (KBr)		12, 88
$Ru_3(CO)_8(PPh_3)_2(NO)_2$	1470, 1445 (KBr)		88
$Os_3(CO)_{10}(NO)_2$	1503, 1484 (KBr)		12, 13
$Os_3(CO)_9(L)(NO)_2$	1470, 1455 (C$_6$H$_{12}$)		13, 126
L = P(OCH$_3$)$_3$[c], PPh$_3$, py, N(CH$_3$)$_3$			
$Os_3(CO)_9(NO)_2$	1731, 1705 (C$_6$H$_{12}$)		13
$Os_3(CO)_8(L)(NO)_2$	1705, 1672 (C$_6$H$_{12}$)		13, 14
L = P(OCH$_3$)$_3$[c], PPh$_3$			
$[Co_3(CY)(CO)_7(NO)]^-$, Y = CH$_3$[c],	1716 (THF)		37
Ph, COOH, (C$_5$H$_5$)Fe(C$_5$H$_4$)			
$HFeRu_2(NH)(CO)_8(NO)$	1775 (C$_6$H$_{14}$)		89
$Fe_4N(CO)_{11}(NO)$	1789 (C$_6$H$_{14}$)	366 (CH$_2$Cl$_2$)	20
$Fe_4N(CO)_{11-n}[P(OCH_3)_3]_n(NO)$	1776 (C$_6$H$_{14}$)		89
n = 1[c], 2			

Compound	IR ν(NO)		Ref.
[Fe$_4$C(CO)$_{11}$(NO)]$^-$	1736 (THF)		89
[FeRu$_3$(CO)$_{12}$(NO)]$^-$	1736 (THF)	391 (CH$_2$Cl$_2$)	42, 43
FeRu$_3$N(CO)$_{11}$(NO)	1784 (C$_6$H$_{14}$)	369 (CH$_2$Cl$_2$)	113
L = P(OCH$_3$)$_3$,^{c} dppe	1753 (C$_6$H$_{14}$)		113
HOs$_4$(CO)$_{13}$(NO)	1603 (CH$_2$Cl$_2$)		38
H$_3$Os$_4$(CO)$_{12}$(NO)	1725 (CH$_2$Cl$_2$)		22
[Fe$_4$(S)$_3$(NO)$_7$]$^-$	1790 (CH$_2$Cl$_2$)		29, 55
Fe$_4$S$_4$(NO)$_4$	1720 (CH$_3$CN)		49, 50
[Fe$_4$S$_4$(NO)$_4$]$^-$	1722 (KBr)		50
Co$_4$(NCMe$_3$)$_4$(NO)$_4$			48
Fe$_4$(S)$_2$(NCMe$_3$)$_2$(NO)$_4$	1792 w, 1760 vs, 1745 s (KBr)		49, 51
[Fe$_4$(S)$_2$(NCMe$_3$)$_2$(NO)$_4$]$^-$	1762 w, 1732 vw, 1682 vs, 1664 s (KBr)		51
Mo$_4$(OH)$_4$(CO)$_8$(NO)$_4$			127
(η^5-C$_5$H$_4$Me)$_2$V$_2$Fe$_2$(S)$_4$(NO)$_2$	1765, 1740 (CCl$_4$)		52
(η^5-C$_5$H$_4$Me)$_2$V$_2$Co$_2$(S)$_4$(NO)$_2$	1785, 1762 (CCl$_4$)		52
(η^5-C$_5$H$_4$Me)$_2$V$_2$Fe(S)$_4$(NO)$_2$	1758, 1722 (CCl$_4$)		52
(η^5-C$_5$H$_4$Me)$_2$V$_2$FeCo(S)$_4$(NO)$_2$	1785, 1745 (CCl$_4$)		52
(η^5-C$_5$Me$_5$)$_2$Cr$_2$Fe$_2$(S)$_4$(NO)$_2$	1738, 1712 (KBr)		53
(η^5-C$_5$Me$_5$)$_2$Mo$_2$Fe$_2$(S)$_4$(NO)$_2$	1734, 1708 (KBr)		53
{Na[Mo$_3$(CO)$_6$(NO)$_3$(OCH$_3$)$_3$(O)]$_2$}$^{3-}$	1595 (CH$_3$CN)		31
[Mo$_3$(CO)$_6$(NO)$_3$(OCH$_3$)$_4$]$^-$	1628 (CH$_3$CN)		31
{Na[W$_3$(CO)$_6$(NO)$_3$(OCH$_3$)$_3$(O)]$_2$}$^{3-}$	1585 (CH$_3$CN)		31
[Re$_7$C(CO)$_{20}$(NO)]$^{2-}$			21
[Os$_{10}$C(CO)$_{24}$(NO)]$^-$	1603 (CH$_2$Cl$_2$)		19
[Os$_{10}$C(CO)$_{23}$(NO)]$^-$	1760 (CH$_2$Cl$_2$)		19

^{a} Ph, phenyl; THF, tetrahydrofuran; dppe, Ph$_2$PCH$_2$CH$_2$PPh$_2$; py, pyridine.
^{b} (NH$_3$) denotes chemical shift relative to NH$_3$.
^{c} Denotes derivative for which spectral data are included.

TABLE II. STRUCTURAL DATA

Type	Compound	M—N distance (Å)
Terminal	$Os_3(CO)_8[P(OCH_3)_3](NO)_2$	1.77(4) av.[a] 1.65(4)
Terminal	$[FeRu_3(CO)_{12}(NO)]^-$	1.643(4)
Terminal	$Fe_4N(CO)_{11}[P(OCH_3)_3](NO)$	1.753(4)
Terminal	$Ru_6C(CO)_{14}(NO)_2$	av. 1.78(1)
Terminal	$Ru_6C(CO)_{14}(NO)(AuPPh_3)$	av. 1.72(2)
Terminal	$Fe_4S_4(NO)_4$	av. 1.663(6)
Terminal	$[Fe_4S_4(NO)_4]^-$	av. 1.659(5)
Terminal	$Fe_4S_2(NCMe_3)_2(NO)_4$	av. 1.661(8)
Terminal	$[Fe_4S_2(NCMe_3)_2(NO)_4]^-$	av. 1.61(2)
Terminal	$Co_4(NCMe_3)(NO)_4$	av. 1.65(1)
Terminal	$[Fe_4S_3(NO)_7]^-$	av. 1.64(1)
Terminal	$(Ph_3P)_2Pt(\mu_3-S)_2Fe_2(NO)_4$	av. 1.63(3) (×2) av. 1.69(3) (×2)
Terminal	$\{Na[Mo_3(CO)_6(NO)_3(OCH_3)_3(O)]_2\}^{3-}$	av. 1.82(5)
Terminal	$[Mo_3(CO)_6(NO)_3(OCH_3)_4]^-$	1.815(3)
Terminal	$Mo_4(OH)_4(CO)_8(NO)_4$	1.89(2)[b]
Doubly bridging (M—M bond)	$[Os_3(CO)_{10}(NO)]^-$	(M1)2.014(6) (M2)1.973(5)
Doubly bridging and triply bridging (M—M bond)	$(\eta^5-C_5H_5)_3Mn_3(NO)_4$	av. 1.848(8) (×6) av. 1.93(1) (×3)
Doubly bridging (M—M bond)	$[(\eta^5-C_5H_5)_3Mn_3(NO)_3(NOH)]^+$	av. 1.856(4)
Doubly bridging (M—M bond)	$[(\eta^5-C_5H_5)_3Mn_3(NO)_3(NH)]^+$	av. 1.860(3)
Doubly bridging (No M—M bond)	$Ru_3(CO)_{10}(NO)_2$	av. 2.03(3)
Doubly bridging (No M—M bond)	$Os_3(CO)_9(NMe_3)(NO)_2$	av. 2.05(1)
Doubly bridging (No M—M bond)	$H_3Os_4(CO)_{12}(NO)$	av. 2.10(2)
Doubly bridging (No M—M bond)	$[Os_{10}C(CO)_{24}(NO)]^-$	(M1)2.10(7) (M2)2.10(7)

[a] av., Average.

N—O distance (Å)	M—N—O angle (degrees)	M—M distance (Å)	M—N—M angle (degrees)	References
(N1)1.16(5)	(N1)171(4)			14
(N2)1.21(5)	(N2)165(4)			
1.174(5)	178.1(4)			43
1.131(4)	175.7(3)			89
av. 1.14(2)				34
av. 1.20(2)				34
av. 1.15(1)	av. 117.6(8)			49, 50
av. 1.168(6)	av. 177.5(9)			50
av. 1.166(8)	av. 178.7(8)			49, 51
av. 1.209(5)	av. 177(2)			51
av. 1.16(1)	av. 171(4)			48
av. 1.17	176.3(9) (apical)			55
	171.0(9)			
	165.1(10)			
	166.3(8)			
	164.6(9)			
	167.5(9)			
	166.1(8)			
	av. 166(2)			
	av. 172(8) (×2)			54
	av. 150.3(6) (×2)			
av. 1.19(2)	av. 178(2)			31
1.170(6)	173.6(11)	2.751(1)		31
1.22(3)[b]				127
1.232(10)	(M1)135.6(6)	2.816(2)	90.6(3)	69
	(M2)135.2(6)			
1.212(6) (×3)	av. 137.0(4) (×6)	av. 2.506(3)	av. 85.4(4) (×3)	
1.247(5)	av. 131.4(9) (×3)		av. 81.0(2) (×3)	40
av. 1.207(5)		2.508(1)		68
av. 1.205(4)		2.503(1)		68
(N1)1.20(1)	av. 129(3)	3.150(1)	(N1)101.3(3)	
(N2)1.24(1)			(N2)102.5(4)	12
		3.197(2)	av. 102.4(5)	126
				22
1.23(11)				19

[b] Average of disordered CO and NO.

are (1) the N—O bond distance increases from terminal [1.17 (2) Å] to doubly bridging [1.22(2) Å] to triply bridging [1.247(5) Å] bonding modes, and (2) there is a sharp increase in the M—N bond length of ~0.25 Å in going from a terminal NO to a bridging NO (with an M—M present). Clusters with the NO bridging two metals with an M—M bond have longer M—N distances by about 0.07 Å relative to those that do not have M—M bonds.

The M—N—O angles for the terminal NO ligands average 176(2)°. A noteworthy deviation from this value occurs in the structurally character-ized clusters that contain the $M(NO)_2$ group, $Os_3(CO)_8[P(OCH_3)_2](NO)_2$ (14), and $[Fe_4S_3(NO)_7]^-$ (55), and $(PPh_3)_2 Pt (\mu_3-S)_2 Fe_2(NO)_4$ (54). The M—N—O angles average 168° for the osmium cluster and 166(2)° for the tetrairon cluster. As just mentioned, the $PtFe_2$ cluster contains one linear and one bent nitrosyl ligand. This behavior is also observed in mono-nuclear metal dinitrosyls and has been previously attributed to a strong electronic coupling between the two NO ligands (56, 57).

A problem that is always faced in the structural analysis of compounds containing both CO and NO is the assignment of the nitrogen atom. If the spectroscopic data (discussed in Section II,B,2) do not yield a unique position for the NO, a crystallographic test can be applied. After complete refinement of the proposed model the atom being labeled as N is switched to C and refined again. If the original assignment was correct, then the thermal parameters of the nitrogen should have decreased. In cases where this test has been applied to ordered systems the observed decrease in the isotropic thermal parameter is on the order of 30%. Further, the difference in the M—C and M—N distances is usually around 0.05 Å. However, this cannot be used by itself as proof of the nitrogen location. Due to the similarities of the M—NO and M—CO groups partial or complete disorder is not uncommon, and care must be taken in analyzing these structures.

The most unambiguous approach to the assignment of C or N is the use of neutron diffraction. The neutron scattering power of C and N differs enough that even in disordered compounds the occupancies of C and N can be assigned. The only report of this technique for nitrosyl carbonyl complexes is for $HW_2(CO)_9(NO)$ (58).

2. Infrared and ^{15}N-NMR Spectroscopy of NO-Containing Clusters

The N—O stretching frequencies of cluster coordinated nitrosyl ligands are useful for differentiating the possible bonding modes. The tabulation of the available data in Table I shows that terminal nitrosyl ligands absorb in the range 1789–1664 cm^{-1}. Obviously, for highly charged species such as $\{Na[Mo_3(CO)_6(NO)_3(OCH_3)_3(O)]_2\}^{3-}$ the ν_{NO} will appear at lower energy (1585 cm^{-1} in this example). Doubly bridging nitrosyls absorb in the range 1603–1445 cm^{-1}, and there are no significant differences attributable to the presence or absence of an M—M bond. The one example of a triply bridging NO occurs in $(\eta^5-C_5H_5)_3Mn_3(NO)_4$ which absorbs at 1340 cm^{-1}.

An inherent experimental difficulty that is often encountered results from the low energies of bridging nitrosyl stretching frequencies. Absorbances due to solvent, counterions, and other ligands such as PPh$_3$ can mask the region making it difficult to observe *and* assign the ν_{NO}. In cases where some interference is present the use of isotopic substitution of either the N (the easier and less costly element to substitute) or the oxygen is required. This method has been used to establish the ν_{NO} for PPN[Ru$_3$(CO)$_{10}$(NO)] ($\nu_{^{14}NO} = 1479$ cm^{-1}; $\nu_{^{15}NO} = 1451$ cm^{-1}) (*33*).

Another trend that has been observed with low-valent mononuclear nitrosyl complexes is the dramatic shift of ν_{NO} to lower energy in moving from the first to the second row of the transition elements. This trend is not generally observed in M—CO spectra. In moving from [Fe(CO)$_3$(NO)]$^-$ to [Ru(CO)$_3$(NO)]$^-$ the ν_{CO} peaks actually move $\sim$5cm^{-1} higher in energy, while the ν_{NO} drops $\sim$50 cm^{-1} to lower energy (*45*). Apparently the improved energy match of the π^*NO orbitals with the d_π metal orbitals allows the nitrosyl ligand to absorb all of the additional electron density in going from the first to the second row. The trend is also observed in the series M(PPh$_3$)$_2$(NO)$_2$ for M = Fe (*59*), Ru, and Os (*60*). This series further shows that little difference exists between the second and third rows of the transition elements.

One method that can be used to surmount the problems of observing low-energy nitrosyl absorbances is to use ^{15}N-NMR spectroscopy. Although this typically requires ^{15}N-enriched material, the difference between a terminal linear nitrosyl and a bridging NO is far more dramatic. Until recently only a few ^{15}N chemical shifts for nitrosyl complexes were known; however, several recent studies have substantially increased the data base and have broadened the bonding modes observed (*61–66, 33*). To date the data have been found to fall into four general regions. Terminal, linear, mononitrosyl complexes appear from 350 to 450 ppm [downfield from NH$_3$(l), 25°C]. As in the structural studies metal dinitrosyls are seen to give rise to a region separate from mononitrosyls appearing from 510 to 570 ppm. Complexes containing bent NO ligands are substantially deshielded because of the lone pair on the N, and they appear at 740–870 ppm. Unfortunately this region overlaps directly with the region associated with μ_2-NO ligands from 750 to 815 ppm. Due to the rarity of carbonyl nitrosyl complexes containing a bent NO ligand, it is not likely that the overlap of the bent NO and bridging NO regions will cause much of a problem.

C. Reactivity of Nitrosyl Clusters

As in mononuclear nitrosyl complexes, it is convenient to separate the reactivity into two sections: (1) reactions directly involving change at the NO, and (2) reactions of the metals that are enhanced by the presence of the nitrosyl ligand. The reaction of overwhelming importance that occurs with cluster coordinated nitric oxide is deoxygenation. Depending on the conditions this can ultimately yield NH, NH_2, NCO, or simply N atoms coordinated to the cluster.

1. N Formation

The formation of nitrido clusters from coordinated NO will be discussed first, since it is likely that the other products (NH, etc.) form via coordinated N atoms. In this section we will consider only the reactions of well-characterized nitrosyl clusters. The general condensation of mononuclear nitrosyl complexes to yield nitrido clusters will be discussed in Section III,A,2.

The conversion of $[FeRu_3(CO)_{12}(NO)]^-$ into $HFeRu_3N(CO)_{12}$ via protonation (42) has been found to occur via the nitrido cluster $[FeRu_3N(CO)_{12}]^-$ (43). Equation (31), which is essentially quantitative,

$$[FeRu_3(CO)_{12}(NO)]^- + CO \xrightarrow{THF} [FeRu_3N(CO)_{12}]^- + CO_2 \qquad (32)$$

has been studied in detail. A kinetic analysis of this reaction in the temperature range of 25–65°C yielded $\Delta H^{\ddagger} = +26.4(4)$ kcal/mol and $\Delta S^{\ddagger} = +3.9(1.3)$ eu. These results were consistent with either CO dissociation or M—M bond cleavage being the important step. Since declusterification of the starting material itself occurs under excess CO, inhibition studies were not possible. Scheme 1 shows two possible pathways for the NO to N conversion. Since the kinetic analysis revealed data for the first step only, the intermediates shown must be considered speculative at this time.

The tetranuclear cluster $H_3Os_4(CO)_{12}(NO)$, which contains a nitrosyl ligand bridging an open metal–metal bond (4) reacts at room temperature over several days to form $HOs_4N(CO)_{12}$ [Eq. (33)] (38). Formally the

$$H_3Os_4(CO)_{12}(NO) \xrightarrow[\text{several days}]{\text{THF or } CH_2Cl_2} HOs_4N(CO)_{12} + \text{``}H_2O\text{''} \qquad (33)$$

molecule loses the elements for water, but no details concerning the direct observation of H_2O have been presented. It is noteworthy that the

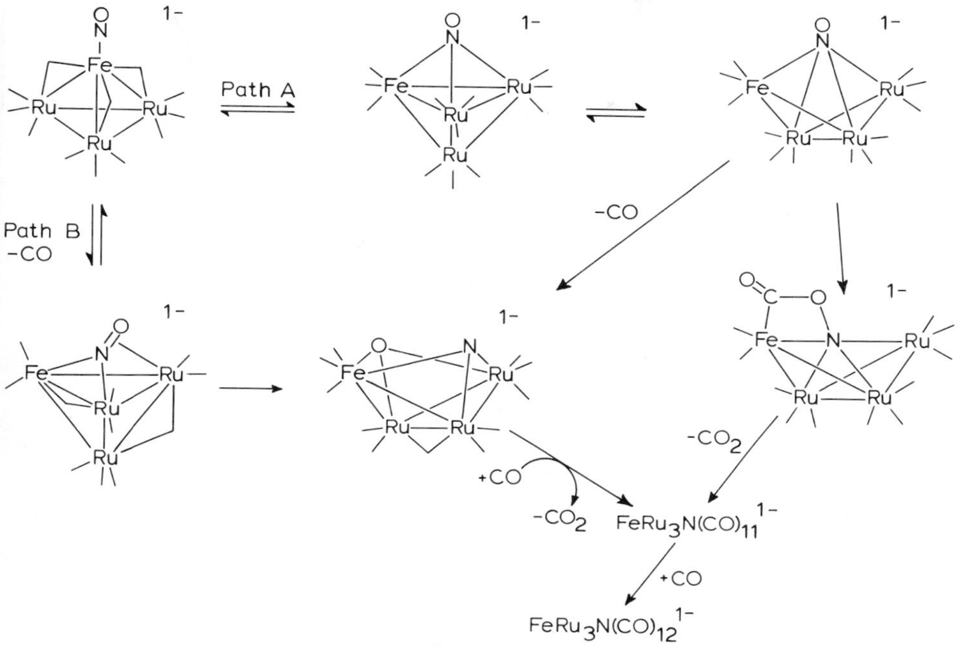

SCHEME 1

preparation of $H_3Os_4(CO)_{12}(NO)$ results from the reaction of NO^+ with $[H_3Os_4(CO)_{12}]^-$, and that the analogous reaction with Ru leads directly to nitrido clusters. One of the two ruthenium clusters, $HRu_4N(CO)_{12}$, apparently is formed in a route similar to the Os derivative (22), i.e., loss of the elements of H_2O. However, $H_3Ru_4N(CO)_{11}$ is also formed in this reaction presumably by loss of "CO_2" (38).

2. Reaction with Electrophiles

O-Methylation (67) and O-protonation (67, 68) of coordinated NO have been reported [Eqs. (34)–(36)]. In each of these reactions the NO ligand is

$$[Ru_3(CO)_{10}(NO)]^- + CF_3SO_3CH_3 \rightarrow Ru_3(CO)_{10}(NOCH_3) \quad (34)$$

$$(60\% \text{ yield})$$

$$[Ru_3(CO)_{10}(NO)]^- + CF_3SO_3H \rightarrow Ru_3(CO)_{10}(NOH) \quad (35)$$

$$[\eta^5\text{-}C_5H_5Me]_3Mn_3(NO)_4 + HBF_4 \rightarrow [(\eta^5\text{-}C_5H_4Me)_3Mn_3(NO)_3(NOH)]^+ \quad (36)$$

$$(12)$$

$$(66\% \text{ yield})$$

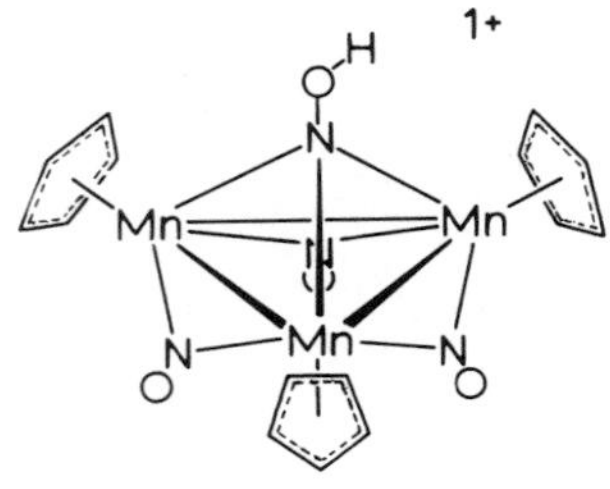

(12)

bridging either two [for $[Ru_3(CO)_{10}(NO)]^-$, which is isostructural with $[Os_3(CO)_{10}(NO)]^-$ (5) (69), or three [for $(\eta^5\text{-}C_5H_4Me)_3Mn_3(NO)_4$ (7) (40)] metals. This same type of reactivity has been observed for carbonyl systems (70–80). In both cases the oxygen of the bridging ligand is considered to be more nucleophilic as a result of increased back-bonding. The Mn system, which is neutral, contains both μ_2 and μ_3 nitrosyls and the structure of the O-protonated product has a μ_3-NOH (12) (68). Although unambiguous evidence proving that protonation occurs only on the μ_3-NO is not available, it certainly would appear to be the most basic site of the molecule. The ν_{NO} for the μ_3-NO appears at 1340 cm^{-1} compared to 1530 and 1480 cm^{-1} for the μ_2-NO ligands, and the μ_3-N—O bond is substantially longer than found for the μ_2-NO ligands.

Protonation of $[Ru_3(CO)_{10}(NO)]^-$ [Eq. (35)] exhibits an interesting dependence on the nature of the acid. While CF_3SO_3H yields the O-protonated product (13), CF_3CO_2H generates the known tautomer $HRu_3(CO)_{10}(NO)$ (14). In inert solvents such as hexane, ether, and

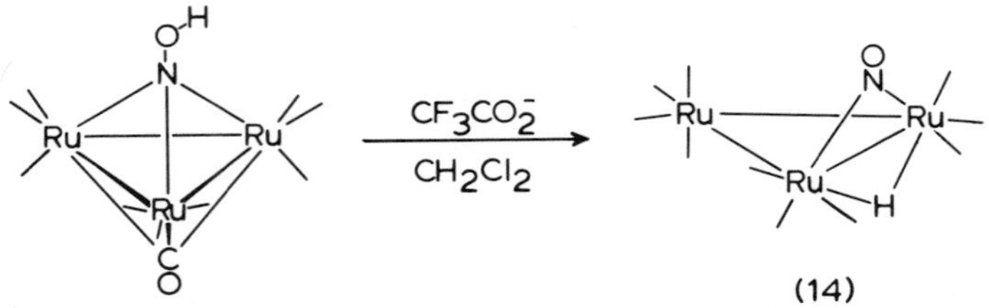

$$\xrightarrow[\text{CH}_2\text{Cl}_2]{\text{CF}_3\text{CO}_2^-} \tag{37}$$

(13)   (14)

CH_2Cl_2 the two compounds do not interconvert. However, as shown in Eq. (37), the addition of a catalytic amount of $[CF_3CO_2]^-$ (or other anions such as NO_3^-) rapidly catalyzes the O—H to M—H tautomerization (67). Two conclusions from this are (1) $HRu_3(CO)_{10}(NO)$ is thermodynamically more stable than $Ru_3(CO)_{10}(NOH)$, and (2) the kinetic site of protonation is on the oxygen.

Further treatment of $[(\eta^5\text{-}C_5H_4Me)_3Mn_3(NO)_3(NOH)]^+$ with two equivalents of acid leads to cleavage of the NO bond [Eq. (38)] and

$$[(\eta^5\text{-}C_5H_4Me)_3Mn_3(NO)_3(NOH)]^+ + 2H^+ \rightarrow [(\eta^5\text{-}C_5H_4Me)_3Mn_3(NO)_3(NH)]^+ + H_2O$$

$$\text{(14\% yield)} \qquad (38)$$

formation of the imido cluster (68). The lower yield must result from oxidation of some Mn species to provide the required two electrons for this reaction to occur. The only other structurally characterized example of the μ_3-NH ligand is $FeCo_2(NH)(CO)_9$, which was formed upon reaction of $[FeCo_3(CO)_{12}]^-$ with NO^+ (81).

The use of protons and electrons to reduce coordinated NO has also been shown with mononuclear complexes. The remarkable transformation of coordinated NO_2^- all the way to NH_3 on $[Ru(trpy)(bpy)L]^{2+}$ (trpy = 2,2′,2″-terpyridine and bpy = 2,2′-bipyridine) proceeds through the nitrosyl complex $[Ru(trpy)(bpy)NO]^{3+}$ (82). This chemistry models the reactions of the enzyme nitrite reductase. The reduction of NO with hydridic reagents is uncommon and the only case reported for multinuclear systems involves the reduction of the dinuclear complex $(\eta^5\text{-}C_5H_5)_2Cr_2(NO)_4$ to $(\eta^5\text{-}C_5H_5)_2Cr_2(NO)_3(NH_2)$ in 15% yield using $Na[H_2Al(OCH_2CH_2OCH_3)_2]$ (83). In mono- or polynuclear complexes with both NO and CO as ligands, the latter is observed to be more electrophilic. Therefore the attack on NO will be anticipated only for complexes without CO or other ligands such as allyl or olefins that are susceptible to attack by nucleophiles.

3. Reduction with H_2

Two reports of the use of molecular hydrogen to reduce the N—O bond have appeared for the trinuclear cluster $HRu_3(CO)_{10}(NO)$ (84, 85). Equation (39) represents the results obtained using a batch reactor system

$$HRu_3(CO)_{10}(NO) + H_2 \xrightarrow{75°C} HRu_3(NH)(CO)_9(10\%) \text{ (15)}$$

$$+ HRu_3(NH_2)(CO)_{10}(37\%) \text{ (16)}$$

$$+ H_4Ru_4(CO)_{12}(8\%) + H_3Ru_4(NH_2)(CO)_{12}(\text{trace}) \qquad (39)$$

(84), and similar results were obtained by in situ infrared spectroscopic examination with a high-pressure IR cell (85). The rate of the reaction was shown (86) to be inhibited by added CO and to depend on the H_2 pressure up to 100 atm; it was independent of H_2 pressure at higher pressures.

These data were interpreted in terms of a reversible CO dissociation as the first step followed by H_2 oxidative addition. Monitoring the reaction by high-pressure liquid chromatography revealed that $H_2Ru_3(NH)(CO)_9$ is formed before $HRu_3(NH_2)(CO)_{10}$.

Independent studies of $H_2Ru_3(NH)(CO)_9$ (**15**) show that it converts to the amido cluster **16** [Eq. (40)], under the reaction conditions, but the

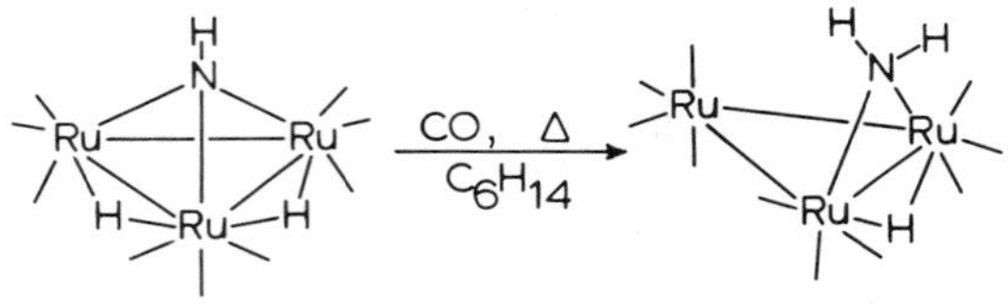

$$ (40) $$

(15)   (16)

reverse reaction does not occur. Under the conditions of the reaction, $HRu_3(NH_2)(CO)_{10}$ slowly forms $H_4Ru_4(CO)_{12}$. The nature of the final nitrogenous product was suggested to be NH_3 but this has not yet been proven.

One mechanism suggested for this reduction involved the intermediacy of a coordinated NOH (*84*). Evidence supporting this conclusion was the similarity of the reaction of $Ru_3(NOCH_3)(CO)_{10}$ and H_2 with the reaction in Eq. (39) (*67*). The initial reaction [Eq. (41)] activates the molecular

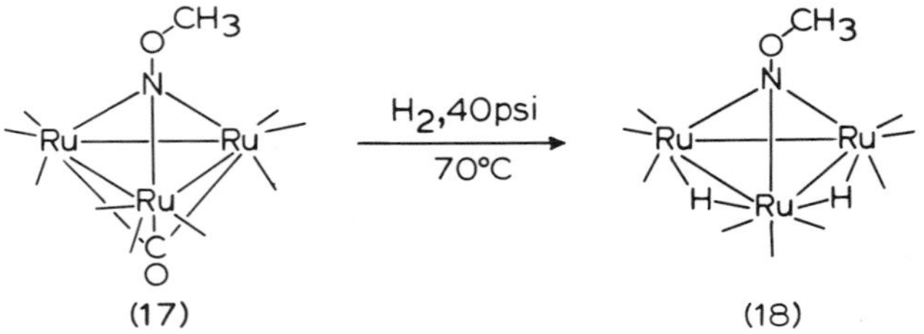

$$ (41) $$

(17)   (18)

hydrogen. This is followed by a reduction that yields the same products in similar ratios as found for the nitrosyl cluster.

Reduction of the NO ligand on $HOs_3(CO)_{10}(NO)$ requires higher temperatures (140°C) than the ruthenium analogue, however, the products are the same with one exception. The unusual hydrogen-rich cluster $H_4Os_4(NH)(CO)_8$ is the major product (*86*). Its structure is similar to $H_2Ru_3(NH)(CO)_9$ (**15**) with the third M—M bond bridged by one hydrogen and a terminal hydrogen replacing one CO. It is similar to the structurally characterized analog $H_4Os_3(NCH_2CF_3)(CO)_8$ (*87*).

4. *Metal-Centered Reactions*

The kinetics of the phosphine substitution of $Ru_3(CO)_{10}(NO)_2$ were reported to proceed via the dissociative mechanism shown in Eqs. (42) and (43) (*88*). The overall rate was only slightly faster than that observed for

$$Ru_3(CO)_{10}(NO)_2 \rightarrow Ru_3(CO)_9(NO)_2 + CO \qquad (42)$$

$$Ru_3(CO)_9(NO)_2 + PPh_3 \rightarrow Ru_3(CO)_9(PPH_3)(NO)_2 \qquad (43)$$

$Ru_3(CO)_{12}$, which suggests that the bridging nitrosyl ligands do not act as electron wells in the same fashion proposed for mononuclear nitrosyl carbonyl complexes.

The cluster $Os_3(CO)_9(NO)_2$ is a unique and important species (*13, 14*). As shown in **2** the structure of the $P(OCH_3)_3$-substituted derivative contains two $Os(CO)_4$ fragments and an $Os[P(OCH_3)_3](NO)_2$ group having linear terminal NO ligands. The reactivity of $Os_3(CO)_9(NO)_2$ with donors varying from PPh_3 to toluene has been examined (*13*). The general scheme proposed to account for the reactivity is shown in Scheme 2. With $L = NH_3$ or H_2NEt, Step 1 is reversible and the adduct (which has new ν_{NO} bands at 1678 and 1617 cm^{-1} compared to 1731 and 1705 cm^{-1} in the starting cluster) is not stable in the absence of excess ligand. With $L = P(OCH_3)_3$ or PPh_3, no simple adduct is observed at all, rather CO substitution occurs rapidly, presumably by an L association step followed by a CO dissociation. When $L = CO$ a slow association occurs to give a metastable form of $Os_3(CO)_{10}(NO)_2$. At higher temperature (80°C) CO dissociation reforms the nonacarbonyl along with some $Os_3(CO)_{10}(\mu_2\text{-}NO)_2$. With $L = $ pyridine (at reflux for 5 hours) the green cluster $Os_3(CO)_9(py)(NO)_2$ with bridging NO ligands is formed. Finally, ^{13}C-NMR and IR spectral studies show that even toluene itself forms an adduct reversibly. This study has clearly shown the value that can be derived from a conversion of three CO ligands to two NO ligands. The ease with which adduct formation occurs even with weak donors is greatly enhanced relative to $Os_3(CO)_{12}$. The synthesis of additional clusters of this type is clearly an important goal.

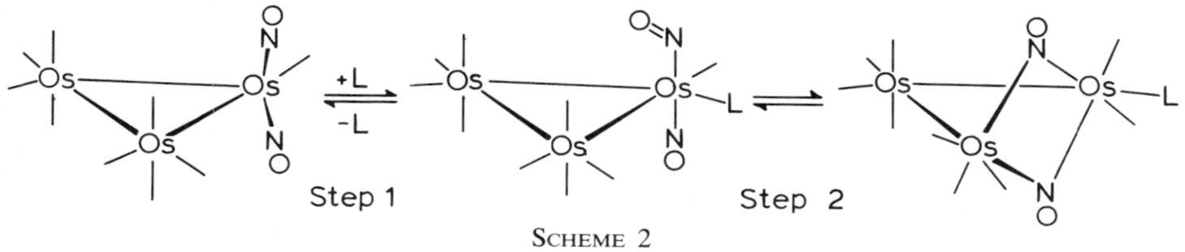

SCHEME 2

A kinetic study of the reaction of $Fe_4N(CO)_{11}(NO)$ **(19)** *(20)* with PPh_3 [Eq. (44)] shows the reaction to be cleanly second order *(89)*. The structural analysis of $Fe_4N(CO)_{10}[P(OCH_3)_3](NO)$ **(20)** shows that the NO

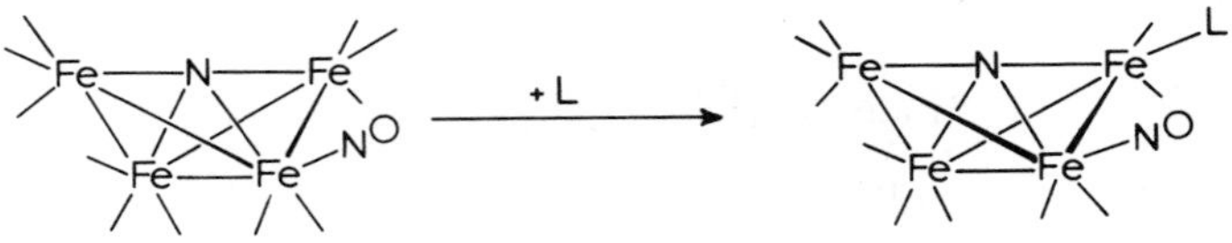

$$(44)$$

(19)   (20)

is located in the symmetric position on a hinge Fe and the $P(OCH_3)_3$ is located on a wing-tip Fe. The rate for the analogous substitution using $P(OCH_3)_3$ was too fast to measure at room temperature. The activation parameters for L = PPh_3 were $\Delta H^{\ddagger} = 12.2$ kcal/mol and $\Delta S^{\ddagger} = -34$ eu. Because this is one of the first kinetic studies involving butterfly clusters it is not clear whether the NO is directly enhancing the rate or not.

III

NITRIDO CLUSTERS

A. *Preparation of Nitrido Clusters*

The primary source of the nitrogen atoms in nitrido clusters is ultimately its monoxide, which is usually coordinated to a transition metal. Therefore, several of the methods used to form nitrido clusters are identical to methods used to form nitrosyl complexes. In this section the synthetic approaches are broken into three categories. The first involves the use of nitrosylating reagents such as NO^+, NO_2^-, or isolated metal nitrosyls themselves. The second category involves the conversion of a coordinated isocyanate into a nitrido cluster, and the last section discusses the interconversion of one nitrido cluster into another. Table III contains a list of the known nitrido clusters.

1. *Via NO: NO, NO^+, or NO_2^-*

The first syntheses of carbonyl clusters containing an interstitial nitrogen atom involved the use of NO^+ [Eqs. (45) and (46)] *(3)*. In the case of rhodium, the major product was suggested to be $[Rh_6(CO)_{14}(NO)]^-$, but no further data have been reported on this. The reaction of NO^+ with the mixture of iron carbonyls, $Fe(CO)_5$ and $[Fe_2(CO)_8]^{2-}$, yielded two clusters

TABLE III

KNOWN NITRIDO CLUSTERS AND ^{15}N CHEMICAL SHIFTS

Cluster[a]	δ^{15}N[ppm(NH$_3$)][b]	References
$(\eta^5\text{-}C_5H_5)_3Mo_3(N)(O)(CO)_4$		92, 93
$(\eta^5\text{-}C_5H_5)_3Mo_2W(N)(O)(CO)_4$		92
$(\eta^5\text{-}C_5H_5)_3MoW_2(N)(O)(CO)_4$		92
$(\eta^5\text{-}C_5H_5)_3W_3(N)(O)(CO)_4$		92
$[Fe_4N(CO)_{12}]^-$	618 (CH$_2$Cl$_2$)	20, 90
$HFe_4N(CO)_{12}$	591 (CH$_2$Cl$_2$)	20, 90
$Fe_4N(CO)_{11}(NO)$	596 (CH$_2$Cl$_2$)	20
$Fe_4N(CO)_{11-n}[P(OCH_3)_3]_n(NO)$, $n = 1^c,2$		89
$[FeRu_3N(CO)_{12}]^-$	559, 520 (CH$_2$Cl$_2$)	43
$\{FeRu_3N(CO)_{12-n}[P(OCH_3)_3]_n\}^-$, $n = 1^c,2$	545, 508 (CH$_2$Cl$_2$)	113
$HFeRu_3N(CO)_{12}$	531 (CH$_2$Cl$_2$)	43
$HFeRu_3N(CO)_{12-n}[P(OCH_3)_3]_n$, $n = 1,2^c,3$	514, 574 (J_{PN} = 6.1 Hz) (CH$_2$Cl$_2$)	113
$FeRu_3N(CO)_9(L)_2(NO)$, L = P(OCH$_3$)$_3$, dppe		113
$CoRu_3N(CO)_{12}$		81
$[Ru_4N(CO)_{12}]^-$	519 (CH$_2$Cl$_2$)	91
$HRu_4N(CO)_{12}$		22, 91
$HRu_4N(CO)_{11}[P(OCH_3)_3]$		22
$H_3Ru_4N(CO)_{11}$		38
$[Os_4N(CO)_{12}]^-$		38
$HOs_4N(CO)_{12}$		38
$[Fe_5N(CO)_{14}]^-$		90, 101
$HFe_5N(CO)_{14}$		90
$[HFe_5N(CO)_{13}]^{2-}$		90
$[FeRu_4N(CO)_{14}]^-$		43
$[Ru_5N(CO)_{14}]^-$	465 (CH$_2$Cl$_2$)	67, 91, 94
$(\eta^5\text{-}C_5Me_5)_2Mo_2Co_3N(CO)_{10}$		128
$[Ru_6N(CO)_{16}]^-$	559 (CH$_2$Cl$_2$)	91, 94
$[Co_6N(CO)_{15}]^{1-}$	196 (acetone-d_6)	3
$[Rh_6N(CO)_{15}]^{1-}$	108 (acetone-d_6)	3
$[Rh_{10}PtN(CO)_{22}]^{3-}$		96

[a] Ph, phenyl; dppe, Ph$_2$PCH$_2$CH$_2$PPh$_2$.

[b] (NH$_3$) denotes chemical shift relative to NH$_3$; solvent is given in parentheses.

[c] Denotes derivative for which spectral data are included.

$$[Co_6(CO)_{15}]^{2-} + NOBF_4 \xrightarrow{THF} Co(II) + Co(CO)_3(NO) + Co_4(CO)_{12} + [Co_6N(CO)_{15}]^-$$

(45)

(**21**)

(40–50% yield)

$$[Rh_6(CO)_{15}]^{2-} + NOBF_4 \xrightarrow{THF} [Rh_6N(CO)_{15}]^-$$

(< 10% yield)   (46)

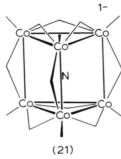

(21)

$[Fe_4N(CO)_{12}]^-$ and $[Fe_5N(CO)_{14}]^-$ (*90*). After 1 hour at 130°C the tetranuclear cluster is isolated in 3.3% yield, while the pentanuclear species, which forms after longer reaction times (2 hours) at a slightly higher temperature (145°C), is produced in 66% yield. This anion could be protonated and the structure of $HFe_5N(CO)_{14}$ is shown in **22**. It was noted that below 130°C a nitrosyl complex was formed.

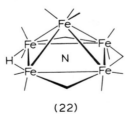

(22)

The reaction of the tetranuclear Ru and Os anionic clusters, $[H_3M_4(CO)_{12}]^-$, with NO^+ has been reported to give both nitrosyl and nitrido clusters (*22, 38*). With M = Ru only nitrido products are isolated in low yield [Eq. (47)]. The dodecacarbonyl cluster is similar in structure to

$$[H_3Ru_4(CO)_{12}]^- + NO^+ \rightarrow HRu_4N(CO)_{12} + H_3Ru_4N(CO)_{11}$$ (47)

the iron analogue (discussed in Section III,A,2) while $H_3Ru_4N(CO)_{11}$ has two additional bridging hydrogen atoms in place of one CO (**23**).

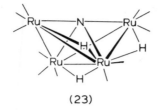

(23)

Nitrosylation of $[CoRu_3(CO)_{13}]^-$ with NO^+ produces the nitrido cluster $CoRu_3N(CO)_{12}$ in low yield (81). The major products from this reaction $[Ru_3(CO)_{12}$ and $Co(CO)_3(NO)]$ arise from cluster fragmentation.

The use of nitrite (NO_2^-) as a source of metal nitrosyls is now well established as discussed in Section II,A3. One study of the reaction of $PPN(NO_2)$ with carbonyl clusters of the Co triad revealed an alternative synthesis of $[Co_6N(CO)_{15}]^-$ and $[Rh_6N(CO)_{15}]^-$ (35). As shown in Eqs. (48) and (49), both the tetranuclear and hexanuclear starting materials for Rh lead to the same nitrido cluster. The reasonable yields for Rh by either route suggest that the product is somewhat of a thermodynamic sink. Consistent with this idea, Eq. (50) shows the use of gaseous nitric oxide itself to form the Rh nitrido cluster directly (3).

$$M_4(CO)_{12} + PPN(NO_2) \xrightarrow[\text{room temp.}]{\text{THF}} [M_6N(CO)_{15}]^- \tag{48}$$

$$(M = Co, 12.5\%)$$

$$(M = Rh, 30\%)$$

$$M_6(CO)_{16} + PPN(NO_2) \rightarrow [M_6N(CO)_{15}]^- \tag{49}$$

$$(M = Rh, 74\%)$$

$$[Rh_7(CO)_{16}]^{3-} + NO/CO\ (1:1) \xrightarrow{CH_3OH} [Rh_6N(CO)_{15}]^- \tag{50}$$

$$(40\text{–}60\%\ \text{yield})$$

It is highly probable that all of the above reagents act by formation of an intermediate metal nitrosyl. In the following examples the source of nitrogen is a coordinated nitric oxide.

2. Via NO: Metal Nitrosyls

The first observation of the formation of a nitrido ligand from a nitrosyl ligand involved the homonuclear Fe reaction shown in Eq. (51) (20, 42). There are numerous examples of reactions involving the conversion of a coordinated nitrosyl ligand into a nitrido cluster. In some cases [Eqs. (51) (20), (52) (91), and (53) (92, 93)], the NO is bound to a mononuclear

$$2[Fe(CO)_3(NO)]^- + 3Fe_3(CO)_{12} \rightarrow 2[Fe_4N(CO)_{12}]^- + 3Fe(CO)_5 + 2CO + CO \quad (51)$$

(**24**)

(50% yield)

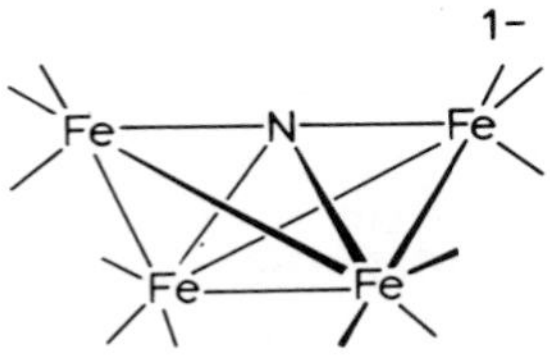

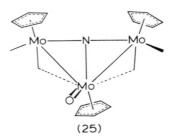

(**24**)

$$[Ru(CO)_3(NO)]^- + Ru_3(CO)_{12} \rightarrow [Ru_4N(CO)_{12}]^- + CO_2 + 2CO \quad (52)$$

(40% yield)

$$M(\eta^5\text{-}C_5H_5)(CO)_2(NO) + [M'(\eta^5\text{-}C_5H_5)(CO)_3]_2 \xrightarrow{200°C,\ 1\ hour}$$

$$MM_2' \ (\eta^5\text{-}C_5H_5)_3(N)(O)(CO)_4 \quad (53)$$

(**25**)

(low yield)

(M = Mo, W; M' = Mo, W)

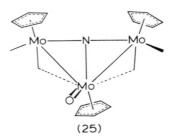

(**25**)

complex which is condensed with a di- or polynuclear species to give the product. For all these reactions CO_2 has been detected using mass spectrometry, and in the study of Eq. (53), ^{18}O from $N^{18}O$ was found in the CO_2 rather than in the oxide ligand (*92*). The compound $Mo_3(\eta^5\text{-}C_5H_5)_3(N)(O)(CO)_4$ was first isolated in low yield from the reaction of $[Mo_2(\eta^5\text{-}C_5H_5)(CO)_2]_2$ with ethyl diazoacetate (*93*).

In the above studies the overall complexity of the transformation renders it difficult to comment on the details of the NO bond cleavage step. Two studies involving the direct conversion of a coordinated NO into a nitrido ligand without a change in nuclearity of the cluster are simpler reactions and were discussed in Section II,C,1. The only other example of a conversion of a cluster coordinated nitrosyl into a nitrido ligand is the pyrolysis of $[Ru_3(CO)_{10}(NO)]^-$ to form low yields of $[Ru_5N(CO)_{14}]^-$ (67).

3. Via NCO

The observation of the conversion of a coordinated isocyanate into a nitrido cluster opens a new pathway to these species (91, 94). Equations (54) (94) and (55) (91) both proceed in high yield.

$$2Ru_3(CO)_{12} + PPN(N_3) \rightarrow PPN[Ru_6N(CO)_{16}] + CO \qquad (54)$$

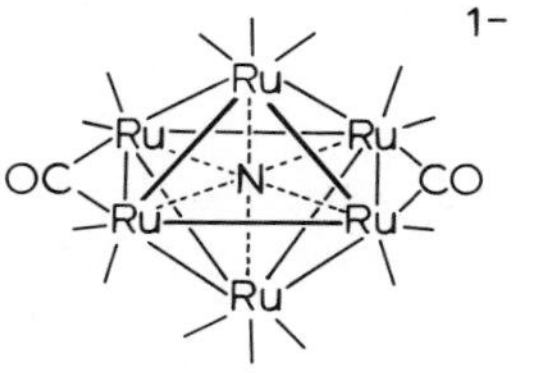

(26)

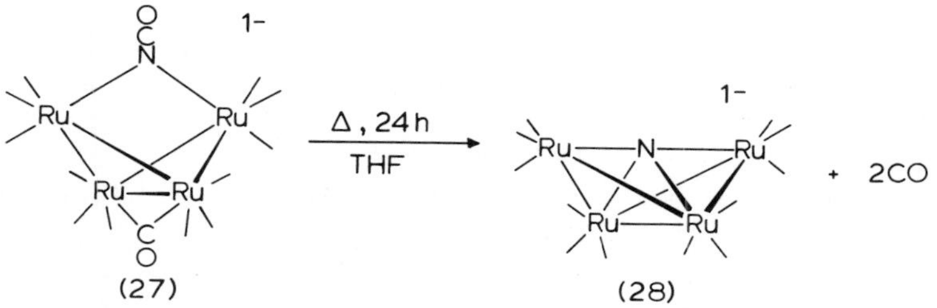

(90% yield)

(26)

(55)

The reaction of N_3^- with $Ru_3(CO)_{12}$ has been studied and shown to produce in a stepwise fashion $[Ru_3(NCO)(CO)_{11}]^-$, which loses a CO to form $[Ru_3(\mu\text{-}NCO)(CO)_{10}]^-$, which slowly loses more CO at room temperature to form $[Ru_4(NCO)(CO)_{13}]^-$ (27) (95). In Eq. (54) the formation of the trinuclear isocyanato clusters occurs upon mixing with the azide. These eventually react with the excess $Ru_3(CO)_{12}$ to form the product. Mechanistic details of Eq. (54) will be difficult to establish, once again, because of the change in nuclearity of the cluster. However, the simplicity of Eq. (55) offers some hope that the mechanistic details of the N—C bond breaking step may be unraveled, and this work is underway. The reverse of this reaction can occur under high CO pressure and will be discussed in Section III,F.

4. *Interconversions of Nitrido Clusters*

Although the conversion of a particular carbido cluster into another with a different metal framework is an important synthetic method for carbides (*1*), the analogous process is essentially untapped for nitrido clusters. It has been used [Eq. (56)] to form the largest known nitrido cluster, $[PtRh_{10}N(CO)_{21}{}^-]$ (**29**) (*96*), and it proved successful in adding a fifth vertex in the formation of the mixed-metal cluster $[FeRu_4N(CO)_{14}]^-$ [Eq. (57)] (*43*).

$$[Rh_6N(CO)_{15}]^- + [PtRh_4(CO)_4]^{2-} \rightarrow PtRh_{10}N(CO)_{21}{}^{3-} + \text{other products} \quad (56)$$

$$\text{(10–20\% yield)}$$

$$[FeRu_3N(CO)_{12}]^- + Ru_3(CO)_{12} \rightarrow [FeRu_4N(CO)_{14}]^- \quad (57)$$

$$\text{(10\% yield)}$$

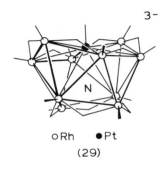

3-

oRh   ●Pt

(29)

One reaction that moves in both directions (from larger to smaller number of metals, and vice versa) occurs with $[Ru_6N(CO)_{16}]^-$ [Eq. (58)] (*91*). Under higher pressures of CO, $[Ru_5N(CO)_{14}]^-$ releases $Ru(CO)_5$ and forms $[Ru_4N(CO)_{12}]^-$. This reaction can also be reversed.

$$[Ru_6N(CO)_{16}]^- + 2CO \rightleftharpoons [Ru_5N(CO)_{14}]^- + \tfrac{1}{3}Ru_3(CO)_{12} \quad (58)$$

$$\textbf{(30)}$$

$$\text{(95\% yield)}$$

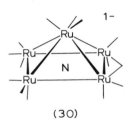

1-

(30)

The analogous reaction with the isoelectronic carbido cluster, $Ru_6C(CO)_{17}$, requires far more vigorous conditions (80 atm CO, 70°C, 3 hours) (97).

B. *Structural Features of Nitrido Clusters*

The structures of most of the nitrido clusters are related by an octahedral (or part thereof) geometry about the nitrogen (Scheme 3). Specific data for the nitrido clusters are listed in Table IV. At the top end the only cluster in which the nitrogen is surrounded by six metals in a proposed octahedral fashion is $[Ru_6N(CO)_{16}]^-$ (26). Although an X-ray crystallographic analysis has not been reported, the 86 electron count is the same as $Ru_6C(CO)_{17}$ (98) and $[Ru_6C(CO)_{16}]^{2-}$ (99), both of which adopt an octahedral metal atom arrangement.

The notable exceptions to Scheme 3 are the hexanuclear nitrido clusters of the cobalt triad. These adopt the trigonal prismatic framework shown in **21** for $[Co_6N(CO)_{15}]^{1-}$ (3). The difference between these clusters and $[Ru_6N(CO)_{16}]^{1-}$ is their 90 electron count. The trigonal prismatic structure can be rationalized as an *arachno*-square antiprism using skeletal electron pair theory. Once again related carbido clusters are known with the formula $[M_6C(CO)_{15}]^{2-}$ for Co and Rh (100). In all of these structures the M—M bond distance between the two triangular planes is slightly longer than within the planes.

The three M_5N structures that have been characterized exhibit square pyramidal geometry with the nitrogen situated slightly *below* or in the square face. For $HFe_5N(CO)_{14}$ (**22**) (90), the nitrogen is 0.093 Å below the face, and in $[Ru_5N(CO)_{14}]^-$ (**30**) (91) the corresponding value is 0.21 Å. However, in the iron cluster $[Fe_5N(CO)_{14}]^-$, the nitrogen atom lies in the basal plane of metal atoms (101). For comparison, in the neutral ruthenium carbido cluster, $Ru_5C(CO)_{15}$ (97), the carbon atom is 0.11 Å below the square face. The remaining parameters such as M—N(C) distances exhibit the expected decrease in going from C to N [with the exception of the apical M—N(C) distance].

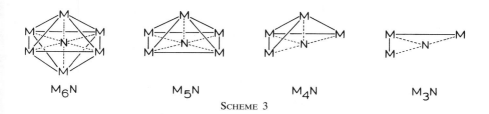

M_6N M_5N M_4N M_3N

SCHEME 3

TABLE IV

STRUCTURAL DATA FOR NITRIDO CLUSTERS

Type	Formula	Geometry	M—N distances (Å)	M—N—M angles (degrees)	References
M_3N	$Mo_3(N)(O)(\eta^5\text{-}C_5H_5)_3(CO)_4$	T-Shaped	1.972(4)	171.8(2)	93
M_4N	$[Fe_4N(CO)_{12}]^-$	Butterfly	1.921(4) (×2) 1.900(5) (×2) 1.771(5) (×2)	94.1(2) (×2) 179.0(3) 90.4(4) (×4) 82.8(2)	20
M_4N	$HFe_4N(CO)_{12}$	Butterfly	1.92(2) (×2) 1.77(1) (×2)	178.4(6)	90
M_4N	$[Os_4N(CO)_{12}]^-$	Butterfly	2.12(2) (×2) 1.96(2) (×2)		38
M_4N	$HRu_4N(CO)_{11}[P(OCH_3)_3]$	Butterfly	2.110(4) (×2) 1.93(2) (×2)	173.2(3)	22
M_4N	$H_3Ru_4N(CO)_{11}$	Butterfly	2.097(5) (×2) 2.017(5) (×2) 1.960(9) (×4)	173.7(3)	38
M_4N	$Fe_4N(CO)_{10}[P(OCH_3)_3](NO)$	Butterfly	1.906(3) (×2) 1.774(3) (×2)	178.2(2) 90(1) (×4)	89
M_4N	$[FeRu_3N(CO)_{12}]^-$	Butterfly	2.059(8) (×2) 1.86(1) (×2)	176.6(2) 89(1) (×4) 80.1(1)	43

M$_4$N	[FeRu$_3$N(CO)$_{10}$[P(OCH$_3$)$_3$]$_2$]$^-$	Butterfly	2.03(1) 1.94(1)	172.0(8) 86.9(8) (×2) 79.9(5)	113
M$_5$N	HFe$_5$N(CO)$_{14}$	Square pyramidal	1.913(2) 1.836(3) (×4)		90
M$_5$N	[Fe$_5$N(CO)$_{14}$]$^-$	Square pyramidal	1.92(2) 1.85(3) (×4)		101
M$_5$N	[Ru$_5$N(CO)$_{14}$]$^-$	Square pyramidal	2.14(2) 2.03(5) (×4)		91, 94
M$_6$N	[Co$_6$N(CO)$_{15}$]$^-$	Trigonal prismatic	1.94(1) (×6)	82 (×6) 83 (×3)	3
M$_{11}$N	[Rh$_{10}$PtN(CO)$_{21}$]$^{3-}$	Trigonal bipyramidal (surrounding N)	2.12(1) (×2 Rh eq.)[a] 2.05(1) (×2 Rh ax.) 1.92(2) (Pt eq.)	150 (ax.)[a] 120 (×3 eq.)	96

[a] Eq., Equatorial; ax., axial.

71

Although its formula contains more than five metal atoms the nitrogen atom is surrounded by five metals in $[Rh_{10}PtN(CO)_{21}]^{3-}$ (**29**). The nitrogen can be considered to be in a distorted trigonal bipyramidal (tbp) environment (**31**) where the Pt—Rh5—Rh5′ atoms form a nearly

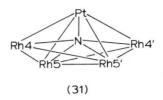

(**31**)

equilateral triangle with the nitrogen slightly closer to the platinum (1.92 vs 2.12 Å) (*96*). Atoms Rh4 and Rh4′ form the axial positions and it is here that the greatest distortion from tbp geometry occurs. The Rh4—N—Rh4′ angle is ~150°. As shown in **29**, the metals in this cluster exist in two layers: (1) a six-membered, puckered ring of rhodium atoms with a Pt in the center of the ring, and (2) a puckered four-membered ring.

The M_4N butterfly is the most prevalent structure for nitrido clusters. $[Fe_4N(CO)_{12}]^-$ (**24**) and $H_3Ru_4N(CO)_{11}$ (**23**) are two examples which emphasize the basic $[M_4N(L)_{12}]$ structure. In each case the metals surrounding the nitrogen are in nearly perfect octahedral geometry with two cis sites vacant. The interaction of the nitrogen with the wing-tip sites is always much stronger than with the hinge sites, as evidenced by the shortened (0.16 Å) bond lengths.

In the only M_3N cluster that has been structurally characterized, $(\eta^5\text{-}C_5H_5)_3Mo_3(N)(O)(CO)_4$ (**25**) (*93*), the Mo—N bonds of the linear Mo—N—Mo fragment are shorter than the unique Mo—N bond. This structure is interesting in that it contains both low (+2) and high (+4) formal oxidation states of Mo, making this cluster a connection between the organometallic nitrido clusters and the high-valent nitrido complexes (*11*).

Some comments regarding the electron counting rules are warranted. In every nitrido cluster prepared to date the nitrogen atom should be considered as a five-electron donor. In M_6N clusters or others where the nitrogen is completely surrounded by metals there is less ambiguity regarding the number of electrons donated. However, for the clusters such as the M_4N butterfly type it is not obvious how all five of the electrons are contributed. The alternative to contributing five electrons would be a scheme where N donated three electrons to the cluster and maintained a lone pair of electrons pointing radially away from the center of the cluster. In essence this means the nitrogen atom would be a vertex of the cluster, meaning that a species such as $[Fe_4N(CO)_{12}]^-$ would be considered as a five-vertex cluster. We believe the structural evidence argues against such a formulation. In *every* example of a structure with an exposed N atom the M—N distances are much shorter between the metals located at the exposed edge of the nitrogen. We have proposed that these metals are able to stabilize the extra electrons on nitrogen via π-bonds (20). These π-interactions are then responsible for allowing all five electrons on nitrogen to be contributed to the cluster. Such an explanation would mean that the reactivity of N atoms as donors would be greatly moderated in these clusters, which appears to be true. This is not to say that a three-electron donor pyramidal N atom is not accessible during reactions, and, in fact, this has been proposed in the isomerization of $[FeRu_3N(CO)_{12}]^-$ discussed in Section III,D.

Three papers have discussed the details of the bonding in the closely related carbido clusters (102–104). The results of the calculations confirm the importance of the π-overlap between the metals and the carbon atom along the exposed edge of the cluster. Another important point relevant to the question of the number of electrons donated by the carbon atom was addressed in one of the papers. It was found that although the C $2s$ orbital lies at much lower energy, than the metal orbital with which it interacts, the interaction is sizable due to the very large overlap (102). While the energy difference will be even greater with the more electronegative nitrogen atom, we again note that the M—N bond distances are shorter than the M—C bond distances, and this should enhance the overlap.

C. Nitrogen vs Carbon

Enough data on these clusters are now available to make a preliminary comparison of the similarities and differences of carbide and nitride ligands. Advances in understanding of the perturbations caused by the shift from C to N will help draw a clearer picture of the complex bonding of atomic species to clusters. In this light, there are three observations that seem to point to the same conclusion.

1. In the high nuclearity cluster, $[Rh_{10}PtN(CO)_{21}]^{3-}$, nitrogen is found to be five coordinate. It was pointed out that in large carbido clusters carbon usually occupies sites with coordination numbers of six to eight (96). In $[Rh_{10}PtN(CO)_{21}]^{3-}$ the nitrogen has forced the reduction of the overall compactness of the cluster to achieve five coordination. Again this is not observed for high-nuclearity carbido clusters.

2. Currently there are no reported three-coordinate carbido clusters, while the family of compounds with the formula $(\eta^5\text{-}C_5H_5)_3M_3\text{-}(N)(O)(CO)_4$, where M = Mo and W, does contain a three-coordinate

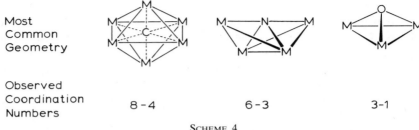

Most Common Geometry		
Observed Coordination Numbers		
8 - 4	6 - 3	3-1

SCHEME 4

nitrogen (*92*). In cases where the synthesis of a M_3C cluster has been attempted the stable product contains the ketenylidene (CCO) ligand (*105*).

3. While interstitial elements are generally believed to exert a stabilizing influence on a molecule by acting as "atomic glue," the cluster $[Ru_6N(CO)_{16}]^-$ converts to $[Ru_5N(CO)_{14}]^-$ within minutes under CO (1 atm, 25°C). The analogous reaction with $Ru_6C(CO)_{17}$ requires 80 atm CO at 70°C for 3 hours.

All of these observations point to the preference of nitrogen for lower coordination numbers relative to carbon. Certainly the smaller size of the atom makes it increasingly difficult to fit larger numbers of metals around it, while maintaining a good bonding distance to each of the metals. An extension of this trend to the next element over, oxygen, also seems to follow this pattern. Several examples of clusters with tricoordinate, pyramidal oxygen are known (*106–110*), whereas no examples of higher coordination numbers exist for low-valent clusters. Scheme 4 seems also to show that the ability of the more electronegative element to stabilize the lone pair of electrons pointing radially away from the cluster contributes to the trend. It will be interesting to view future results to see if this trend holds true.

D. *Nitrogen NMR of Nitrido Clusters*

The first measurements of the nitrogen NMR spectra of nitrido clusters were reported for $[Co_6N(CO)_{16}]^-$ and $[Rh_6N(CO)_{15}]^-$ and posed somewhat of an enigma (*3*). The chemical shifts were 196.2 and 107.6 ppm, respectively (downfield from NH_3), which were far upfield from the anticipated position based on the related carbido clusters. As a first approximation, the chemical shifts for nitrogen (measured relative to NH_4^+) are *twice* the values for carbon (measured relative to CH_4) for isoelectronic molecules (*61*). This relationship arises from the comparison of the radial factor ($\langle r^{-3} \rangle_{2p}$) for nitrogen and carbon as it appears in the paramagnetic shielding term. Since the ^{13}C resonance for $[Co_6C(CO)_{15}]^{2-}$ appears at 332.8 ppm (relative to CH_4) (111) the unusual nature of the upfield shift in $[Co_6N(CO)_{15}]^-$ is particularly apparent.

More recent work with nitrido clusters within the iron triad revealed chemical shifts in the "expected" region. For instance, $[Ru_6N(CO)_{16}]^-$ appears at 538 ppm (NH_4^+) (91) which can be compared to the carbon resonance of $[Ru_6C(CO)_{16}]^{2-}$ which is at 461.2 ppm (CH_4) (112). Although the $\delta^{15}N$ value is not twice as large as $\delta^{13}C$, it is at least larger than the carbon resonance. Of the approximately 15 compounds measured, a range from ~450 to ~620 ppm is observed. Because many of the compounds are closely related some useful trends have appeared.

1. For the same structure the chemical shift of the compound with the first row transition element is ~90–100 ppm downfield from the shift for the second row transition element (3, 43, 91).

2. The effect of protonation of a monoanionic cluster causes the chemical shift to decrease by ~20–30 ppm. This has been tested only with butterfly clusters at this point. In one case the addition of NO^+ to a monoanionic cluster had the same effect on the chemical shift of the nitrido ligand (43, 113).

3. In butterfly clusters the chemical shift is dictated by the nature of the wing-tip metals. $[Ru_4N(CO)_{12}]^-$ appears at 519 ppm, while $[Fe_4N(CO)_{12}]^-$ is located at 619 ppm. In the mixed-metal cluster $[FeRu_3N(CO)_{12}]^-$ two isomeric forms (32 and 33) exist and interconvert [Eq. (59)] (43).

(59)

(32)   (33)

The chemical shift for 33 which has both wing-tip sites occupied by Ru, is 520 ppm, nearly identical to that for $[Ru_4N(CO)_{12}]^-$. For isomer 32 with one Fe in the wing-tip position the chemical shift is 40 ppm downfield from 33. The magnitude and direction of this difference are also reflected in the two isomeric forms of $\{FeRu_3N(CO)_{11}[P(OCH_3)_3]\}^-$ (113).

Returning to $[Co_6N(CO)_{15}]^-$ and $[Rh_6N(CO)_{15}]^-$, both of these are 90-electron hexanuclear clusters. As such they adopt a more open structure than octahedral clusters containing 86 electrons, i.e., $[Ru_6N(CO)_{16}]^-$.

Several reasons for these unusual chemical shifts have been suggested. These include the differing electronegativities of C and N (114), the different ΔE contributions in the denominator of the paramagnetic term (115), as well as differences in site symmetry (116). In view of the more recent data discussed above, it appears that $\delta^{15}N$ is indeed linked to the ΔE factor. In line with this the color of $[Rh_6N(CO)_{15}]^-$ is orange compared to dark red for $[Rh_6C(CO)_{15}]^{2-}$ (115). While this proposal is reasonable, it is clear that more data are necessary to establish the importance of low-lying excited states on the chemical shift of the nitrogen.

E. Dynamic Properties of Nitrido Clusters

The basic fluxional properties of the carbonyl ligands bound to nitrido clusters are the same as those found for related carbido clusters (1). The difference between C and N has no effect on the mechanism of CO migrations and apparently little effect on the activation barriers for these processes. This latter point, however, has not been quantitatively addressed. In M_6N systems the dynamic properties depend on the structure. Both $[Co_6N(CO)_{15}]^-$ and $[Rh_6N(CO)_{15}]^-$ exhibit static ^{13}C-NMR spectra at 25°C consistent with their solid-state structures (3). The octahedral cluster $[Ru_6N(CO)_{16}]^-$ exhibits one sharp resonance at 207.5 ppm down to −87°C indicative of rapid CO exchange over the entire molecule (94). For the square pyramidal structures $[Fe_5N(CO)_{14}]^-$ (90) and $[FeRu_4N(CO)_{14}]^-$ (43), two resonances are observed in a 3:11 ratio. This has been interpreted as involving intrametal CO exchange on the apical metal and intermetal exchange around the basal plane of the cluster.

In all of the M_4N butterfly clusters intrametal CO exchange is the only CO scrambling process observed (Scheme 5). The activation barrier is a function of the metal site (wing-tip or hinge) and whether or not a proton bridges the hinge M—M bond. Also, as in most other CO scrambling processes the activation barrier increases down the triad. For the simplest systems, such as $[Fe_4N(CO)_{12}]^-$, the rearrangements occur rapidly even at −80°C giving rise to two equal intensity singlets, one due to the wing-tip $Fe(CO)_3$ groups and the other due to the hinge sites (90). Protonation of this cluster gives $HFe_4N(CO)_{12}$ which at −80°C exhibits a set of four

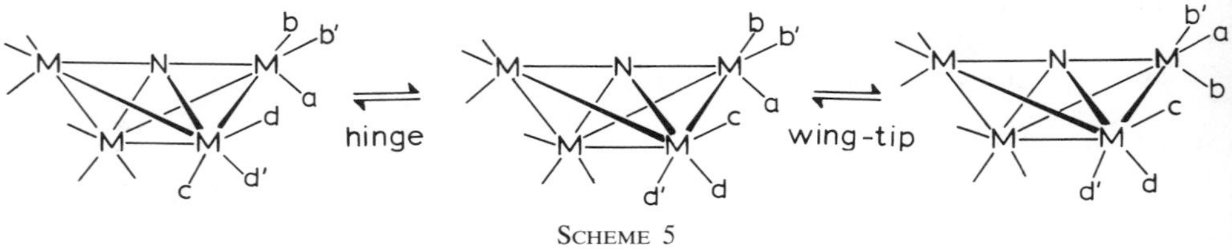

SCHEME 5

resonances in a 4:4:2:2 intensity ratio (90). This is consistent with the solid-state structure **34**; however, no detailed assignment of the resonances

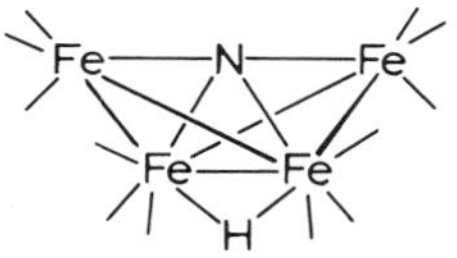

(34)

to the various carbonyls was made. In the closely related clusters $HFe_4(CH)(CO)_{12}$ (117) and $HFeRu_3N(CO)_{12}$ (43), assignments of the spectra were possible at low temperature, and as the temperature was increased the carbonyls on the wing-tip sites scrambled before those on the hinge sites. In the anionic cluster $[FeRu_3N(CO)_{12}]^-$ the opposite was found; the hinge sites had the lower activation barriers (43). The differences are attributable to the bridging hydrogen. When the hydrogen is present the hinge metals appear similar to a six-coordinate metal. Removing the hydrogen reduces the metal to a pseudo-five-coordinate geometry which should lower the activation barrier for CO scrambling. A similar effect has been noted in trinuclear clusters (69).

A recent study of $[FeRu_3N(CO)_{12}]^-$ has elucidated another dynamic process that occurs *slowly on the laboratory time scale*. The cluster exists in two isomeric forms **32** and **33** [Eq. (59)], which are in equilibrium with one another (43). This represents the first examples where the wing-tip and hinge positions of a butterfly cluster with a μ_4 atom have been shown to interconvert. The thermodynamic parameters for **32** $\rightleftharpoons$ **33** are $\Delta H = -3.5 \pm 1.0$ kcal/mol and $\Delta S = -13 \pm 2$eu. The large value of ΔS was attributed to ion-pairing effects. The rate of conversion of **32** $\rightarrow$ **33** is $4.2 \pm 0.2 \times 10^{-7}$ sec^{-1}, and possible mechanisms for this rearrangement are shown in Scheme 6. It will be interesting to see how such rearrangements vary with structure and in particular as a function of the μ_4 atom.

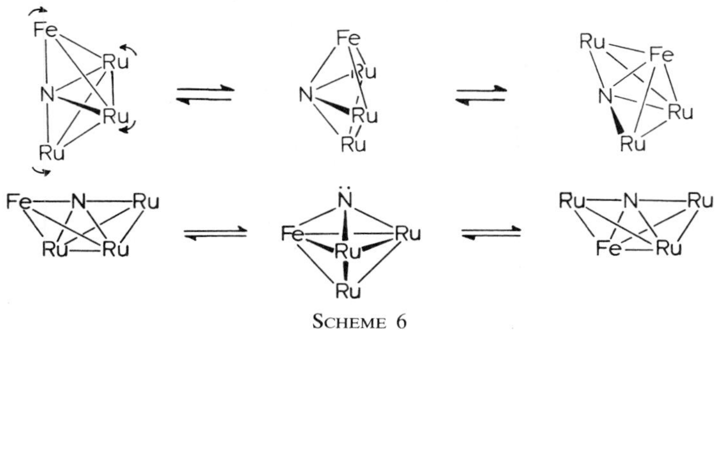

SCHEME 6

F. Reactivity of Nitrido Clusters

This section will be limited to reactions involving the nitrogen atom, specifically N protonation and isocyanate formation.

1. N Protonation

The formation of N—H bonds by protonation of a nitrido cluster was first reported for the reaction of $[Fe_4N(CO)_{12}]^-$ with aqueous acid. The trinuclear clusters $Fe_3(NH)(CO)_{10}$ and $Fe_3(NH)_2(CO)_9$ were both isolated in very low yield (20). Experiments with related compounds have led to more satisfying results (118). Intensely colored intermediates were observed upon reacting CF_3SO_3H with the following butterfly nitrido clusters: $[FeRu_3N(CO)_{12}]^-$, $[Ru_4N(CO)_{12}]^-$, and $\{FeRu_3N(CO)_{10}[P(OCH_3)_3]_2\}$. Although the final products formed had the proton bridging the hinge metal–metal bond, the intermediates were conclusively shown to contain a direct N—H bond.

The most valuable evidence for this came from using the structurally characterized disubstituted derivative $\{FeFu_3N(CO)_{10}[P(OCH_3)]_2\}^-$ (**35**), which contains iron only in the hinge position and which undergoes a quantitative rearrangement upon protonation to (**40**) (Scheme 7). The reaction proceeds through an intensely green colored solution. A ^1H- and ^{15}N-NMR spectral study of the protonation of ^{15}N-enriched cluster revealed that two species containing NH ligands are formed and are in equilibrium with one another. The first was suggested to have a μ_4-NH (**36**), and the second was proposed to have a μ_3-NH (**37**). Structure **36** is related to the 60-electron species $[Fe_4(CCO_2CH_3)(CO)_{12}]^-$ ($119, 120$), while **37** is isostructural with $Os_4(NCH_3)(CO)_{12}$ (121). Scheme 7 offers a rationale for the final stereochemistry observed in **40**.

Once the nitrogen is protonated (**36**) it can move to one of two different capping μ_3-NH species depending on toward which side of the cluster the N migrates. From **38** the product can be formed directly. This is not possible from **37**, which must proceed back through the high-energy intermediate before going on to the product.

Also shown in Scheme 7 is the structure of the NH product (**41**) resulting from trapping the intermediate with CO. The CO also quenches the intense color in the protonations of $[FeRu_3N(CO)_{12}]^-$ and $[Ru_4N(CO)_{12}]^-$. Analysis of the products in the mixed-metal cluster reaction gives $FeRu_2(NH)(CO)_{10}$, $[FeRu_4N(CO)_{14}]^-$, $HFeRu_3N(CO)_{12}$, and $Ru_3(CO)_{12}$. Likewise, the same reaction with $[Ru_4N(CO)_{12}]^-$ produces $Ru_3(NH)(CO)_{10}$, $[Ru_5N(CO)_{14}]^-$, $HRu_4N(CO)_{12}$, and $Ru_3(CO)_{12}$. Equa-

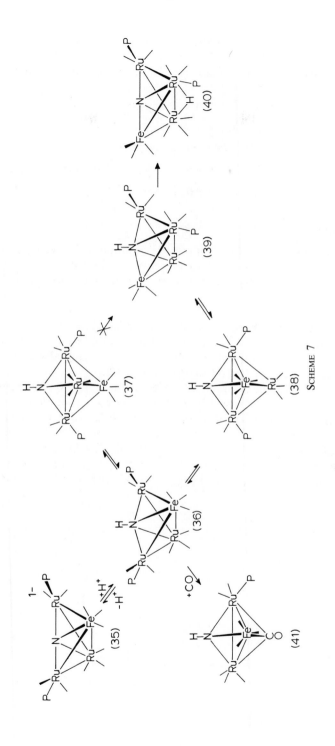

SCHEME 7

tions (60)–(64) offer a reasonable explanation for these product distributions.

$$[M_4N(CO)_{12}]^- + H^+ \rightleftharpoons M_4(NH)(CO)_{12} \tag{60}$$

$$M_4(NH)(CO)_{12} + 3CO \rightarrow M_3(NH)(CO)_{10} + M(CO)_2 \tag{61}$$

$$M_4(NH)(CO)_{12} \rightarrow HM_4N(CO)_{12} \tag{62}$$

$$[M_4N(CO)_{12}]^- + M(CO)_5 \rightarrow 3CO + [M_5N(CO)_{14}] \tag{63}$$

$$3M(CO)_5 \rightarrow 3CO + M_3(CO)_{12} \tag{64}$$

2. Isocyanate Formation

The scission of the N—C bond of coordinated isocyanates has been discussed as a useful method to prepare nitrido clusters. Under higher pressures of CO, Eq. (65) can be reversed. When $[Ru_6N(CO)_{16}]^-$ is placed

$$M_n(NCO) \rightleftharpoons M_n(N) + CO \tag{65}$$

under a CO atmosphere it rapidly forms $[Ru_5N(CO)_{14}]^-$, but further reaction under CO does not occur until the temperature and pressure are increased. At higher pressures, $[Ru_4N(CO)_{12}]^-$ is initially formed and is slowly converted to $[Ru_3(NCO)(CO)_{11}]^-$ (91).

Equation (65) is an example of a process that may be readily promoted by a cluster but not necessarily with a mononuclear complex. Scheme 8 shows how the multiple-coordination capabilities of a trinuclear cluster could reduce the N—C bond and eventually cleave it. To date, examples of the NCO ligand bridging more than two metals in either a mono- or polyhapto fashion are unknown. The reverse process represents a unique method for forming carbon–nitrogen bonds. Although the above chemis-

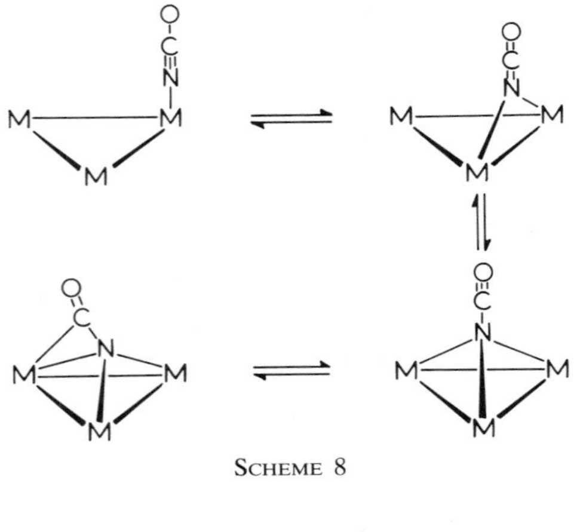

SCHEME 8

try is the only example of NCO formation from a characterized nitrido ligand, reports of NCO formation in Eqs. (66) (*122*) and (67) (*35*) are proposed to proceed via nitrido intermediates.

$$(\eta^5\text{-}C_5Me_5)_2Mo_2(N_2CMe_2)(CO)_4 \xrightarrow[\Delta]{toluene} (\eta^5\text{-}C_5Me_5)_2Mo_2(NCO)(NCMe_2)(CO)_3$$

(66)

$$[Ru_3(CO)_{10}(NO)]^- + 3CO \xrightarrow{h\nu} [Ru_3(NCO)(CO)_{11}]^- + CO_2$$

(67)

IV

OVERVIEW

Much of the chemistry discussed in this review can be summarized in Scheme 9. Although details regarding most of these reactions are lacking, a general picture of the formation and reactivity of coordinated nitrogen atoms is beginning to emerge. The formation and subsequent reactions of nitrido clusters bear a striking resemblance to the heterogeneous catalytic chemistry of NO plus CO or H_2. On the surface of catalysts used to promote the reduction of NO with CO nitrogen atoms are known to reversibly form bound NCO (*123, 124*). When H_2 and NO are mixed, NH_3 is the final product. Once again N—H bond formation has been observed with nitrosyl and nitrido clusters. The fact that catenation is not particularly important in nitrogen chemistry significantly reduces the complexity of these reactions relative to CO reduction processes. This simplifying feature should continue to promote valuable synthetic and mechanistic studies aimed at developing a better understanding of the reactions and interconversions of nitrogen atoms and small nitrogenous ligands on the surface of metal cluster compounds.

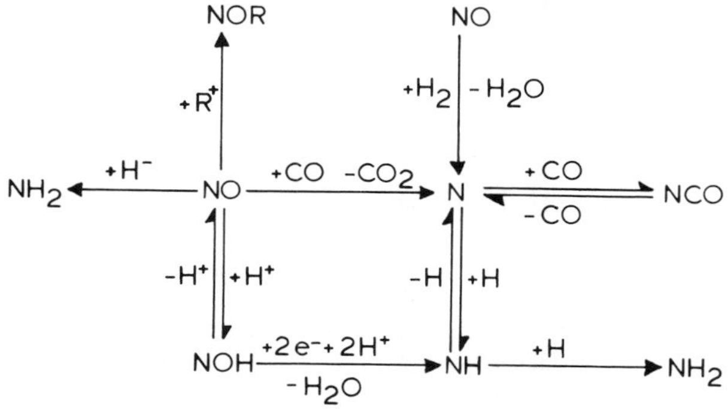

Scheme 9

There also exist structural similarities between the M_6N clusters and the class of solid-state compounds known as transition metal nitrides. In these refractory compounds the nitrogen exists in either an octahedral or a trigonal prismatic array of transition metals (*125*). The interest in using transition metal nitrides for superconducting thin films as well as chemically inert coatings (*125*) should promote studies in which nitrido clusters could be used as starting materials for the synthesis of new refractory materials.

ACKNOWLEDGMENTS

I want to thank the graduate and undergraduate students who have contributed to our efforts in this area, the National Science Foundation for supporting this work, and the Alfred P. Sloan Foundation for a fellowship (1983–1985). I would also like to thank Professors L. F. Dahl, J. E. Ellis, J. Mason, T. B. Rauchfuss, and J. R. Shapley for preprints and preliminary communication of their work in this area.

REFERENCES

1. M. Tachikawa and E. L. Muetterties, *Prog. Inorg. Chem.* **28**, 203 (1981).
2. C. K. Rofer-Deporter, *Chem. Rev.* **81**, 447 (1981).
3. S. Martinengo, G. Ciani, A. Sironi, B. T. Heaton, and J. Mason, *J Am. Chem. Soc.* **101**, 7095 (1979).
4. A. Ozaki and K. Aika, *Catal. Sci. Technol.* **3**, 87 (1981).
5. R. Eisenberg and D. E. Hendriksen, *Adv. Catal.* **28**, 79 (1979).
6. B. J. Savatsky and A. T. Bell, *ACS Symp. Ser.* **178**, 105 (1982).
7. J. A. McCleverty, *Chem. Rev.* **79**, 53 (1979).
8. K. G. Caulton, *Coord. Chem. Rev.* **14**, 317 (1975).
9. J. H. Enemark and R. D. Feltham, *Coord. Chem. Rev.* **13**, 339 (1974).
10. N. G. Connelly, *Inorg. Chim. Acta Rev.* **6**, 47 (1972).
11. W. P. Griffith, *Coord. Chem. Rev.* **8**, 369 (1972).
12. J. R. Norton, J. P. Collman, G. Dolcetti, and W. T. Robinson, *Inorg. Chem.* **11**, 382 (1972).
13. S. Bhaduri, B. F. G. Johnson, J. Lewis, D. J. Watson, and C. Zuccaro, *J. Chem. Soc., Dalton Trans.*, p. 557 (1979).
14. A. V. Rivera and G. M. Sheldrick, *Acta Crystallogr. Sect. B* **34**, 3372 (1978).
15. R. L. Mond and A. F. Wallis, *J. Chem. Soc.* p. 32 (1922).
16. E. O. Fischer, O. Beckert, W. Hafner, and H. O. Stahl, *Z. Naturforsch. B: Anorg. Chem., Org. Chem.* **10**, 598 (1955).
17. L. O. Brockway and J. S. Anderson, *Trans. Faraday Soc.* **33**, 1233 (1937).
18. B. F. G. Johnson, P. R. Raithby, and C. Zuccaro, *J. Chem. Soc., Dalton Trans.*, p. 99 (1980).
19. D. Braga, K. Henrick, B. F. G. Johnson, M. McPartlin, W. J. H. Nelson, and J. Puga, *J. Chem. Soc., Chem. Commun.* p. 1083 (1982).
20. D. E. Fjare and W. L. Gladfelter, *Inorg. Chem.* **20**, 3533 (1981).
21. J. R. Shapley, unpublished observations.
22. D. Braga, B. F. G. Johnson, J. Lewis, J. M. Mace, M. McPartlin, J. Puga, W. J. H. Nelson, P. R. Raithby, and K. H. Whitmire, *J. Chem. Soc., Chem. Commun.*, p. 1081 (1982).

23. B. F. G. Johnson, J. Lewis, W. J. H. Nelson, J. Puga, P. R. Raithby, M. Schoeder, and K. H. Whitmire, *J. Chem. Soc., Chem. Commun.*, p. 610 (1982).

24. B. F. G. Johnson, J. Lewis, W. J. H. Nelson, J. Puga, K. Henrick, and M. McPartlin, *J. Chem. Soc., Dalton Trans.*, p. 1203 (1983).

25. B. F. G. Johnson, J. Lewis, W. J. H. Nelson, J. Puga, P. R. Raithby, and K. H. Whitmire, *J. Chem. Soc., Dalton Trans.*, p. 1339 (1983).

26. B. F. G. Johnson, J. Lewis, P. R. Raithby, and C. Zuccaro, *J. Chem. Soc., Dalton Trans.*, p. 716 (1980).

27. B. F. G. Johnson and J. A. Segal, *J. Chem. Soc., Dalton Trans.*, p. 478 (1973).

28. R. E. Stevens, M. L. Blohm, and W. L. Gladfelter, unpublished observations.

29. F. Seel, in "Handbook of Preparative Inorganic Chemistry" (G. Brauer, ed.), p. 1764. Academic Press, New York, 1965.

30. W. Hieber and H. Z. Beutner, *Z. Anorg. Allg. Chem.* **320,** 101 (1963).

31. S. W. Kirtley, J. P. Chanton, R. A. Love, D. L. Tipton, T. V. Sorrell, and R. Bau, *J. Am. Chem. Soc.* **102,** 3451 (1980).

32. R. E. Stevens, T. J. Yanta, W. L. Gladfelter, *J. Am. Chem. Soc.* **103,** 4981 (1981).

33. R. E. Stevens and W. L. Gladfelter, *Inorg. Chem.* **22,** 2034 (1983).

34. B. F. G. Johnson, J. Lewis, W. J. H. Nelson, J. Puga, P. R. Raithby, D. Braga, M. McPartlin, and W. Clegg, *J. Organomet. Chem.* **243,** C13 (1983).

35. R. E. Stevens and W. L. Gladfelter, unpublished observations.

36. B. F. G. Johnson, J. Lewis, W. J. H. Nelson, J. N. Nicholls, and M. D. Vargas, *J. Organomet. Chem.* **249,** 255 (1983).

37. S. B. Colbran, B. H. Robinson, and J. Simpson, *J. Organomet. Chem.* **265,** 199 (1984).

38. M. A. Collins, B. F. G. Johnson, J. Lewis, J. M. Mace, J. Morris, M. McPartlin, W. J. H. Nelson, J. Puga, and P. R. Raithby, *J. Chem. Soc., Chem. Commun.*, p. 689 (1983).

39. R. B. King and M. B. Bisnette, *Inorg. Chem.* **3,** 791 (1964).

40. R. C. Elder, *Inorg. Chem.* **13,** 1037 (1974).

41. B. W. S. Kolthammer and P. Legzdins, *J. Chem. Soc., Dalton Trans.*, p. 31 (1978).

42. D. E. Fjare and W. L. Gladfelter, *J. Am. Chem. Soc.* **103,** 1572 (1981).

43. D. E. Fjare and W. L. Gladfelter, *J. Am. Chem. Soc.* **106,** 4799 (1984).

44. Y. Chen and J. E. Ellis, *J. Am. Chem. Soc.* **105,** 1689 (1983).

45. D. Mantell, R. E. Stevens, and W. L. Gladfelter, unpublished observations.

46. J. E. Ellis and K. L. Fjare, unpublished observations.

47. W. L. Gladfelter, R. E. Stevens, and D. E. Fjare, *Int. Symp. Homo. Catal. 3rd*, p. 115, 1982.

48. R. S. Gall, N. G. Connelly, L. F. Dahl, *J. Am. Chem. Soc.* **96,** 4017 (1974).

49. R. S. Gall, C. T.-W. Chu, L. F. Dahl, *J. Am. Chem. Soc.* **96,** 4019 (1974).

50. C. T.-W. Chu, F. Y.-K Lo, L. F. Dahl, *J. Am. Chem. Soc.* **104,** 3409 (1982).

51. C. T.-W Chu, R. S. Gall, L. F. Dahl, *J. Am. Chem. Soc.* **104,** 737 (1982).

52. T. B. Rauchfuss, T. D. Weatherill, S. R. Wilson, and J. P. Zebrowski, *J. Am. Chem. Soc.* **105,** 6508 (1983).

53. H. Brunner, H. Kauermann, and J. Wachter, *Angew. Chem., Int. Ed. Engl.* **22,** 549 (1983).

54. A. M. Mazany, J. P. Fackler, Jr., M. K. Gallagher, and D. Seyferth, *Inorg. Chem.* **22,** 2593 (1983).

55. C. T.-W. Chu and L. F. Dahl, *Inorg. Chem.* **16,** 3245 (1977).

56. Y. S. Yu, R. A. Jacobson, and R. J. Angelici, *Inorg. Chem.* **21,** 3106 (1982).

57. B. J. Morris-Sherwood, B. W. S. Kolthammer, and M. B. Hall, *Inorg. Chem.* **20,** 2771 (1981).

58. J. P. Olsen, T. F. Koetzle, S. W. Kirtley, M. Andrews, D. L. Tipton, and R. Bau, *J. Am. Chem. Soc.* **96,** 6621 (1974).

59. D. W. McBride, S. L. Stafford, and F. G. A. Stone, *Inorg. Chem.* **1,** 386 (1962).
60. K. R. Grundy, K. R. Laing, and W. R. Roper, *J. Chem. Soc., Chem. Commun.,* p. 1500 (1970).
61. J. Mason, *Chem. Rev.* **81,** 205 (1981).
62. L. K. Bell, J. Mason, D. M. P. Mingos, and D. G. Tew, *Inorg. Chem.* **22,** 3497 (1983).
63. D. H. Evans, D. M. P. Mingos, J. Mason, and A. Richards, *J. Organomet Chem.* **249,** 293 (1983).
64. J. Mason, *Chemistry in Britain,* p. 654 (1983).
65. J. Mason, D. M. P. Mingos, D. Sherman, and R. W. M. Wardle, *J. Chem. Soc., Chem. Commun.,* in press.
66. R. E. Botto, B. W. S. Kolthammer, P. Legzdins, and J. D. Roberts, *Inorg. Chem.* **18,** 2049 (1979).
67. R. E. Stevens and W. L. Gladfelter, *J. Am. Chem. Soc.* **104,** 645 (1982).
68. P. Legzdins, C. R. Nurse, and S. J. Rettig, *J. Am. Chem. Soc.* **105,** 3727 (1983).
69. B. F. G. Johnson, J. Lewis, J. M. Mace, P. R. Raithby, R. E. Stevens, and W. L. Gladfelter, *Inorg. Chem.* **23,** 1600 (1984).
70. D. F. Shriver, D. Lehman, and D. Strope, *J. Am. Chem. Soc.* **97,** 1594 (1975).
71. P. D. Gavens and M. J. Mays, *J. Organomet. Chem.* **162,** 389 (1978).
72. J. B. Keister, *J. Chem. Soc., Chem. Commun.,* p. 214 (1979).
73. B. F. G. Johnson, J. Lewis, A. G. Orpen, P. R. Raithby, and G. Süss, *J. Organomet. Chem.* **173,** 187 (1979).
74. E. M. Holt, K. Whitmire, and D. F. Shriver, *J. Chem. Soc., Chem. Commun.,* p. 778 (1980).
75. K. Whitmire, D. F. Shriver, and E. M. Holt, *J. Chem. Soc., Chem. Commun.,* p. 780 (1980).
76. P. A. Dawson, B. F. G. Johnson, J. Lewis, and P. R. Raithby, *J. Chem. Soc., Chem. Commun.,* p. 781 (1980).
77. H. A. Hodali, D. F. Shriver, and C. A. Ammlung, *J. Am. Chem. Soc.* **100,** 5239 (1978).
78. G. Fachinetti, *J. Chem. Soc., Chem. Commun.,* p. 397 (1979).
79. J. B. Keister, *J. Organomet. Chem.* **190,** C36 (1980).
80. K. H. Whitmire and D. F. Shriver, *J. Am. Chem. Soc.* **103,** 6754 (1981).
81. D. E. Fjare, D. G. Keyes, and W. L. Gladfelter, *J. Organomet. Chem.* **250,** 383 (1983).
82. W. R. Murphy, Jr., K. J. Takeuchi, and T. J. Meyer, *J. Am. Chem. Soc.* **104,** 5817 (1982).
83. B. W. Hames, P. Legzdins, and J. C. Oxley, *Inorg. Chem.* **19,** 1565 (1980).
84. W. L. Gladfelter, *in* "Organometallic Compounds: Synthesis, Structure, and Reactivity" (B. L. Shapiro, ed.), p. 281. Texas A & M University Press, College Station, Texas, 1983.
85. B. F. G. Johnson, J. Lewis, and J. M. Mace, *J. Chem. Soc., Chem. Commun.,* p. 186 (1984).
86. J. A. Smieja, R. E. Stevens, D. E. Fjare, and W. L. Gladfelter, *Inorg. Chem.,* in press.
87. Z. Dawoodi, M. J. Mays, and K. Henrick, *J. Chem. Soc., Dalton Trans.,* p. 433 (1984).
88. J. R. Norton and J. P. Collman, *Inorg. Chem.* **12,** 476 (1973).
89. F. R. Furuya, D. E. Fjare, and W. L. Gladfelter, unpublished observations.
90. M. Tachikawa, J. Stein, E. L. Muetterties, R. G. Teller, M. A. Beno, E. Gebert, and J. M. Williams, *J. Am. Chem. Soc.* **102,** 6648 (1980).
91. M. L. Blohm and W. L. Gladfelter, *Organometallics* **4,** 45 (1985).

92. N. D. Feasey and S. A. R. Knox, *J. Chem. Soc., Chem. Commun.*, p. 1062 (1982).
93. N. D. Feasey, S. A. R. Knox, and A. G. Orpen, *J. Chem. Soc., Chem. Commun.*, p. 75 (1982).
94. M. L. Blohm, D. E. Fjare, and W. L. Gladfelter, *Inorg. Chem.* **22**, 1004 (1983).
95. D. E. Fjare, J. A. Jensen, and W. L. Gladfelter, *Inorg. Chem.* **22**, 1774 (1983).
96. S. Martinengo, G. Ciani, and A. Sironi. *J. Am. Chem. Soc.* **104**, 328 (1982).
97. D. H. Farrar, P. F. Jackson, B. F. G. Johnson, J. Lewis, J. N. Nicholls, and M. McPartlin, *J. Chem. Soc., Chem. Commun.*, p. 415 (1981).
98. A. Sirigu, M. Bianchi, and E. Beneditti, *J. Chem. Soc., Chem. Commun.*, 596 (1969).
99. J. S. Bradley, G. B. Ansell, and E. W. Hill, *J. Organomet. Chem.* **184**, C33 (1980).
100. V. G. Albano, M. Sansoni, P. Chini, and S. Martinengo, *J. Chem. Soc., Dalton Trans.*, p. 651 (1973).
101. A. Gourdon and R. Jeanin, *C. R. Acad. Sc. Paris* **295**, 1101 (1982).
102. S. D. Wijeyesekera and R. Hoffmann, *Organometallics* **3**, 949 (1984).
103. S. D. Wijeyesekera, R. Hoffmann, and C. N. Wilker, *Organometallics* **3**, 962 (1984).
104. S. Harris and J. S. Bradley, *Organometallics* **3**, 1086 (1984).
105. J. W. Kolis, E. M. Holt, and D. F. Shriver, *J. Am. Chem. Soc.* **105**, 7307 (1983).
106. M. A. Goudsmit, B. F. G. Johnson, J. Lewis, P. R. Raithby, and K. H. Whitmire, *J. Chem. Soc., Chem. Commun.*, p. 246 (1983).
107. A. Ceriotti and L. Resconi, *J. Organomet. Chem.* **249**, C35 (1983).
108. V. A. Uchtman and L. F. Dahl, *J. Am. Chem. Soc.* **91**, 3763 (1969).
109. A. Bertolucci, M. Freni, P. Romiti, G. Ciani, A. Sironi, and V. G. Albano, *J. Organomet. Chem.* **113**, C61 (1976).
110. J. R. Shapley, J. T. Park, M. R. Churchill, J. W. Ziller, and L. R. Beanan, *J. Am. Chem. Soc.* **106**, 1144 (1984).
111. V. G. Albano, P. Chini, G. Ciani, M. Sansoni, D. Strumolo, B. T. Heaton, and S. Martinengo, *J. Am. Chem. Soc.* **98**, 5027 (1976).
112. J. S. Bradley, *Adv. Organomet. Chem.* **22**, 1 (1983).
113. D. E. Fjare and W. L. Gladfelter, unpublished observations.
114. B. T. Heaton, L. Strona, and S. Martinengo, *J. Organomet. Chem.* **215**, 415 (1981).
115. J. Mason in Discussion in: B. T. Heaton, *Philos. Trans. R. Soc. London Ser. A* **308**, 95 (1982).
116. J. Mason, private communication.
117. M. A. Beno, J. M. Williams, M. Tachikawa, and E. L. Muetterties, *J. Am. Chem. Soc.* **103**, 1485 (1981).
118. M. L. Blohm, D. E. Fjare, and W. L. Gladfelter, submitted for publication.
119. J. S. Bradley, G. B. Ansell, and E. W. Hill, *J. Am. Chem. Soc.* **101**, 7417 (1979).
120. J. S. Bradley, *Philos. Trans. R. Soc. London Ser. A* **308**, 103 (1982).
121. Y. C. Lin, C. B. Knobler, and H. D. Kaesz, *J. Organomet. Chem.* **213**, C41 (1981).
122. W. A. Herrmann, L. K. Bell, M. L. Ziegler, H. Pfisterer, and C. Pahl, *J. Organomet. Chem.* **247**, 39 (1983).
123. M. L. Unland, *J. Phys. Chem.* **77**, 1952 (1973).
124. R. J. H. Voorhoeve and L. E. Trimble, *J. Catal.* **54**, 269 (1978).
125. L. E. Toth, "Transition Metal Carbides and Nitrides," Academic Press, New York, 1971.
126. B. F. G. Johnson, J. Lewis, P. R. Raithby, and C. Zuccaro, *J. Chem. Soc., Chem. Commun.*, p. 916 (1979).
127. V. Albano, P. Bellon, G. Ciani, and M. Manassero, *J. Chem. Soc., Chem. Commun.*, p. 1242 (1969).
128. L. F. Dahl, unpublished observation.

NOTE ADDED IN PROOF

Listed below are nitrosyl and nitrido clusters that have been recently reported.

$\{[H_3Re_3(CO)_{10}]_2(\mu_4\text{-}\eta^2\text{-}NO)\}^{1-}$ [T. Beringhelli, G. Ciani, G. D'Alfonso, H. Molinari, A. Sironi, and M. Freni, *J. Chem. Soc., Chem.Commun.*, p. 1327 (1984)].

$Fe_6C(CO)_{11}(NO)_4$ and $[Fe_6C(CO)_{15}(NO)]^{1-}$ [A. Gourdon and Y. Jeannin, *J. Organomet. Chem.* **282,** C39 (1985)].

$(C_5Me_5)_3Mo_3Co_2(N)(NH)(CO)_8$, $(C_5Me_5)_3Mo_3(N)(O)(CO)_4$, $(C_5Me_5)_3Mo_3(NCO)(O)\text{-}(CO)_4$, $[C_5H_5)_3Co_3(NO)(NH)]^+$, $(C_5H_4Me)_3Co_3(NO)(NH)$ [L. F. Dahl, R. Bedard, and C. Gibson, preliminary communication].

ADVANCES IN ORGANOMETALLIC CHEMISTRY, VOL. 24

The Electron-Transfer Reactions of Polynuclear Organotransition Metal Complexes

WILLIAM E. GEIGER

Department of Chemistry
University of Vermont
Burlington, Vermont

and

NEIL G. CONNELLY

Department of Inorganic Chemistry
University of Bristol
Bristol, England

I
INTRODUCTION

During the past two decades, the chemistry of polynuclear organometallic complexes has become increasingly important, particularly in aiding

ISBN 0-12-031124-0

our understanding of the relationship between structure, bonding, and reactivity. Metal clusters have generated particular interest, as catalyst precursors, as models of metal surfaces with chemisorbed small molecules, and as models for biological species such as the larger iron–sulfur proteins.

More recently, it has become clear that many polynuclear compounds undergo electron-transfer reactions, and several reviews devoted wholly or partly to this subject have appeared (1–4). Studies of the redox properties of polymetallic species can provide information on the nature of the highest occupied molecular orbital (HOMO)[1] or lowest unoccupied molecular orbital (LUMO), on the possible cooperativity between metal sites, on the existence of mixed-valence compounds, and on the ways in which one-electron (or, more rarely, two-electron) changes affect structure and reactivity.

In a general sense, metal clusters have been regarded as "electron reservoirs" which can gain or lose electrons at will without molecular disruption. However, few real species possess this property, and the addition or subtraction of more than one electron usually leads to structural changes, and a concomitant increase in reactivity. Detailed studies are being made, therefore, of redox-induced changes in structure and reactivity, studies which promise to provide rational routes to new polymetallic complexes and information about electronic structure.

We have previously surveyed the electron-transfer reactions of mononuclear organotransition metal complexes (5), and we now focus our attention on related species containing two or more metals. As in our earlier article, the results presented are dominated by, but by no means limited to, electrochemical investigations.

Multimetallic sandwiches, including polymetallocenes, bis(fulvalene)-dimetal compounds, multi-decker sandwiches, and compounds in which two metals are bonded to a common unsaturated cyclic hydrocarbon are discussed in Section II. Sections III and IV are subdivided according to the metal group, and describe ligand-bridged bimetallics and metal clusters, respectively. Where appropriate, redox potentials have been included. Unless otherwise stated they are referenced versus the aqueous saturated calomel electrode (sce).

---

[1] Abbreviations: ave, average; cot, cyclooctatetraene; Cp, η^5-cyclopentadienyl; DME, 1,2-dimethoxyethane; DMF, N,N-dimethylformamide; dppe, $Ph_2PCH_2CH_2PPh_2$; ESR, electron spin resonance; Fc, ferrocenyl; HOMO, highest occupied molecular orbital; LUMO, lowest unoccupied molecular orbital; sce, saturated calomel electrode; TCNQ, 7,7,8,8-tetracyano-p-quinodimethane; THF, tetrahydrofuran.

II

MULTIPLE SANDWICH COMPOUNDS

A. Sandwich Compounds Containing Linked Cyclopentadienyl or Aryl Groups

A great deal of effort has been devoted to the study of linked metallocene units in order to probe the redox properties of compounds containing two or more identical electron-transfer sites; much of the interest in these systems stems from the fact that one-electron changes lead formally to mixed-valence species. Iron compounds, including biferrocene **1** (M = Fe), bridged biferrocenes **2** [X = $(CH_2)_n$, S, Hg, etc.], bis(fulvalene)diiron **3** (M = Fe), ferrocenophane **4** (M = M′ = Fe), and their derivatives, have attracted the most attention.

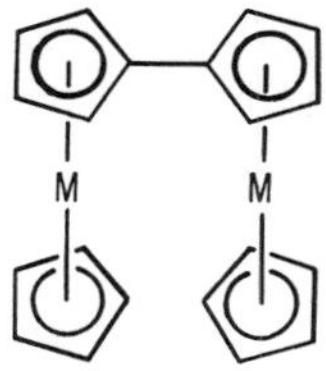

(1) 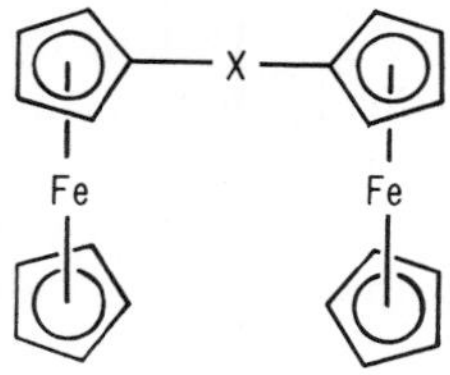   (2)

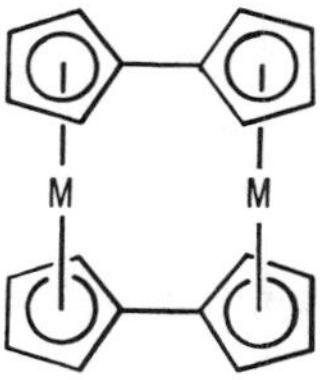

(3) 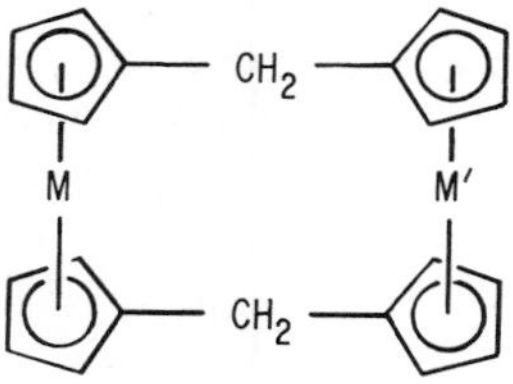   (4)

Biferrocene **1** (M = Fe) oxidizes in two separate one-electron steps (6, 7) ($E_{1/2}$ = 0.31 and 0.64 V), of which only the first is chemically reversible (i.e., the monocation is stable but the dication is not); mild oxidation yields the monocation $[1]^+$ (M = Fe) as a mixed-valence Fe(II)Fe(III) species (8–10). Bis(fulvalene)diiron **3** (M = Fe) similarly undergoes two successive one-electron oxidations (the dication is also stable), but in this case the Mössbauer spectrum of the monocation shows the iron atoms to be equivalent, implying a delocalized electronic structure and an oxidation state of 2.5 for each metal (9, 11). The separation of E^0

values, ΔE^0, for the two oxidations of **3** (M = Fe, $\Delta E^0 = 0.59$ V) is greater than that of **1** (M = Fe, $\Delta E^0 = 0.33$ V) (*12*), apparently due to increased electronic interactions in the rigid cation $[\mathbf{3}]^+$ (M = Fe) (*7*). Similarly, the difference in successive one-electron oxidation potentials for [1.1]-ferrocenophane **4** (M = M′ = Fe) is larger than that of its nonrigid analogue diferrocenylmethane **2** (X = CH$_2$) (*13*).

In bridged biferrocenes having poor delocalization through the bridging group, the two ferrocene halves act more like independent, noninteracting units. The redox potentials reflect this in that the E^0 values of the two oxidation processes move closer together. Thus, the ΔE^0 value of 0.33 V for **1** (M = Fe), which has no bridging unit, is lowered to 0.17 V for **2** (X = CH$_2$), and 0.04 V for **2** (X = CH$_2$CH$_2$) (*7*). The E^0 values of the C$_2$H$_4$-bridged compound are so close that separate waves are not resolved in the rotating electrode voltammogram. Instead, a single wave of two-electron height is observed. Similar behavior was found for three other bridged biferrocenes **2** (X = Hg, C$_2$Me$_4$, or divinylbenzene), which yielded only dications when chemically oxidized (*7*). The question of delocalization through the bridge has been addressed in further studies (*14–16*), and a great deal has been learned about the mechanisms of charge-transfer interactions in bimetallic systems from electrochemical, spectroscopic, and magnetic data (*10, 16, 17*).

The vinylferrocene polymers [{Fc(CH$_2$CH$_2$)}$_n$] (Fc = ferrocenyl) constitute a good model system for understanding multiple electron-transfer reactions involving mutually isolated redox sites (*18*). An important conclusion arising from theoretical and experimental studies is that multiple, noninteracting redox centers exhibit voltammetric wave *shapes* that match exactly those of a corresponding single redox center. Only the wave *height* is different. For example, a polymer having approximately 75 vinylferrocene units gave a cyclic voltammogram with a peak separation of about 60 mV in N,N'-dimethylformamide (DMF), the same as a reversible one-electron process but with a wave height about 75 times that of an equimolar solution of the vinylferrocene monomer. The E^0 potentials of the polymers were slightly more negative (by 80–120 mV) than that of the monomer (*18*). It is expected, therefore, that a binuclear compound such as **2** (X = Hg), which has little or no delocalization through the bridge, should exhibit a cyclic voltammogram not with the 30 mV peak separation of a reversible two-electron wave, but rather with a separation of about 60 mV and with a wave height approximately twice that of ferrocene. If the number of ferrocenyl units were increased by joining longer bridge-insulated chains, the peak current per mole (proportional to the apparent n value) would continue to rise in proportion to the number of iron atoms, but the peak separation would continue to be close to 60 mV (assuming nearly Nernstian charge transfer).

Unfortunately, many of the published voltammetric data on molecules of the types **1–4** are not reported in sufficient detail to compare with theory. However, Rieke *et al.* (*19*) successfully analyzed the voltammetric behavior of the Group IV compounds $[EMe_{4-n}\{(\eta^6\text{-}C_6H_5)Cr(CO)_3\}_n]$ (E = Si, Ge, Sn or Pb; n = 1–3), in which the aryl groups are σ-bonded to E, and π-bonded to electroactive $Cr(CO)_3$ units. Single waves with similar peak separations, but having wave heights corresponding to n values of 1, 2, and 3, were observed for $[SnMe_3\{(\eta^6\text{-}C_6H_5)Cr(CO)_3\}]$, $[SnMe_2\{(\eta^6\text{-}C_6H_5)Cr(CO)_3\}_2]$, and $[SnMe\{(\eta^6\text{-}C_6H_5)Cr(CO)_3\}_3]$, respectively. It was concluded that the arenetricarbonylchromium groups are essentially noninteracting and each is the site of a one-electron oxidation. Unfortunately, the di- and trications were too unstable to be isolated, and their magnetic properties could not be studied (*19*).

The electronic absorption spectrum of $[BFc_4]$ shows a broad band at 2200 nm, assigned to a charge-transfer transition and supporting the formulation of the complex as a zwitterion containing three neutral Fe(II) ferrocenyl groups, one cationic Fe(III) center, and, at least formally, a negatively charged boron atom (*20*).

The cyclic voltammogram of $[BFc_4]$ shows four separate waves, implying that delocalization through the central boron is greater than through the central tin atom in the chromium compounds mentioned above (*19*). The reported electrochemical preparation (*20*) of the monocation $[BFc_4]^+$ is, however, hard to reconcile with the published cyclic voltammetric data, which show that several of the waves have shifted in potential after electrolysis.

The cyclobutadiene derivatives $[Co(\eta^4\text{-}C_4Fc_nPh_{4-n})Cp]$ (n = 1–3) (Cp = η^5-cyclopentadienyl) undergo one-electron oxidation to stable cations (*21*) which show charge-transfer bands in the near infrared region. An analysis of the spectra led to the conclusion that the cations are Class II mixed-valence compounds and that the charge-transfer transition arises from a weak interaction between Co(I) and Fe(III) rather than between Fe(II) and Fe(III). Each compound showed further oxidations at more positive potentials, including an irreversible process ascribed to electron loss from the cobalt center.

When ferrocenyl groups are linked together without insulating bridges, one oxidation wave per iron atom is observed. Thus, 1,1′-terferrocene **5** and 1,1′-quaterferrocene **6** show three and four oxidation waves, respectively (*12*). Increasing the chain length leads to a more facile first oxidation [E^0 values of 0.40 V for ferrocene, 0.31 V for **1** (M = Fe), 0.22 V for **5**, and 0.16 V for **6**], and smaller separations between the first two E^0 values [ΔE^0 values of 0.34 V for **1** (M = Fe), 0.22 V for **5**, and 0.20 V for **6**], consistent with electronic delocalization between metal centers despite their formally trapped valences.

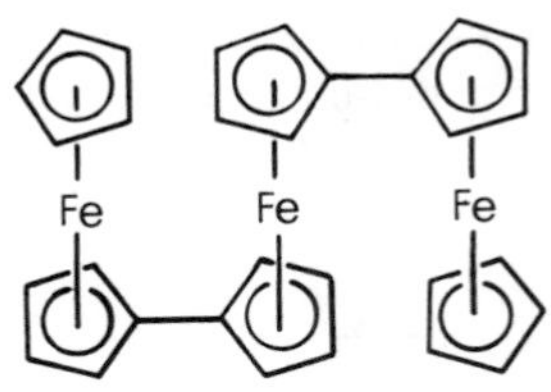

(5) 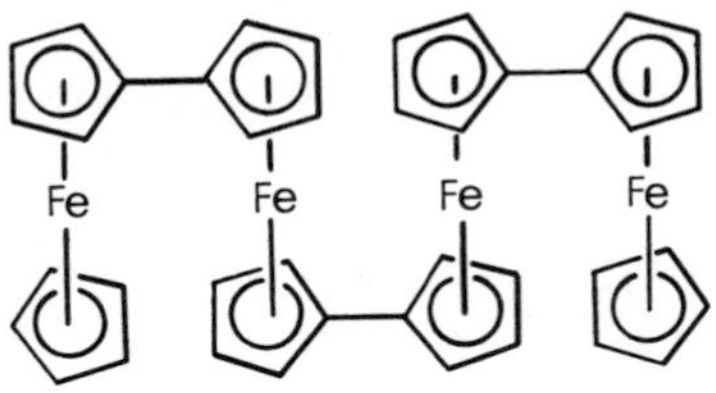   (6)

The monocations of **5** and **6** were prepared and their near infrared intervalence charge-transfer bands studied; the dication of **6** was also obtained but, as suggested qualitatively by cyclic voltammetry, the tri-cation was too unstable to be isolated (12).

When one iron atom in biferrocene is replaced by ruthenium, the second oxidation moves to a more positive potential (by ~400 mV), reflecting the greater difficulty in oxidizing ruthenocene (22). A similar effect is observed in the mixed-metal metallocenophane **4**, (M = Fe, M′ = Ru); the more positive wave (peak potential E_{pk} = 0.94 V) is due to an irreversible two-electron process presumably localized on the ruthenocene unit. Interestingly, [1.1]-ruthenocenophane **4** (M = M′ = Ru) oxidizes in a chemically reversible two-electron step (E^0 = 0.38 V) to a persistent di-cation (23).

The dication $[Fe_2(\eta^5,\eta'^5\text{-}C_{10}H_8)(\eta^6\text{-arene})_2]^{2+}$ (**7**), a mixed-sandwich analogue of biferrocene, shows two *reduction* waves, assigned to the formation of the Fe(II)Fe(I) cation and the Fe(I)Fe(I) neutral compound (24). Unlike biferrocene, no oxidations were observed, probably due to the effect of the high positive charge on **7**, which would be expected to destabilize the Fe(III) oxidation state. When the sandwich moieties are "insulated" by an aliphatic connecting group, as in **8**, a single two-electron

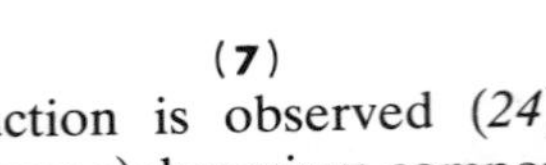

(7) 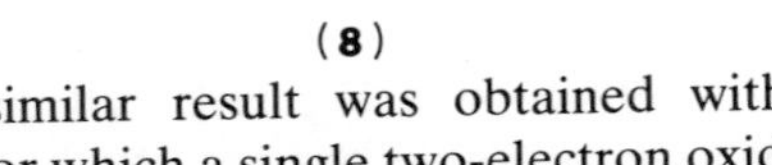   (8)

reduction is observed (24). A similar result was obtained with the bis(arene)chromium compound **9** for which a single two-electron oxidation wave was observed at virtually the same potential as that of the related mononuclear complex $[Cr(\eta\text{-}C_6H_5R)_2]$ [R = $(CH_2)_4C_6H_5$] (25).

Directly linked, bimetallic bis(arene)metal complexes such as $[M_2(\eta^6,\eta'^6\text{-}C_{12}H_{10})(\eta^6\text{-}C_6H_6)_2]^Z$ (**10**) (M = V, Z = 0; M = Cr, Z = 2),

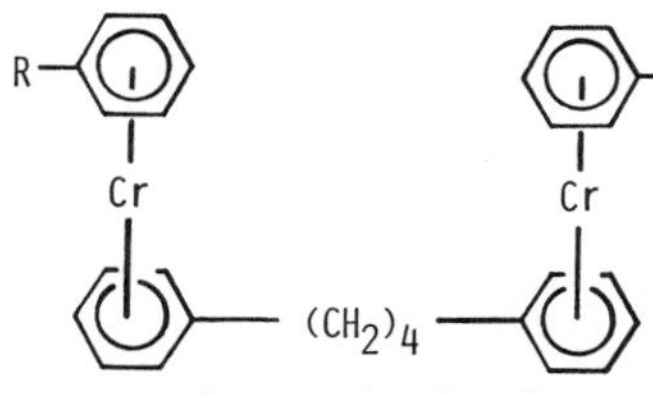

$[9, \ R = (CH_2)_4Ph]$

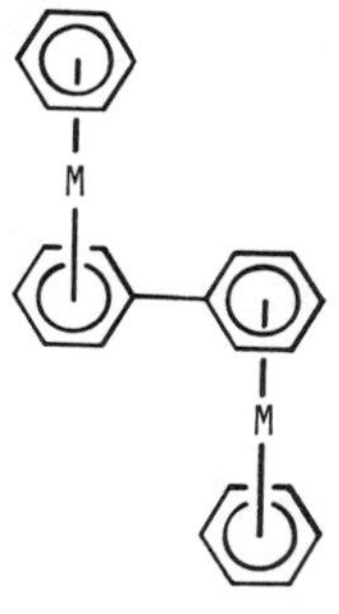

(10)

are paramagnetic and have been studied by ESR spectroscopy (26, 27). The divanadium triplet radical gives a beautifully resolved spectrum both in fluid solution and in a rigid glass. From the zero-field splitting, an average distance between the two paramagnetic centers of 540 pm was calculated, consistent with the assignment of a trans structure to the complex (27).

The electrochemical behavior of two types of linked half-sandwich complexes appears to be influenced by structural changes which accompany the electron-transfer steps. The biphenyl complex $[Cr_2(CO)_6(\eta^6,\eta'^6\text{-}C_{12}H_{10})]$ (11) is reduced in a reversible two-electron process at -1.61 V [versus Ag/AgCl, in tetrahydrofuran (THF) at 0°C] to a persistent dianion (28). The cyclic voltammetric peak separation of ~35 mV argues against an interpretation involving two one-electron reductions to isolated redox sites (which would give a 60 mV peak separation) [vide ante (18)]. Rather, E^0 for the second electron transfer must be more positive than E^0 for the first, and the second reduction is, therefore, more facile than the first.

In order to account for these observations, structural changes were proposed as shown in Scheme 1; the attraction of this scheme lies in the avoidance of 19-electron metal centers in the monoanion 12 and dianion 13. Data on the bis(chromiumtricarbonyl) complex of stilbene were interpreted similarly (28).

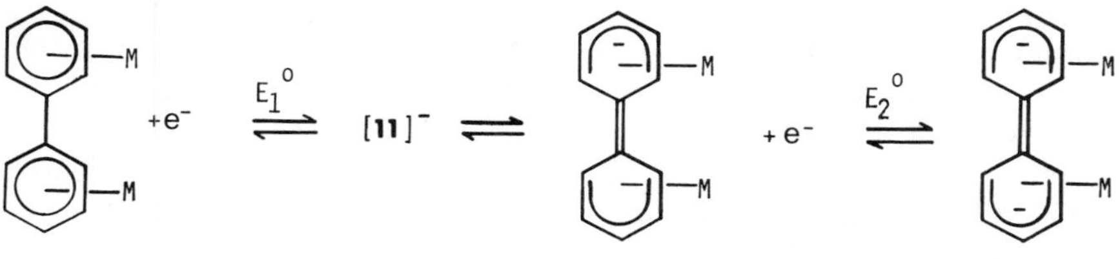

SCHEME 1. $M = Cr(CO)_3$

The fulvalenedirhodium complex **14** and the corresponding dication **15** are similarly linked by a two-electron transfer process with a small cyclic

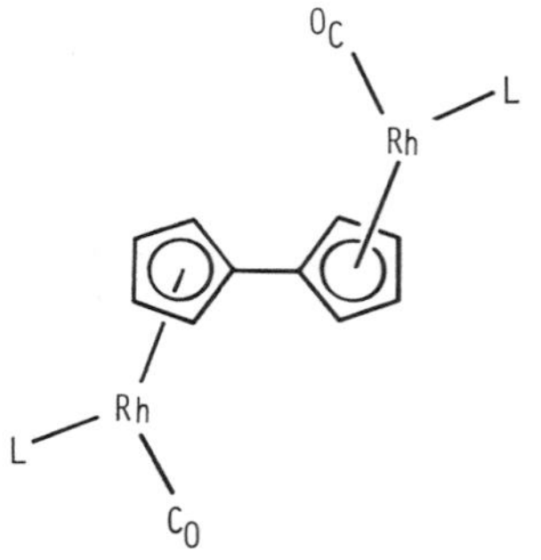

(**14**, L = PPh$_3$) 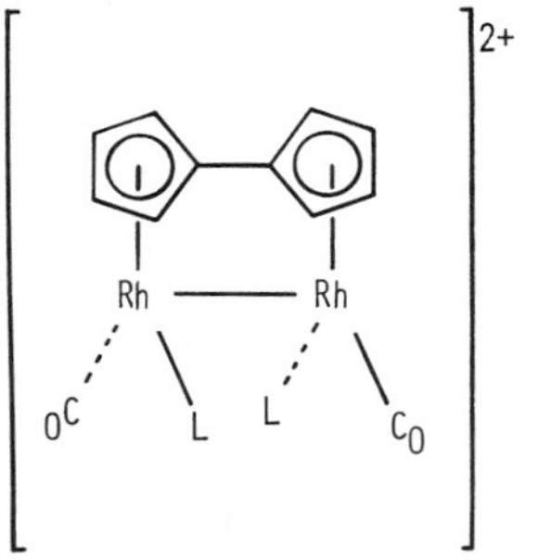   (**15**, L = PPh$_3$)

voltammetric peak separation (~38 mV). Again the second electron-transfer seems to be easier, in the thermodynamic sense, than the first (29). X-Ray structural studies have shown that the rhodium atoms are mutually cis in the dication but trans disposed in the neutral compound (30).

Cobalt analogues of biferrocene and bis(fulvalene)diiron have not been investigated as extensively as the iron compounds, despite the information on electronic interactions which could be gained from observing cobalt hyperfine splittings by ESR spectroscopy. The bis(fulvalene)dicobalt di-cation [**3**]$^{2+}$ (M = Co) can be reduced in two successive one-electron steps at mild potentials (E^0 = −0.07 V and −0.95 V), and [Co$_2$(η^5,η'^5-C$_{10}$H$_8$)Cp$_2$]$^{2+}$, [**1**]$^{2+}$ (M = Co), is reduced at −0.53 V and −0.88 V (31); the preparation of the monocation [**3**]$^+$ (M = Co) has been briefly reported (32, 33).

Bis(fulvalene)divanadium complexes in three different overall oxidation states have been prepared (34). The dication [V$_2$(η^5,η'^5-C$_{10}$H$_8$)$_2$]$^{2+}$, [**3**]$^{2+}$ (M = V) is reversibly reduced to the monocation (E^0 = 0.14 V), but the reduction to the neutral complex (E_{pk} = −0.09 V) is irreversible in acetonitrile. Since **3** (M = V) can be synthesized separately by the reaction of the fulvalene dianion with VCl$_2$·2THF in THF, the irreversibility of the one-electron reduction of [**3**]$^+$ (M = V) in acetonitrile is probably due to the reaction of the neutral complex with the solvent.

The electrochemistry of the bis(fulvalene)divanadium derivatives [V$_2$L$_2$(η^5,η'^5-C$_{10}$H$_8$)$_2$]^Z (**16**) (L = CO, Z = 0; L = MeCN, Z = 2) is ill defined. In acetonitrile, the cyclic voltammogram of **16** (L = CO, Z = 0) shows two oxidation waves, at about −0.7 V and −1.2 V (34). The first was ascribed to the oxidation of the monocation to the dication, and is not fully reversible; the second is totally irreversible yet assigned to the oxidation of the neutral compound to the monocation. The chemical studies on this system are also puzzling. Both monocation and dication are said to be isolable by oxidizing **16** (L = CO, Z = 0) with the ferrocenium ion in THF, yet the infrared spectra of the products are not in agreement with the formulations given [**16**, L = CO: Z = 0, $\bar{\nu}$(CO) = 1970sh, 1960s, 1900s, and 1870w cm^{-1}; Z = 1, $\bar{\nu}$(CO) = 2055, 2010, 1960, and 1915 cm^{-1}; Z = 2, $\bar{\nu}$(CO) = 2055 and 2005 cm^{-1}].

The cyclic voltammogram of **16** (L = MeCN, Z = 2) is also not easily interpreted. The E^0 value for the reduction of the dication, reported as 0.30 V, would be expected to be far more negative if the reduction of **16** (L = CO, Z = 2) does indeed occur at −0.7 V.

The bis(fulvalene)dinickel complex [**3**]$^{2+}$ (M = Ni) is reversibly reduced to the corresponding monocation (E^0 = 0.12 V) and neutral complex (E^0 = −0.29 V) (*35*). The unusual (fulvalene)diniobium compounds [Nb$_2$(μ-NR)$_2$(η^5,η'^5-C$_{10}$H$_8$)Cp$_2$] **17** (R = C$_6$H$_5$ or *p*-CH$_3$OC$_6$H$_4$) are chemically oxidized, for example by AgBF$_4$ in THF, to the paramagnetic monocations; the 19-line ESR spectra (A_{Nb} = 49 G) of [**17**]$^+$ arise from the interaction of the unpaired electron with two equivalent metal atoms (^{93}Nb, $I = \frac{9}{2}$). X-Ray structural studies on **17**^{Z} (Z = 0 and 1, R = OMe) have revealed a lengthening of the Nb—Nb bond, from 283.4(5) to 292.1(1) pm, and a slight increase in the folding about that bond on oxidation (*36*).

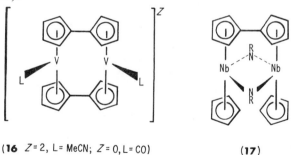

(**16** Z = 2, L = MeCN; Z = 0, L = CO) (**17**)

B. *Triple-Decker Sandwich and Related Compounds*

Although recent reviews of metal–carborane electrochemistry (*37, 38*) obviate the need to discuss multimetallic carborane clusters here, it seems appropriate to mention the results of studies on triple-decker sandwich compounds. The only structurally confirmed (*39*) all-hydrocarbon triple-decker, [Ni$_2$Cp$_3$]$^+$ (**18**), has apparently not been investigated electrochemically, but carborane analogues containing two or three boron atoms in the bridging five-membered ring have an extensive electrochemistry. The dicobalt compound $[Co_2(\eta^5\text{-}C_2B_3H_5)Cp_2]$ (**19**) is both oxidized and reduced in two separate reversible one-electron steps; cyclic voltammetry has shown that all of the complexes **19**^{Z}, ($Z = +2, +1, 0, -1, -2$) are stable on the voltammetric time scale with lifetimes of at least 5 seconds (*40*). A more extensive set of data exists for the complexes $[MM'(\mu\text{-}C_3B_2R_5)Cp_2]$ (**20**) (MM' = FeCo, Co_2, CoNi, or Ni_2); for each, only well-separated, one-electron waves are observed, consistent with a high degree of delocalization (*41*). The monocations $[20]^+$ (M = Co, M' = Fe, Co, or Ni) have been isolated, and the ESR spectra of the iron and nickel derivatives showed cobalt hyperfine splitting.

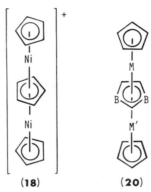

(**18**)   (**20**)

In agreement with the accepted molecular orbital treatment (*42*), which predicts that the total electron count around the two metals will not exceed 34, none of the complexes **19** and **20** can be reduced beyond the 34-electron state (in the potential range available at the dropping mercury electrode in nonaqueous solvents); $[20]^-$ (MM' = Ni_2) has 34 valence electrons. The heterogeneous charge-transfer rates, measured by alternating current polarography, confirm that the electron-transfer reactions of these compounds are very rapid (*41*).

One possible approach to the preparation of an all-hydrocarbon triple-decker compound, with the necessary 34-electron configuration, is to attempt to oxidize or reduce a "near-miss" complex (*42*), that is, one containing the correct number of ligands and metals but with an excess or deficiency of electrons. Two such attempts, involving complexes with bridging cyclooctatetraene (cot) ligands, have been reported.

The 32-electron dititanium compound **21** reacts with two equivalents of potassium in THF to give a green solution with one single line in the ^{1}H-NMR spectrum (*43*). However, the published data are insufficient to convince one that the structure assigned to the dianion, **22**, is correct.

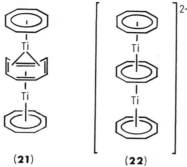

(21)   (22)

The oxidation of the electron-rich complexes $[M_2(\mu\text{-}C_8H_8)Cp_2]$ (23) (M = Co or Rh), which contain 36 electrons, provides definitive evidence for a structural change in a bridging cot ligand during electron transfer. Each of these complexes is oxidized in a single, two-electron process to a dication. The dicobalt compound $[23]^{2+}$ (M = Co), was too unstable to be structurally characterized (44), but the dirhodium analogue has been isolated and examined crystallographically (45). X-Ray studies have shown that the cot ring is only slightly twisted from planarity (Fig. 1). Nevertheless $[23]^{2+}$ (M = Rh) cannot be considered a true triple-decker complex.

Compounds related to 23, but with two cyclopentadienylmetal groups mutually cis with respect to the bridging cot ligand, also undergo electron-transfer reactions. For example, $[Rh_2(\mu\text{-}C_8H_8)Cp_2]^{2+}$, which contains a single metal–metal bond, is reduced to $[Rh_2(\mu\text{-}C_8H_8)Cp_2]$ by sodium amalgam (46).

When reduced with potassium metal, $[V_2(\mu\text{-}C_8H_8)Cp_2]$ (24) forms a radical anion, the ESR spectrum of which (15 lines, $A_V = 11.6$ G) suggests

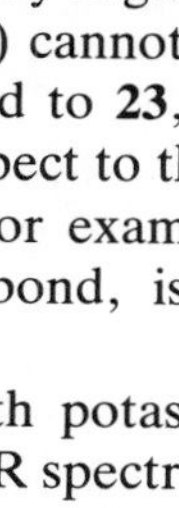

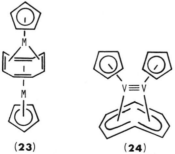

(23) 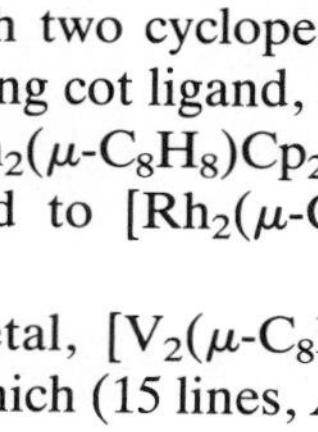   (24)

an orbitally nondegenerate ground state (47). The dichromium complex $[Cr_2(\mu\text{-}C_8H_8)Cp_2]$, structurally analogous to 24 but with a metal–metal double bond, is electrochemically oxidized to a radical cation isoelectronic with $[24]^-$ The ESR spectrum of this species shows properties similar to those of the divanadium radical anion (48).

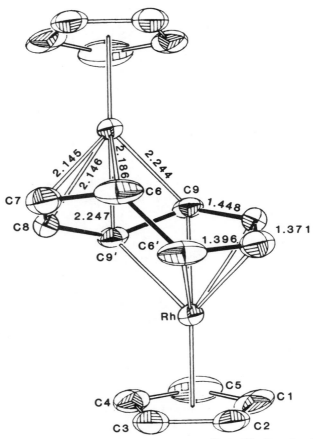

Fig. 1. The molecular structure of $[Rh_2(\mu\text{-}C_8H_8)Cp_2]^{2+}$, $[23]^{2+}$. [Reprinted with permission from *J. Am. Chem. Soc.* **106**, 3052 (1984).]

III

BRIDGED BIMETALLIC COMPOUNDS

A. Titanium

The imido-bridged complexes $[Ti_2(\mu\text{-}NR)_2X_2Cp_2]$ (**25**) (X = Cl or OCH$_3$, R = Me or Ph) are reduced in a reversible one-electron process (e.g., X = Cl, R = Ph, $E_{1/2}$ = 0.39 V versus Ag/AgI in CH$_2$Cl$_2$). The stable anion radicals, which can be prepared from **25** and sodium amalgam or $[Fe(CO)_4]^{2-}$, show fluid solution ESR spectra with appreciable hyperfine coupling to the bridging nitrogen atoms (e.g., X = Cl, R = Ph, A_N = 6.7 G) (49).

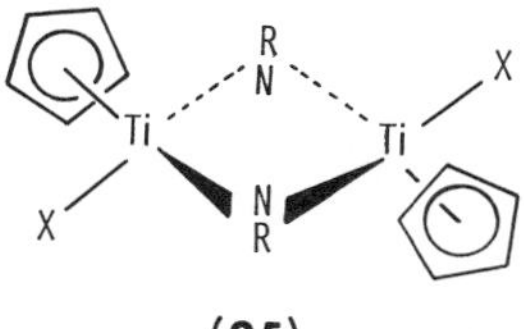

(25)

B. *Vanadium and Niobium*

The complexes $[\{V(CO)_4(\mu\text{-}EMe_2)\}_2]$ (E = P or As) are reduced in two, successive one-electron steps in benzonitrile (e.g., E = P, E^0 = −0.35 and −1.04 V versus Ag/AgCl). The monoanion (E = P), prepared by sodium naphthalenide reduction of the neutral compound and isolable as a sodium salt, has an ESR spectrum consisting of 15 triplets; the hyperfine couplings to two equivalent metal atoms and to two equivalent phosphorus atoms are 24.9 and 9.3 G, respectively (50). The redox potential for the first reduction of $[V_2(CO)_{8-n}L_n(\mu\text{-}PMe_2)_2]$ (n = 1 or 2, L = P- or As-donor) correlates with the donor–acceptor ability and the steric requirements of the ligand L.

The hydride complexes $[Nb_2H_2(\sigma,\eta^5\text{-}C_5H_3X)_2(\eta\text{-}C_5H_4Y)_2]$ (26) [X = Y = H (51) or XY = —SiMe_2OSiMe_2— (52)], originally thought to

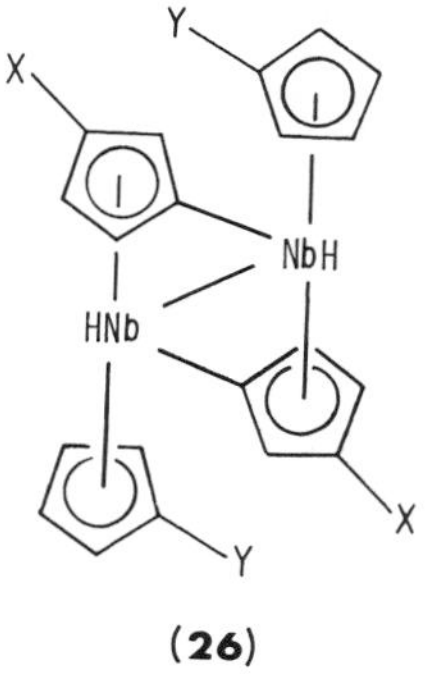

(26)

be niobocenes, are reduced by sodium naphthalenide in THF to the corresponding dianions (52). X-Ray structural studies have shown that the metal–metal bond of 26 (X = Y = H, Nb—Nb = 310.5 pm) (51) is absent in $[26]^{2-}$ (XY = —SiMe_2OSiMe_2—; Nb—Nb = 393.2 pm) (52).

The reversible oxidation of the fulvalene complex $[Nb_2(\mu\text{-}NR)_2(\eta^5,\eta'^5\text{-}C_{10}H_8)Cp_2]$ **(17)** is described in Section II,A.

C. Chromium, Molybdenum, and Tungsten

Dimethylphosphido-bridged complexes of the type $[\{M(CO)_4(\mu\text{-}PMe_2)\}_2]$ **(27)** (M = Cr, Mo, and W) undergo reversible two-electron reduction to the corresponding dianions *(53, 54)*; NMR spectroscopy showed that $[27]^{2-}$, prepared by allowing **27** to react with sodium–potassium alloy, are fluxional *(55)*. The monoanions $[27]^-$ were prepared

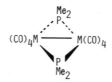

(27; M = Cr, Mo, or W)

by mixing **27** and $[27]^{2-}$, and studied by ESR spectroscopy. The moderate ^{31}P hyperfine splitting in $[27]^-$ (e.g., M = Cr, $A_P = 13.0$ G), was originally interpreted in terms of a half-filled orbital having high bridging-ligand character *(55)*. Later, this and NMR spectral data were reinterpreted in terms of a half-filled metal–metal antibonding orbital *(56)*, and a detailed theoretical treatment *(57)* established that the reduction of **27** results in the occupancy of a σ^* in-plane molecular orbital and effectively reduces the metal–metal bond order in the dianion to zero.

The redox behavior of singly bridged dimetal compounds having metal–metal bonds *(58)* has been compared to that of those which do not *(59)*. Nineteen mixed-metal compounds $[MM'(\mu\text{-}AsMe_2)]$ [M = $Cr(CO)_5$, $W(CO)_5$, $Fe(CO)_4$, and $Co(CO)Cp$; M' = $Cr(CO)_2Cp$, $Mo(CO)_2Cp$, $W(CO)_2Cp$, $Mn(CO)_4$, $Re(CO)_4$, and $Co(CO)_3$], and some of their phosphine and phosphite derivatives, undergo one-electron oxidations between 0.3 and 1.0 V (vs Ag/AgCl) in dichloroethane, and are also reduced *(58)*.

The potential for the reduction process correlates with the optically measured excitation energy of the σ–σ^* transition, and the presence of heavier metals, or greater carbonyl substitution, results in more negative $E_{1/2}$ values.

Twenty-seven dimetal compounds, bridged by one dimethylarsenide group but containing no direct metal–metal bond, were found to undergo irreversible oxidation and/or reduction *(59)*, usually giving mononuclear products. Binuclear [{Mo(CO)$_5$}$_2$(μ-AsMe$_2$AsMe$_2$)] was isolated from both the sodium–potassium alloy reduction and the iodine oxidation of [Cp(CO)$_3$Mo(μ-AsMe$_2$)Mo(CO)$_5$].

The nitrosyl complex [{Cr(μ-NO)(NO)Cp}$_2$] (28) is reversibly reduced ($E^0 = -1.00$ V) to [28]$^-$ in CH$_2$Cl$_2$. The monoanion [28]$^-$ is much more stable than that derived from the isoelectronic iron carbonyl compound [{Fe(μ-CO)(CO)Cp}$_2$] (vide infra), but exhaustive electrochemical reduction of 28 yields [Cr$_2$(μ-NO)(μ-NH$_2$)(NO)$_2$Cp$_2$] and [Cr$_2$(μ-NH$_2$)$_2$(NO)$_2$Cp$_2$]. The oxidation of 28 is irreversible, and yields [Cr(NO)$_2$Cp]$^+$ (60).

The sulfur-bridged dimolybdenumnitrosyl complexes [{Mo(μ-SR)(NO)Cp}$_2$] (R = Me, Et, Pr, Bu, or benzyl) undergo reversible oxidation at ~0.7 V, but attempts to measure the ESR spectra of the resulting cations were unsuccessful (61).

The complexes [{Mo(μ-SR)$_2$Cp}$_2$] (29) and [{Fe(μ-SR)(CO)Cp}$_2$] (30), which have quite similar oxidation potentials (e.g., R = Me, $E^0 \approx -0.1$ V) (62, 63), have provided an interesting insight into the effect of electron transfer on molecular structure. The molybdenum derivative (R = Me) shows virtually no change on one-electron oxidation (Mo—Mo $\approx$ 261 pm) (64), possibly because of steric restraints imposed by the presence of four bridging ligands. The iron–iron distance of 339 pm in the neutral complex (30, R = Ph) (65) is shortened to 293 pm in the cation [30]$^+$ (R = Me), a change consistent with the formation of a one-electron metal–metal bond on oxidation (66). Surprisingly, the greater structural change accompanying the oxidation of 30 does not lead to a substantial lowering of the heterogeneous charge-transfer rate compared to that for the oxidation of 29 (63).

Dithiolate- or disulfide-bridged analogues of 29 oxidize reversibly to mono- and dications. The Mo—Mo distance (259.9 pm) in [Mo$_2$(μ-SC$_3$H$_6$S)$_2$Cp$_2$]$^+$ is very similar to that of [29]$^+$, (R = Me); preliminary magnetic data on the dications, isolated from the reaction of the neutral species with ceric ion, suggest the presence of two unpaired electrons (67).

η^6-Arene analogues of 29 undergo oxidatively induced reactions at the aromatic ring (Scheme 2). In propylene carbonate, [Mo$_2$(μ-SMe)$_4${η^5-C$_6$H$_5$R(CN)}(η^6-C$_6$H$_5$R)]$^+$ [31]$^+$ is oxidized ($E^0 = 1.17$ V) to a persistent dication which slowly loses the radical CN· to give the bis(arene) complex [Mo$_2$(μ-SMe)$_4$(η^6-C$_6$H$_5$R)$_2$]$^{2+}$, [32]$^{2+}$. The oxidation of [31]$^+$ at a more positive potential ($E^0 = 1.9$ V) results in the loss of H$^+$ rather than CN·, giving [Mo$_2$(μ-SMe)$_4$(η^6-C$_6$H$_5$R)(η^6-C$_6$H$_5$CN)]$^{2+}$ [33]$^{2+}$ (68).

The mixed-valence complexes [Mo$_2$(μ-X)$_3$(η^7-C$_7$H$_7$)$_2$] (69) are readily oxidized by arenediazonium ions to the diamagnetic monocations. The dependence of E^0 on X (X = Cl, $E^0 = -0.15$ V; X = Br, $E^0 = -0.12$ V;

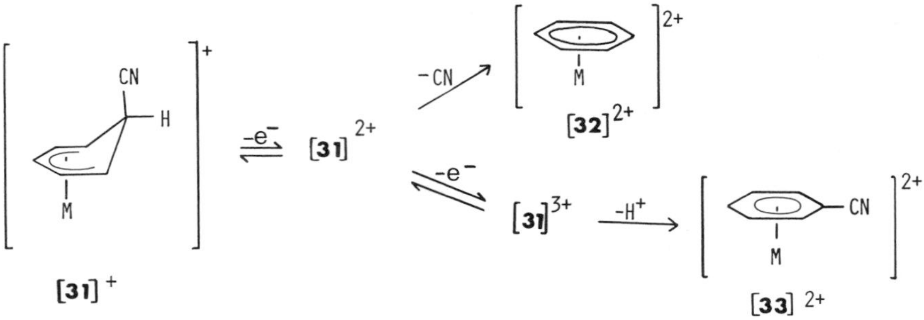

$$[31]^+ \quad \overset{-e^-}{\underset{}{\rightleftharpoons}} \quad [31]^{2+} \quad \overset{-CN}{\nearrow} \quad [32]^{2+}$$

$$\overset{-e^-}{\searrow} \quad [31]^{3+} \quad \overset{-H^+}{\longrightarrow} \quad [33]^{2+}$$

SCHEME 2. $M = Mo_2(\mu\text{-SMe})_4(\eta\text{-}C_6H_5R)$

$X = I$, $E^0 = -0.11$ V) is such that the electron is probably removed from an orbital antibonding with respect to the metal–metal bond (70).

D. Manganese

The ESR spectrum observed when $[Mn_2(CO)_{10-n}(PR_3)_n]$ (34) ($n = 0$ or 2) is γ-irradiated has been the source of some dispute. On irradiating a frozen solution of 34 in methyltetrahydrofuran, Symons and co-workers assigned the spectrum to the anion radical [34]$^-$, formed by simple electron transfer (71–73). The same spectrum is apparently observed when single crystals of $[Mn_2(CO)_{10}]$ are γ-irradiated. However, the directional dependence of the manganese hyperfine splitting allowed Morton, Preston, and co-workers to show that the half-occupied orbital has antibonding $3d_{z^2}$ (Mn) character and is inclined at an angle of 118° to the metal–metal vector (74). Thus, the second authors deduced the formula $[Mn_2(CO)_9]^-$ for the radical and assigned the carbonyl-bridged structure 35.

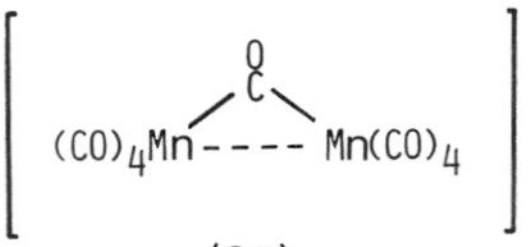

$$\left[(CO)_4Mn \cdots Mn(CO)_4 \right]^-$$

(35)

Iodine oxidation of $[\{Mn(CO)_4(\mu\text{-SR})\}_2]$ (R = H or SnMe$_3$) did not give a stable cation, but instead yielded the cluster $[Mn_4S_4(CO)_{15}]$ (75).

The dibridged nitrosyls $[\{Mn(\mu\text{-SR})(NO)Cp\}_2]$ (R = Pr^i, Bu^i, or Bu^t) are reversibly oxidized in CH$_2$Cl$_2$ to monocations ($E^0 \approx -0.2$ V) and dications ($E_0 \approx 0.5$ V). The former are readily prepared by air or iodine oxidation of the neutral species, and the ESR spectra show coupling ($A_{55Mn} \approx 38$ G) to two equivalent metal atoms (76).

The singly bridged complex $[Mn_2(\mu\text{-SEt})(CO)_4(\eta\text{-}C_5H_4Me)_2]^+$ (77) shows two reversible, one-electron reduction waves ($E^0 = 0.17$ and

−0.22 V vs Ag/AgCl) (*78*); the neutral complex is said to be indefinitely stable in cold acetone. Electron addition is thought likely to lead to metal–metal bond rupture (*77*).

The reduction of $[Mn_2(\mu\text{-}CH_2)(CO)_4Cp_2]$ is irreversible, but the reversible oxidation ($E^0 = 0.53$ V) gave a persistent radical cation with an ESR spectrum showing equal coupling ($A_{Mn} = 38$ G) to two manganese atoms (*63*).

The one-electron reduction of $[Mn_2(NO)_3Cp_3]$ has been mentioned briefly (*62*).

E. *Iron and Ruthenium*

The phosphido-bridged diiron complexes $[\{Fe(CO)_3(\mu\text{-}PR_2)\}_2]$ (**36**) have short metal–metal distances (e.g., R = Ph, Fe—Fe = 262 pm), consistent with the presence of a single iron–iron bond and a concomitant 18-electron configuration for each metal. Sodium reduction of **36** gives the dianion $[\mathbf{36}]^{2-}$, which has an iron–iron distance (R = Ph) of 363 pm indicating the absence of a direct metal–metal interaction. The structural data suggest that the LUMO of **36** is antibonding between the metals (*79*), supporting a bonding model arrived at earlier through molecular orbital calculations (*80*).

Details of the electrochemical reduction of **36** and the possible existence of the monoanionic intermediate (**36**)⁻ (R = Ph) are not yet settled. The neutral dimer is reported to undergo a single, chemically reversible, two-electron reduction at a mercury electrode ($E^0 = -1.26$ V in CH_3CN) (*79*), although the process was less reversible (R = Ph or Me) in propylene carbonate (*81*) and totally irreversible at platinum (*79*). The peak separation of 45 mV (at mercury), which contrasts with the expected value of 30 mV for a reversible two-electron reduction, is reported to be essentially independent of the cyclic voltammetric scan rate (*79*). Although this seems to imply that the overall process may consist of two closely spaced one-electron waves, mixing solutions of $[\mathbf{36}]^{2-}$ (R = Ph) and **36** (R = Ph) did not produce a sufficient concentration for $[\mathbf{36}]^{-}$ (R = Ph) to be detected by ESR spectroscopy. Thus, the high disproportionation constant for the reaction:

$$2\ [\mathbf{36}]^{-} \rightleftharpoons \mathbf{36} + [\mathbf{36}]^{2-} \qquad (R = Ph)$$

implied by ESR spectroscopy, is inconsistent with the electrochemical data.

Although $[\{Fe(CO)_3(\mu\text{-}PMe_2)\}_2]$ (**36**) (R = Me) also reduces in a reversible two-electron step in 1,2-dimethoxyethane (DME) (*55, 57*), the monoanion $[\mathbf{36}]^{-}$ (R = Me), *can* be generated by allowing **36** (R = Me) to react with $[36]^{2-}$ (R = Me). The ESR spectrum of the monoanion, and NMR studies on the dianion, were originally interpreted in favor of a bridging-ligand localized LUMO for 36 (R = Me) (54, 55), but it is now known that this orbital is metal–metal antibonding in character (56, 57, 79). The observation of appreciable ^{31}P hyperfine splitting in the ESR spectrum of $[36]^-$ (R = Me), which led to the original misinterpretation, can be understood by analogy with data on $[\{Co(\mu\text{-}PPh_2)Cp\}_2]^+$. The phosphido bridge contributes less than 10% to the half-filled orbital of this radical cation, and yet the isotropic phosphorus splitting is ~11 G (82).

The anion $[36]^{2-}$ (R = Ph) oxidatively adds alkyl halides, RX, to give the anionic acyl complexes 37 (83).

The complexes $[\{Fe(CO)_3(\mu\text{-}SR)\}_2]$ are reduced in a single two-electron step (R = Me, $E^0 = -1.22$ V) much like the isoelectronic phosphido-bridged compounds 36. The resulting dianions are apparently more stable in DME (62) than in propylene carbonate (81). The mixed-bridged compound $[Fe_2(CO)_6(\mu\text{-}SMe)(\mu\text{-}PMe_2)]$ behaves similarly (81), but two one-electron reductions, separated by 400 mV, were observed for $[\{Fe(NO)_2(\mu\text{-}PPh_2)\}_2]$ (53).

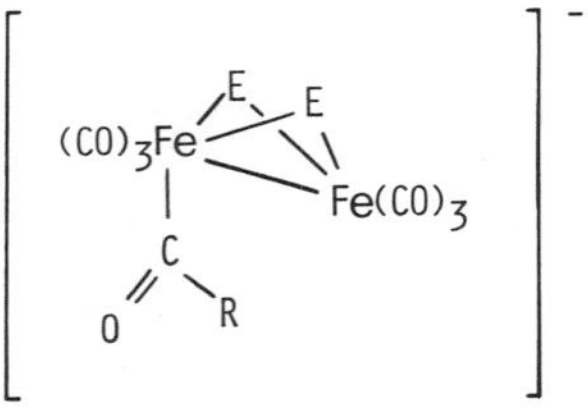

(37, E = PPh_2) 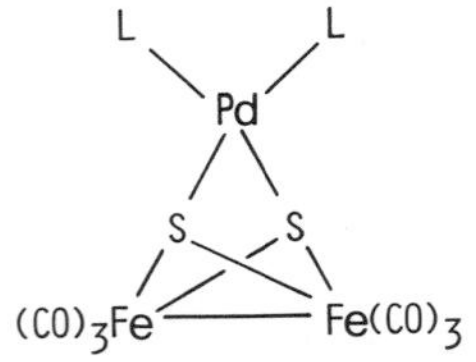   (38, L = PPh_3)

The disulfide-bridged complex $[Fe_2(CO)_6(\mu\text{-}S_2)]$, which contains iron–iron and sulfur–sulfur bonds, may be chemically reduced to a reactive dianion. This species, in which the disulfide linkage has presumably been cleaved (84, 85), is stable at $-78°C$ in THF when generated with LiBHEt$_3$. It reacts with a variety of electrophiles, such as organic halides, to provide a general route to complexes with bridging organosulfur ligands, and with metal complexes such as $[PdCl_2(PPh_3)_2]$ to give sulfur-bridged heteropoly-nuclear compounds such as 38 (84–87).

The reductions of the sulfur-bridged complexes 39 and 40 (L = CO) are reversible, giving monoanions stable on the time scale of cyclic voltamme-try. In the presence of P(OMe)$_3$, electrocatalytic carbonyl substitution occurs since E^0 for the substitution product is more negative than that of the carbonyl precursor. Thus, for example, 40 (L = CO) gave 88% yield of 40 [L = P(OMe)$_3$] when only 0.05 $\mathscr{F}$ of charge was passed (88).

(39)

(39)

(40)

The carbonyl-bridged dimers $[\{M(\mu\text{-CO})(CO)Cp\}_2]$ **(41)** [M = Fe (*89, 90*) or Ru (*89*)] are reductively cleaved to the monoanions $[M(CO)_2Cp]^-$. In acetonitrile, **41** (M = Fe) is also cleaved by electrochemical or chemical oxidation to give $[Fe(NCMe)(CO)_2Cp]^+$ or, in the presence of iodine, $[FeI(CO)_2Cp]$ (*91–93*). Although the cation $[41]^+$ (M = Fe) is more persistent in other solvents, and the oxidation of **41** (M = Fe) is reversible in CH_2Cl_2 (*60, 94*) and THF (*95*), bulk electrolysis at 0.67 V in CH_2Cl_2 with $[NBu^n_4][PF_6]$ as supporting electrolyte gave $[Fe(CO)_2Cp][PF_6]$ (*60*).

The ditertiaryphosphine-bridged derivatives of **41** (M = Fe) are readily oxidized by iodine or silver(I) salts to the isolable monocations $[Fe_2(\mu\text{-}CO)_2(\mu\text{-L})Cp_2]^+$; further oxidation with Ag^+ gives $[Fe(CO)_2LCp]^+$ **[42]**$^+$ (L = Ph_2PRPPh_2, R = CH_2, C_2H_2, C_2H_4, or NEt), probably via the unstable dications **[42]**$^{2+}$ (*95*). Indeed, electrochemical studies on **42** [L = $Ph_2PCH_2CH_2PPh_2$ (dppe)] show a reversible one-electron oxidation ($E^0 = 0.10$ V) followed by a second, irreversible process at 0.95 V (*96*) in CH_2Cl_2.

X-Ray diffraction studies on $[\{Fe(CO)(\mu\text{-SR})Cp\}_2]^Z$ (Z = 0, R = Ph; Z = 1, R = Me) (*66*) were referred to in Section III,C in connection with related work on $[\{Mo(\mu\text{-SMe})_2Cp\}_2]^Z$ (Z = 0 and 1) (*64*). The dications $[\{Fe(CO)(\mu\text{-SR})Cp\}_2]^{2+}$ have also been prepared (*61, 63, 97*) but not structurally characterized.

By comparison, the geometries of $[\{Fe(CO)(\mu\text{-PPh}_2)Cp\}_2]^Z$ (Z = 0, 1, and 2) are all well defined. The successive shortening of the iron–iron distance, from 350 pm (Z = 0) to 314 pm (Z = 1) and then to 276 pm (Z = 2), suggests electron removal from an HOMO which is antibonding between the metals, and the sequential formation of metal–metal bonds of order 0.5 and 1.0 (*98*). It is interesting to note here the brief report that the one-electron oxidation of $[Fe_2(\mu\text{-CO})_2(\mu\text{-Ph}_2PCH_2PPh_2)Cp_2]$ is accompanied by an increase in the metal–metal bond length of less than 2 pm (*99*).

The disulfide bridge has a profound effect on the course of the redox reactions of $[Fe_2(\mu\text{-S}_2)(\mu\text{-SEt})_2Cp_2]$ **(43)**. Reversible one-electron oxidation ($E^0 = 0.21$ V) (*100*) gives an isolable monocation **[43]**$^+$ (*101*) with an Fe—Fe distance (306 pm) approximately 25 pm shorter than that of **43** (*102*). Further oxidation, at 0.9 V in acetonitrile, gave the transient

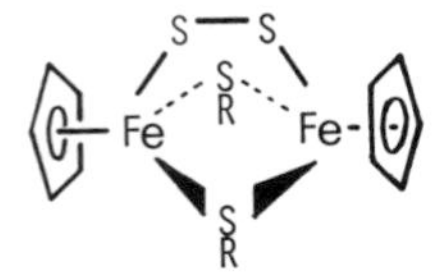

(43)

dication $[43]^{2+}$, which reacted with the solvent to give $[\{Fe(NCMe)(\mu\text{-}SR)Cp\}_2]^{2+}$ (44). Presumably, the additional shortening of the iron–iron distance required to form $[43]^{2+}$ results in a strained Fe_2S_2 ring and, therefore, in the expulsion of the S_2 bridge (103).

Complex 44 is also prepared by the chemical oxidation of 43 with $[NO][PF_6]$. By contrast, the reaction with Ag^+ gave $[Fe_2Ag_2(\mu\text{-}S_2)(\mu\text{-}SEt)_2(NCMe)_4Cp_2]^{2+}$, in which the silver atoms are assumed to be coordinated to sulfur and nitrile ligands (103).

The cyclic voltammogram of the dimeric nitrosyl $[\{Fe(\mu\text{-}NO)Cp\}_2]$ shows a reversible reduction wave at -1.38 V (vs $Ag/AgNO_3$), and sodium in DME affords the highly reactive salt $Na[\{Fe(\mu\text{-}NO)Cp\}_2]\cdot1.5$ DME; the anion reacts with methyl iodide to give $[\{FeMe(\mu\text{-}NO)Cp\}_2]$ (104).

The diiron metallacycle 45 is reported to undergo an *irreversible* reduction to an anion which nevertheless regenerates the neutral complex on reoxidation (105); detailed studies of this potentially interesting result have not appeared. The dimetallacycle 46 undergoes irreversible reduction

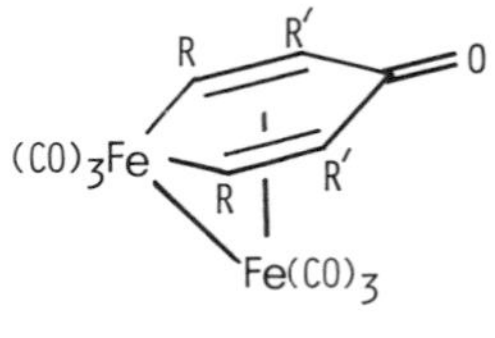

(45; R = Me, R′ = Ph 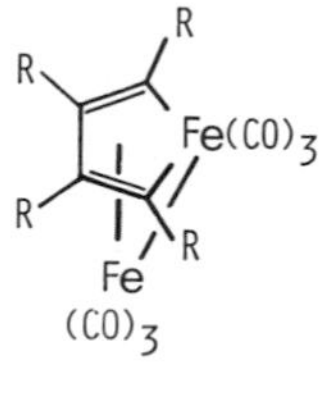)   (46, R = C_6H_4Cl-p)

(105), but analogous benzoferrole derivatives give stable monoanions which have been studied by ESR spectroscopy (106). Further discussion of these species may be found in Section VI of our earlier article (5).

F. Cobalt and Rhodium

The synthetically important dimeric anions $[\{Co(\mu\text{-}CO)(\eta\text{-}C_5H_4R)\}_2]^-$ $[47]^-$ (R = H, $SiMe_3$, or $SiMePh_2$) are prepared by sodium amalgam reduction of the appropriate monomer $[Co(CO)_2(\eta\text{-}C_5H_4R)]$ (107, 108). The ESR spectra of the monoanions showed equal hyperfine coupling to

TABLE I

Compound	Co—Co distance (pm)	Electronic configuration	References
$[\{Co(\mu\text{-}CO)(\eta\text{-}C_5Me_5)\}_2]$	233.8	d^8, d^8	110, 111
$[\{Co(\mu\text{-}CO)(\eta\text{-}C_5Me_5)\}_2]^-$	237.2	d^8, d^9	110, 111
$[\{Co(\mu\text{-}CO)Cp\}_2]^-$	237.2	d^8, d^9	107
	236.4		110
$[Co_2(\mu\text{-}CO)(\mu\text{-}NO)Cp_2]$	237.2	d^8, d^9	112
$[\{Co(\mu\text{-}NO)Cp\}_2]$	237.0	d^9, d^9	112
$[\{Co(\mu\text{-}NO)Cp\}_2]^+$	234.8	d^8, d^9	113

both cobalt nuclei (107, 108) but did not allow unambiguous assignment of the molecular orbital containing the unpaired electron.

One-electron oxidation of $[47]^-$ [e.g., R = H, $E^0 = -1.05$ V vs Ag/AgClO$_4$ (109) or -0.75 V vs sce (63)] gives the neutral complex $[\{Co(\mu\text{-}CO)(\eta\text{-}C_5H_4R)\}_2]$ (47). Since the formal cobalt–cobalt bond order increases from 1.5 to 2.0 when $[47]^-$ is oxidized to 47, an appreciable shortening of the metal–metal bond length might be expected. However, a comparison of the Co—Co distance in $[\{Co(\mu\text{-}CO)(\eta\text{-}C_5Me_5)\}_2]$ with that in the corresponding monoanion showed a difference of only 3.4 pm (Table I) (110, 111).

The nitrosyl complexes $[Co_2(\mu\text{-}CO)(\mu\text{-}NO)Cp_2]$ (48) and $[\{Co(\mu\text{-}NO)Cp\}_2]^+$ $[49]^+$ are isoelectronic with $[47]^-$, and $[\{Co(\mu\text{-}NO)Cp\}_2]$ is isoelectronic with the hypothetical dianion $[47]^{2-}$. Once again, only small increases in the cobalt–cobalt bond length are observed as electrons are added to the dimeric system (112, 113).

The structural results presented in Table I are in disagreement with one theoretical study which predicted an increase of about 8 pm in the cobalt–cobalt distance for each electron added to the complex (114). The smaller observed difference has been ascribed to the fact that as the Co—Co bond length increases the Co—CO bond length decreases; the equilibrium geometry appears to reflect the balancing of these two effects (110, 115).

The one-electron oxidation of $[Co_2(\mu\text{-}CO)(\mu\text{-}NO)Cp_2]$ (48) is irreversible in acetonitrile (112), probably because the monocation reacts with the solvent. Thus, the oxidation of $[Co_2(\mu\text{-}CO)(\mu\text{-}NO)(\eta\text{-}C_5Me_5)_2]$ ($E^0 = -0.2$ V) is reversible in CH$_2$Cl$_2$ and the monocation is isolable (116).

The cations $[\{Co(\mu\text{-}NO)(\eta\text{-}C_5R_5)\}_2]^+$ $[49]^+$ (R = H or Me), prepared from the nitrosonium ion and $[Co(CO)_2(\eta\text{-}C_5R_5)]$, are part of a four-membered redox series; cyclic voltammetry reveals the formation of dications $[49]^{2+}$ (R = H, $E^0 = 1.17$ V; R = Me, $E^0 = 0.68$ V), neutral species **49** (R = H, E^0 = 0.34 V; R = Me, E^0 = −0.15 V), and a monoanion [**49**]⁻ (R = Me, E^0 = −1.7 V) (*116*).

The oxidized forms of **49** are susceptible to attack by electron donors. The dication [**49**]²⁺ (R = H), generated *in situ* from [**49**]⁺ and silver(I) ions, reacts with alkenes to give, for example, [Co(η^2-cyclooctene)(NO)Cp]⁺; the cation [**49**]⁺ reacts with PPh₃ to give [Co(PPh₃)(NO)Cp]⁺ but the neutral complex **49** is inert (*116*).

Rhodium complexes related to **48** and **49** also undergo electron-transfer reactions. Thus, [{Rh(μ-NO)Cp}₂] (**50**) is reduced in a reversible one-electron step (E^0 = −1.17 V) to a stable monoanion. Surprisingly, since the dication [{Rh(μ-NO)(η-C₅Me₅)}₂]²⁺ can be isolated from [Rh(CO)₂(η-C₅Me₅)] and [NO][PF₆], the oxidation of **50** is totally irreversible.

The cation [Rh₂(μ-CO)(μ-NO)(η-C₅Me₅)₂]⁺ also undergoes reversible one-electron reduction in CH₂Cl₂ (E^0 = −0.57 V), and electrolysis at −0.7 V gave a very air-sensitive solution of the neutral complex; a second, irreversible reduction process was also detected (*117*).

The sulfur-bridged dimers [{Co(μ-SR)Cp}₂] are reversibly oxidized to monocations (*53, 61*) and, in DME, a reversible one-electron reduction was reported (*53*). A similar reduction was detected for [{Co(μ-PPh₂)(Cp}₂] (*53*), and although the one-electron oxidation was described as irreversible in DME (*53*), reference has been made to the generation of [{Co(μ-PPh₂)Cp}₂]⁺ (*82*) and to its well-resolved ESR spectrum (31 lines, $A_{^{59}Co}$ = 22.66 G, $A_{^{31}P}$ = 11.33 G).

The dirhodium complexes [{Rh(μ-SR)(η-C₅H₄R′)}₂] (R = aryl, R′ = H or Me) exist with syn and anti geometries, and each isomer gives rise to separate, irreversible oxidation waves; the *thermal* syn to anti isomerization was followed by cyclic voltammetry. The radical cations [{Rh(μ-SR)(η-C₅H₄R)}₂]⁺ are highly unstable, and, when generated by [NO][PF₆] oxidation of the neutral dimer, undergo metal–metal bond-insertion reactions. Thus, [Rh₂(μ-SC₆H₄CH₃-*p*)₃Cp₂]⁺ is formed in the presence of di-*p*-tolyldisulfide (*118*).

The potential for the reduction of the acetylene-bridged dicobalt compounds [Co₂(CO)₆(μ-RC₂R)] (**51**) (R = H, Ph, Buᵗ, CF₃, or SiMe₃) is very dependent on R [R = CF₃, E^0 = −0.49 V; R = Buᵗ, E^0 = −1.01 V vs Ag/AgCl], and the stability of the resulting radical anion [**51**]⁻ also varies greatly (R = H, $t_{1/2}$ < 60 seconds at −60°C; R = CF₃, persistent radical) (*119*). ESR spectra indicate that the unpaired electron in [**51**]⁻ occupies an orbital that is antibonding between the two cobalt atoms, and that the Co—Co bond is "bent" (*119, 120*). Except for [**51**]⁻ (R = CF₃), all of the monoanions eventually decompose to mononuclear compounds including [Co(CO)₄]⁻.

When the reduction of **51** is performed in the presence of Lewis bases, phosphine derivatives are formed. For anion radicals [**51**]⁻ with limited lifetimes, mononuclear products such as $[Co(CO)_3(PR_3)]^-$ and paramagnetic $[Co(CO)_2(RC_2R)(PR'_3)]$ result, but, if [**51**]⁻ is more stable, direct substitution to give binuclear species occurs (121, 122).

Because the reduction potential of $[Co_2(CO)_{6-n}L_n(\mu\text{-}RC_2R)]$ (**52**) (L = P-donor) is more negative than that of **51**, the neutral form of the substitution product is thermodynamically favored at the potential required to generate [**51**]⁻. Thus, substitution of **51** affords [**52**]⁻ which is immediately oxidized (at the electrode or by another molecule of neutral **51**) to **52**. The net result is that the reductive substitution of **51** to give **52** is catalytic in the quantity of electricity used.

The real utility of this electron-transfer catalyzed reaction lies in the enormous enhancement of reactivity toward substitution in the anion radicals. For example, thermal substitution of **51** (R = CF_3) by phosphines gives 10–60% product yields after several hours in boiling benzene. By contrast, the monosubstituted complexes **52** ($n = 1$) were produced in greater than 90% yield, in less than 1 minute at room temperature, by adding a catalytic amount of the reductant $[Ph_2CO]^-$ to a 1:1 mixture of **51** and L (121).

In the presence of higher concentrations of phosphines, di- and trisubstituted products may be obtained (121), and interesting examples of the stereospecific formation of isomers by electron-transfer catalysis have been reported (123, 124). The "flyover" complex $[Co_2(CO)_4\{\mu\text{-}C_6(CF_3)_6\}]$ (**53**) forms the bis(phosphite) derivative $[Co_2(CO)_2\{P(OMe)_3\}_2\{\mu\text{-}C_6(CF_3)_6\}]$ when reduced in the presence of excess $P(OMe)_3$, but two different isomers can be exclusively produced depending on the applied electrolysis potential. At the potential of the one-electron reduction of **53** ($E^0 = -0.1$ V vs Ag/AgCl), isomer **54** is formed, but, at a potential appropriate for the reduction of **54** ($E^0 = -0.7$ V), quantitative conversion to isomer **55** is achieved. The reaction is electrocatalytic in both cases,

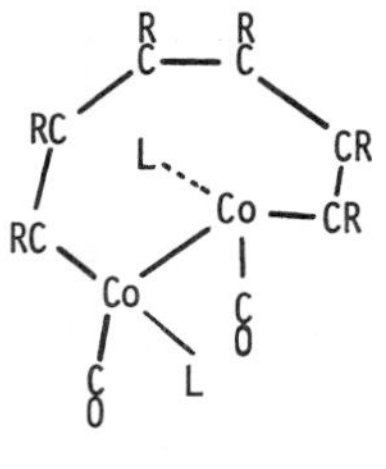

(**54**, R = CF_3) 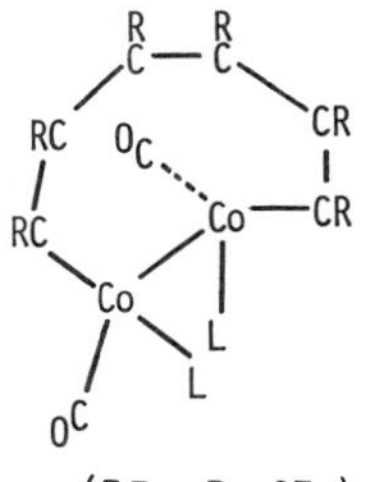   (**55**, R = CF_3)

and stereochemical selectivity seems to depend on the specific conformations of [**54**]$^-$ and [**55**]$^-$ rather than of the neutral analogues (*123*).

The tribridged complex [Co$_2$(CO)$_2${μ-F$_2$PN(Me)PF$_2$}$_3$], is easily reduced to a monoanion radical ($E^0 = -0.59$ V) and to a persistent dianion
($E^0 = -1.07$ V); the three species were characterized by carbonyl bands in
the IR spectrum at 2003, 1940, and 1877 cm^{-1}, respectively. The dianion is
said to react with methyliodide, but no products were isolated (*125*).

The dimers [{Co(CO)(μ-CO)(η^4-diene)}$_2$] (**56**) (diene = butadiene,
isoprene, or cyclohexadiene) react with weak oxidants, such as [CPh$_3$]$^+$, in
the presence of PPh$_3$ to give [Co(CO)$_{3-n}$(PPh$_3$)$_n$(η^4-diene)]$^+$ (n = 1 or
2). Infrared evidence was presented (*126*) for the intermediacy of the
cation radical [{Co(CO)(μ-CO)(η^4-diene)}$_2$]$^+$ [**56**]$^+$, but cyclic voltammetry on **56** (diene = 2,3-dimethylbutadiene) in CH$_2$Cl$_2$ showed a diffusion-controlled, but totally irreversible, one-electron oxidation wave
(*127*).

G. *Nickel and Palladium*

The complexes [{Ni(μ-SR)Cp}$_2$] (**57**), [{Ni(μ-PPh$_2$)Cp}$_2$] (**58**), and
[{Ni(CO)$_2$(μ-PPh$_2$)}$_2$] (**59**) are reduced to persistent anion radicals, and
the carbonyl also forms a dianion (*53*). The sulfur-bridged dimers **57**
(R = alkyl or aryl) are also easily oxidized ($E^0 \approx 0.0$ V) and the monocations, although not isolable, were characterized by ESR spectroscopy (*53*,
61). Second, reversible oxidation waves at about 0.8 V were also observed
for **57** (*61*).

Although structural comparisons have not been made on **58** and [**58**]$^+$,
the nickel–nickel distance in the neutral species is 80 pm longer than that
in [{Co(μ-PPh$_2$)Cp}$_2$]. The cobalt complex, which is isoelectronic with
[**58**]$^{2+}$, contains a direct metal–metal bond; by implication, the HOMO
of **58** is antibonding between the two nickel atoms (*128*).

The alkyne-bridged complexes [Ni$_2$(μ-RC$_2$R)Cp$_2$] are oxidized to reactive radical cations which, in the presence of dienes, cleave to the
mononuclear species [Ni(η^4-diene)Cp]$^+$ (*129*). The analogous dipalladium
compound [Pd$_2$(μ-PhC$_2$Ph)(η-C$_5$Ph$_5$)$_2$] (**60**) is reversibly oxidized in two
one-electron steps (E^0_1 = 0.52 V, E^0_2 = 1.13 V), and [**60**]$^+$ is isolable
from the reaction of **60** with Ag$^+$. The radical cation shows a single
line in the solution ESR spectrum (g = 2.044), and reacts with ligands,
L, to give mononuclear cations [PdL$_2$(η-C$_5$Ph$_5$)]$^+$ (L = PPh$_3$, L$_2$ =
Ph$_2$PCH$_2$CH$_2$PPh$_2$ or diene). The neutral compound **60** is also reduced to
an anionic radical ($E^0 = -1.12$ V) which slowly decomposes releasing
[C$_5$Ph$_5$]$^-$ (*130*).

IV

METAL CLUSTERS

A. *Vanadium and Chromium Groups*

There is very little information available on the redox properties of organometallic clusters of the early transition metals. There are preliminary indications that the triangular cluster $[V_3(CO)_9Cp_3]$ undergoes a series of reversible one-electron transfers, but no details are available (*63*).

X-Ray diffraction studies on the redox-related pair $[Nb_3(\mu\text{-Cl})_6(\eta\text{-}C_6Me_6)_3]^Z$ (**61**) $[Z = 1$ (*131, 132*) and 2 (*133*)] show the structure to be essentially unchanged during the redox process, implying the HOMO of the monocation to be nonbonding with respect to the metal triangle. The Nb—Nb distances are virtually the same ($\sim$334 pm), and the nonplanar hexamethylbenzene ligands show an average folding of approximately 156° in both complexes. The paramagnetic dication was prepared from the monocation by oxidation with 7,7,8,8-tetracyano-*p*-quinodimethane (TCNQ) (*131, 133*), and the resulting salt of the $[\{TCNQ\}_2]^{2-}$ dianion is a semiconductor (*133*).

Cyclic voltammetry shows that $[Nb_3(\mu\text{-Cl})_6(\eta\text{-}C_6Me_6)_3]^+$ undergoes reversible one-electron transfer ($E^0 = -0.21$ V vs Ag/Ag$^+$, in DME), but anodic oxidation, or reaction with ceric ion, gave the hexanuclear product $[Nb_6Cl_{12}(\eta\text{-}C_6Me_6)_6]^{4+}$ (*134*). Taken with the demonstrated stability of $[Nb_3(\mu\text{-Cl})_6(\eta\text{-}C_6Me_6)_3]^{2+}$, mentioned earlier, this seems to indicate that the trinuclear dication undergoes slow dimerization. The monocation is also reported to be reducible ($E_{1/2} \approx -1.8$ V vs Ag/Ag$^+$), but the reduction products were not studied (*134*).

Cyclic voltammetry suggests that the mixed-metal clusters $[\{MM'(\mu_2\text{-}CO)_2(\mu_3\text{-CO})(PPh_3)Cp\}_2]$ (**62**) (M = Cr, Mo, or W; M' = Pd or Pt)

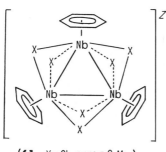

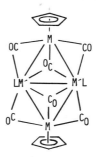

(**61**; X = Cl, arene = C₆Me₆)   (**62**, L = PPh₃)

undergo irreversible two-electron reduction, and two one-electron oxidations the first of which is partially reversible. A partial product analysis suggested that reduction occurs via the simultaneous addition of one electron to each of the palladium or platinum centers; oxidation, which also leads to cluster degradation, is thought to occur at the Group VI metals *(135)*.

B. *Iron*

The trinuclear carbonyls $[M_3(CO)_{12}]$ (M = Fe, Ru, or Os) and the mixed metal clusters $[Fe_2Ru(CO)_{12}]$ and $[FeRu_2(CO)_{12}]$ undergo one-electron reduction; the triruthenium and triosmium compounds are irreversibly reduced, at potentials (M = Ru, $E_{1/2} = -0.82$ V; M = Os, $E_{1/2} = -1.16$ V vs Ag/AgCl) *(136)* considerably more negative than the triiron analogue, which shows two one-electron waves (-0.21 and -0.76 V) *(137)*.

The first reduction of $[Fe_3(CO)_{12}]$ is now established as a diffusion-controlled, fully reversible process *(137–139)* giving $[Fe_3(CO)_{12}]^-$. The second reduction of $[Fe_3(CO)_{12}]$ is variously described as (1) chemically irreversible in acetone or CH_2Cl_2 at a platinum electrode, leading to cluster fragmentation *(137)*; (2) not fully reversible in dry CH_2Cl_2 even at cyclic voltammetric scan rates as high as 8 V sec^{-1} *(139)*; and (3) reversible at a scan rate of 100 mV sec^{-1} *(138)*.

The radical anion $[Fe_3(CO)_{12}]^-$ can be generated by the electrolytic reduction of $[Fe_3(CO)_{12}]$ in CH_2Cl_2, or by chemical methods using alkali metals or the paramagnetic cluster $[Co_3(CO)_9(\mu_3\text{-}CR)]^-$ (Section IV,D). ESR spectroscopy serves to characterize $[Fe_3(CO)_{12}]^-$ *(136, 140)* and $[Fe_{3-n}Ru_n(CO)_{12}]^-$ (n = 1 or 2) *(136)*, and also shows that the triiron anion decomposes in CH_2Cl_2 to other paramagnetic carbonyls including $[Fe_2(CO)_8]^-$ **(63)**, and an Fe(I) monomeric complex. In THF, the same radical species are formed on reduction of $[Fe_3(CO)_{12}]$, but it is noteworthy that **63** and **64** are detectable even in the absence of a reducing agent *(136)*.

The substituted complexes $[Fe_3(CO)_{12-n}L_n]$ (n = 1–3, L = P-donor) also undergo one-electron reduction, with the potential becoming increasingly more negative with the degree of carbonyl substitution [e.g., L = $P(OMe)_3$, n = 1, $E_{1/2} = -0.52$ V; n = 2, $E_{1/2} = -0.84$ V; and n = 3, $E_{1/2} = -1.13$ V vs Ag/AgCl, in acetone]. The anions, including two isomers of $[Fe_3(CO)_{11}\{P(OPh)_3\}]^-$, were detected by ESR spectroscopy but showed a strong tendency to form $[Fe_3(CO)_{12}]^-$ *(136)*.

At low temperatures, $[Fe_3H(CO)_{11}]^-$ undergoes reversible one-electron oxidation at mercury ($E_{1/2} = -1.13$ V vs Ag/AgCl) *(137)*. The violet, neutral radical can be generated at $-80°C$ by Ag^+ or $[FeCp_2]^+$ oxidation in THF, and was characterized by ESR spectroscopy (141). Similarly the chemical oxidation of $[Fe_2(CO)_8]^{2-}$, in the presence of $HBF_4 \cdot OEt_2$, gave a species tentatively formulated as $[Fe_4H(CO)_{13}]$ (141).

The oxidation of $[Fe_3(CO)_9\{P(OMe)_3\}_3]$ in acetone is a reversible one-electron process at Pt or Hg ($E^0 = 0.0$ V vs Ag/AgCl), but the cation radical has not been characterized; cluster degradation leads to mononuclear Fe(I) complexes (137).

The metal framework of trinuclear clusters capped by one or two bridging groups is more robust toward oxidation or reduction. The mixed-metal clusters $[FeCo_2(CO)_9(\mu\text{-}E)]$ (65) (E = S, Se, or PR) reduce in successive one-electron steps, first to a somewhat persistent anion radical [E = S, $t_{1/2} \approx 60$ seconds (142)] at mild potentials (E = S, $E^0 = -0.24$ V vs Ag/AgCl), and then to a very unstable dianion (142, 143).

The frozen solution ESR spectrum of $[65]^-$ (E = S) shows coupling to two equivalent cobalt nuclei with a rhombic cobalt hyperfine tensor of 66, 54, and 16 G. The ESR spectral parameters were compared with those of the isoelectronic species $[Co_3(CO)_9(\mu_3\text{-}CR)]^-$ (Section IV,D) which have axially symmetric cobalt splittings of $A_{\parallel} = 74$ G and $A_{\perp} = 15$ G. It was concluded (1) that the replacement of cobalt by iron results in the mixing of the cobalt $3d_{z^2}$ orbital into one composed primarily of an antibonding combination of cobalt $3d_{xy}$ orbitals, and (2) that approximately 60% of the unpaired electron density of $[65]^-$ resides on the two cobalt nuclei (142).

The single-crystal ESR spectra of $[Co_3(CO)_9(\mu_3\text{-}E)]$ (E = S or Se), which are isoelectronic with 65^-, show definitively that the half-filled orbital has $2a_2$ symmetry (in C_{3v} geometry) and is antibonding with respect to the metal–metal bonds (144, 145). Consistent with this, the metal–metal distances of the diamagnetic complexes 65, which do not contain an extra electron in the antibonding M_3 orbital, are shorter by 4–9 pm than those of $[Co_3(CO)_9(\mu_3\text{-}E)]$ (145, 146).

The mixed-metal complex $[FeCo_2(CO)_7(\mu\text{-}CO)(\mu\text{-}PPh_2)_2]$ is reduced in a one-electron step to a monoanion ($E^0 = -0.68$ V), and there is some evidence for a very short-lived dianion in a further reduction at $E^0 = -1.2$ V (147).

(65)   (66)   (67; ● = MCp, E = bridging group)

Doubly bridged trinuclear clusters may be oxidized or reduced, but the stability of the one-electron product depends on the identity of the bridging group and on the terminal ligands bonded to the metals. The iron complex $[Fe_3(CO)_2(\mu\text{-}S)(\mu\text{-}SR)Cp_3]$, the structure of which has not been fully defined, oxidizes reversibly in two one-electron steps. The monocation was synthesized by iodine oxidation of the neutral species but apparently decomposes to $[\{Fe(CO)(\mu\text{-}SR)Cp\}_2]^+$ (Section III,E) (148).

The tricarbonyliron analogues $[Fe_3(\mu_3\text{-}E)(\mu_3\text{-}E')(CO)_9]$ (66) (E = CO, E' = S; E = E' = S; E = S, E' = SO) are irreversibly oxidized in benzonitrile, and the last two complexes undergo multielectron decomposition reactions. The compounds are reducible, however, at very mild potentials. The disulfide (E = E' = S) forms a persistent monoanion ($E^0 = -0.43$ V vs Ag/AgCl), but 66 (E = S, E' = CO or SO) are irreversibly reduced, albeit more readily ($E^0 = -0.26$ V and -0.31 V, respectively) because of the electron-withdrawing nature of the CO and SO bridges (149).

The reduction of $[Fe_3(\mu_3\text{-}S)_2(CO)_8(\overline{CSCHCHS})]$ in DMF is reversible at fast cyclic voltammetric sweep rates ($E^0 = -0.73$ V). In the presence of $P(OMe)_3$, three new reduction waves are observed, revealing the stepwise substitution of carbonyl ligands to give $[Fe_3(\mu_3\text{-}S)_2(CO)_{8-n}\{P(OMe)_3\}_n(\overline{CSCHCHS})]$ ($n = 1$–3). Under 1 atm carbon monoxide, the rate of substitution was less, leading the authors to favor a mechanism involving CO displacement after the initial reduction step, rather than rupture of the M—M bond before ligand loss. In either case, reduction clearly activates the cluster toward carbonyl substitution (150).

Chemical and electrochemical studies of the redox properties of tetranuclear clusters have been the key to probing the relationship between geometrical structures and electron configuration. The compounds in question have the cubane-like structure 67, in which each metal is bonded to a cyclopentadienyl ring and to three triply bridging ligands. Cyclic voltammetric studies on 67 (E = S, M = Fe) revealed a remarkable series (151) of four reversible electron-transfer reactions. The electrochemical data for 67 (M = Fe, E = S) were originally misinterpreted in terms of one reduction ($E^0 = -0.33$ V) and three oxidations ($E^0 = 0.33, 0.88,$ and 1.41 V) (151, 152), but it is now known (153, 154) that the wave at -0.33 V also corresponds to an oxidation process. Thus, the Fe_4S_4 core remains intact in the *five* cluster oxidation states $[67]^Z$ ($Z = 0$–4).

These cubane-like clusters are apparently very adept at charge delocalization. The mono- and dications of $[\{Fe(\mu_3\text{-}S)Cp\}_4]$ were isolated in good yield after bulk electrolysis, and the crystal structures of $[67]^Z$ [M = Fe, E = S; $Z = 0$ (155, 156), 1 (157), and 2 (152)] have been determined.

The neutral cluster has two short Fe—Fe distances (~264 pm) and four long, nonbonding distances (336 pm) (155, 156). After one-electron oxidation, the short pair remains essentially unchanged, while two of the longer Fe—Fe distances shorten significantly (by 17 pm, to 319 pm), and the other pair shortens only marginally to 332 pm; the overall result is a slight twisting of the cube (157). Further oxidation to the dication gives a flattened tetrahedron of iron atoms, now having four short (283 pm) and two long (325 pm) iron–iron distances (152). The structural changes are consistent with the removal of electrons from a degenerate molecular orbital antibonding with respect to the tetrametal framework (157).

The Mössbauer spectra of $[\{Fe(\mu_3\text{-}S)Cp\}_4]^Z$ ($Z = 0$–2) show that in each cluster the four metal atoms are identical. The fact that the isomer shift for iron does not change appreciably from one complex to the next implies that there is ligand-to-metal charge redistribution after oxidation (158).

The cyclic voltammogram of **67** (M = Fe, E = CO) apparently showed that this cluster can exist with overall charges −1 to +2, and evidence was presented for the generation of both the monoanion and the monocation by bulk electrolysis (151, 159). As in the case of $[\{Fe(\mu_3\text{-}S)Cp\}_4]$, however, an erroneous assignment of the original electrochemical data was made, meaning that the carbonyl cluster $[\{Fe(\mu\text{-}CO)Cp\}_4]^Z$ is actually detectable with the overall charges $Z = 0$ to 3 (153, 154).

The cation $[\{Fe(\mu_3\text{-}CO)Cp\}_4]^+$ can also be produced by irradiating **67** (M = Fe, E = CO) in the presence of halocarbons. Although the mechanism of the photooxidation process is not well understood, the "charge-acceptor" nature of the solvent was thought to be important (160).

Comparative X-ray structural studies on $[\{Fe(\mu_3\text{-}CO)Cp\}_4]^Z$ [$Z = 0$ (161) and 1 (162)] showed that the fully bonded tetrahedron of metal atoms (Fe—Fe = 252 pm) undergoes only minor distortions on oxidation, with a slight shortening of the average iron–iron distance to 248 pm. Electron removal from an orbital largely nonbonding with respect to the tetrametal framework was deduced (161).

The cubane-like cluster $[\{Fe(\mu_3\text{-}S)(NO)\}_4]$ (**68**) is not formally an organometallic compound, but the structural changes observed when it is reduced, and the MO description of its bonding, are relevant to the cyclopentadienyl complexes described above. The nitrosyl complex contains 60 valence electrons, 8 less than $[\{Fe(\mu_3\text{-}S)Cp\}_4]$, and is reduced in two consecutive one-electron steps to an isolable monoanion [**68**]⁻, and to a dianion [**68**]²⁻.

X-Ray studies on [**68**]^Z ($Z = 0$ and −1) show that the iron–iron distances are lengthened on reduction consistent with electron addition to a tetrairon antibonding orbital. The observed increases in the metal–metal bond lengths of **68** are less than those which occur when $[\{Fe(\mu_3\text{-}S)Cp\}_4]^+$ is reduced to $[\{Fe(\mu_3\text{-}S)Cp\}_4]$, implying that the π^* (NO) orbitals absorb some of the effect of the increased electron density in $[\textbf{68}]^-$ (*163*).

The sulfur-rich cluster $[Fe_4S_5Cp_4]$, which contains three triply bridging sulfur atoms and one triply bridging disulfide group, is sequentially oxidized in three one-electron steps ($E^0 = -0.28$, $+0.02$, and $+1.14$ V), and the mono- and dications have been isolated and characterized.

The related complex $[Fe_4S_6Cp_4]$, which contains a second disulfide group, is reversibly reduced to a monoanion ($E^0 = -1.33$ V) and oxidized to a monocation ($E^0 = 0.06$ V); loss of a second electron ($E^0 = 0.37$ V) results in the expulsion of sulfur and the formation of $[Fe_4S_5Cp_4]^{2+}$ (*164*).

One of the disulfide groups of $[Fe_4S_6Cp_4]$ may act as a ligand to a $Mo(CO)_4$ group, and the resulting cluster, $[Fe_4Mo(CO)_4S_6Cp_4]$, is reversibly reduced and oxidized to a monoanion ($E^0 = -1.02$ V) and monocation ($E^0 = 0.28$ V), respectively. A second, irreversible oxidation process occurs at ~ 0.6 V (*165*).

C. Ruthenium and Osmium

The catalytic substitution reactions of metal carbonyl clusters, including $[M_3(CO)_{12}]$ (M = Fe, Ru, or Os), $[Ru_4H_4(CO)_{12}]$, $[Rh_6(CO)_{16}]$, and $[Co_3(CO)_9(\mu\text{-}CCl)]$, with isocyanides or Group V-donor ligands may be induced by either electrochemical or chemical (benzophenone ketyl) reduction. The most favorable conditions for efficient substitution include (1) the formation of a radical anion with a significant lifetime and (2) the use of a ligand which is not reduced by $[Ph_2CO]^-$, and which is less of a π acid than CO (*166*).

In the cases of $[Ru_3(CO)_{12}]$ and $[Ru_4H_4(CO)_{12}]$, which have very unstable anion radicals, Lewis base substitution reactions are stoichiometric rather than catalytic in the reducing agent (*167*); dppe and the triruthenium cluster give $[Ru_3(CO)_{10}(dppe)]$, $[Ru_3(CO)_{11}(dppe)]$, or $[\{Ru_3(CO)_{11}\}_2(\mu\text{-}dppe)]$, depending on the relative stoichiometries of the reactants (*168*). Complexes such as $[Ru_3(CO)_{11}L]$ and $[Ru_4H_4(CO)_{11}L]$ [L = $PPh_2CH_2CH_2Si(OEt)_3$], precursors to surface-anchored species, have also been prepared by reductively induced substitution reactions (*169*).

The hexanuclear cluster $[Os_6(CO)_{18}]$ (**69**) and the corresponding dianion $[\textbf{69}]^{2-}$ are notable since the structural change from the bicapped tetrahedron of **69** (*170*) to the octahedron of osmium atoms in $[\textbf{69}]^{2-}$ (*171*) is exactly that predicted by polyhedral skeletal electron-pair theory (*172*).

The reduction of **69** by iodide ion proceeds by an inner-sphere process involving prereductive association (*173*). The electrochemical reduction involves a single, chemically reversible two-electron transfer at 0.0 V (*174*). The large peak separations observed in the cyclic voltammograms of **69**, at Hg, Pt, or Au electrodes, are due to slow charge-transfer kinetics for the reduction process; the slowness of the electron transfer may arise from the accompanying structural rearrangement (*174*).

The tetranuclear cluster $[Os_4H_3(CO)_{12}]^-$ (**70**) is irreversibly oxidized in acetonitrile ($E_{pk} = 0.44$ V vs Ag/AgNO$_3$), and bulk electrolysis provides $[Os_4H_3(CO)_{12}(NCMe)]^+$ (**71**) in quantitative yield. The cation, which is the first osmium cluster with a "butterfly" structure unsupported by bridging ligands, may also be prepared from **70** and $[NO]^+$ (*175*).

The irreversible reduction of the cation **71** ($E_{pk} = -1.69$ V vs Ag/AgNO$_3$) leads to the regeneration of **70** so that the two clusters are related by the chemically reversible electron-transfer reaction:

$$[Os_4H_3(CO)_{12}]^- + MeCN \rightleftharpoons [Os_4H_3(CO)_{12}(NCMe)]^+ + 2\ e^-$$

D. Cobalt, Rhodium, and Iridium

Tricobalt clusters with one triply bridging ligand, isostructural with **65**, have been extensively investigated. The carbonyls $[Co_3(CO)_9(\mu\text{-E})]$ (**72**) (E = S, Se, or PR) have one electron in excess of that required for an 18-electron configuration about each metal. The half-filled orbital is antibonding with respect to the metal triangle (*144, 145*), and the complexes are easily oxidized (see also, Section IV,B).

The complexes $[Co_3(CO)_7(\mu_2\text{-X})(\mu_3\text{-S})]$ [X = CRNR′, R = Me or Ph, R′ = C$_6$H$_{11}$; X = SCNMe$_2$ or NHC(R)S, R = Me, Ph, or p-OMeC$_6$H$_4$], in which the three-electron donor, X, replaces one carbonyl group on each of two metal atoms, are diamagnetic, and therefore isoelectronic with [**72**]$^+$. Accordingly, they undergo one-electron reduction ($E^0 = -0.4$ to -0.8 V) to persistent radical anions isoelectronic with **72**. Interestingly, the shorter cobalt–cobalt distance in $[Co_3(CO)_7\{\mu_2\text{-NHC(Me)S}\}(\mu_3\text{-S})]$ (Co—Co$_{ave}$ = 263.7 pm) relative to that in **72** is again in accord with the bonding description for the paramagnetic species (*176*).

The alkylidyne-capped clusters $[Co_3(CO)_9(\mu\text{-CR})]$ (**73**), also with one less valence electron than **72**, are easily reduced to anion radicals (*177–180*). The reversibility of the redox couple was verified by a variety of voltammetric methods (*178*); that the E^0 values (e.g., R = F, $E^0 = -0.28$ V; R = Me, $E^0 = -0.43$ V vs Ag/AgCl), and cobalt hyperfine splittings in the ESR spectra of the anions ($A_{^{59}Co} \sim 36$ G)

(179), vary so little with the substituent R is further evidence that the half-filled orbital of [73]$^-$ is predominantly trimetal in character.

Reduction beyond the monoanionic stage for [Co$_3$(CO)$_9$(μ_3-CR)] is irreversible, even at 193 K in acetone or at cyclic voltammetric scan rates up to 1000 V sec^{-1}. The addition of a second electron to the cluster ($E_{1/2} \sim -1.2$ V vs Ag/AgCl) apparently results in metal–metal bond cleavage as [Co(CO)$_4$]$^-$ is formed as one of the reduction products (178).

Oxidation was not observed for [Co$_3$(CO)$_9$(μ_3-CR)], but the phosphine derivatives [Co$_3$(CO)$_{9-n}$L$_n$(μ_3-CR)] (74) ($n = 1$–3) display anodic cyclic voltammetric waves. The cation radicals apparently produced in the oxidation reactions appear to be very reactive, and trisubstitution is necessary before reversible waves were observed (e.g., L = PEt$_2$Ph, $n = 3$, R = Me, $E^0 = 0.44$ V vs Ag/AgCl).

As expected, the E^0 values for the couples 74/[74]$^-$ and 74/[74]$^+$ become more negative as n increases; substitution of CO by PPh$_3$ shifts the potential by ~250 mV (180).

The redox-catalyzed substitution reactions of [Co$_3$(CO)$_9$(μ_3-CR)] (73) closely parallel those of [Co$_2$(CO)$_6$(μ-R$_2$CR)] (Section III,F) and give [Co$_3$(CO)$_{9-n}$L$_n$(μ_3-CR)] efficiently and in high yield (122, 166). However, a significant extension of this method has allowed mixed-metal carbonyl clusters to be prepared. Thus, for example, addition of the reductant [Ph$_2$CO]$^-$ to a mixture of [{Mo(CO)$_3$Cp}$_2$] and 73 (R = Ph) in THF affords good yields of [Co$_2$Mo(CO)$_8$Cp(μ_3-CPh)] in which one basal Co(CO)$_3$ group has been replaced by Mo(CO)$_2$Cp (181). The equivalent thermal substitution reaction gives less product and requires the mixture to be refluxed in benzene for 3 days.

Ferrocenyl derivatives of 73, namely [Co$_3$(CO)$_{9-n}$L$_n$(μ_3-CFc)] ($n = 0$–3, L = phosphine or phosphite) undergo reversible reduction at the tricobalt center to detectable radical anions ($n = 0$), and oxidation at the iron center to give isolable cations (182). For the trisubstituted derivatives [$n = 3$, L = P(OMe)$_3$ or P(OPh)$_3$], the monocations appear to be mixed-valence species in which the two redox sites are weakly interacting; low-energy absorption bands, assigned to intervalence transfer transitions, have been observed. The dications [Co$_3$(CO)$_6${P(OR)$_3$}$_3$(μ_3-CFc)]$^{2+}$ (R = Me or Ph) are also isolable, oxidation having occurred at both possible redox sites (183).

The photoelectron spectra of [Co$_3$(CO)$_9$(μ_3-CR)], and the results of molecular orbital calculations, lead to two conclusions. First, the alkylidyne bridge donates a lone pair to the metal triangle, and second, the presence of this bridge, rather than the metal–metal interactions, is crucial to the stability of the complex (184, 185) and, by inference, its radical anion.

Studies of the reduction of [{Co(CO)Cp}$_3$] (75) tend to support these conclusions. In solution, the structure of 75 contains, at best, one labile, triply bridging carbonyl (186). The reduction of 75, originally thought to yield a stable radical anion (187), is now known (63) to give a short-lived anion ($E^0 \approx -1.0$ V) which undergoes metal–metal bond cleavage. The extrusion of one Co(CO)Cp fragment results in quantitative yields of [{Co(μ-CO)Cp}$_2$]$^-$ (Section III,F), which is apparently responsible for the ESR spectrum originally assigned (87) to ⌊75]$^-$.

Further indications that capping ligands stabilize clusters toward redox reactions is found with [Co$_3$(μ_3-E)(μ_3-E')Cp$_3$] (76), which adopt a structure similar to 66. Two reversible one-electron oxidations and one reduction are found for 76 (E = E' = S), and the monocation and monoanion of 76 (E = S, E' = CS) (149) were detected by cyclic voltammetry.

Iodine and 76 (E = E' = S) readily give the paramagnetic monocation [76]$^+$ (E = E' = S), and comparative X-ray studies of the redox-related pair showed that the equilateral triangle of metal atoms in the neutral species (Co—Co = 268.7 pm) is significantly distorted on oxidation. The presence of one short metal–metal bond (247.4 pm) and two longer distances (264.9 pm) suggests that the electron is removed from a degenerate ground state, and that a Jahn–Teller distortion occurs in the cation (188).

The complex [Co$_3$(μ_3-CPh)$_2$Cp$_3$] (76) (E = E' = CPh) is reduced by potassium metal to a radical anion whose ESR spectrum suggests a symmetrical structure with a half-filled orbital constructed from cobalt $3d$ atomic orbitals. The cation [76]$^+$ (E = E' = CPh), detected by cyclic voltammetry ($E^0 = 0.34$ V) and prepared by electrolytic oxidation, has the unpaired electron in a degenerate orbital, and a structural Jahn–Teller distortion is again expected (189).

The tetracobalt cubanes [{Co(μ_3-S)Cp}$_4$] (67) (M = Co, E = S) have not been studied as extensively as the iron analogues (Section IV,B), but it is to be expected that electrochemical studies will reveal an extensive electron-transfer series. A comparison of the structures of [{Co(μ_3-S)Cp}$_4$]^Z (Z = 0 and 1) showed that the nonbonded tetrahedron of metal atoms in the neutral molecule (Co—Co$_{ave}$ = 329.5 pm) undergoes a tetragonal distortion. The shortening of four of the cobalt–cobalt distances (317.2 pm) is due to the removal of an electron from an orbital antibonding with respect to the four metal atoms (190).

Preliminary data have been reported (154) which show that [{Co(μ_3-As)Cp}$_4$] is oxidized to the mono- and dications. In addition, it is the first cyclopentadienylmetal-cubane complex to be reduced to a genuine monoanion.

On oxidation with $FeCl_3$, the trigonal prismatic cluster $[Co_6C(CO)_{15}]^{2-}$ undergoes rearrangement to the distorted octahedral monoanion; the carbido-carbon atom is retained within the Co_6 cavity. Once again, a metal–metal antibonding orbital is implicated in the redox process (*191*).

Little information exists on the redox properties of rhodium clusters despite the fact that electron transfer is implicated in the redox condensation reactions of these compounds (*192*).

The 12-atom cluster $[Rh_{12}(CO)_{30}]^{2-}$ [**77**]$^{2-}$ undergoes a two-electron reduction in THF giving two equivalents of hexanuclear $[Rh_6(CO)_{15}]^{2-}$; the reduction process appears to be partly reversible at $-40°C$. The dianion [**77**]$^{2-}$ is also oxidized in a two-electron step, yielding a short-lived intermediate presumed, but not proven, to be **77**. Cyclic voltammetric measurements in a pressurized electrochemical cell were also used to monitor the formation of $[Rh_5(CO)_{15}]^-$ ($E_{pk} = -0.8$ V) from [**77**]$^{2-}$ and CO (*193*).

The irreversible one-electron oxidation and reduction reactions of $[Ir_3(CO)_6(\mu\text{-}SBu^t)(\mu\text{-}RC_2R)]$ appear to result in the production of binuclear products (*194*).

E. *Nickel, Platinum, and Gold*

The paramagnetic cluster $[Ni_3(\mu_3\text{-}CO)_2Cp_3]$ (**78**) is reversibly reduced in a one-electron step ($E^0 = -0.8$ V vs Ag/AgCl) but its oxidation is irreversible (*149*). The diamagnetic complexes $[CoNi_2(\mu_3\text{-}CO)_2(\eta\text{-}C_5H_{5-n}Me_n)]$ (**79**) ($n = 0$, 1, or 5) are isoelectronic with [**78**]$^+$, and a comparison of the structure of **79** with that of **78** shows that the average metal–metal distance in the cobalt complex is 3.1 pm shorter. The assignment of a half-filled antibonding trimetal orbital for **78** (*195*) contradicted an earlier ESR spectroscopic study which had led to the deduction of a half-filled bonding orbital (*196*).

The mixed-metal complexes **79** are reduced to stable monoanions ($n = 0$, $E^0 = -1.53$ V vs Ag/Ag$^+$), and [**79**]$^-$ ($n = 5$) was prepared by allowing sodium naphthalenide to react with **79** ($n = 5$). Consistent with the bonding description outlined above, the nickel–nickel and cobalt–nickel distances are increased in the anion, by 6.2 and 2.1 pm, respectively. The larger increase observed in the nickel–nickel bond length implied greater nickel $3d$ character in the half-filled orbital (*197*).

The anion $[Ni_3(\mu_3\text{-}CO)_2Cp_3]^-$ [**78**]$^-$ was prepared by the same method as [**79**]$^-$, and, as expected, a further lengthening of the nickel–nickel bonds was observed. As [**78**]$^-$ is diamagnetic, the HOMO of **78** appears to be nondegenerate (*197*).

The complex $[Ni_3(\mu_3\text{-}CO)_2(\eta\text{-}C_5Me_5)_3]$ (**80**) gives much more stable oxidation products than **78** and two reversible, one-electron oxidations

($E^0 = -0.35$ and 0.08 V vs Ag/Ag$^+$) were observed in THF. The shift of the first oxidation potential of **80** to a more negative value than that of **78** is expected because of the presence of the electron-donating methyl substituents. However, this does not account for the *positive* shift in E^0 reported for the *reduction* of **80**; steric factors may be involved (*197*).

The related sulfide-bridged complexes [Ni$_3$(μ_3-S)$_2$Cp$_3$] undergo reversible reduction and oxidation processes, but the ions have not been characterized (*149*).

The remarkable hexanuclear complex [{NiCp}$_6$] (**81**), prepared by the sodium naphthalenide reduction of nickelocene, undergoes an extensive series of reversible one-electron transfer reactions; cyclic voltammetry shows waves relating the *six* species [{NiCp}$_6$]^Z ($Z = -2$ to 3). Chemical oxidation of **81**, with Ag$^+$, gave the monocation whose structure shows only a small tetragonal distortion from the octahedral array of nickel atoms in the neutral precursor (*198*).

The paramagnetic, mixed-metal cluster [Fe$_3$Pt$_3$(CO)$_{15}$]$^-$ **82**, prepared by iodine oxidation of the corresponding dianion, is surprisingly stable. Its frozen solution ESR spectrum shows axially symmetric ^{195}Pt hyperfine splitting; once again, half filling of an antibonding, in-plane, triplatinum orbital is favored (*199*).

Mixed platinum–cobalt clusters do not reduce reversibly (*200*); the addition of one electron to [Co$_2$Pt(CO)$_6$(μ_3-CO)(dppe)] ($E_{pk} = -1.2$ V) leads to [Co(CO)$_4$]$^-$ extrusion. The analogous complex [Co$_2$Pt(CO)$_7$(μ_3-CO)(PPh$_3$)] (**83**) likewise loses [Co(CO)$_4$]$^-$ on reduction but the tetranuclear species [Co$_2$Pt$_2$(μ-CO)$_3$(CO)$_5$(PPh$_3$)$_2$] (**84**) is also produced in good yield. A logical mechanism for the formation of **84** would involve the production, and subsequent dimerization, of the radical [CoPt(CO)$_4$(PPh$_3$)] (*200*).

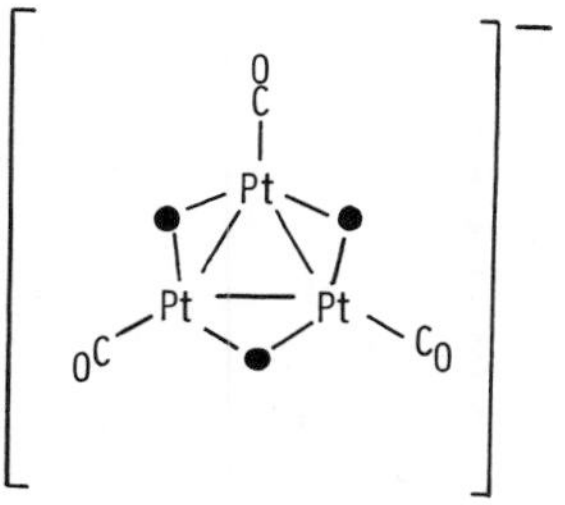

[**82**, ● = Fe(CO)$_4$] 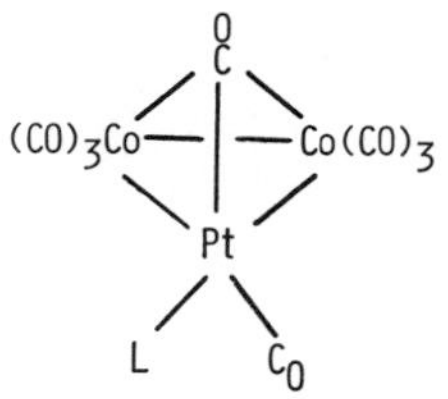   (**83**, L = PPh$_3$) 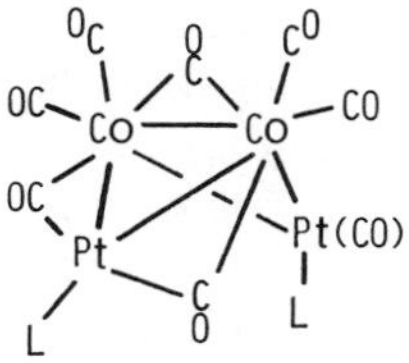   (**84**)

The gold cluster [Au$_9$(PPh$_3$)$_8$]$^{3+}$ [**85**]$^{3+}$ is reduced, apparently in a reversible two-electron process ($E^0 = -0.38$ V vs Ag/AgCl). However, a careful analysis of the voltammetric data suggested that the observed two-electron wave actually corresponds to the superimposition of two one-electron waves ($\Delta E^0 \approx 40$ mV) due to the processes:

$$[85]^{3+} \rightleftharpoons [85]^{2+} \quad \text{and} \quad [85]^{2+} \rightleftharpoons [85]^{+}$$

In support of this hypothesis, mixing solutions of $[85]^{3+}$ and $[85]^{+}$ (the latter isolated from electrolyzed solutions of the former) gave an ESR spectrum ascribed to the dication. Certain aspects of this spectrum, especially the absence of gold hyperfine splittings, have yet to be explained.

A comparison of the crystal structures of $[85]^{3+}$ and $[85]^{+}$ showed that reduction is accompanied by considerable molecular reorganization (201).

V

ADDENDUM

Some recent papers are summarized in this section.

Theoretical studies have shown that the monocations of both biferrocene and bis(fulvalene)diiron show some delocalization; the latter is much more delocalized than the former (202). By contrast, $[Fe_2\{\eta^5,\eta^5\text{-}C_5H_4CH(Me)C_5H_4\}_2]^{+}$ appears to be a trapped valence cation; X-ray structural studies have revealed different environments about the two metal atoms (203) (Section II,A).

The dianions $[\{Mo(CO)_4(\mu\text{-}SR)\}_2]^{2-}$ (R = Bu^t or Ph) are oxidized to the neutral dimers in a chemically reversible two-electron step. The 37 mV peak separation in the cyclic voltammogram is attributed to the simultaneous transfer of two electrons, caused by metal–metal bond cleavage coupled with a molecular rearrangement (204) (Section III,C).

Comments on the redox potentials for the oxidation of $[M_2(CO)_{10}]$ (M = Mn or Re) are included in a discussion of the mechanism of the halogenation reactions of the carbonyls (205) (Section III,D).

Complex 39 may be prepared by the electron-transfer catalyzed isomerization of $[Fe_2(CO)_6\{\mu\text{-}ROC(S)SMe\}]$ (206) (Section III,E).

Sodium and $[\{Fe(NO)_2(\mu\text{-}PPh_2)\}_2]$ afford the corresponding dianion, but an isomeric product, namely $[Fe_2(NO)_3(\mu\text{-}NO)(PPh_2)(\mu\text{-}PPh_2)]^{2-}$, is formed via LiBHEt$_3$ reduction and subsequent treatment with butyllithium (207) (Section III,E).

Steady-state cyclic voltammetric studies in MeCN show that $[\{Fe(\mu\text{-}CO)(CO)Cp\}_2]$ is reduced to $[Fe(CO)_2Cp]^{-}$ via the binuclear anion; the rate constant for the cleavage of $[Fe_2(CO)_4Cp_2]^{-}$ to $[Fe(CO)_2Cp]^{-}$ and $[Fe(CO)_2Cp]$ is 1.060 sec^{-1} at 273 K (208). A brief report (209) suggests that $[\{Fe(\mu\text{-}CO)(CO)(\eta\text{-}C_5Me_5)\}_2]$ is oxidized, in DMF at platinum, in an irreversible two-electron process; in CH_2Cl_2 reversible one-electron oxidation to $[Fe_2(CO)_4(\eta\text{-}C_5Me_5)_2]^+$ occurs (210).

The complexes $[Ru_2(\mu\text{-}CO)(\mu\text{-}CHR)(\mu\text{-}dppm)Cp_2]$ (R = H or Me, dppm = $Ph_2PCH_2PPh_2$) also undergo reversible one-electron oxidation, but loss of a second electron leads to deprotonation of the bridging carbene and isolation of $[Ru_2(\mu\text{-}CO)(\mu\text{-}CX)(\mu\text{-}dppm)Cp_2]^+$ (X = H or $CHCH_2$) (211). A similar double-oxidation deprotonation mechanism accounts for the synthesis of $[Ru_3(\mu_2\text{-}CO)_3(\mu_3\text{-}CCH_2)(\eta\text{-}C_5Me_5)_3]^+$ from $[Ru_3\text{-}(\mu_2\text{-}CO)_3(\mu_3\text{-}CMe)(\eta\text{-}C_5Me_5)_3]$ (212) Section III,E).

The oxidative cleavage of $[\{CoX_2(\eta\text{-}C_5Me_5)\}_2]$ (X = Cl or Br) in water to give $[Co(aq)_n(\eta\text{-}C_5Me_5)]^{2+}$ has been studied by cyclic voltammetry (213). Details (214) have been presented of the reactions of $[\{Co(\mu\text{-}CO)Cp\}_2]^-$ with alkyl halides; thermolysis of the products, namely $[\{CoR(CO)Cp\}_2]$, leads to carbon–carbon coupling (Section III,F).

The ligand-bridged complexes $[Rh_2(CO)_{4-n}(PPh_3)_n(\mu\text{-}RNXNR)_2]$ (n = 1 or 2, R = aryl, X = N or CMe) undergo reversible one-electron oxidation to isolable monocations (215), which appear to be fully delocalized mixed valence complexes (216). Oxidation also leads to enhanced susceptibility to carbonyl substitution (n = 1, X = CMe), and a drastic shortening in the metal–metal bond distance implies electron loss from an anti-bonding dimetal orbital (215) (Section III,F).

The pyrazolyl-bridged complex $[\{Ir(\mu\text{-}pz)(\eta^4\text{-}cod)\}_2]$ is readily oxidized (0.25 V) (217), but the excited state of the neutral compound is expected to be a powerful reducing agent and is implicated in photo-induced oxidative addition reactions (218) (Section III,F).

Gas-phase clustering reactions involving $[Fe(CO)_4]^-$ and $[Fe(CO)_5]$ have been compared (219) to those occurring during the reduction of $[Fe(CO)_5]$ (Section IV,B).

One-electron reduction of $[Fe_3(CO)_9(\mu_3\text{-}CO)(\mu_3\text{-}CMe)]^-$ gives the unusual cluster $[Fe_3(CO)_9(\mu_3,\eta^2\text{-}MeC\equiv CO)]^{2-}$, possibly via the coupling of oxo–carbyne and ethylidyne ligands (220, 221) (Section IV,B).

Photochemically produced organometallic radicals act as reducing agents toward metal clusters. For example, irradiation of $[Ru_3(CO)_{12}]$ and PMe_2Ph in the presence of $[\{Mo(CO)_3Cp\}_2]$ yields $[Ru_3(CO)_{11}(PMe_2Ph)]$ via the reduction of the ruthenium cluster by $[Mo(CO)_3Cp]$; substitution of $[Ru_3(CO)_{12}]^-$, but not the neutral precursor, is apparent (222) (Section IV,C).

Further detailed studies of the electrochemistry of monocapped triangular clusters have been presented (223–225). The redox behavior depends on the capping group and the number of disparate metal atoms in the heteronuclear clusters. Species such as $[FeCo_2(CO)_9(\mu\text{-}E)]$ (E = S, Se, PBu^t, etc.), and $[FeCoM(CO)_8(\mu\text{-}E)Cp]$ (M = Mo or W), can be prepared from a redox reaction between $[Co_3(CO)_9(\mu\text{-}E)]$ and $Na_2[Fe(CO)_4]$, followed by the electron-transfer catalyzed reaction between $[FeCo_2(CO)_9(\mu\text{-}E)]$ and $[M(CO)_3Cp]^-$ (226).

The ferrocenyl substituent modifies the electrochemistry of $[Co_3(CO)_7(NO)(\mu\text{-}CR)]^-$ (227) and $[Co_3(\mu\text{-}CR)_2Cp_3]$ (R = Fc) (228). For example, the redox series $[Co_3(\mu\text{-}CH)(\mu\text{-}CFc)Cp_3]^Z$ (Z = 2 to −1) is extended in $[Co_3(\mu\text{-}CFc)_2Cp_3]^Z$ to Z = 3 to −1; the bis(ferrocenylcarbyne) cluster cations are Class II mixed valence species (Section IV,D).

The electrochemistry of $[Ru_3H(\mu_3\text{-}R)(CO)_9]$ (R = allyl, allenyl, or alkynyl) (229), $[Ag\{Os_3H(CO)_{10}\}_2]^-$ (230), and $[CoPt_2(\mu\text{-}PPh_2)(\mu\text{-}CO)_2(CO)(PPh_3)_3]$ (231) have also been reported.

REFERENCES

1. T. J. Meyer, *Prog. Inorg. Chem.* **19**, 1 (1975).
2. P. J. Vergamini and G. J. Kubas, *Prog. Inorg. Chem.* **21**, 261 (1976).
3. P. Lemoine, *Coord. Chem. Rev.* **47**, 55 (1982).
4. J. A. McCleverty, *in* "Reactions of Molecules at Electrodes" (N. S. Hush, ed.) p. 460. Wiley (Interscience), New York, 1971.
5. N. G. Connelly and W. E. Geiger, *Adv. Organomet. Chem.* **23**, 1 (1984).
6. T. Matsumoto, M. Sato, and A. Ichimura, *Bull. Chem. Soc. Jpn.* **44**, 1720 (1971).
7. W. H. Morrison, Jr., S. Krogsrud, and D. N. Hendrickson, *Inorg. Chem.* **12**, 1998 (1973).
8. D. O. Cowan, R. L. Collins, and F. Kaufman, *J. Phys. Chem.* **75**, 2025 (1971).
9. W. H. Morrison, Jr. and D. N. Hendrickson, *Inorg. Chem.* **14**, 2331 (1975).
10. D. O. Cowan, C. LeVanda, J. Park, and F. Kaufman, *Acc. Chem. Res.* **6**, 1 (1973).
11. C. LeVanda, K. Bechgaard, D. O. Cowan, U. T. Mueller-Westerhoff, P. Eilbracht, G. A. Candela, and R. L. Collins, *J. Am. Chem. Soc.* **98**, 3181 (1976).
12. G. M. Brown, T. J. Meyer, D. O. Cowan, C. LeVanda, F. Kaufman, P. V. Roling, and M. D. Rausch, *Inorg. Chem.* **14**, 506 (1975).
13. J. E. Gorton, H. L. Lentzner, and W. E. Watts, *Tetrahedron* **27**, 4353 (1971).
14. P. Shu, K. Bechgaard, and D. O. Cowan, *J. Org. Chem.* **41**, 1849 (1976).
15. C. LeVanda, D. O. Cowan, C. Leitch, and K. Bechgaard, *J. Am. Chem. Soc.* **96**, 6788 (1974).
16. J. A. Kramer, F. H. Herbstein, and D. N. Hendrickson, *J. Am. Chem. Soc.* **102**, 2293 (1980), and references therein.
17. T. J. Meyer, *Acc. Chem. Res.* **11**, 94 (1978).
18. J. B. Flanagan, S. Margel, A. J. Bard, and F. C. Anson, *J. Am. Chem. Soc.* **100**, 4248 (1978).
19. R. D. Rieke, I. Tucker, S. N. Milligan, D. R. Wright, B. R. Willeford, L. J. Radonovich, and M. W. Eyring, *Organometallics* **1**, 938 (1982).
20. D. O. Cowan, P. Shu, F. L. Hedberg, M. Rossi, and T. J. Kistenmacher, *J. Am. Chem. Soc.* **101**, 1304 (1979).
21. J. Kotz, G. Neyhart, W. J. Vining, and M. D. Rausch, *Organometallics* **2**, 79 (1983).

22. M. M. Sabbatini and E. Cesarotti, *Inorg. Chim. Acta* **24**, L9 (1977).
23. A. F. Diaz, U. T. Mueller-Westerhoff, A. Nazzal, and M. Tanner, *J. Organomet. Chem.* **236**, C45 (1982).
24. W. H. Morrison, Jr., E. Y. Ho, and D. N. Hendrickson, *Inorg. Chem.* **14**, 500 (1975).
25. L. P. Yureva, S. M. Peregudova, L. N. Nekrasov, A. P. Korotkov, N. N. Zaitseva, N. V. Zakurin, and A. Yu. Vasilkov, *J. Organomet. Chem.* **219**, 43 (1981).
26. C. Elschenbroich and J. Heck, *J. Am. Chem. Soc.* **101**, 6773 (1979).
27. C. Elschenbroich and J. Heck, *Angew. Chem., Int. Ed. Engl.* **20**, 267 (1981).
28. S. N. Milligan and R. D. Rieke, *Organometallics* **2**, 171 (1983).
29. N. G. Connelly, A. R. Lucy, J. D. Payne, A. M. R. Galas, and W. E. Geiger, *J. Chem. Soc., Dalton Trans.*, p. 1879 (1983).
30. M. J. Freeman, A. G. Orpen, N. G. Connelly, I. Manners, and S. J. Raven, *J. Chem. Soc., Dalton Trans.* (1985), in press.
31. A. Davison and J. C. Smart, *J. Organomet. Chem.* **49**, C43 (1973).
32. J. C. Smart, Ph.D. Dissertation, Massachusetts Institute of Technology (1973).
33. C.-P. Lau, P. Singh, S. J. Cline, R. Seiders, M. Brookhart, W. E. Marsh, D. J. Hodgson, and W. E. Hatfield, *Inorg. Chem.* **21**, 208 (1982).
34. J. C. Smart and B. L. Pinsky, *J. Am. Chem. Soc.* **102**, 1009 (1980).
35. P. R. Sharp, K. N. Raymond, J. C. Smart, and R. J. McKinney, *J. Am. Chem. Soc.* **103**, 753 (1981).
36. D. A. Lemenovskii, V. P. Fedin, Yu. L. Slovohotov, and Yu. T. Struchkov, *J. Organomet. Chem.* **228**, 153 (1982).
37. W. E. Geiger *in* "Metal Interactions with Boron Clusters" (R. N. Grimes, ed.), p. 239. Plenum, New York, 1982.
38. H. Gysling and J. Morris, *Chem. Rev.* (in press).
39. A. Salzer and H. Werner, *Angew. Chem., Int. Ed. Engl.* **11**, 930 (1972).
40. D. E. Brennan and W. E. Geiger, *J. Am. Chem. Soc.* **101**, 3399 (1979).
41. J. Edwin, M. Bochmann, M. C. Böhm, D. E. Brennan, W. E. Geiger, C. Krüger, J. Pebler, H. Pritzkow, W. Siebert, W. Swiridoff, H. Wadepohl, J. Weiss, and U. Zenneck, *J. Am. Chem. Soc.* **105**, 2582 (1983).
42. J. W. Lauher, M. Elian, R. H. Summerville, and R. Hoffmann, *J. Am. Chem. Soc.* **98**, 3219 (1976).
43. S. P. Kolesnikov, J. E. Dobson, and P. S. Skell, *J. Am. Chem. Soc.* **100**, 999 (1978).
44. J. Moraczewski and W. E. Geiger, *J. Am. Chem. Soc.* **100**, 7429 (1978).
45. W. E. Geiger, J. Edwin, and A. L. Rheingold, *J. Am. Chem. Soc.* **106**, 3052 (1984).
46. A. Salzer, personal communication, 1983.
47. C. Elschenbroich, J. Heck, W. Massa, E. Nun, and R. Schmidt, *J. Am. Chem. Soc.* **105**, 2905 (1983).
48. C. Elschenbroich, J. Heck, W. Massa, and R. Schmidt, *Angew. Chem., Int. Ed. Engl.* **22**, 330 (1983).
49. C. T. Vroegop, J. H. Teuben, F. van Bolhuis, and J. G. M. van der Linden, *J. Chem. Soc., Chem. Commun.*, p. 550 (1983).
50. T. Madach and H. Vahrenkamp, *Chem. Ber.* **114**, 513 (1981).
51. L. J. Guggenberger, *Inorg. Chem.* **12**, 294 (1973).
52. D. A. Lemenovskii, V. P. Fedin, A. V. Aleksandrov, Yu. L. Slovohotov, and Yu. T. Struchkov, *J. Organomet. Chem.* **201**, 257 (1980).
53. R. E. Dessy, R. Kornmann, C. Smith, and R. Haytor, *J. Am. Chem. Soc.* **90**, 2001 (1968).
54. R. E. Dessy and L. Wieczorek, *J. Am. Chem. Soc.* **91**, 4963 (1969).
55. R. E. Dessy, A. L. Rheingold, and G. D. Howard, *J. Am. Chem. Soc.* **94**, 746 (1972).

56. R. E. Dessy and L. A. Bares, *Acc. Chem. Res.* **5,** 415 (1972).
57. B.-K. Teo, M. B. Hall, R. F. Fenske, and L. F. Dahl, *J. Organomet. Chem.* **70,** 413 (1974).
58. U. Honrath and H. Vahrenkamp, *Z. Naturforsch., B: Anorg. Chem., Org. Chem.* **34,** 1190 (1979).
59. T. Madach and H. Vahrenkamp, *Z. Naturforsch., B: Anorg. Chem. Org. Chem.* **34,** 1195 (1979).
60. P. Legzdins and B. Wassink, personal communication, 1983; *Organometallics* **3,** 1811 (1984).
61. D. P. Frisch, M. K. Lloyd, J. A. McCleverty, and D. Seddon, *J. Chem. Soc., Dalton Trans.*, p. 2268 (1973).
62. R. E. Dessy, F. E. Stary, R. B. King, and M. Waldrop, *J. Am. Chem. Soc.* **88,** 471 (1966).
63. T. Gennett, Ph.D. Dissertation, University of Vermont (1983).
64. N. G. Connelly and L. F. Dahl, *J. Am. Chem. Soc.* **92,** 7470 (1970).
65. G. Ferguson, C. Hannaway, and K. M. S. Islam, *J. Chem. Soc., Chem. Commun.*, p. 1165 (1968).
66. N. G. Connelly and L. F. Dahl, *J. Am. Chem. Soc.* **92,** 7472 (1970).
67. M. Rakowski Dubois, R. C. Haltiwanger, D. J. Miller, and G. Glatzmaier, *J. Am. Chem. Soc.* **101,** 5245 (1979).
68. W. E. Silverthorn, *J. Am. Chem. Soc.* **102,** 842 (1980).
69. G. C. Allen, M. Green, B. J. Lee, H. P. Kirsch, and F. G. A. Stone, *J. Chem. Soc., Chem. Commun.*, p. 794 (1976).
70. M. Bochmann, M. Green, H. P. Kirsch, and F. G. A. Stone, *J. Chem. Soc., Dalton Trans.*, p. 714 (1977).
71. O. P. Anderson and M. C. R. Symons, *J. Chem. Soc., Chem. Commun.*, p. 1020 (1972).
72. S. W. Bratt and M. C. R. Symons, *J. Chem. Soc., Dalton Trans.*, p. 1314 (1977).
73. M. C. R. Symons, J. Wyatt, B. M. Peake, J. Simpson, and B. H. Robinson, *J. Chem. Soc., Dalton Trans.*, p .2037 (1982).
74. T. Lionel, J. R. Morton, and K. F. Preston, *Inorg. Chem.* **22,** 145 (1983).
75. V. Küllmer, E. Rottinger, and H. Vahrenkamp, *Z. Naturforsch., B: Anorg. Chem., Org. Chem.* **34,** 224 (1979).
76. P. Hydes, J. A. McCleverty, and D. G. Orchard, *J. Chem. Soc. A*, p. 3660 (1971).
77. J. C. T. R. Burckett-St. Laurent, M. R. Caira, R. B. English, R. J. Haines, and L. R. Nassimbeni, *J. Chem. Soc., Dalton Trans.*, p. 1077 (1977).
78. G. T. Hefter and G. A. Heath, unpublished results (quoted in ref. 77).
79. R. E. Ginsberg, R. K. Rothrock, R. G. Finke, J. P. Collman, and L. F. Dahl, *J. Am. Chem. Soc.* **101,** 6550 (1979).
80. B. K. Teo, M. B. Hall, R. F. Fenske, and L. F. Dahl, *Inorg. Chem.* **14,** 3103 (1975).
81. R. Mathieu, R. Poilblanc, P. Lemoine, and M. Gross, *J. Organomet. Chem.* **165,** 243 (1979).
82. B.-K. Teo and L. F. Dahl, unpublished esr spectral data (quoted in ref. 80).
83. J. P. Collman, R. K. Rothrock, R. G. Finke, and F. Rose-Munch, *J. Am. Chem. Soc.* **99,** 7381 (1977).
84. D. Seyferth and R. S. Henderson, *J. Am. Chem. Soc.* **101,** 508 (1979).
85. D. Seyferth, R. S. Henderson, and L.-C. Song, *J. Organomet. Chem.* **192,** C1 (1980).
86. D. Seyferth, R. S. Henderson, and M. K. Gallagher, *J. Organomet. Chem.* **193,** C75 (1980).
87. D. Seyferth, R. S. Henderson, and L.-C. Song, *Organometallics* **1,** 125 (1982).
88. E. K. Lhadi, C. Mahe, H. Patin, and A. Darchen, *J. Organomet. Chem.* **246,** C61 (1983); A. Darchen, E. K. Lhadi, and H. Patin, *J. Organomet. Chem.* **259,** 189 (1983).

89. R. E. Dessy, P. M. Weissman, and R. L. Pohl, *J. Am. Chem. Soc.* **88**, 5117 (1966).
90. D. Miholova and A. A. Vlcek, *Inorg. Chim. Acta* **41**, 119 (1980).
91. R. J. Haines and A. L. du Preez, *J. Am. Chem. Soc.* **91**, 769 (1969); *J. Chem. Soc. A*, p. 2341 (1970).
92. E. C. Johnson, T. J. Meyer, and N. Winterton, *J. Chem. Soc., Chem. Commun.*, p. 934 (1970); *Inorg. Chem.* **10**, 1673 (1971).
93. J. A. Ferguson and T. J. Meyer, *Inorg. Chem.* **10**, 1025 (1971).
94. N. El Murr, paper delivered at Meeting of American Chemical Society, Washington, D.C., August 1983.
95. R. J. Haines and A. L. du Preez, *Inorg. Chem.* **11**, 330 (1972).
96. J. A. Ferguson and T. J. Meyer, *J. Chem. Soc., Chem. Commun.*, p. 1544 (1971); *Inorg. Chem.* **11**, 631 (1972).
97. J. A. DeBeer, R. J. Haines, R. Greatrex, and J. A. van Wyk, *J. Chem. Soc., Dalton Trans.*, p. 2341 (1973).
98. J. D. Sinclair, N. G. Connelly, and L. F. Dahl, unpublished results; J. D. Sinclair, Ph.D. Dissertation, University of Wisconsin, 1972; *Diss. Abstr. B* **33**, 4716 (1972–73).
99. R. Mason, unpublished results (quoted in ref. 77).
100. G. J. Kubas, P. J. Vergamini, M. P. Eastman, and K. B. Prater, *J. Organomet. Chem.* **117**, 71 (1976).
101. P. J. Vergamini, R. R. Ryan, and G. J. Kubas, *J. Am. Chem. Soc.* **98**, 1980 (1976).
102. A. Terzis and R. Rivest, *Inorg. Chem.* **12**, 2132 (1973).
103. G. J. Kubas and P. J. Vergamini, *Inorg. Chem.* **20**, 2667 (1981).
104. M. D. Seidler and R. G. Bergman, *Organometallics* **2**, 1897 (1983).
105. R. E. Dessy and R. L. Pohl, *J. Am. Chem. Soc.* **90**, 1995 (1968).
106. G. Zotti, R. D. Rieke, and J. S. McKennis, *J. Organomet. Chem.* **228**, 281 (1982).
107. C. S. Ilenda, N. E. Schore, and R. G. Bergman, *J. Am. Chem. Soc.* **98**, 255 (1976); N. E. Schore, C. S. Ilenda, and R. G. Bergman, *J. Am. Chem. Soc.* **99**, 1781 (1977).
108. N. E. Schore, *J. Organomet. Chem.* **173**, 301 (1979).
109. N. E. Schore, C. S. Ilenda, and R. G. Bergman, *J. Am. Chem. Soc.* **98**, 256 (1976).
110. R. E. Ginsburg, L. M. Cirjak, and L. F. Dahl, *J. Chem. Soc., Chem. Commun.*, p. 468 (1979); L. M. Cirjak, R. E. Ginsburg, and L. F. Dahl, *Inorg. Chem.* **21**, 940 (1982).
111. W. I. Bailey, Jr., D. M. Collins, F. A. Cotton, J. C. Baldwin, and W. C. Kaska, *J. Organomet. Chem.* **165**, 373 (1979).
112. I. Bernal, J. D. Korp, G. M. Reisner, and W. A. Herrmann, *J. Organomet. Chem.* **139**, 321 (1977).
113. F. Wochner, E. Keller, and H. H. Brintzinger, *J. Organomet. Chem.* **236**, 267 (1982).
114. A. R. Pinhas and R. Hoffmann, *Inorg. Chem.* **18**, 654 (1979).
115. F. Bottomley, *Inorg. Chem.* **22**, 2656 (1983).
116. N. G. Connelly, J. D. Payne, and W. E. Geiger, *J. Chem. Soc., Dalton Trans.*, p. 295 (1983).
117. S. Clamp, N. G. Connelly, and J. D. Payne, *J. Chem. Soc., Chem. Commun.*, p. 897 (1981); S. Clamp, N. G. Connelly, J. A. K. Howard, I. Manners, J. D. Payne and W. E. Geiger, *J. Chem. Soc., Dalton Trans.*, p. 1659 (1984).
118. N. G. Connelly and G. A. Johnson, *J. Chem. Soc., Dalton Trans.*, p. 1375 (1978).
119. R. S. Dickson, B. M. Peake, P. H. Rieger, B. H. Robinson, and J. Simpson, *J. Organomet. Chem.* **172**, C63 (1979).
120. B. M. Peake, P. H. Rieger, B. H. Robinson, and J. Simpson, *J. Am. Chem. Soc.* **102**, 156 (1980).
121. C. M. Arewgoda, B. H. Robinson, and J. Simpson, *J. Am. Chem. Soc.* **105**, 1893 (1983).
122. G. J. Bezems, P. H. Rieger, and S. Visco, *J. Chem. Soc., Chem. Commun.*, p. 265

(1981); C. M. Arewgoda, P. H. Rieger, B. H. Robinson, J. Simpson, and S. Visco, *J. Am. Chem. Soc.* **104,** 5633 (1982).

123. C. M. Arewgoda, B. H. Robinson, and J. Simpson, *J. Chem. Soc., Chem. Commun.,* p. 284 (1982).
124. R. G. Cunninghame, A. J. Downard, L. R. Hanton, S. D. Jensen, B. H. Robinson, and J. Simpson, *Organometallics* **3,** 180 (1984).
125. A. Chaloyard, N. El Murr, and R. B. King, *J. Organomet. Chem.* **188,** C13 (1980).
126. F. M. Chaudhary and P. L. Pauson, *J. Organomet. Chem.* **69,** C31 (1974).
127. N. G. Connelly and E. J. Young, unpublished results, 1983.
128. J. M. Coleman and L. F. Dahl, *J. Am. Chem. Soc.* **89,** 542 (1967).
129. N. G. Connelly, G. A. Lane, and W. E. Geiger, unpublished results.
130. K. Broadley, G. A. Lane, N. G. Connelly, and W. E. Geiger, *J. Am. Chem. Soc.* **105,** 2486 (1983).
131. M. R. Churchill and S. W.-Y. Chang, *J. Chem. Soc., Chem. Commun.,* p. 248 (1974).
132. F. Stollmaier and U. Thewalt, *J. Organomet. Chem.* **222,** 227 (1981).
133. S. Z. Goldberg, B. Spivack, G. Stanley, R. Eisenberg, D. M. Braitsch, J. S. Miller, and M. Abkowitz, *J. Am. Chem. Soc.* **99,** 110 (1977).
134. R. B. King, D. M. Braitsch, and P. N. Kapoor, *J. Am. Chem. Soc.* **97,** 60 (1975).
135. R. Jund, P. Lemoine, M. Gross, R. Bender, and P. Braunstein, *J. Chem. Soc., Chem. Commun.,* p. 86 (1983).
136. P. A. Dawson, B. M. Peake, B. H. Robinson, and J. Simpson, *Inorg. Chem.* **19,** 465 (1980).
137. A. M. Bond, P. A. Dawson, B. M. Peake, B. H. Robinson, and J. Simpson, *Inorg. Chem.* **16,** 2199 (1977).
138. N. El Murr and A. Chaloyard, *Inorg. Chem.* **21,** 2206 (1982).
139. D. Miholova, J. Klima, and A. A. Vlcek, *Inorg. Chim. Acta* **27,** L67 (1978).
140. P. J. Krusic, J. S. Filippo, Jr., B. Hutchinson, R. L. Hance, and L. M. Daniels, *J. Am. Chem. Soc.* **103,** 2129 (1981).
141. P. J. Krusic, *J. Am. Chem. Soc.* **103,** 2131 (1981).
142. B. M. Peake, P. H. Rieger, B. H. Robinson, and J. Simpson, *Inorg. Chem.* **20,** 2540 (1981).
143. H. Beurich, T. Madach, F. Richter, and H. Vahrenkamp, *Angew. Chem., Int. Ed. Engl.* **18,** 690 (1979).
144. C. E. Strouse and L. F. Dahl, *Faraday Discuss. Chem. Soc.* **47,** 93 (1969).
145. C. E. Strouse and L. F. Dahl, *J. Am. Chem. Soc.* **93,** 6032 (1971).
146. D. L. Stevenson, C. H. Wei, and L. F. Dahl, *J. Am. Chem. Soc.* **93,** 6027 (1971).
147. D. A. Young, *Inorg. Chem.* **20,** 2049 (1981).
148. R. J. Haines, J. A. DeBeer, and R. Greatrex, *J. Organomet. Chem.* **55,** C30 (1973).
149. T. Madach and H. Vahrenkamp, *Chem. Ber.* **114,** 505 (1981).
150. A. Darchen, C. Mahe, and H. Patin, *J. Chem. Soc., Chem. Commun.,* p. 243 (1982).
151. J. A. Ferguson and T. J. Meyer, *J. Chem. Soc., Chem. Commun.,* p. 623 (1971).
152. Trinh-Toan, B. K. Teo, J. A. Ferguson, T. J. Meyer, and L. F. Dahl, *J. Am. Chem. Soc.* **99,** 408 (1977).
153. L. F. Dahl and R. Bedard, personal communication, 1983.
154. R. Bedard, paper delivered at Meeting of American Chemical Society, Washington, D.C., August, 1983.
155. R. A. Schunn, C. J. Fritchie, Jr., and C. T. Prewitt, *Inorg. Chem.* **5,** 892 (1966).
156. C. H. Wei, G. R. Wilkes, P. M. Treichel, and L. F. Dahl, *Inorg. Chem.* **5,** 900 (1966).
157. Trinh-Toan, W. P. Fehlhammer, and L. F. Dahl, *J. Am. Chem. Soc.* **99,** 402 (1977).
158. H. Wong, D. Sedney, W. M. Reiff, R. B. Frankel, T. J. Meyer, and D. Salmon, *Inorg. Chem.* **17,** 194 (1978).

159. J. A. Ferguson and T. J. Meyer, *J. Am. Chem. Soc.* **94,** 3409 (1972).
160. C. R. Bock and M. S. Wrighton, *Inorg. Chem.* **16,** 1309 (1977).
161. M. A. Neuman, Trinh-Toan, and L. F. Dahl, *J. Am. Chem. Soc.* **94,** 3383 (1972).
162. Trinh-Toan, W. P. Fehlhammer, and L. F. Dahl, *J. Am. Chem. Soc.* **94,** 3389 (1972).
163. C. T.-W. Chu, F. Y.-K. Lo, and L. F. Dahl, *J. Am. Chem. Soc.* **104,** 3409 (1982).
164. G. J. Kubas and P. Vergamini, *Inorg. Chem.* **20,** 2667 (1981).
165. G. J. Kubas, personal communication, 1982.
166. M. I. Bruce, D. C. Kehoe, J. G. Matisons, B. K. Nicholson, P. H. Rieger, and M. L. Williams, *J. Chem. Soc., Chem. Commun.,* p. 442 (1982).
167. M. I. Bruce, J. G. Matisons, and B. K. Nicholson, *J. Organomet. Chem.* **247,** 321 (1983).
168. M. I. Bruce, T. W. Hambley, B. K. Nicholson, and M. R. Snow, *J. Organomet. Chem.* **235,** 83 (1982).
169. J. Evans and B. P. Gracey, *J. Chem. Soc., Chem. Commun.,* p. 247 (1983).
170. R. Mason, K. M. Thomas, and D. M. P. Mingos, *J. Am. Chem. Soc.* **95,** 3802 (1973).
171. M. McPartlin, C. R. Eady, B. F. G. Johnson, and J. Lewis, *J. Chem. Soc., Chem. Commun.,* p. 883 (1976).
172. K. Wade, *Adv. Inorg. Chem. Radiochem.* **18,** 1 (1976).
173. G. R. John, B. F. G. Johnson, J. Lewis, and A. L. Mann, *J. Organomet. Chem.* **171,** C9 (1979).
174. B. Tulyathan, Ph.D. Dissertation, University of Vermont, 1981.
175. B. F. G. Johnson, J. Lewis, W. J. H. Nelson, J. Puga, P. R. Raithby, M. Schröder, and K. H. Whitmire, *J. Chem. Soc., Chem. Commun.,* p. 610 (1982).
176. A. Benoit, A. Darchen, J.-Y. Le Marouille, C. Mahé, and H. Patin, *Organometallics* **2,** 555 (1983).
177. J. C. Kotz, J. V. Petersen, and R. C. Reed, *J. Organomet. Chem.* **120,** 433 (1976).
178. A. M. Bond, B. M. Peake, B. H. Robinson, J. Simpson, and D. J. Watson, *Inorg. Chem.* **16,** 410 (1977).
179. B. M. Peake, B. H. Robinson, J. Simpson, and D. J. Watson, *Inorg. Chem.* **16,** 405 (1977).
180. A. M. Bond, P. A. Dawson, B. M. Peake, P. H. Rieger, B. H. Robinson, and J. Simpson, *Inorg. Chem.* **18,** 1413 (1979).
181. S. Jensen, B. H. Robinson, and J. Simpson, *J. Chem. Soc., Chem. Commun.,* p. 1081 (1983).
182. S. B. Colbran, B. H. Robinson, and J. Simpson, *Organometallics* **2,** 943 (1983).
183. S. B. Colbran, B. H. Robinson, and J. Simpson, *Organometallics* **2,** 952 (1983).
184. P. T. Chesky and M. B. Hall, *Inorg. Chem.* **20,** 4419 (1981).
185. G. Granozzi, E. Tondello, D. Ajo, M. Casarin, S. Aime, and D. Osella, *Inorg. Chem.* **21,** 1081 (1982).
186. F. A. Cotton and J. D. Jamerson, *J. Am. Chem. Soc.* **98,** 1273 (1976); W. I. Bailey, Jr., F. A. Cotton, J. D. Jamerson, and B. W. S. Kolthammer, *Inorg. Chem.* **21,** 3131 (1982).
187. R. E. Dessy, R. B. King, and M. Waldrop, *J. Am. Chem. Soc.* **88,** 5112 (1966).
188. P. D. Frisch and L. F. Dahl, *J. Am. Chem. Soc.* **94,** 5082 (1972).
189. S. Enoki, T. Kawamura, and T. Yonezawa, *Inorg. Chem.* **22,** 3821 (1983).
190. G. L. Simon and L. F. Dahl, *J. Am. Chem. Soc.* **95,** 2164 (1973).
191. V. G. Albano, P. Chini, G. Ciani, M. Sansoni, and S. Martinengo, *J. Chem. Soc., Dalton Trans.,* p. 163 (1980).
192. P. Chini, *J. Organomet. Chem.* **200,** 37 (1980).
193. A. Bonny, T. J. Crane, and N. A. P. Kane-Maguire, *Inorg. Chim. Acta* **65,** L83 (1982).
194. P. Lemoine, M. Gross, D. de Montauzon, and R. Poilblanc, *Inorg. Chim. Acta* **71,** 15 (1983).

195. L. R. Byers, V. A. Uchtman, and L. F. Dahl, *J. Am. Chem. Soc.* **103,** 1942 (1981).

196. H. C. Longuet-Higgins and A. J. Stone, *Mol. Phys.* **5,** 417 (1962).

197. J. M. Maj, A. D. Rae, and L. F. Dahl, *J. Am. Chem. Soc.* **104,** 3054 (1982).

198. M. S. Paquette and L. F. Dahl, *J. Am. Chem. Soc.* **102,** 6621 (1980).

199. G. Longoni and F. Morazzoni, *J. Chem. Soc., Dalton Trans.*, p. 1735 (1981).

200. P. Lemoine, A. Giraudeau, M. Gross, R. Bender, and P. Braunstein, *J. Chem. Soc., Dalton Trans.*, p. 2059 (1981).

201. J. G. M. van der Linden, M. L. H. Paulissen, and J. E. J. Schmitz, *J. Am. Chem. Soc.* **105,** 1903 (1983).

202. D. R. Talham and D. O. Cowan, *Organometallics* **3,** 1712 (1984).

203. M. F. Moore, S. R. Wilson, D. N. Hendrickson, and U. T. Mueller-Westerhoff, *Inorg. Chem.* **23,** 2918 (1984).

204. B. Zhuang, J. W. McDonald, F. A. Schultz, and W. E. Newton, *Organometallics* **3,** 943 (1984).

205. S. P. Schmidt, W. C. Trogler, and F. Basolo, *J. Am. Chem. Soc.* **106,** 1308 (1984).

206. E. K. Lhadi, H. Patin, and A. Darchen, *Organometallics* **3,** 1128 (1984).

207. Y.-F. Yu, C.-N. Chau, A. Wojcicki, M. Calligari, G. Nardin, and C. Balducci, *J. Am. Chem. Soc.* **106,** 3704 (1984).

208. S. G. Davies, S. J. Simpson, and V. D. Parker, *J. Chem. Soc., Chem. Commun.*, p. 352 (1984).

209. D. Catheline and D. Astruc, *J. Organomet. Chem.* **266,** C11 (1984).

210. N. G. Connelly and E. J. Young, unpublished data, 1983.

211. N. G. Connelly, N. J. Forrow, B. P. Gracey, S. A. R. Knox, and A. G. Orpen, *J. Chem. Soc., Chem. Commun.*, p. 14 (1985).

212. N. G. Connelly, N. J. Forrow, S. A. R. Knox, K. A. Macpherson, and A. G. Orpen, *J. Chem. Soc., Chem. Commun.*, p. 16 (1985).

213. U. Kölle and B. Fuss, *Chem. Ber.* **117,** 753 (1984).

214. N. E. Schore, C. S. Ilenda, M. A. White, H. E. Bryndza, M. G. Matturo, and R. G. Bergman, *J. Am. Chem. Soc.* **106,** 7451 (1984).

215. N. G. Connelly, C. J. Finn, M. J. Freeman, A. G. Orpen, and J. Stirling, *J. Chem. Soc., Chem. Commun.*, p. 1025 (1984).

216. G. C. Allen and N. G. Connelly, unpublished data.

217. J. L. Marshall, S. R. Stobart, and H. B. Gray, *J. Am. Chem. Soc.* **106,** 3027 (1984).

218. J. V. Caspar and H. B. Gray, *J. Am. Chem. Soc.* **106,** 3029 (1984).

219. J. Wronka and D. P. Ridge, *J. Am. Chem. Soc.* **106,** 67 (1984).

220. F. Dahan and R. Mathieu, *J. Chem. Soc., Chem. Commun.*, p. 432 (1984).

221. D. de Montauzon and R. Mathieu, *J. Organomet. Chem.* **252,** C83 (1983).

222. A. E. Stiegman, A. S. Goldman, D. B. Leslie, and D. R. Tyler, *J. Chem. Soc., Chem. Commun.*, p. 632 (1984).

223. P. N. Lindsay, B. M. Peake, B. H. Robinson, J. Simpson, U. Honrath, H. Vahrenkamp, and A. M. Bond, *Organometallics* **3,** 413 (1984).

224. U. Honrath and H. Vahrenkamp, *Z. Naturforsch. Ser. B* **39,** 545 (1984).

225. U. Honrath and H. Vahrenkamp, *Z. Naturforsch Ser. B* **39,** 555 (1984).

226. U. Honrath and H. Vahrenkamp, *Z. Naturforsch Ser. B* **39,** 559 (1984).

227. S. B. Colbran, B. H. Robinson, and J. Simpson, *J. Organomet. Chem.* **265,** 199 (1984).

228. S. B. Colbran, B. H. Robinson, and J. Simpson, *Organometallics* **3,** 1344 (1984).

229. P. Zanello, S. Aime, and D. Osella, *Organometallics* **3,** 1374 (1984).

230. M. Fajardo, M. P. Gomez-Sal, H. D. Holden, B. F. G. Johnson, J. Lewis, R. C. S. McQueen, and P. R. Raithby, *J. Organomet. Chem.* **267,** C25 (1985).

231. R. Bender, P. Braunstein, B. Metz, and P. Lemoine, *Organometallics* **3,** 381 (1984).

ADVANCES IN ORGANOMETALLIC CHEMISTRY, VOL. 24

Organometallic Lanthanide Chemistry

WILLIAM J. EVANS

Department of Chemistry
University of California, Irvine
Irvine, California

I

INTRODUCTION

Organolanthanide chemistry currently is one of the most rapidly developing areas of organometallic chemistry. In the last few years, this field has undergone dramatic transformation from an area involving a narrow range of ionic, cyclopentadienyl species such as $(C_5H_5)_3Ln$ (*1*), $[(C_5H_5)_2LnCl]_2$ (*2*), and $[(C_5H_5)_2LnR]_2$ (*3*) (Ln = lanthanide metal; R = alkyl or aryl), for which only limited reactivity had been demonstrated, to an area encompassing a wide variety of new classes of organometallic complexes which often have unique structures and spectacular reactivity. Organolanthanide chemistry has grown from a minor field of little general interest to an area making major contributions to our understanding of organometallic chemistry in a broad sense.

Organolanthanide chemistry was exhaustively reviewed in Volume 9 of this Series in 1970 in an article that required only 15 pages (4). More recent surveys (5–7) have required much more space: the latest review in *Comprehensive Organometallic Chemistry* (7) was nearly six times longer than the 1970 report. Although the latest reviews are quite recent, many new results in organolanthanide chemistry have occurred since they were written. This article will focus only on these recent results and how they relate to trends currently developing in the organolanthanide area. Even with this limited scope, it will not be possible to be exhaustive in this coverage.

Many of the major developments in organolanthanide chemistry in the past few years have involved the synthesis and structural characterization of new complexes. As is appropriate for any young, undeveloped area of organometallic chemistry, these synthetic and structural investigations were an essential first step in establishing what combinations of metals, ligands, and reagents were compatible and significant. Recent synthetic and structural results in organolanthanide chemistry fall into three major categories, each of which comprises a section of this article: synthesis and reactivity of Ln—C bonds, synthesis and reactivity of Ln—H bonds, and low oxidation state chemistry. The advances made in these three areas have provided a basic set of complexes and reactivity patterns upon which future more detailed studies of reaction mechanisms and unstable intermediates can be based. An additional area of recent organolanthanide research involves the C_5Me_5 ligand. Since these C_5Me_5 complexes have rather special chemistry, they are described in a separate section on pentamethylcyclopentadienyllanthanide chemistry. Before any of the recent experimental results are discussed, a brief section on the properties of the lanthanides and the traditional generalizations applied to organolanthanide chemistry, as well as a section on the unique aspects of the lanthanide elements, are presented.

II

BACKGROUND

The lanthanide elements are differentiated from other metallic elements by the fact that their valence electrons are in $4f$ orbitals. Calculations on the $[Xe]4f^n$ electron configurations of the lanthanides indicate that the $4f$

orbitals do not have significant radial extension beyond that filled $5s^2 5p^6$ orbitals of the xenon inert gas core (8, 9). As a result, the external appearance of a lanthanide ion is like that of a closed shell inert gas with a tripositive charge. In this sense, the lanthanides are similar to alkali or alkaline earth ions and lanthanide chemistry traditionally has been found to be quite ionic (10). Electrostatic factors and steric considerations appear to be more important in determining the stability, structure, and chemistry of lanthanide complexes than interactions between the metal and ligand orbitals.

A further consequence of the limited radial extension of the $4f$ orbitals is that a given class of lanthanide complexes often displays similar chemistry regardless of the particular metal or $4f^n$ configuration involved. For example, all members of the class $(C_5H_5)_3Ln$ are known for Ln = La to Lu and are chemically alike. Compare this to the variability in the chemistry of the transition metal complexes, $(C_5H_5)_2M$, as M varies from Ti to Cu. Another similarity across the series is that the +3 oxidation state is the most stable for every element in the series (10). A consequence of the chemical conformity found for these metals is that a single symbol, Ln, has been used in the past to describe rather accurately the chemistry of all 14 elements in the series.

Small differences in the chemistry of the lanthanides are observed due to the gradually changing radial size of the elements, which decreases from 1.061 Å for La^{3+} to 0.848 Å for Lu^{3+} (11). Differences in chemistry also occur for the four elements in the series which have nontrivalent oxidation states accessible under normal reaction conditions: Ce^{4+} $(4f^0)$, Eu^{2+} $(4f^7)$, Yb^{2+} $(4f^{14})$, and Sm^{2+} $(4f^6)$. Since Ce^{4+} is a strong oxidizing agent and the divalent lanthanides are good reducing agents (12), these alternative oxidation states become important primarily under strongly oxidizing or reducing conditions, respectively. Note that in none of these cases are both the +4 and +2 oxidation states available to the same metal. Hence, the chemistry of the lanthanides has not been found to involve two-electron redox processes on a single metal center.

The properties described above for the lanthanide elements have led to the development of two generalizations for stabilizing organometallic complexes of these metals. First, in these ionic systems, the +3 charge must be balanced and electrostatic interactions must be optimized. This is best accomplished using stable organic anions as ligands. Second, the stability of a complex can be enhanced by filling the coordination environment of the metal with bulky ligands to sterically block decomposition reactions. Since the lanthanide ions are large and can accommodate high coordination numbers, 8–12, this latter requirement is not always easy to fulfill. One reason that the later lanthanides, Lu, Yb, and Er, are studied more often than the other members in the series is that it is less difficult to sterically saturate these smaller metal centers. Two ligands that satisfy both the electrostatic and steric needs of a lanthanide ion are cyclopentadienyl monoanions and cyclooctatetraenyl dianions. Traditionally, these ligands have been the most common coordinating groups in organolanthanide chemistry.

Since the chemistry of organolanthanide complexes is rather invariant to the number of $4f^n$ electrons, a metal of similar size and charge but with no f electrons could have similar chemical properties. Such is the case for yttrium and it will be included in this article. Although yttrium is not formally a lanthanide element, it is congeneric with lanthanum. More importantly, the radial size of the trivalent yttrium ion is nearly identical with that of Er^{3+}. As such, its chemistry, at least so far, has proven to be very similar to that of the late lanthanides. Since Y^{3+} is diamagnetic (cf. Er^{3+}, $\mu = 9.4$–9.7 B.M.) with $I = \frac{1}{2}$, it provides NMR-accessible complexes whose spectra can often contain considerable extra structural information via Y–H and Y–C coupling.

III

THE POTENTIAL FOR UNIQUE CHEMISTRY

The properties of the lanthanide elements and their organometallic complexes described in the previous section explain in part why organometallic chemists in the past found lanthanide chemistry much less interesting than transition metal chemistry. The highly ionic, trivalent organolanthanide complexes appeared to have little potential to interact with the small-molecule substrates that provide such a rich chemistry for the transition metals: neutral unsaturated hydrocarbons, H_2, CO, phosphines, etc. The two-electron oxidation $\rightleftharpoons$ reduction cycles so important in catalytic transition metal chemistry in $18 \rightleftharpoons 16$ electron complexes seemed inaccessible to the lanthanides. The wide range of chemical reactivity found for transition metals as the metal changes across the series or as the oxidation state of a given metal changes was not found in the lanthanide series, where all of the metals have similar chemistry and a single oxidation state predominates. These features, coupled with a general but untrue perception that the "rare earth" elements must be rare and expensive [in fact they are more abundant than Hg or Ag (13)] and the fact that organolanthanide chemistry is experimentally demanding (all the complexes discussed in this article are extremely air and moisture sensitive), encouraged organometallic chemists to concentrate their efforts in other parts of the periodic table.

More recently, however, it has been realized that the lanthanide elements have the potential for unique chemistry (14). Indeed, some of the very properties just discussed which appear disadvantageous or limiting in comparison with transition metal chemistry are in fact advantageous in developing new organometallic chemistry. For example, the traditionally observed similarity of the chemistry of the 14 lanthanides suggests that the lanthanides will not have as diverse a chemistry as a row of transition metals. On the other hand, there is no comparable series of metals in the periodic table that has similar chemistry, but a gradually changing radial dimension. This aspect of the lanthanide elements offers the potential to vary reactivity precisely; by selecting the appropriately sized metal from the 14 choices, reactivity can be altered quite substantially (14).

The lack of extensive metal–ligand orbital overlap in organolanthanide complexes suggested that some valuable covalent interactions such as π backbonding may not be possible. On the other hand, with no orbital constraints, certain types of "orbitally forbidden" reactions may be quite easy to accomplish at a lanthanide metal center. Furthermore, ligand rearrangement and reversible modification of coordination geometry may be relatively facile in the absence of orbital factors as long as electrostatic demands are met. In a catalytic cycle, some nonessential intermediate steps, which are necessitated by favorable orbital interactions in transition metal complexes, may be absent in a lanthanide system, which, consequently, could have a faster reaction rate. Hence, organolanthanide complexes may provide the optimum geometrical environment for unique or uniquely rapid reaction chemistry.

The large size and ionic, electropositive character of the lanthanides, properties which make the chemistry experimentally difficult, also make the lanthanides strongly electrophilic and oxophilic. These properties can also impart unusual chemistry.

In summary, the special combination of physical properties of the lanthanides should translate into novel and potentially useful chemical behaviour. As stated in an earlier summary of organolanthanide chemistry (14), "the challenge in the lanthanide area, therefore, is to place the lanthanide metals in chemical environments which allow exploitation of their chemical uniqueness." In the past 5 years, organometallic environments beyond the simple, original, tris(cyclopentadienyl) and bis(cyclopentadienyl) chloride and alkyl types have been explored and some remarkable chemistry has resulted.

IV

SYNTHESIS AND REACTIVITY OF Ln—C BONDS

The synthesis, structure, and physical and chemical properties of organo-lanthanide complexes containing only polyhaptocyclopentadienyl and cyclooctatetraenyl metal–carbon bonds are fully covered in other reviews (5–7). This section will focus on lanthanide–carbon single bonded species only. Ln—C single bonds are found almost exclusively in only two types of complexes: homoleptic complexes, $LnR_x^{(x-3)-}$, and heteroleptic complexes containing two cyclopentadienyl rings, $[(C_5H_5)_2LnR]_2$ and $(C_5H_5)_2LnR(solvent)$, where R = alkyl or aryl. The chemistry of monocyclopentadienyl complexes $(C_5R_5)LnR_2(solvent)_x$ is an unexplored area which will undoubtedly lead to interesting chemistry in the future. The first report of the synthesis of a new class of mono-ring alkyl species, the monocyclooctatetraenylalkyl complexes, $C_8H_8LnR(thf)_2$ (thf = tetrahydrofuran), has appeared (15).

A. Homoleptic Ln—C Complexes

For many years, the only fully characterized homoleptic organolantha-nide complex and the only structurally characterized Ln—C species of any type was the tetrakis(2,6-dimethylphenyl) complex $\{Lu[C_6H_3(CH_3)_2]_4\}$ $\{Li(thf)_4\}$ (16). The four bulky 2,6-dimethylphenyl ligands were ar-ranged in a tetrahedral fashion around the metal with Lu—C bond lengths of 2.4–2.5 Å. The ionic metathesis reaction used to generate this complex [Eq. (1)] is the classic method for organolanthanide synthesis (6). This study in general was the prototype demonstration of the import-

$$LnCl_3 + 4\ LiC_6H_3Me_2 \rightarrow [Ln(C_6H_3Me_2)_4][Li(thf)]_4 + 3\ LiCl \tag{1}$$

ance of saturating the coordination sphere of the metal to obtain stable complexes: the reaction in Eq. (1) gave isolable products only for Ln = Yb and Lu and attempts to use larger metals or less substituted phenyl groups failed (17).

Recent advances in homoleptic organolanthanide chemistry have cen-tered primarily on the synthesis of new compounds. One major class of ligands recently explored involves the bulky, trimethylsilyl-substituted methyl ligands, $(CH_3)_3SiCH_2$— and $[(CH_3)_3Si]_2CH$—. Neutral complexes of formula $Ln(CH_2SiMe_3)_3(thf)_x$ (Ln = Y, Tb, Er, Tm, Yb, Lu; x = 0, 2, 3) (18–20) and $Y[CH(SiMe_3)_2]_3$ were reported, as well as anionic species such as $[Ln(CH_2SiMe_3)_4][Li(L)_4]$ (Ln = Y, Er, Yb, Lu; L = thf, Et$_2$O; L_2 = $Me_2NCH_2CH_2NMe_2$ = tmeda) (19–22) and the heteroleptic

{Ln[CH(SiMe₃)₂]₃Cl}{Li(thf)₄} (*19*). The ytterbium derivative of the latter complex was crystallographically shown to have a pseudotetrahedral arrangement of ligands around the metal with Yb—C bond lengths of 2.37–2.39 Å.

The CH_2SiMe_3 and $CH(SiMe_3)_2$ ligands were chosen because they lacked hydrogen on the β carbon and hence decomposition by β-hydrogen elimination was prevented. α-Hydrogen reactivity is possible, however, and elimination of $Si(CH_3)_4$ from some of these complexes has been reported to generate $CH(SiMe_3)^{2-}$ ligands from CH_2SiMe_3 species (*22*) [Eq. (2)]. None of the proposed alkylidene complexes have yielded to

$$Ln(CH_2SiMe_3)_3(thf)_2 \rightarrow SiMe_4 + [Ln(CH_2SiMe_3)(CHSiMe_3)]_n + 2 \, thf \qquad (2)$$

X-ray structural analysis, but the area of lanthanide alkylidene and alkylidyne complexes is likely to develop significantly in the next few years (*23*).

The choice of CH_2SiMe_3 and $CH(SiMe_3)_2$ ligands for organolanthanide synthesis was logical based on the facile β-hydrogen elimination reactivity observed in transition metal complexes. However, it was not known whether the lanthanide metals would have the same reactivity as the transition metals in this regard, since β-hydrogen elimination had not been demonstrated for organolanthanides. Indeed, when the *tert*-butyl group was examined as a bulky alkyl ligand, isolable complexes were obtainable [Eq. (3), Ln = Sm, Er, Y, Yb, Lu] (*24, 25*). These complexes displayed

$$LnCl_3 + 4 \, t\text{-}C_4H_9Li \xrightarrow{\text{THF}} [Ln(t\text{-}C_4H_9)_4][Li(thf)_4] + 3 \, LiCl \qquad (3)$$

remarkable stability considering that 36 β-hydrogens were present on a single metal center. For example, the erbium complex begins to slowly decompose only at 60°C in the solid state. Even more surprising were the results of studying the thermal decomposition of the least stable complex of the series, $[Sm(t\text{-}C_4H_9)_4][Li(thf)_4]$. After 16 hours in THF (tetrahydrofuran) at 40°C, ¹H-NMR spectroscopy indicated that complete decomposition had occurred with the formation of 3.25 equivalents of $(CH_3)_3CH$ per mole of $[Sm(t\text{-}C_4H_9)_4][Li(thf)_4]$ (*24*). The absence of equal amounts of $(CH_3)_2C{=}CH_2$ and $(CH_3)_3CH$, the expected β-hydrogen elimination products, suggests that β-hydrogen elimination is not the preferred mode of decomposition for this organolanthanide.

This was a clear example of the potential of organolanthanides to display unique reactivity patterns. The *tert*-butyl results also demonstrated that one need not necessarily expect the principles of organotransition metal chemistry to apply exactly to organolanthanide compounds. Indeed, one of the exciting aspects of this young organolanthanide area is that the principles of reactivity are still being established, i.e., current experiments are providing the data from which reactivity generalizations will ultimately be derived.

The *tert*-butyl ligands in $[Ln(t-C_4H_9)_4][Li(thf)_4]$ are good metallating agents and the reaction of these complexes with terminal alkynes was used to generate new homoleptic species [Eq. (4), R = C_6H_5, n-C_4H_9, t-C_4H_9] (26). This metallation provides a convenient halide-free synthesis of

$$[Ln(t\text{-}C_4H_9)_4][Li(thf)_4] + 4\ HC\equiv CR \rightarrow [Ln(C\equiv CR)_4][Li(thf)_4] + 4\ HC(CH_3)_3 \quad (4)$$

Ln—C bonds and has been usefully extended to a variety of other organolanthanide syntheses (see Sections IV,B, V,C, and VII,A).

In addition to the formally four-coordinate homoleptic complexes already described, six-coordinate species have been obtained using methyl ligands. Treatment of lanthanide trichlorides with a greater than sixfold excess of CH_3Li in the presence of tetramethylethylenediamine (TMEDA) forms complexes containing an octahedral $Ln(CH_3)_6^{3-}$ unit [Eq. (5), Ln = Pr, Nd, Sm, Er, Tm, Yb, Lu] (27–29). X-ray crystallographic studies

$$LnCl_3 + 6\ LiCH_3 + 3\ TMEDA \xrightarrow{Et_2O} [Ln(CH_3)_6][Li(tmeda)]_3 + 3\ LiCl \quad (5)$$

of the erbium and holmium derivatives revealed that each Li(tmeda) unit bridges two methyl groups to give an $Ln[(\mu\text{-}CH_3)_2Li(tmeda)]_3$ structure with 2.57(2) Å Ln—C distances in the case of erbium (28, 29).

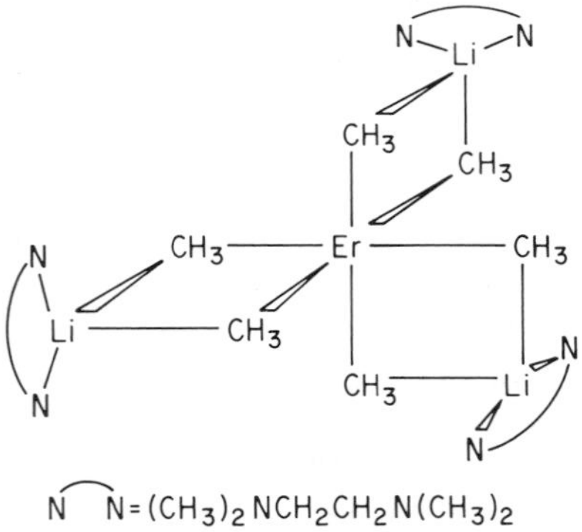

$$\stackrel{\frown}{N\quad N} = (CH_3)_2\ NCH_2CH_2\ N(CH_3)_2$$

Related to the six-coordinate methyl complexes are the ylide complexes $Ln[(CH_2)_2PMe_2]_3$ prepared according to Eqs. (6) and (7) (Ln = La, Pr, Nd, Sm, Gd, Ho, Er, and Lu) (30). Although structural data have not

$$LnCl_3 + 3\ Me_3P{=}CH_2 \rightarrow Ln(CH_2PMe_3)_3Cl_3 \quad (6)$$

$$Ln(CH_2PMe_3)_3Cl_3 + 3\ t\text{-}C_4H_9Li \rightarrow Ln[(CH_2)_2PMe_2]_3 + 3\ LiCl + 3\ (CH_3)_3CH \quad (7)$$

been reported, a trischelate structure similar to the structure of the $Ln[(\mu\text{-}CH_3)_2Li(tmeda)]_3$ complexes is likely.

With the exception of the metallation reaction in Eq. (4), little chemistry has been reported for homoleptic lanthanide complexes. With four to six reactive alkyl sites per metal center, these compounds are potentially much more reactive than the bis(cyclopentadienyl)alkyl compounds described in Section IV,B. This is an area of future growth in the organolanthanide field.

B. Heteroleptic Ln—C Complexes

The first examples of bis(cyclopentadienyl)alkyllanthanide complexes were synthesized nearly 10 years ago by allowing $[(C_5H_5)_2LnCl]_2$ to react with LiR (R = CH_3, C_6H_5, $C\equiv CC_6H_5$) (3). The form in which the $(C_5H_5)_2LnR$ unit is isolated, i.e., as an electron-deficient bridged dimer, $[(C_5H_5)_2Ln(\mu\text{-}R)]_2$ (31), or as a monomeric solvate, $(C_5H_5)_2LnR(solvent)$ (32–34), depends on whether the solvent of isolation will strongly coordinate to the metal and whether the R group is a good bridging ligand. From THF, monomeric THF adducts, $(C_5H_5)_2LnR(thf)$, are routinely isolated except when R is a group that forms particularly strong bridges, such as $C\equiv CR'$ (35, 36). From toluene, bridged dimers are generally isolated except for alkyl groups such as tert-butyl that have little tendency to bridge (32).

Recent advances in $(C_5H_5)_2LnR$ chemistry have involved primarily structural characterization and reactivity studies. The synthesis of new Ln—C complexes starting from lanthanide hydrides is described in Section V,C. The chemistry of $(C_5Me_5)_2LnR$ complexes is described in Section VII.

1. Structural Results

The X-ray crystal structures are now known for four dimeric bis(cyclopentadienyl)alkyl complexes: $[(C_5H_5)_2Yb(\mu\text{-}CH)_3]_2$ (31), $[(C_5H_5)_2Y(\mu\text{-}CH_3)]_2$ (31), $[(C_5H_5)_2Er(\mu\text{-}C\equiv CCMe_3)]_2$ (35), and $[(CH_3C_5H_4)_2Sm(\mu\text{-}C\equiv CCMe_3)]_2$ (36). The role of the hydrogen atoms in the $\mu\text{-}CH_3$ bridges was carefully studied in the first two examples and a simple, electron-deficient carbon-centered bridge was found with an average Yb—C distance of 2.511(35) Å (Scheme 1). The bridged alkynide structures proved to be unusual in that the $\mu\text{-}C\equiv CCMe_3$ bridge was asymmetrical with unequal A and B angles (Scheme 2) of 149(2)° and 115(2)° for the erbium complex and 151(1)° and 122(1)° for the samarium derivative. From simple steric and electrostatic considerations, the bulky $C\equiv CCMe_3$ bridges might be expected to reside symmetrically between the $(C_5H_4R)_2Ln$ units (R = H, CH_3) (Scheme 2).

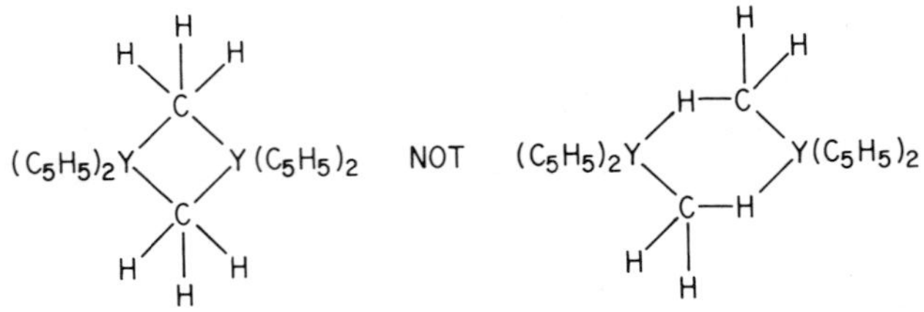

$(C_5H_5)_2Y$ ... $Y(C_5H_5)_2$ NOT $(C_5H_5)_2Y$ ... $Y(C_5H_5)_2$

SCHEME 1

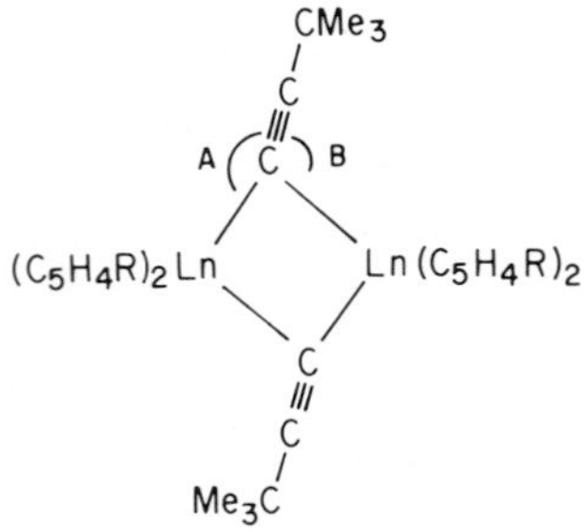

CMe$_3$

A (C) B

$(C_5H_4R)_2Ln$ ... $Ln(C_5H_4R)_2$

Me$_3$C

SCHEME 2

Four monomeric $(C_5H_5)_2LnR(thf)$ complexes have been character-ized by X-ray crystallography: $R = t\text{-}C_4H_9$ (*32*), CH_2SiMe_3 (*33*), and $4\text{-}CH_3C_6H_4$ (*33*) for Ln = Lu; and $R = CH_3$ for Ln = Yb (*34*). In each complex, the two cyclopentadienyl ring centroids, the THF oxygen atom, and the carbon bound to the metal roughly describe a tetrahedral geometry around the metal. The metal–carbon distance in $(C_5H_5)_2Lu(t\text{-}C_4H_9)(thf)$, 2.47(2) Å, is significantly longer than that in the other three monomeric alkyls, 2.34–2.38 Å, and more like the Lu—C distances in $Lu[C_6H_3(CH_3)_2]_4^-$, 2.4–2.5 Å (*16*). The longer distances may reflect increased steric crowding. How this lengthening translates into reactivity remains to be determined. The longer bond could be more reactive, but a sterically more saturated ligand environment may have diminished reac-tivity.

2. Reactivity

The *tert*-butyl complex $(C_5H_5)_2Lu(t\text{-}C_4H_9)(thf)$ has played a prominent role in studies of the reactivity of heteroleptic alkyllanthanides (*37*). One example is that this complex was the precursor to the first tractable products isolable from an organolanthanide CO reaction system (*38*). Hence, in contrast to the reaction of $(C_5H_5)_2LuCH_3(thf)$ with CO, which gives a complex product mixture, $(C_5H_5)_2Lu(t\text{-}C_4H_9)(thf)$ reacts cleanly with CO in a two-step process as shown in Eqs. (8) and (9) $(R = t\text{-}C_4H_9)$

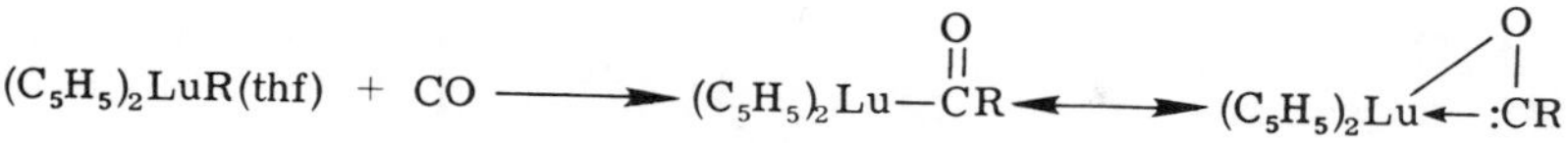

$$(C_5H_5)_2LuR(thf) + CO \longrightarrow (C_5H_5)_2Lu-\overset{\overset{O}{\|}}{C}R \longleftrightarrow (C_5H_5)_2Lu \leftarrow :CR \overset{\diagup O}{\underset{|}{}} \quad (8)$$

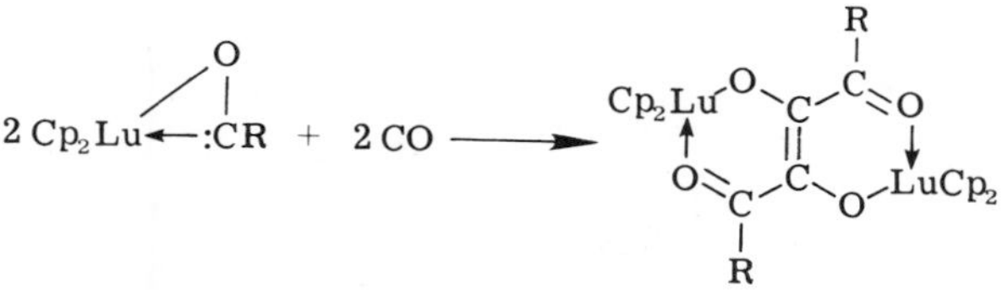

$$\text{(9)}$$

(38). The product isolated from the reaction of a 1:1 stoichiometric amount of CO with the lutetium complex was characterized as a mono-insertion product, an acyl. From spectroscopic data, a dihaptoacyl structure involving significant Lu–O interaction was postulated. With excess CO, a dimeric complex was isolated in which four CO molecules were coupled via one C=C double bond and two C—C single bonds to form an enedionediolate moiety. The condensation of these four CO molecules can be rationalized based on the carbenoid character of the dihaptoacyl carbon atom. This multiple coupling of CO has precedent in early transition metal (39) and actinide chemistry (40), but this was the first instance in which the C_5Me_5 ligand was not needed as a co-ligand.

A second major reaction in which $(C_5H_5)_2Ln(t\text{-}C_4H_9)(thf)$ was a crucial reactant was hydrogenolysis of the Ln—C bond (41). This reaction was important not only because it provided the first crystallographically characterized organolanthanide hydride complexes (see Section V), but also because it revealed some general principles of Ln—R reactivity. Since none of the lanthanide metals have both an $n+$ and $(n + 2)+$ oxidation state readily available, the reaction of Ln—R with hydrogen is unlikely to occur by an oxidative addition of H_2 followed by reductive elimination of RH. Instead, a "four-center" reaction [Eq. (10)] is more likely (42). Hence, hydrogenolysis provided an opportunity to study four-center reactions with the lanthanide metals.

$$\begin{array}{c} H\cdots H \\ \vdots \quad \vdots \\ Ln - R + H_2 \rightarrow Ln\cdots R \rightarrow Ln - H + RH \end{array} \quad (10)$$

The first organolanthanide hydrogenolysis reaction described in the literature involved the reaction of the dimeric methyl bridged complex, $[(CH_3C_5H_4)_2Yb(\mu\text{-}CH_3)]_2$, with H_2 in toluene (43). The reaction was found to be quite slow requiring approximately 2 weeks to be complete. During this time Yb^{3+}-H^- conversion to Yb^{2+} occurred [Eq. (11), $Cp' = CH_3C_5H_4$] and the reaction proved to be a good route to the divalent $(CH_3C_5H_4)_2Yb(thf)$ (Section VI), but not to a stable trivalent hydride. In contrast, the reaction of $(C_5H_5)_2Lu(t\text{-}C_4H_9)(thf)$ with H_2 in toluene occurs rapidly [Eq. (12), $Cp = C_5H_5$, $CH_3C_5H_4$] (40). The

$$[Cp'YbCH_3]_2 + 2 \ H_2 \xrightarrow[\text{weeks}]{2} 2 \ CH_4 + 2 \ [Cp'_2YbH] \rightarrow 2 \ Cp'_2Yb \qquad (11)$$

$$2 \ Cp_2Ln(t\text{-}C_4H_9)(thf) + 2 \ H_2 \xrightarrow{\text{toluene}} 2 \ (CH_3)_3CH + [Cp_2Ln(\mu\text{-}H)(thf)]_2 \qquad (12)$$

toluene-insoluble hydride complex, which is described in Section V, starts to precipitate soon after hydrogen addition and an 85% yield of 2-methylpropane and a 70–80% isolated yield of the hydride can be achieved within 24 hours. Reaction of the *tert*-butyl complex in THF, however, gave no hydrogenolysis over a 6-hour period. As described in the following paragraph, these hydrogenolysis reactions provided some of the first data on general patterns of reactivity of the Ln—C bond.

The enhanced reactivity of the monomeric *tert*-butyl complex compared to the dimeric methyl complex in the toluene reactions is consistent with the expected greater reactivity of a terminal alkyl ligand compared to a bridged alkyl. This trend is well established in alkyllithium chemistry where chelating nitrogen bases such as TMEDA are used routinely to enhance reactivity by converting alkyl bridged $(RLi)_n$ clusters to monomeric (tmeda)LiR species (44). The enhanced reactivity of $(C_5H_5)_2Lu(t\text{-}C_4H_9)(thf)$ in toluene versus THF suggests that even higher reactivity can be achieved with a terminal alkyl if it is attached to a sterically unsaturated, unsolvated metal center. Presumably, the toluene reaction differs from the THF reaction in that in toluene some dissociation of THF occurs to provide access to a highly reactive species like $(C_5H_5)_2$ $Lu(t\text{-}C_4H_9)$.

Obviously, a terminal alkyl on a coordinatively unsaturated, unsolvated metal will have favorable reactivity for steric reasons. Such a system is also likely to be reactive for electronic reasons. The electropositive lanthanide is very electrophilic and the rather ionic alkyl anion is very nucleophilic. The combination of two adjacent sites of high nucleophilicity and high electrophilicity, both with strong tendencies to coordinate/react further, should be exceptionally reactive, even with strong bonds with little dipolar character such as H—H or C—H (see Section VII).

The possibility of converting the methyl-bridged dimers $[(C_5H_5)_2Ln\text{-}(\mu\text{-}CH_3)]_2$ to more reactive species was investigated by attempting hydrogenolysis is mixed ether–alkane or ether–arene solvent systems which would generate more reactive $(C_5H_5)_2LnCH_3$ and $(C_5H_5)_2Ln(CH_3)(ether)$

species (45). In contrast to the reaction in pure arene solvent, hydrogenolysis is efficient in these mixed solvent systems. However, for $(C_5H_4R)_2Y(CH_3)(thf)$ (R = H, CH_3) hydrogenolysis is also efficient in THF, in contrast to the situation with the bulkier *tert*-butyl ligand. Hence, reactivity in bis(cyclopentadienyl)alkyl complexes is a sensitive function of ligand size, degree of oligomerization, solvation, and solvent. The basic principles of Ln—C bond reactivity suggested by these preliminary results in reactions with the H—H bond appear to be followed by other lanthanide-based reactions which may occur by "four-center" mechanisms, but the area is too young to assume these generalizations to be definitive.

A third major reaction of bis(cyclopentadienyl)alkyllanthanide complexes is metallation [Eq. (13) (35)], analogous to the reactions described

$$2 (C_5H_5)_2Er(CH_3)(thf) + 2 HC{\equiv}CR \rightarrow [(C_5H_5)_2Er(\mu\text{-}C{\equiv}CR)]_2 + 2 CH_4 \quad (13)$$

earlier for homoleptic complexes [Eq. (4)]. Whether or not these reactions occur by four-center mechanisms related to hydrogenolysis remains to be determined. An unusual variation of the metallation reaction is the reaction of $(C_5H_5)_3Yb$ with terminal alkynes $HC{\equiv}CR$ to form C_5H_6 and $[(C_5H_5)_2YbC{\equiv}CR]_x$ complexes (46). If this reaction does occur by a metallation mechanism, it goes in a direction opposite that predicted by the acidity of C_5H_6 and $HC{\equiv}CR$. Since ytterbium has an accessible divalent state, alternative mechanisms may be possible.

A fourth major reaction of $[(C_5H_5)_2LnR]_2$ complexes is the catalytic polymerization of alkenes [Eq. (14), R = H, Me, $SiMe_3$; $n = 1$ or 2] (47). Metallation of cyclopentadienyl rings [Eq. (15)] was observed as a

$$[(C_5H_4R)_2ErCH_3]_n + x\, H_2C{=}CH_2 \rightarrow [(C_5H_4R)_2Er(CH_2CH_2)_xCH_3]_n \rightarrow \text{polymer} \quad (14)$$

$$[(C_5H_4R)_2Ln(CH_2CH_2)_xCH_3]_2 \rightarrow$$
$$(C_5H_4R)Ln(\mu\text{-}C_5H_3R)_2Ln(C_5H_4R) + 2\, CH_3CH_2(CH_2CH_2)_{x-1}CH_3$$
$$(15)$$

deleterious side reaction leading to termination of the polymerization process. By using peralkylated cyclopentadienyl complexes, this ring metallation could be avoided (47) (also see Section VII). The $AlMe_3$ adduct, $(C_5H_4R)_2Ln(\mu\text{-}Me)_2AlMe_2$, was also found to initiate ethene polymerization (47, 48).

C. Summary

Many synthetic and structural aspects of the Ln—C bond are now well established. In regard to reactivity, the majority of reactions of the Ln—C bond in homoleptic or heteroleptic $[(C_5H_5)_2LnR]_n$ complexes fall into three categories: (a) formal insertion, demonstrated with CO and alkenes; (b) metallation of C—H bonds; and (c) hydrogenolysis. Since these three reactions have been demonstrated with only a few substrates, each area of reactivity should prove fertile for further development. The bulky, non-bridging *tert*-butyl group has proven to be a key ligand in the development of the principles of Ln—C chemistry. *tert*-Butyl complexes have demonstrated (a) that the traditional reactivity patterns of organotransition metal chemistry may not always be directly applicable in the lanthanide area; and (b) that reactivity of the Ln—C bond appears to be correlated with the availability of a nonbridging alkyl group on a sterically unsaturated, unsolvated metal center. As described in Section VII, examination of Ln—C reactions of the type described in this section but with C_5Me_5 groups as co-ligands tend to follow these principles with spectacular results.

V

SYNTHESIS AND REACTIVITY OF Ln—H BONDS

One of the most recent and rapidly developing areas of organolanthanide chemistry involves species containing Ln—H bonds. This area is so new that the information in this section is not presented in other review articles (5–7). Nevertheless, the area has developed so quickly that not only are synthetic and structural results available, but information on reactivity is also known.

Although binary, solid state, lanthanide hydrides (LnH_x) were known for many years (49), results suggesting the existence of molecular lanthanide hydrides were not obtained until 1978. The first evidence for the existence of organolanthanide hydride molecules arose from a study of catalytic hydrogenation by lanthanide metal vapor reaction products (50) (see Section VI). Homogeneous catalytic hydrogenation of alkynes was observed and the presence of lanthanide hydride intermediates was suggested (see Section V,C). These results and further lanthanide metal vapor reactions involving terminal alkynes (51) indicated that molecular lanthanide hydrides were accessible and initiated specific synthetic studies toward that goal.

A. *Dimeric Organolanthanide Hydrides*

As described in Section IV,B, hydrogenolysis of lanthanide—alkyl bonds in $(C_5H_4R)_2LnR$ complexes provided a synthetic route to organolanthanide hydrides [Eqs. (10)–(12)]. The structure of the first crystallographically characterized organolanthanide hydrides, $[(C_5H_4R)_2Ln(\mu\text{-}H)(thf)]_2$ (R = H, CH$_3$; Ln = Lu, Er, Y, Tb) (*41*), is shown in Fig. 1. Each metal is surrounded by two cyclopentadienyl rings, two bridging hydride ligands, and a terminal THF for a formal coordination number of nine electron pairs. X-Ray crystal structures were obtained for R = CH$_3$ and Ln = Er and Y, and in the yttrium case the hydride ligands were located crystallographically. Both the ^{1}H- and ^{89}Y-NMR spectra of $[(CH_3C_5H_4)_2Y(\mu\text{-}H)$ (thf)]$_2$ contain triplets (J_{YH} = 27.2 Hz) indicating that the dimeric structure involving formally electron deficient hydride bridges is retained in solution (*41, 52*).

The dimeric hydrides, $[(C_5H_4R)_2Ln(\mu\text{-}H)(thf)]_2$, could be obtained from $(C_5H_4R)_2Ln(t\text{-}C_4H_9)(thf)$ not only by hydrogenolysis but also by β-hydrogen elimination [Eq. (16)].

$$2\ (C_5H_4R)_2Ln(t\text{-}C_4H_9)(thf) \rightarrow [(C_5H_4R)_2Ln(\mu\text{-}H)(thf)]_2 + 2\ H_2C{=}C(CH_3)_2 \quad (16)$$

Hence, in contrast to the homoleptic *tert*-butyl complexes, $[Ln(t\text{-}C_4H_9)_4]$-$[Li(thf)_4]$ (Section IV,A), these heteroleptic species β-hydrogen eliminate. $(C_5H_5)_2Lu(t\text{-}C_4H_9)(thf)$ requires temperatures of 75°C to decompose according to Eq. (16) and the reaction is quite complex. The $(C_5H_5)_2$-$Ln(t\text{-}C_4H_9)(thf)$ complexes of the larger metals, Er and Y, are more reactive, since they are sterically less saturated, and β-elimination occurs at room temperature. Interestingly, the methylcyclopentadienyl derivatives, $(CH_3C_5H_4)_2Ln(t\text{-}C_4H_9)(thf)$, are less stable, i.e., more prone to β-hydrogen elimination, than their C$_5$H$_5$ counterparts even though they are formally more sterically saturated. This apparent exception to the principle

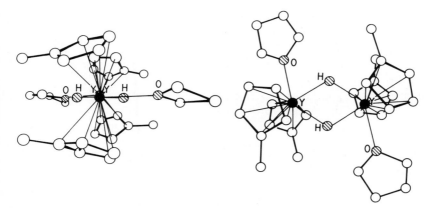

Fig. 1. Top view (right) and view down the Y—Y vector (left) of $[(CH_3C_5H_4)_2Y$-$(\mu\text{-}H)(thf)]_2$.

that steric saturation confers stability may result because the coordination environments in the methylcyclopentadienyl complexes are *sterically oversaturated*. As suggested by comparing Ln—C bond distances in $(C_5H_5)_2Lu(t\text{-}C_4H_9)(thf)$ and other $(C_5H_5)_2LnR(thf)$ species (Section IV,B), the bis(cyclopentadienyl)-*tert*-butyl complex is already sterically crowded. In the $CH_3C_5H_4$ analogue, further steric crowding may cause THF to dissociate. This would generate a sterically unsaturated, unsolvated terminal alkyl complex which, as described in Section IV,B, is one of the most reactive (i.e., unstable) forms of an organolanthanide. The relative stabilities of the heteroleptic *tert*-butyl complexes demonstrate an additional principle of organolanthanide reactivity: steric oversaturation of the metal coordination environment can lead to less stable species (cf. ref. 53). This has important consequences for the C_5Me_5 complexes described in Section VII.

B. *Trimeric Organolanthanide Hydrides*

If the $(C_5H_5)_2Ln(t\text{-}C_4H_9)(thf)$ complexes β-hydrogen eliminate in the presence of LiCl, a new type of organolanthanide complex is formed [Eq. (17), Cp = C_5H_5, Ln = Er,Y] (54). Since LiCl is a by-product in the

$$3\,Cp_2Ln[C(CH_3)_3]\,(thf) \xrightarrow{\text{LiCl}} [Li(thf)_4] \begin{bmatrix} Cp_2 \\ Ln \\ H \diagup | \diagdown H \\ | \diagdown H \diagup | \\ Cp_2Ln \diagdown_{Cl} \diagup LnCp_2 \end{bmatrix} \quad (17)$$

synthesis of $(C_5H_5)_2Ln(t\text{-}C_4H_9)(thf)$ from $[(C_5H_5)_2LnCl]_2$ and $t\text{-}C_4H_9Li$, and since the 50% yield of Eq. (17) requires only one LiCl per six lanthanides, it is rather easy to generate the trimer hydrides which incorporate LiCl. In fact, great care must be taken to remove all of the LiCl from the alkylorganolanthanide hydrogenolysis substrates if pure dimer hydrides, $[(C_5H_5)_2Ln(\mu\text{-}H)(thf)]_2$, are desired according to Eq. (12).

Inspection of the metal coordination environment in the crystallographically characterized $\{[(C_5H_5)_2ErH]_3Cl\}\{Li(thf)_4\}$ (Scheme 3) (54) reveals that it is similar to the environment in the dimeric hydrides discussed above (41). Each metal is formally nine-coordinate with two cyclopentadienyl groups, two bridging ligands (in this case two hydrides or a hydride and a chloride), and one additional ligand (in this case a bridging hydride or chloride instead of terminal THF) in the coordination sphere. Although the local metal coordination environment in the trimer has precedent in the

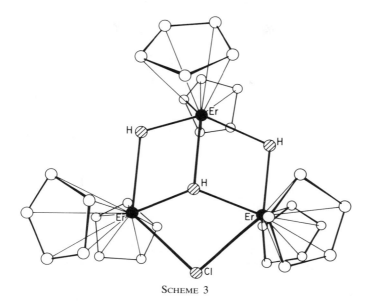

SCHEME 3

dimer structure, the overall result gives an unprecedented structure with respect to the central hydrogen. The position of this triply coordinated hydrogen coplanar with the three metal atoms is unique in metal hydride chemistry. μ_3-Hydrides in trimeric transition metal clusters must reside above the plane of the metals because generally there is insufficient room in the center of the triangle of metal–metal bonds to accommodate a hydrogen atom (e.g., ref. 55). Hence, organolanthanide chemistry provides an unusual, new, metallic coordination environment for hydrogen.

After the crystallographic establishment of the structure of $\{(C_5H_5)_2ErH]_3Cl\}\{Li(thf)_4\}$, the remarkable $(\mu_3\text{-H})Ln_3$ unit was found in another type of organolanthanide hydride species, the tetrahydride $\{[(C_5H_5)_2LuH]_3H\}\{Li(thf)_4\}$ (54). This complex was obtained from an experiment designed to explore the β-hydrogen elimination chemistry of $(C_5H_5)_2Ln(t\text{-}C_4H_9)(Et_2O)$. The synthesis of this complex was pursued in order to get a heteroleptic t-butyllutetium derivative which would β-hydrogen eliminate cleanly at a lower temperature than the THF complex, $(C_5H_5)_2Lu(t\text{-}C_4H_9)(thf)$, which required 75°C for decomposition. It was anticipated that Et_2O would not solvate as strongly as THF, and dissociation to form a highly reactive unsolvated, sterically unsaturated terminal alkyl, $(C_5H_5)_2Lu(t\text{-}C_4H_9)$, would result. This strategy was successful and the postulated $(C_5H_5)_2Lu(t\text{-}C_4H_9)(Et_2O)$ was too unstable to isolate. Decomposition of the *tert*-butyl species gave the trimer tetrahydride $\{[(C_5H_5)_2LuH]_3H\}\{Li(thf)_4\}$ according to Eq. (18), where Ln = Lu.

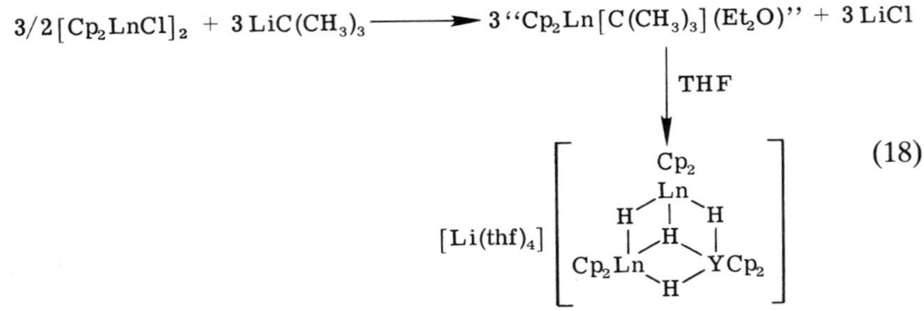

$$3/2\,[Cp_2LnCl]_2 + 3\,LiC(CH_3)_3 \longrightarrow 3\,\text{``}Cp_2Ln[C(CH_3)_3](Et_2O)\text{''} + 3\,LiCl$$

$$\Big\downarrow THF$$

$$[Li(thf)_4]\left[\begin{array}{c} Cp_2 \\ Ln \\ H\diagup\;\big|\;\diagdown H \\ |\quad\diagup H\diagdown\quad| \\ Cp_2Ln\diagdown\;\diagup YCp_2 \\ H \end{array}\right] \qquad (18)$$

The lutetium complex can also be obtained by hydrogenolysis in the presence of lithium reagents starting from $(C_5H_5)_2Lu(t\text{-}C_4H_9)(thf)$. The origin of the LiH incorporated into this trimer was more difficult to explain than the origin of LiCl in Eq. (17). Perhaps related to this is the fact that the yield in Eq. (18) was 12%.

Since the μ_3-hydrides may have interesting chemical and physical properties in addition to their remarkable structure, it was desirable to have a high-yield synthesis of these species. Investigation of the reactivity of the yttrium complexes $[(C_5H_5)_2Y(\mu\text{-}H)(thf)]_2$ with a variety of lithium reagents has led to a synthesis of $\{[(C_5H_5)_2YH]_3H\}\{Li(thf)_4\}$ in 75% yield according to Eq. (19) (56).

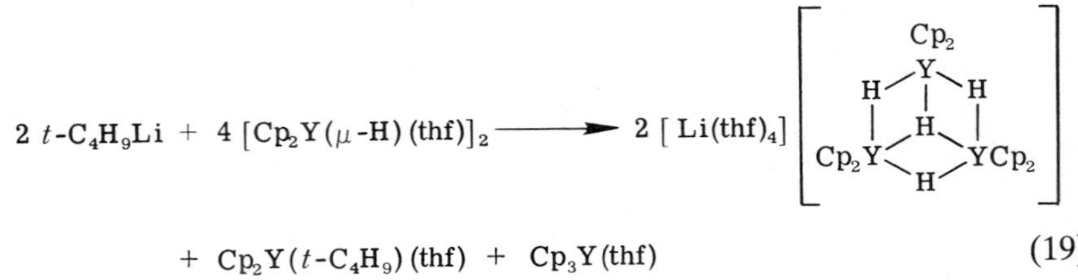

$$2\,t\text{-}C_4H_9Li + 4\,[Cp_2Y(\mu\text{-}H)(thf)]_2 \longrightarrow 2\,[Li(thf)_4]\left[\begin{array}{c} Cp_2 \\ Y \\ H\diagup\;\big|\;\diagdown H \\ |\quad\diagup H\diagdown\quad| \\ Cp_2Y\diagdown\;\diagup YCp_2 \\ H \end{array}\right]$$

$$+ Cp_2Y(t\text{-}C_4H_9)(thf) + Cp_3Y(thf) \qquad (19)$$

Scheme 4 provides one set of reactions consistent with known organolanthanide chemistry which may explain the observed synthesis. This mechanistic rationale is also consistent with the reaction of $t\text{-}C_4H_9Li$ with $[(C_5H_5)_2Y(\mu\text{-}D)(thf)]_2$, which forms the tetradeuteride $\{[(C_5H_5)_2YD]_3D\}\{Li(thf)_4\}$ exclusively, and with the reaction of CH_3Li with $[(C_5H_5)_2Y(\mu\text{-}H)(thf)]_2$, which forms a complex reaction mixture which may contain methyl-bridged trimers such as $\{[(C_5H_5)_2Y(\mu\text{-}CH_3)][(C_5H_5)_2Y(\mu\text{-}H)]_2(\mu_3\text{-}H)\}\{Li(thf)_4\}$. The *tert*-butyl group is a key ligand in Eq. (19) and Scheme 4 since alternative reaction pathways and products involving

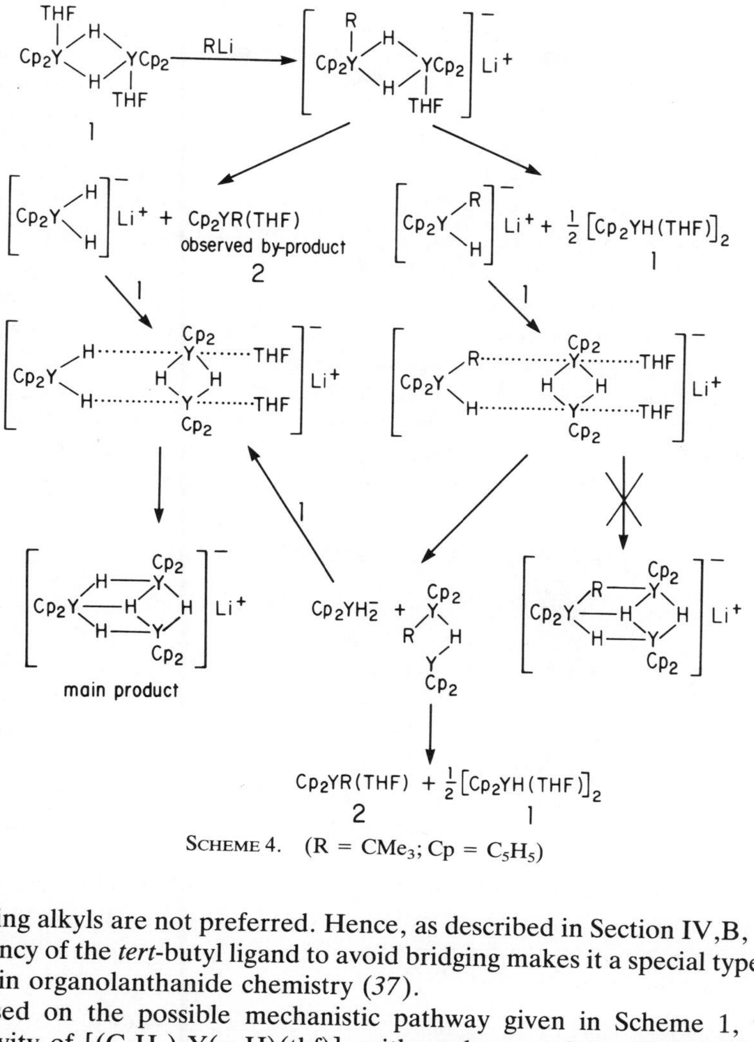

SCHEME 4. $(R = CMe_3; Cp = C_5H_5)$

bridging alkyls are not preferred. Hence, as described in Section IV,B, the tendency of the *tert*-butyl ligand to avoid bridging makes it a special type of alkyl in organolanthanide chemistry (37).

Based on the possible mechanistic pathway given in Scheme 1, the reactivity of $[(C_5H_5)_2Y(\mu\text{-H})(\text{thf})]_2$ with analogues of the $(C_5H_5)_2YH_2^-$ reagent was examined. Using a precursor to $(CH_3C_5H_4)_2ZrH_2$, a new class of intermetallic trimer tetrahydrides was obtained [Eq. (20), $Cp' = CH_3C_5H_4$] (56).

$$[Cp'_2ZrH_2]_2 + 2\left[Cp'_2Y(\mu\text{-H})(\text{thf})\right]_2 \longrightarrow 2 \begin{array}{c} Cp'_2 \\ Zr \\ H \overset{|}{\underset{|}{}} H \\ \overset{|}{\underset{H}{}} \\ Cp'_2Y \overset{}{\underset{H}{\diagup}} YCp'_2 \end{array} \qquad (20)$$

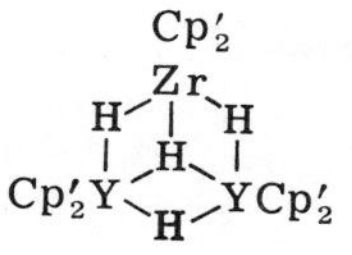

The synthesis of the trimer hydrides discussed above suggests that the $(\mu_3\text{-}H)Ln_3$ unit may constitute a common structural feature in organolanthanide chemistry. It should be possible to synthesize trimers with a wide variety of doubly bridging substituents. Other oligomeric structures may also be possible as long as similar coordination environments around the metal are maintained. Related to this, the high pressure hydrogenolysis of $(C_5H_5)_2Lu(CH_2SiMe_3)(thf)$ in THF (57), the reaction of $(C_5H_5)_2LuCl(thf)$ with NaH (57), and the reaction of $[(C_5H_5)_2Y(\mu\text{-}H)(thf)]_2$ with LiH (56) generate still other forms of $(C_5H_5)_2LnH$ species that differ from those discussed above based on IR and NMR spectroscopy, but whose structures have not yet been determined by X-ray diffraction (see also ref. 58). The polymetallic polyhydride complexes discussed in this section may prove to be prototypes for an extensive class of polynuclear lanthanide complexes. These organopolylanthanides may constitute a new class of polynuclear complexes distinct from metal–metal bonded transition metal clusters in that their polynuclear structure is maintained primarily by hydride bridges.

C. Reactivity of Cyclopentadienyllanthanide Hydrides

The hydrogen ligand in the organolanthanide complexes discussed in Sections V,A,B, is very hydridic. These complexes react rapidly with D_2O and CH_3I to generate HD and CH_4, respectively, in high quantitative yield (41). The basic organometallic reactivity patterns of the Ln—H bond have been investigated by surveying the reactivity of $[(C_5H_4R)_2Y(\mu\text{-}H)(thf)]_2$ with unsaturated hydrocarbons (45, 59). These reactions not only have provided information about the Ln—H bond, but also have generated new heteroleptic Ln—C species, extending the scope of material discussed in Section IV.

As expected, organolanthanide hydrides react with hydrocarbons containing acidic hydrogen by metallation [Eq. (21)] (45). This metallation reaction is analogous to reactions of organolanthanide alkyls with acidic hydrocarbons [Eqs. (4) and (13)].

$$[(C_5H_4R)_2Ln(\mu\text{-}H)(thf)]_2 + 2\,HC\equiv CR \rightarrow [(C_5H_4R)_2Ln(\mu\text{-}C\equiv CR)]_2 + 2\,H_2 \quad (21)$$

The most common mode of reactivity of Y—H with unsaturated hydrocarbons in the absence of acidic hydrogens is 1,2-addition as shown in Scheme 5. The hydrides react with alkenes to form alkyl complexes, with 1,2-propadiene to form allyl systems, and with alkynes to form cis-alkenyl complexes (45). Crystallographic confirmation of the 1,2-addition mode of Y—H with unsaturated hydrocarbons was obtained using tert-butylnitrile as a substrate. The structure of the alkylideneamido product, $[(C_5H_5)_2Y(\mu\text{-}N\!=\!CHCMe_3)]_2$, was determined by X-ray crystallography.

Cp₂Y⟨THF, CH₂CH₂R′

Cp₂Y⟨THF, (η³-CH₂CHCH₂)

[Cp₂Y(μ-H)(NC₅H₅)]₂

H₂C=CHR′

H₂C=C=CH₂

NC₅H₅ pentane

CH₃CH₂C≡CCH₂CH₃

[Cp₂Y(μ-H)(THF)]₂ — NC₅H₅/THF → Cp₂Y-N... → Cp₂Y-N...

Cp₂Y⟨THF, C=C, CH₃CH₂ CH₂CH₃, H

RC≡N

RNC

Cp₂Y-N⟨C=R,H / N-C... YCp₂

SCHEME 5. (Cp = C₅H₅ or CH₃C₅H₄; R = CMe₃)

Unfortunately, since the complex dimerized, the question of cis or trans addition could not be answered (45). With *tert*-butyl isocyanide as a substrate (59), the Y—H moiety again adds to the multiple bond, but in this case the structure of the product is such that the metal center is within bonding distance of both the carbon and the nitrogen atoms. Some structural parameters and three descriptions of possible bonding schemes for this species are shown in Scheme 6. This η^2-formimidoyl structure is the first crystallographically characterized example of η^2-coordination of a multiply bonded organic moiety within bonding distance of a metal of this type. The cyclopentadienyl- and methylcyclopentadienyl-substituted lanthanide hydrides, $[(C_5H_4R)_2Y(\mu\text{-H})(\text{thf})]_2$, do not react with unactivated

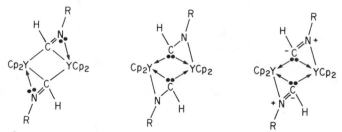

SCHEME 6. Average bond lengths: Y—C, 2.553(5) Å; C—N, 1.275(6) Å; Y—N, 2.325(4)Å. [Cp = C₅H₅, R = C(CH₃)₃]

benzene rings under mild conditions. Pyridine reacts in nonpolar solvents by displacement of THF to form pyridine solvates $[(C_5H_4R)_2Y(\mu\text{-H})(C_5H_5N)]_2$. In polar solvents, 1,2-addition is observed to occur (45).

If the 1,2-addition reaction of Ln—H to an alkyne to form an alkenyl complex is combined with a hydrogenolysis reaction, a cycle for catalytic hydrogenation of unsaturated substrates results [Eq. (22), R = H,

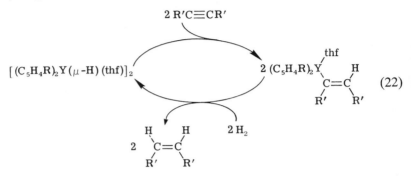

$$(22)$$

CH_3, $R' = C_2H_5$]. Indeed, $[(C_5H_4R)_2Y(\mu\text{-H})(thf)]_2$ and $(C_5H_4R)_2Y$-$[C(R')\!=\!CHR'](thf)$ can be interconverted in this way and both initiate catalytic hydrogenation of alkynes to cis alkenes (45). Hence, one plausible rationale exists for the lanthanide-based catalytic hydrogenation reaction which originally stimulated the study of molecular lanthanide hydrides (50, 60).

D. Summary

Since metal hydride complexes are crucial to a wide range of stoichiometric and catalytic organometallic reactions, the synthesis and characterization of organolanthanide hydride complexes were important to the overall development of organolanthanide chemistry. The Ln—H moiety is now solidly established in organolanthanide chemistry. The recent structural results indicate that the Ln—H unit will provide new coordination environments for hydrogen and new classes of polynuclear complexes for the lanthanides in general. The syntheses of molecular lanthanide hydride complexes have provided important information on organolanthanide reactivity. Hydrogenolysis reactions have demonstrated how terminal alkyl ligands on sterically unsaturated, unsolvated metal centers lead to significantly higher reactivity even in reactions with small substrates such as H_2. β-Hydrogen elimination reactions have shown how steric oversaturation can lead to enhanced reactivity. The systematic high-yield syntheses of the trimer tetrahydrides demonstrate how synthetic organolanthanide chemistry is evolving from exploratory synthesis to designed synthesis. All of these developments suggest that much interesting chemistry will be derived from the Ln—H bond in the future.

VI

LOW OXIDATION STATE CHEMISTRY

One of the major growth areas in organolanthanide chemistry involves the chemistry of the metals in lower oxidation states. Prior to 1978, with the exception of a few, brief synthetic reports on divalent species, organolanthanide chemistry involved almost exclusively the +3 oxidation state (5). In the late 1970s, major research efforts in low-valent lanthanide chemistry were initiated (14, 60) and they served to stimulate the development of the entire organolanthanide area.

There are two low-valent oxidation states available to the lanthanides under normal conditions: the +2 oxidation state and the formally zero oxidation state found in the elemental metals. The zero oxidation state is available to all the lanthanides, but only three members of the series have +2 oxidation states accessible under common organometallic reaction conditions: $Eu^{+2}(4f^7)$, $Yb^{+2}(4f^{14})$, and $Sm^{+2}(4f^6)$. The Ln^{+3}/Ln^{+2} reduction potentials [vs. normal hydrogen electrode (NHE)] (12), -0.34 V for Eu, -1.04 V for Yb and -1.50 V for Sm, indicate that Eu^{+2} is the most stable and Sm^{+2} the most reactive of these divalent ions. Sm^{2+} is also the most reactive based on radial size considerations, since it is the largest and most difficult to stabilize by steric saturation.

A. Zero Oxidation State Chemistry

1. Room Temperature, Bulk Metal Reactions

Some of the earliest studies of organolanthanide chemistry described reactions of the elemental metals with alkyl and aryl iodide reagents (RI) (61). Analysis of the soluble products obtained for Ln = Eu, Yb, Sm indicated a formula of primarily "RLnI" although it was acknowledged that this could represent a number of different species in equilibrium. The amount of contamination of the divalent product with trivalent species was observed to follow the order of stability of the divalent states: the Eu system was the cleanest, while the Sm system had only 50% of the metal in the divalent state. These species reacted like Grignard reagents. The reaction of RI with La and Ce metals gave trivalent products with formulas postulated to be $R_{3-x}LnI_x$ based on iodide analysis. A recent study of the reactions of RI with Nd, Pr, Gd, and Ho similarly gave a mixture of $R_{3-x}LnI_x$ species too complex to yield a single, fully characterizable complex (62).

A second, early approach to elemental lanthanide chemistry involved transmetallation reactions of Yb, Eu, and Sm with R_2Hg reagents [Eq. (23)] (63, 64).

$$Yb + (C_6F_5)_2Hg \xrightarrow{THF} [(C_6F_5)_2Yb(thf)_4] + Hg \qquad (23)$$

In addition to divalent perfluoroaryl products, divalent alkynide complexes, $Yb(C\equiv CR)_2$, were prepared by this route (65, 66).

2. Metal Vapor Reactions

The room temperature reactions of the bulk lanthanide metals discussed above were limited in scope. Transmetallation was demonstrated only for Ln = Yb, Eu, and Sm and the number of reagent classes which oxidized the elemental metals under mild conditions was limited. Furthermore, in the latter reactions, the reagents which oxidized the metals generally provided not just an organic ligand, but also some other ligating species (in the above cases, a halide) which invariably led to complex product mixtures from which it was difficult to isolate a single, well-characterized compound. These experimental limitations could be avoided by using the metal in the vapor state, however, and lanthanide metal vapor chemistry has proven to be a general method for developing new organolanthanide chemistry.

The metal vapor technique, in which a metal is vaporized from a resistively heated tungsten container under high vacuum and is co-condensed with a potential ligand at −125 to −196°C, had proven useful in the synthesis of a variety of unusual low-valent transition metal complexes (67–71). With lanthanide metals, this method not only has generated low oxidation state species, but it has also provided the opportunity to study zero-valent lanthanide chemistry on an atomic/molecular basis for the first time. These studies have been important in identifying new directions in organolanthanide chemistry.

Consideration of the chemical nature of zero-valent lanthanide metals raises some intriguing questions. The stability of zero oxidation state transition metal complexes depends in large part on the capacity of the metal to transfer its excess electron density back to the ligands via backbonding. Given the limited radial extension of the 4f orbitals (8, 9), the possibility of stabilizing a zero valent lanthanide complex by back-bonding of $4f$ electron density is remote. However, the atomic spectra of the lanthanides show that *in low oxidation states*, the $5d$ orbitals are close in energy to the $4f$ levels and that a variety of mixed $4f$ $5d$ $6s$ electron configurations are low in energy (72). Hence, it is possible that the valence electrons of a low valent lanthanide metal would possess $5d$ as well as $4f$ character, an electronic situation which would be novel among metals. For example, zero valent erbium in a neutral complex could have configurations such as $4f^{13}5d^1$, $4f^{12}5d^2$, $4f^{11}5d^26s^1$ as well as the $4f^{12}6s^2$ configuration of the elemental metal. A complex containing an uncharged $5d^1$ or $5d^2$ metal would be unprecedented.

One of the earliest studies of zero valent lanthanide metal vapor chemistry involved the matrix isolation reaction of the metals with CO (73, 74). Based on infrared data, a variety of zero valent metal carbonyl complexes, $Ln(CO)_n$, $n = 1$–6, were postulated. These were not preparative scale experiments, however, and the reaction products were not stable except at very low temperature. Hence, unambiguous confirmation of the formula and structure of these complexes could not be obtained.

A major effort to study the chemistry of the zero oxidation state lanthanides on a preparative scale involved their reactivity with neutral unsaturated hydrocarbons (14, 60). This class of reagents was of interest because reactions of unsaturated hydrocarbons with metals constitute such an important component of organometallic chemistry and because species such as alkenes and alkynes were not common as ligands or reactants in organolanthanide chemistry at that time.

Lanthanide metal vapor reactions with unsaturated hydrocarbons such as C_5H_6 (75) or C_8H_8 (76), which are readily reduced by electropositive metals to common anionic ligands, give products arising from the reduction of the ligand and oxidation of the metal [Eqs. (24)–(26)]. The products of

$$Ln + C_5H_6 \rightarrow Ln(C_5H_5)_n \quad (Ln = Yb, n = 2; Ln = Nd, n = 3) \quad (24)$$

$$Yb + C_8H_8 \rightarrow YbC_8H_8 \quad (25)$$

$$Ln + C_8H_8 \rightarrow Ln_2(C_8H_8)_3 \quad (Ln = La, Ce, Nd, Er) \quad (26)$$

Eqs. (24) and (25) are identical to those made by solution means (77). The trivalent cyclooctatetraenyl complexes, $Ln_2(C_8H_8)_3$, contain $Ln(C_8H_8)_2{}^-$ anions identical to species made from $LnCl_3$ and $K_2C_8H_8$ in solution, but the countercation is the new species, $C_8H_8Ln(thf)_2{}^+$. These form a tight ion pair as shown in Scheme 7.

In contrast to the above systems, lanthanide metal vapor reactions with unsaturated substrates *not* readily convertible to stable organic anions common as ligands generate some of the most unusual organolanthanide

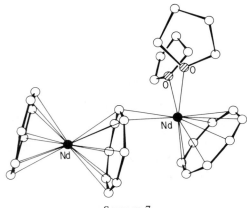

SCHEME 7

SCHEME 7

species known to date. Some typical reactions are shown in Eqs. (27) and (28) (78), (29) (79), and (30) and (31) (50).

$$Ln + H_2C{=}CH{-}CH{=}CH_2 \rightarrow Ln(C_4H_6)_3 \qquad (Ln = Nd, Sm, Er) \qquad (27)$$

$$Ln + H_2C{=}C(CH_3){-}C(CH_3){=}CH_2 \rightarrow Ln(C_6H_{10})_2 \qquad (Ln = La, Nd, Sm, Er) \quad (28)$$

$$Ln + CH_3CH{=}CH_2 \rightarrow Ln(C_3H_6)_3 \qquad (Ln = Er) \qquad (29)$$

$$Ln + CH_3CH_2C{\equiv}CCH_2CH_3 \rightarrow Ln_2(C_6H_{10})_3 \qquad (Ln = Nd, Er) \qquad (30)$$

$$Ln + CH_3CH_2C{\equiv}CCH_2CH_3 \rightarrow Ln(C_6H_{10}) \qquad (Ln = Sm, Yb) \qquad (31)$$

These metal vapor products differ from traditional organolanthanides in terms of stoichiometry, optical and magnetic properties, and solution behavior (14, 60). For example, in contrast to typical organolanthanide trends, the formal ligand-to-metal stoichiometries in these products are quite low, they vary from metal to metal, and they vary depending on the amount of methyl substitution on the ligand. These complexes are intensely colored compared to the normal pale colors of organolanthanides (10) and their magnetic moments are often outside the normal range of "free ion" values usually found for organolanthanides. These compounds are highly associated in solution, they are more highly associated in coordinating solvents like THF, and they oligomerize rather than crystallize in concentrated solution. The magnetic properties have prevented definitive characterization by NMR spectroscopy and the solution behavior so far has precluded X-ray diffraction studies. Hence, the precise nature of the metal ligand interaction in these complexes remains to be determined.

These studies clearly showed that (a) the lanthanide elements have interesting chemistry with unsaturated hydrocarbons, (b) zero-valent lanthanide atoms have variable reactivity within the series (cf. the similarity of the trivalent metals), and (c) zero-valent lanthanide reactivity differs from zero-valent transition metal chemistry (e.g., transition metal alkyne co-condensations result in polymerization of the ligand and rarely allow isolation of an organometallic complex) (14, 60). In addition, these metal vapor complexes were found to be the first f element complexes capable of homogeneous catalytic activation of hydrogen in hydrogenation reactions (14, 50). This stimulated further studies in organolanthanide chemistry since this was the first evidence that f elements had the capacity to function in homogeneous catalytic reactions involving small molecule transfer and provided the first evidence for the existence of soluble, molecular lanthanide hydrides, a class of complexes unknown at that time (see Section V).

The interaction of a lanthanide metal with a substrate such as 3-hexyne could occur in several ways (60): by π complex formation, by oxidative addition into a C—H bond, or by reduction involving radical species. Subsequent lanthanide metal vapor studies were designed to test some of these possibilities. Co-condensation of lanthanide metal vapor with reagents containing acidic hydrogen atoms, e.g., terminal alkynes, demonstrated that oxidative addition of C—H was a viable reaction [Eqs. (32) and (33)] (51). These reactions also provided access to a new class of

$$Yb + HC{\equiv}CR \rightarrow [HYb_2(C{\equiv}CR)_3)_n \tag{32}$$

$$Ln + HC{\equiv}CR \rightarrow [HLn(C{\equiv}CR)_2]_n \qquad (Ln = Er, Sm) \tag{33}$$

organolanthanide hydrides, complexes which were oligomerized presumably by strong alkynide bridges (35).

Lanthanide metal vapor co-condensation reactions with small hydrocarbons such as $H_2C{=}CH_2$, $H_2C{=}C{=}CH_2$, $CH_3CH{=}CH_2$ and $\overline{CH_2CH_2CH_2}$ were also studied to examine the importance of substituents (or lack of substituents) on the reactions (79). The reactivity of lanthanide metals with these smaller substrates is extensive. A variety of sites on the hydrocarbon are evidently attacked and the resulting organolanthanide products are not soluble, in contrast to the previously described lanthanide metal vapor reaction products. Characterization of the organolanthanide products by hydrolysis indicated the following reactions were occurring: oxidative addition into C—H bonds, cleavage of carbon–carbon multiple bonds, homologation, oligomerization, dehydrogenation, and ring opening. One of the more remarkable reactions in this study was the reaction of erbium with the rather inert C—H bonds of cyclopropane. Although the study of small hydrocarbon reactions did not yield crystallizable products, it did provide important information about the substantial reactivity accessible with the lanthanide elements. As stated in the conclusion of the paper on this research (*79*), these reactions "defined a set of conditions under which a variety of hydrocarbon activation reactions take place in the presence of the lanthanide metals. Obviously, the challenge in this area is to control this reactivity so that it can be used selectively." Subsequent research has suggested this is possible (see Section VII).

B. *Divalent Chemistry*

Early divalent organolanthanide chemistry involved the synthesis of the following complexes: $(C_5H_5)_2Ln$, $(C_5H_5)_2LnB$, and C_8H_8Ln, where Ln = Eu, Yb; B = NH_3, thf, and $[(C_5H_5)_2Sm]_n$. The syntheses all involved rather strongly reducing conditions. The Eu and Yb complexes were prepared by dissolving the metal in liquid ammonia and adding C_5H_6 (*80, 81*) or C_8H_8 (*82*). $(C_5H_5)_2Yb$ was also prepared from $[(C_5H_5)_2YbCl]_2$ or $(C_5H_5)_3Yb$ by Na or Yb reduction (*83*). $[(C_5H_5)_2Sm]_n$ was prepared from $(C_5H_5)_3Sm$ by reduction with potassium (*84*). Since $[(C_5H_5)_2Sm]_n$ and the C_8H_8Ln complexes were insoluble in all solvents with which they did not react, their chemistry was not investigated. Early research on $(C_5H_5)_2Yb$ and $(C_5H_5)_2Eu$ focused largely on their color and optical spectra, properties which were sensitive to the nature of the complexed base (*83*).

Recent developments in the chemistry of divalent lanthanide complexes of simple cyclopentadienyl ligands have involved new syntheses and structural characterization. As described in Section VII, by using the C_5Me_5 ligand, advances in reactivity as well as synthesis and structure have been made.

Equations (34)–(36) (*43*) show three new syntheses for $(CH_3C_5H_4)_2Yb$ which involve conditions less drastic than the alkali metal reductions

$$[(CH_3C_5H_4)_2YbCH_3]_2 \xrightarrow[\substack{toluene \\ 5°C}]{h\nu} (CH_3C_5H_4)_2Yb \tag{34}$$

$$[(CH_3C_5H_4)_2YbCH_3]_2 + H_2 \rightarrow CH_4 + [(CH_3C_5H_4)_2YbH] \rightarrow (CH_3C_5H_4)_2Yb \tag{35}$$

$$[(CH_3C_5H_4)_2YbCH_3]_2 \xrightarrow{80°C} (CH_3C_5H_4)_2Yb \tag{36}$$

previously employed. These reactions are significant for any catalytic cycle involving an Ln^{+3}/Ln^{+2} couple. In such a cycle, the critical reactive component is the divalent species. Facile access to Yb^{+2} via photolysis, hydrogenolysis, or thermolysis of Yb^{+3}–alkyl complexes extends the range of possibilities for cycles involving Ln^{+2}-based reactions. These results also indicate that Yb^{3+} reactions, under the appropriate conditions, may involve Yb^{2+} species as well.

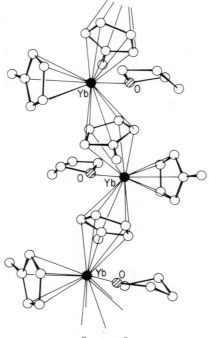

SCHEME 8

Although the first divalent organolanthanides were synthesized in 1965, it was not until 1980 that the first structure of a Ln(II) organometallic complex was determined. In that year, structures of three Yb(II) complexes were reported: $(CH_3C_5H_4)_2Yb(thf)$ *(43)*, $[C_5H_3(SiMe_3)_2]_2Yb(thf)$ *(85)*, and $(C_5Me_5)_2Yb(thf)(toluene)_{0.5}$ *(86)*. The latter two structures with the bulky substituted cyclopentadienyl rings crystallize as monomers. In contrast, $(CH_3C_5H_4)_2Yb(thf)$ oligomerizes in the solid state via bridging cyclopentadienyl rings which are coordinated to metals on both sides. Three monomeric units of the extended structure are shown in Scheme 8. Recently, the structure of $(C_5H_5)_2Yb(CH_3OCH_2CH_2OCH_3)$ has been described. With the bidentate ether, a monomeric formally ten-coordinate structure is observed *(87)*.

C. Summary

The exploration of low-valent lanthanide chemistry opened up a wide range of possibilities in organolanthanide chemistry. These studies demonstrated that unsaturated hydrocarbons were clearly viable reactants/substrates, that a variety of reactions with C—H and C—C bonds were possible, that small molecule homogeneous catalysis occurred with these elements, that Ln^{+3}/Ln^{+2} reaction cycles were conceivable, and that low-valent species were not so ligand deficient and highly oligomerized that they could not be structurally characterized with common ligands. In addition, these low-valent results stimulated a variety of other organometallic studies involving the trivalent metals. Additional low-valent chemistry involving the C_5Me_5 ligand will be described in Section VII,B.

VII

PENTAMETHYLCYCLOPENTADIENYLLANTHANIDE CHEMISTRY

The pentamethylcyclopentadienyl ligand, C_5Me_5, has proven to be of great importance in organometallic chemistry (88–91). Of particular relevance here is the research of Bercaw and co-workers, who demonstrated how this ligand could confer stability, solubility, and crystallinity to titanium and zirconium systems difficult to fully characterize using simple unsubstituted C_5H_5 ligands (90). Subsequently, Marks used this ligand similarly to develop an extensive chemistry for uranium and thorium (91). Since steric saturation is so important in organolanthanide chemistry, the sterically bulky C_5Me_5 ligand clearly had something to offer to the $4f$ elements. This ligand has had a major impact on all the areas discussed previously in this chapter and is intimately involved in many of the major recent advances in the area. To date, C_5Me_5-lanthanide chemistry has involved primarily preparing the C_5H_5 analogues and examining the special properties of these modified derivatives.

A. Trivalent C_5Me_5 Chemistry

Since three C_5Me_5 rings are predicted to be too large to fit around a lanthanide metal (92), C_5Me_5-lanthanide chemistry has involved primarily bis- and mono-ring species. One of the first C_5Me_5-organolanthanide studies involved the use of this ligand to extend the $[(C_5H_5)_2LnCl]_2$ system to metals larger than Sm. For Ln = La, Ce, Pr, and Nd, $[(C_5H_5)_2LnCl]_2$ is unstable with respect to disproportionation to the more sterically saturated $(C_5H_5)_3Ln$ complexes (2). Using C_5Me_5, a bis(ring)halide complex of Nd was isolated as a LiCl adduct, $(C_5Me_5)_2NdCl_2Li(thf)_2$ [Eq. (37)] (93).

$$NdCl_3 + 2\ LiC_5Me_5 \rightarrow (C_5Me_5)_2NdCl_2Li(thf)_2 + LiCl \qquad (37)$$

Subsequently, the alkali metal free systems $(C_5Me_5)_2NdCl(thf)$ and

$(C_5Me_5)_2NdN(SiMe_3)_2$ were isolated (92) along with $(C_5Me_5)_2SmX_2ML_y$ (92) and an extensive series of ytterbium derivatives $(C_5Me_5)_2YbX_2ML_y$, where X = halide, M = alkali metal, L = ethers or amines (94). X-Ray crystal structures of $(C_5Me_5)_2Yb(\mu\text{-}I)_2Li(Et_2O)_2$, $(C_5Me_5)_2Yb(\mu\text{-}Cl)_2Li$-$(Et_2O)_2$, and $(C_5Me_5)_2Yb(\mu\text{-}Cl)_2AlCl_2$ were obtained (94) and had the general structure shown in Scheme 9. Comparison of the first two members of this series indicated that the $(C_5Me_5)_2Yb(\mu\text{-}X)_2$ unit was rather sterically crowded when X was larger than Cl. As described below, the fact that the large C_5Me_5 ligands caused steric congestion in $(ring)_2Ln(bridge)_2$ structures of the later, smaller lanthanides proved to have important consequences on reactivity. X-Ray crystal structures of $(C_5H_4SiPh_2Me)_2Yb(\mu\text{-}Cl)_2Li(Et_2O)_2$ (94), $\{[C_5H_3(SiMe_3)_2]_2PrCl\}_2$ (95), $[C_5H_3(SiMe_3)_2]_2Nd(\mu\text{-}Cl)_2Li(thf)_2$ (96), and $\{[C_5H_3(SiMe_3)_2]_2NdCl_2\}$-$\{AsPh_4\}$ (97) indicated that mono- and disubstitution of cyclopentadienyl rings with bulky R_3Si groups could give the same effect as pentamethyl substitution. The $(C_5Me_5)_2LnX_2M(Et_2O)_2$ complexes have been used as precursors to $(C_5Me_5)_2Yb(O_2CR)$ $(R = Me_3C$ and $CF_3)$ and to $(C_5Me_5)_2Ln(S_2CNEt_2)(Ln = Yb, Nd)$ via ionic metathesis reactions (98). The mono-ring complexes $[(C_5Me_5)NdCl_3][Na(Et_2O)_2]$ (92), $(C_5Me_5)Nd[N(SiMe_3)_2]_2$ (92), and $(C_5Me_5)YbX_3LiL_y$ (94) have also been reported.

Following the precedent set in C_5H_5-lanthanide chemistry, the next step to developing the chemistry of these C_5Me_5 complexes was the conversion of the halides to alkyls and then the alkyls to hydrides via hydrogenolysis. This goal was initially approached using Ln = Yb and Lu, since these are the smallest lanthanides and hence the easiest to sterically saturate (99). Based on the traditional principles of organolanthanide chemistry, this choice of small metals and large ligand should have given the most stable alkyl and hydride derivatives. It did not. The first indication that this combination would not work in a traditional fashion was the complex synthesis needed to obtain a simple methyl derivative from a chloride

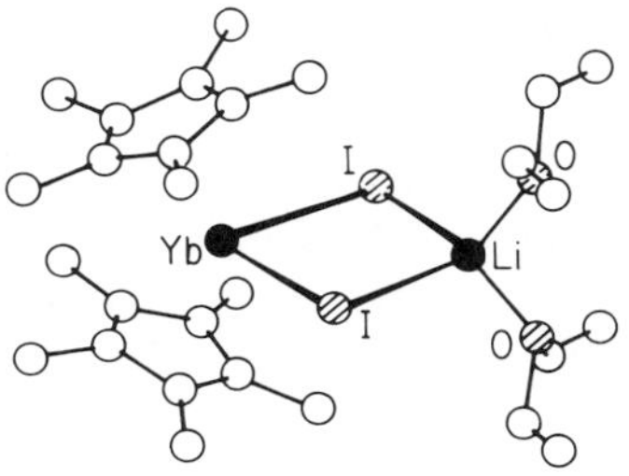

SCHEME 9

precursor. Although CH_3Li reacts with $(C_5Me_5)_2Lu(\mu\text{-}Cl_2)Li(Et_2O)_2$ in THF below $-20°C$ to form $(C_5Me_5)_2LuCH_3(thf)$ (99), the preparative scale synthesis used for $(C_5Me_5)_2LuCH_3$ complexes has been the sequence in Eq. (38) (100, 101). These reactions are evidently rather dependent on

$$(C_5Me_5)_2Lu(\mu\text{-}Cl)_2Li(Et_2O)_2 + 2\ CH_3Li \rightarrow (C_5Me_5)_2Lu(CH_3)_2Li(thf)_3 \xrightarrow[\substack{\text{vacuum} \\ 1-2\ \text{days}}]{75°C}$$

$$[(C_5Me_5)_2Lu(CH_3)_2]Li \xrightarrow{AlMe_3} (C_5Me_5)_2LuMe_2AlMe_2 \xrightarrow{Et_2O}$$

$$(C_5Me_5)_2LuCH_3(Et_2O) \xrightarrow{THF} (C_5Me_5)_2LuCH_3(thf) \quad (38)$$

solvent, alkali metal, halide, and temperature. For example, although $(C_5Me_5)_2Lu(CH_3)_2Li(Et_2O)_2$ is reported to react with $AlMe_3$ to form $(C_5Me_5)_2LuCH_3(thf)$ and $AlMe_4^-$, it will not react with excess $AlMe_3$ to form $(C_5Me_5)_2LuMe_2AlMe_2$ (99), which is accessible from $[(C_5Me_5)_2Lu\text{-}(CH_3)_2]Li$ in Eq. (38) (100). Many details of this chemistry remain to be described.

Much of the important chemistry of $(C_5Me_5)_2LuCH_3(Et_2O)$ was elucidated by studying its alkene polymerization reactions (100). Earlier studies of the reactivity of $[(C_5H_4R)_2LnCH_3]_2$ complexes ($R = H$, CH_3, $SiMe_3$) had shown that alkyllanthanides could polymerize ethene (47). As described in Section IV,B, catalyst deactivation occurred by metallation of the cyclopentadienyl rings to form $\eta^5,\eta^1\text{-}C_5H_3R$ species. In that study, $(C_5Me_4Et)_2LnR$ complexes were examined to prevent this and were found to be longer lived (47).

Consistent with the earlier work, $(C_5Me_5)_2LuCH_3(Et_2O)$ polymerizes ethene and oligomerizes propene (100). NMR analysis of the kinetics of the propene reactions was consistent with dissociation of Et_2O followed by insertion of the alkene into the lanthanide alkyl bond [Eqs. (39) and (40)] (100). The rates were such that the first few oligomerization products

$$(C_5Me_5)_2LuCH_3(Et_2O) \rightarrow (C_5Me_5)_2LuCH_3 + Et_2O \quad (39)$$

$$(C_5Me_5)_2LuCH_3 + CH_3CH{=}CH_2 \rightarrow (C_5Me_5)_2LuCH_2CH(CH_3)_2 \quad (40)$$

$$(C_5Me)_2LuCH_2CH(CH_3)_2 + CH_3CH{=}CH_2 + (C_5Me_5)_2LuCH_2CH(CH_3)CH_2CH(CH_3)_2$$

$$(41)$$

could be observed by NMR spectroscopy [Eqs. (40) and (41)]. Hence, this lanthanide-based system provided an excellent example of direct alkene insertion into metal–carbon bonds. As such, this system constitutes one of the few clear-cut experimental models for direct insertion as the primary mechanism of alkene polymerization by Ziegler–Natta catalysis (102). The result was particularly important because the possibility of alternative reaction pathways for this insertion involving oxidation state changes could be excluded for lutetium. Hence, the fact that the lanthanide metal had a limited redox chemistry could be used to advantage.

Further studies (*103*) of this system showed that the latter reaction was reversible, i.e., β-alkyl elimination was occurring [Eqs. (42) and (43)].

$$(C_5Me_5)_2LuCH_2CH(CH_3)_2 \rightarrow (C_5Me_5)_2LuCH_3 + CH_2{=}CHCH_3 \qquad (42)$$

$$(C_5Me_5)_2LuCH_2CH(CH_3)CH_2CH_2CH_3 \rightarrow (C_5Me_5)_2LuCH_3 + CH_2{=}CHCH_2CH_2CH_3$$

$$(43)$$

Except for a 1960 report on one alkylaluminum system (*104*), no β-alkyl elimination reactions of this type had ever been observed. Thus, this reaction provided another example in which lanthanide systems demonstrated special organometallic chemistry.

The propene oligomerization reaction was complicated by a variety of side reactions and by-products. Some of these were attributed to β-hydrogen elimination reactions leading to $(C_5Me_5)_2LuH$. In contrast to the crystallographically characterized species $[(CH_3C_5H_4)_2Ln(\mu\text{-}H)(thf)]_2$ (*41*), $\{[(C_5H_5)_2LnH]_3H\}\{Li(thf)_4\}$ (*54*) (Section V), and $[(C_5Me_5SmH]_2$ (*105*) (see section VII,B), this species could not be readily isolated (*100*). Attempts to independently synthesize it from $(C_5Me_5)_2LuCH_3(Et_2O)$ by hydrogenolysis, the route which generated the $[(C_5H_4R)_2Ln(\mu\text{-}H)(thf)]_2$ hydrides (*41*), failed. Instead, $(C_5Me_5)_2LuOC_2H_5$ was isolated (*100*). The reactivity of $(C_2H_5)_2O$ in this system was avoided by making the ether-free methyl complex $(C_5Me_5)_2LuCH_3$ by desolvation of $(C_5Me_5)_2LuCH_3(NEt_3)$ [Eq. (44)] (*101*). Hydrogenolysis of $(C_5Me_5)_2LuCH_3$ generated in this way

$$(C_5Me_5)_2LuCH_3(Et_2O) \xrightarrow{\text{NEt}_3 \text{ vacuum}} (C_5Me_5)_2LuCH_3 \qquad (44)$$

forms $(C_5Me_5)_2LuH$ (*103*). Both of these complexes are reported to exist in solution in a monomer/asymmetrical dimer equilibrium: $2\ (C_5Me_5)_2\text{-}LuZ \rightleftharpoons (C_5Me_5)_2ZLu(\mu\text{-}Z)Lu(C_5Me_5)_2$ (Z = H, CH$_3$) (*103, 106*). These dimers are postulated to adopt the asymmetrical structure, because a symmetrically bridged dimer is too sterically crowded when the bulky C_5Me_5 rings are present. Hence, this system constitutes another example of steric oversaturation (see Section V,A). This has important consequences on reactivity. As a monomer, the $(C_5Me_5)_2LuZ$ complexes have a nonbridged alkyl (or hydride) group on a sterically unsaturated metal center. As demonstrated in Ln—R hydrogenolysis studies (*41, 45*) and as is common in alkyllithium chemistry (*44*), this leads to high reactivity (see Section IV,B,2). The asymmetrical dimer also has a reactive terminal alkyl (or hydride) group which is adjacent to a sterically unsaturated metal center. This, too, would be expected to have high reactivity. Due to the electropositive nature of the metals and the ionic nature of the complexes, these $(C_5Me_5)_2LuZ$ complexes have adjacent sites of high nucleophilicity

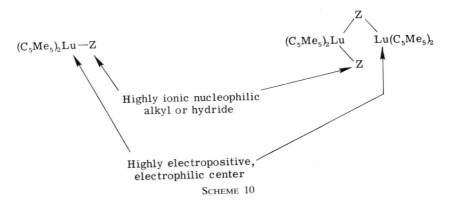

Highly ionic nucleophilic
alkyl or hydride

Highly electropositive,
electrophilic center

SCHEME 10

and electrophilicity and should be ideally suited for high reactivity, even with molecules with little dipolar character (Scheme 10) (*45*). Indeed, the $(C_5Me_5)_2LuZ$ complexes metallate a wide range of substrates (*107*) that ordinarily can be metallated only by the most active alkyllithium reagents in the presence of coordinating bases like TMEDA (*44*). Examples are shown in Eqs. (45)–(50). Although these metallation reactions have

$$(C_5Me_5)_2LuZ + SiMe_4 \rightarrow (C_5Me_5)_2LuCH_2SiMe_3 + ZH \tag{45}$$

$$(C_5Me_5)_2LuZ + NC_5H_5 \longrightarrow (C_5Me_5)_2Lu \overset{\displaystyle H}{\underset{\displaystyle H}{\overbrace{}}} + ZH \tag{46}$$

$$(C_5Me_5)_2LuZ + C_6H_6 \rightarrow (C_5Me_5)_2LuC_6H_5 + ZH \tag{47}$$

$$(C_5Me_5)_2LuC_6H_5 + (C_5Me_5)_2LuZ$$

$$(C_5Me_5)_2Lu - C\overset{HC-CH}{\underset{HC=CH}{\diagdown}}C - Lu(C_5Me_5)_2 + HZ \tag{48}$$

$$(C_5Me_5)_2LuH + D_2 \rightarrow (C_5Me_5)_2LuD + HD \tag{49}$$

$$(C_5Me_5)_2LuH + Et_2O \rightarrow (C_5Me_5)_2LuOEt + C_2H_6 \tag{50}$$

precedent with other metals (*44*), the reaction shown in Eq. (51) does not:

$$(C_5Me_5)_2LuZ + {}^{13}CH_4 \rightarrow (C_5Me_5)_2Lu^{13}CH_3 + ZH \tag{51}$$

This reaction was "the first well-characterized example of the reaction of methane with a homogeneous organometallic complex" (*106*) and provided a clear-cut example of the unique chemistry possible with the lanthanide metals.

B. Low-Valent C_5Me_5 Chemistry

1. Ytterbium and Europium Complexes

The C_5Me_5 ligand was initially used in the area of low-valent lanthanide chemistry to aid in elucidating the syntheses and structures of $(C_5Me_5)_xYb(halide)_y$ species (*99*). In a reaction related to the early studies of Yb plus alkyl halides (*61*), Yb metal was allowed to react with C_5Me_5I in the presence of LiI. With a 15-hour reaction time, $C_5Me_5YbI_3Li(Et_2O)_2$ was isolated in 30% yield. A 39-hour reaction period gave a product with a different distribution of ligands, $(C_5Me_5)_2Yb(\mu\text{-}I)_2Li(Et_2O)_2$, again in 30% yield (*99*). As described in the previous section, this complex was characterized by X-ray crystallography (*94*). Hence, in contrast to the alkyl and aryl halide oxidation reactions, by using the sterically bulky C_5Me_5 ligand as the organic constituent in these oxidations, single crystallizable products could be isolated, albeit in relatively low yield.

Starting from $YbBr_2(thf)_x$, which is conveniently prepared from Yb and $BrCH_2CH_2Br$, divalent $(C_5Me_5)_2YbL_y$ derivatives can be prepared directly by ionic metathesis [Eq. (52)] (*99*).

$$YbBr_2 + KC_5Me_5 \rightarrow [KYb(C_5Me_5)Br_2] \xrightarrow{KC_5Me_5} (C_5Me_5)_2YbL_y + 2 KBr \quad (52)$$

Interestingly, one C_5Me_5 ring of $(C_5Me_5)_2YbL_y$ can be displaced by reaction with LiI or $LiCH_3$ [Eq. (53), Q = I or CH_3] (*99*).

$$(C_5Me_5)_2YbL_y + 2 LiQ \rightarrow [(C_5Me_5)YbQ_2]LiL_y + LiC_5Me_5 \quad (53)$$

The ionic metatheses between lanthanide dihalides and alkali metal salts of C_5Me_5 are quite sensitive to reaction conditions. Although $(C_5Me_5)_2YbL_y$ can be prepared from $YbBr_2$ and LiC_5Me_5 and $YbCl_2$ and NaC_5Me_5 in THF (*86*), the reactions of $YbCl_2$ and NaC_5Me_5 in Et_2O and $YbCl_2$ and LiC_5Me_5 in THF fail to give isolable compounds (*86*). The europium system is more complex. The divalent product $(C_5Me_5)_2Eu(thf)$ is best prepared from trivalent $EuCl_3$ and three equivalents of NaC_5Me_5 in THF. The reaction fails in toluene and when LiC_5Me_5 is used as the reagent. The $EuCl_2/NaC_5Me_5/THF$ reaction analogous to the successful synthesis of $(C_5Me_5)_2Yb(thf)$ fails to give an isolable compound (*86*).

X-Ray crystal structure determinations of $(C_5Me_5)_2Yb(thf) \cdot (toluene)_{0.5}$ (*86*) and $(C_5Me_5)_2Yb(pyridine)_2$ (*108*) revealed monomeric complexes in contrast to the oligomeric structure of the less highly substituted derivative $(CH_3C_5H_4)_2Yb(thf)$ (43) (Section VI,B).

The $(C_5Me_5)_2Yb(solvent)$ complexes have been subjected to a wide range of reactions. Oxidation of $(C_5Me_5)_2Yb(dme)$ (dme = 1,2-dimethoxyethane) with $[(C_5H_5)_2Fe][PF_6]$ gave the cationic trivalent organolanthanide complex $[(C_5Me_5)_2Yb(dme)][PF_6]$. This species should be an ideal precursor to new trivalent products although attempts to make trivalent hydride and alkyl derivatives from this species using KH and $(CH_3)_3CCH_2Li$ regenerated the divalent product $(C_5Me_5)_2Yb(dme)$ (99).

The reaction of $Me_2PCH_2CH_2PMe_2$ with $(C_5Me_5)_2Yb(OEt_2)$ gave an insoluble phosphine complex, but using $Me_2PCH_2PMe_2$ the hydrocarbon-soluble $(C_5Me_5)_2Yb(Me_2PCH_2PMe_2)$ was obtained and structurally characterized (109). This product was the first well-characterized divalent organolanthanide phosphine derivative and demonstrated that a phosphorus ligand could displace an oxygen donor ligand from these traditionally oxophilic metals. The europium analogue was made similarly. In contrast to the phosphine reaction, neither CO, NO, $CH_2{=}CH_2$, nor butadiene were found to react with $(C_5Me_5)_2Yb(thf)$ (99), although these are more difficult displacements, since THF generally coordinates better than Et_2O. Reaction of $(C_5Me_5)_2Yb(Me_2PCH_2PMe_2)$ with $YbCl_3$ in toluene has been used to make the trivalent phosphine complex $(C_5Me_5)_2YbCl(Me_2PCH_2PMe_2)$, in which the phosphine is monodentate (109).

Transition metal carbonyl complexes also react with $(C_5Me_5)_2Yb(OEt_2)$ to produce interesting products. The divalent organolanthanide reduces $Co_2(CO)_8$ to the $Co(CO)_4^-$ anion to produce the complex, $(C_5Me_5)_2$-$Yb(thf)[(\mu\text{-}OC)Co(CO)_3]$ (110). The crystallographically determined structure shows that the $Co(CO)_4^-$ unit is coordinated to the ytterbium by a Yb—O—C—Co linkage. With $Fe(CO)_5$, $(C_5Me_5)_2Yb(OEt_2)$ is thought to form a related reduction product involving the $Fe(CO)_4^{2-}$ anion (110). These products are consistent with reducing ability of divalent lanthanides and the oxophilicity of the lanthanide metals (cf. the CO reaction, section IV,B).

The reaction of $(C_5Me_5)_2Yb(OEt_2)$ with $Fe_3(CO)_{12}$ or $Fe_2(CO)_9$ is more complex, giving a product of formula $[(C_5Me_5)_2Yb\ (\mu\text{-}OC)_2Fe\text{-}(CO)_3]_2Fe(CO)$ (111). A schematic of the structure (Scheme 11) shows that an Fe—Fe bond of the $Fe_3(CO)_{12}$ precursor has been broken to form a metalloacetonylacetonate-like ligand. The reaction of this product with H_2 and CO at 18 atm left the complex unchanged. $Mn_2(CO)_{10}$ also reacts with $(C_5Me_5)_2Yb(OEt_2)$ via reductive metal–metal bond cleavage. A complex structure involving dimeric $[(C_5Me_5)_2Yb][(\mu\text{-}CO)_2Mn(CO)_3]$ units and polymeric $[(C_5Me_5)_2Yb][(\mu\text{-}CO)_3Mn(CO)_2]$ units is formed (112).

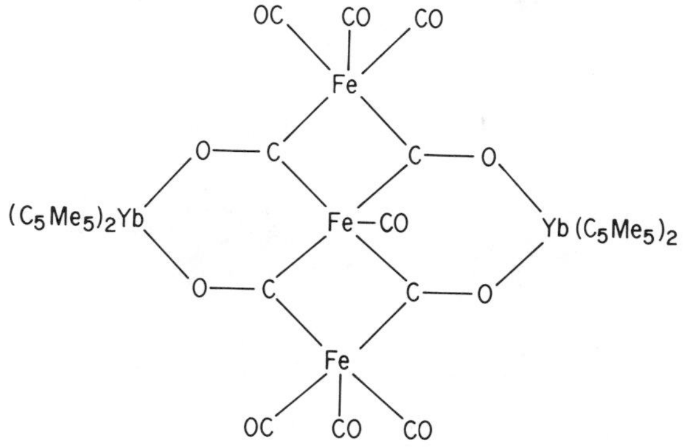

SCHEME 11

2. Samarium Complexes

The greatest impact of the C_5Me_5 ligand on divalent organolanthanide chemistry has been with samarium. In contrast to Eu and Yb, the unsubstituted cyclopentadienyl derivatives of samarium, $[(C_5H_5)_2Sm]_n$ (84) and $[(CH_3C_5H_4)_2Sm]_n$ (113), are insoluble in common solvents. Hence, prior to the synthesis of the alkane-soluble $(C_5Me_5)_2Sm(thf)_2$ (114), there was no opportunity to explore the organometallic chemistry of Sm(II), the most reactive of the divalent lanthanides.

The initial synthesis of $(C_5Me_5)_2Sm(thf)_2$ was accomplished by the metal vapor method [Eq. (54)] (114). This approach provided a halide-free route to the complex and also provided another example of oxidative

$$\text{Sm} + C_5Me_5H \xrightarrow{\text{THF}} [(C_5Me_5)SmH(thf)_2] \rightarrow (C_5Me_5)_2Sm(thf)_2 \qquad (54)$$

addition of C—H to a lanthanide metal (see Section VI,A). The intermediate divalent hydride, "$C_5Me_5SmH(thf)_2$," was identified by elemental analysis and deuterolysis (which formed HD and D_2), but it was not obtainable in pure form. Given the high reactivity of $(C_5Me_5)_2Sm(thf)_2$, it is not surprising that the much more coordinatively unsaturated hydride would be difficult to isolate and would readily convert to the bis(ring) complex. The related complex $(C_5Me_4Et)_2Sm(thf)_2$ was prepared in an analogous manner (115). An X-ray crystal structure of $(C_5Me_5)_2Sm(thf)_2$ revealed a monomeric, disolvated, bent metallocene structure. The complex was also characterized by NMR spectroscopy, despite a magnetic moment of ~3.6 B.M., thereby demonstrating for the first time that Sm(II) organometallics are NMR-accessible species (114).

$(C_5Me_5)_2Sm(thf)_2$ subsequently has been prepared by ionic metathesis from SmI_2 and KC_5Me_5 (116). This reaction, like the other lanthanide

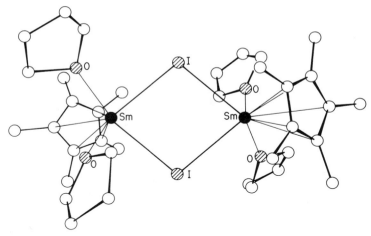

SCHEME 12

dihalide reactions, is sensitive to reaction conditions. In the course of studying this reaction, a new divalent samarium complex, $[(C_5Me_5)$-$SmI(thf)_2]_2$, was isolated and structurally characterized (Scheme 12). This complex completes the series $SmI_2(thf)_x$ (*117*), $[(C_5Me_5)SmI$-$(thf)_2]_2$, $(C_5Me_5)_2Sm(thf)_2$, which should provide substantial variation in coordination environment for the reactive Sm(II) center.

$(C_5Me_5)_2Sm(thf)_2$ reacts with CO and with NO, polymerizes ethene, functions as a catalyst precursor for homogeneous catalytic hydrogenation, and reacts with HgR_2 to form trivalent $(C_5Me_5)_2SmR(thf)$ complexes (*114, 115*). The CO and NO reactions are complex and full elucidation of all of the products has yet to be accomplished. One fully characterized product of these reactions is the bridged oxide complex $[(C_5Me_5)_2Sm]_2$-$(\mu\text{-}O)$. This complex (Scheme 13) can be obtained more directly from epoxides [Eq. (55)], and it forms readily whenever trace oxygen contaminants contact $(C_5Me_5)_2SmZ$ systems where Z = alkyl or hydride (see below). The Sm—O—Sm unit is rigorously linear and the Sm—O bond is comparatively short (*118*).

$$2\ (C_5Me_5)_2 Sm(thf)_2 + CH_3CH_2\overline{CHCH_2O} \rightarrow [(C_5Me_5)_2Sm]_2(\mu\text{-}O) \tag{55}$$

Since $(C_5Me_5)_2Sm(thf)_2$ was a crystallographically characterized low-valent lanthanide complex with catalytic activity in hydrogenation reactions, it was an ideal complex to study with respect to the catalyses which were initiated by metal vapor products as discussed earlier. This system was of further interest since it initiated catalytic hydrogenation of alkynes at rates 1000 times faster than the previously studied organolanthanide

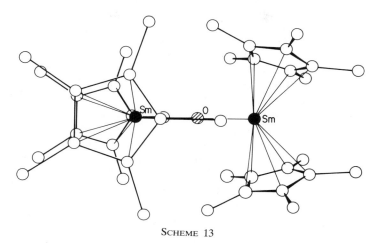

Scheme 13

systems, rates which are the fastest known for an f element complex and which are comparable to rhodium-based hydrogenations. Stoichiometric addition of internal alkynes, $RC\equiv CR$, to the purple $(C_5Me_5)_2Sm(thf)_2$ generated an intensely colored black material which analyzed for $(C_5Me_5)_2SmCR$ [Eq. (56)] *(105)*. Spectroscopic and hydrolytic data were

$$2(C_5Me_5)_2Sm(thf)_2 + C_6H_5C\equiv CC_6H_5 \underset{+THF}{\overset{-THF}{\rightleftharpoons}} [(C_5Me_5)_2Sm]_2C_2(C_6H_5)_2 \quad (56)$$

consistent with but not definitive for an enediyl structure $[(C_5Me_5)_2Sm]$ $(C_6H_5)C{=}C(C_6H_5)[Sm(C_5Me_5)_2]$. Although the magnetic moment of this complex was indicative of a Sm(III) species, the complex was unusual for a Sm(III) compound in that (a) it was black and (b) addition of THF regenerated $(C_5Me_5)_2Sm(thf)_2$ and $C_6H_5C\equiv CC_6H_5$ [Eq. (56)]. Considering the strong tendency of Sm(II) to oxidize to Sm(III), this reverse transformation occurs under surprisingly mild conditions.

$[(C_5Me_5)_2Sm]_2C_2R_2$ complexes react with hydrogen to form $[(C_5Me_5)_2SmH]_2$ [Eq. (57)] *(105)*. This is an unusual hydride in several

$$[(C_5Me_5)_2Sm]_2C_2R_2 + 3\,H_2 \rightarrow [(C_5Me_5)_2SmH]_2 + RCH_2CH_2R \quad (57)$$

respects. Compared to the formally nine-coordinate $[(C_5H_4R)_2Ln(\mu\text{-H})\text{-}(thf)]_2$ hydrides of the smaller metals (Ln = Y, Er, Lu) *(41)*, the formally eight-coordinate samarium system involving the much larger metal was very sterically unsaturated. Comparison of $[(C_5Me_5)_2SmH]_2$ with the nine-coordinate $[(C_5Me_5)_2ThH(\mu\text{-H})]_2$ *(119)* which has the same size rings and a metal of similar radius, leads to the same conclusion. Although this hydride is very reactive due to its steric unsaturation, it could be crystallographically characterized. The structure contains two bent metallocene units, $(C_5Me_5)_2Sm$, skewed by 87° with respect to each other, in an arrangement very similar to that shown previously for $[(C_5Me_5)_2Sm]_2(\mu\text{-}O)$. This is an unusual orientation based on the normal bonding requirements of bent transition metal metallocenes (120), which have their additional ligands in a plane bisecting the ring centroid–metal–ring centroid angle. Since the analogous planes of the two bent metallocene units in $[(C_5Me_5)_2SmH]_2$ are nearly perpendicular, the hydride ligands (which were not located crystallographically) cannot lie in a single plane which bisects both ring centroid–metal–ring centroid angles. As described in Section III, due to the limited radial extension of the $4f$ orbitals, organolanthanides may be able to adopt unusual structures which would be "orbitally forbidden" for transition metal complexes; $[(C_5Me_5)_2SmH]_2$ may be such an example.

Although the sterically unsaturated $[(C_5Me_5)_2SmH]_2$ is highly reactive, it is not as reactive as the sterically oversaturated $(C_5Me_5)_2LuH$ system described in Section VII,A (106, 107). For example, although $[(C_5Me_5)_2SmH]_2$ reacts with THF and Et_2O, it is relatively stable to alkanes and arenes, allowing its chemistry to be studied in these solvents without the complicating metallation chemistry observed for $(C_5Me_5)_2LuH$. These two $(C_5Me_5)_2LnH$ systems, Ln = Sm and Lu, and the cyclopentadienyllanthanide hydrides in general, constitute a powerful example of how one can take advantage of the special lanthanide property of gradually changing radial size in a series of chemically similar systems. The $[(C_5H_4R)_2Ln(\mu\text{-}H)(thf)]_2$ hydrides (Ln = Lu, Er, Y, and R = H, CH_3) are stable, isolable species which have a good match of ligand size and metal size. This combination provides a fully characterizable Ln—H prototype with which the basic properties of the Ln—H bond can be established. Increasing the ligand size to C_5Me_5 in complexes of the small metals causes steric oversaturation. When this generates a complex with terminal hydride sites, it leads to an exceptionally reactive species such as $(C_5Me_5)_2LuH$, which is so reactive that it is difficult to fully characterize. By increasing the size of the metal from Lu to Sm, while keeping the large C_5Me_5 ligand, one obtains a species sterically saturated enough to characterize fully and study cleanly, but still sterically unsaturated enough to be highly reactive. These hydride complexes demonstrate how extensively organometallic reactivity/stability can be varied within the lanthanide series. For (cyclopentadienyl ring)$_2$Ln(reactive ligand) complexes, three parameters can be varied: ring size, metal size, and reactive ligand size. By manipulating these factors, a variety of reactivity patterns can be turned on and off. At present, this is done with individual complexes. As the area develops, this may be possible in the context of catalytic cycles.

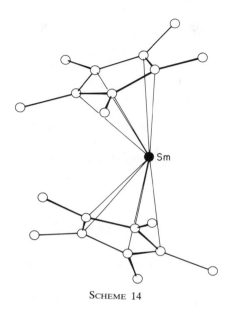

SCHEME 14

A recent development in the low-valent lanthanide area is the synthesis and X-ray structural determination of the unsolvated complex $(C_5Me_5)_2Sm$ (*121*). This species is the first structurally characterized bis(cyclopentadienyl)lanthanide species which has no other ligands in the metal coordination sphere. As such, it is the closest lanthanide analogue of the bis(ring)metallocene sandwich compounds like ferrocene. Samarocene was obtained by desolvation of $(C_5Me_5)_2Sm(thf)_2$ under high vacuum and sublimation of the product [Eq. (58)]. The sublimed crystals have a bent

$$(C_5Me_5)_2Sm(thf)_2 \xrightarrow[\text{vacuum}]{125°C} (C_5Me_5)_2Sm + 2\ THF \qquad (58)$$

metallocene structure (Scheme 14) with angles similar to those in $(C_5Me_5)_2Sm(thf)_2$. This exceedingly sterically unsaturated divalent samarium complex should display a rich and varied chemistry.

C. Summary

As was found in early transition metal and actinide chemistry, the C_5Me_5 ligand has allowed the isolation of complexes difficult to obtain by other means. More importantly, however, following the principle of steric oversaturation, the C_5Me_5 ligand has provided access to some of the most highly reactive organolanthanide complexes known to date. In addition, as demonstrated by the samarium and lutetium complexes $[(C_5Me_5)_2LnH]_2$, this ligand has allowed the variation in lanthanide metal size to be used productively to vary reactivity. In the past, it was known that organolanthanide reactivity could be increased by increasing the size of the metal to make a complex sterically less saturated. The resulting complex was highly reactive, but rather nonselective as it sought to become sterically saturated and a variety of products often formed. Using the oversaturation principle to generate highly reactive terminal ligands on sterically unsaturated metal centers provides a reactive site which is prevented by the size of the C_5Me_5 ring from being so nonselective. As a result, the enhanced reactivity obtained is more controllable and hence more valuable.

VIII

CONCLUSION

Recent advances in organolanthanide synthesis, structural characterization, and reactivity clearly have demonstrated that an extensive, distinctive chemistry is available to the lanthanide elements. The view that the lanthanides have only a limited, ionic chemistry is no longer valid. Not only has a much wider range of ligands and reagents been established as viable in organolanthanide systems, but the recent results also indicate that no class of molecules, not even saturated hydrocarbons like CH_4, should be excluded from consideration.

The thesis that the lanthanide elements offer something unique to organometallic chemistry has been proven in a variety of ways. New classes of complexes, unusual structural types, and novel reactivity patterns have been observed with these elements. In addition, the special properties of the lanthanide elements have allowed major contributions to be made to our knowledge of a variety of fundamental organometallic reactions of general interest including the polymerization of alkenes, the activation of C—H and H—H bonds, and the reduction of CO.

Since organolanthanide chemistry is still a relatively young, unexplored area of organometallic chemistry, the field has considerable potential for generating further contributions to our general understanding of organometallic chemistry in the future. Much remains to be learned about the special chemical properties of the lanthanides before they can be exploited fully. The traditional principles of organolanthanide stability/reactivity involving optimization of electrostatic and steric factors have only recently been examined in detail. These studies have led to significant refinements and variations including the principle of steric oversaturation as a route to high reactivity, the importance to reactivity of terminal rather than bridging ligands on sterically unsaturated, unsolvated metal centers, and the variability in reaction chemistry attainable by varying the size of the lanthanide. As these features of organolanthanide chemistry are experimentally developed to full advantage, other equally important variations of the reactivity principles are likely to be revealed.

As our basic knowledge of organolanthanide chemistry increases, the full potential of these elements in organometallic chemistry should be realized. Consistent with this bright future, organolanthanide chemistry is expanding very rapidly. Just as the early transition metals provided much of the excitement in organometallic chemistry in the 1970s, the lanthanide elements may be the metals of the 1980s.

Acknowledgments

The preparation of this article was aided by grants from the Department of Energy, the National Science Foundation, and the Alfred P. Sloan Foundation. Molecular orientations of some of the figures were developed using the computer graphics program of Professor Warren J. Hehre.

References

1. J. M. Birmingham and G. Wilkinson, *J. Am. Chem. Soc.* **78,** 42 (1956).
2. R. E. Maginn, S. Manastyrskyj, and M. Dubeck, *J. Am. Chem. Soc.* **85,** 672 (1963).
3. N. M. Ely and M. Tsutsui, *Inorg. Chem.* **14,** 2680 (1975).
4. H. Gysling and M. Tsutsui, *Adv. Organomet. Chem.* **9,** 361 (1970).
5. T. J. Marks, *Prog. Inorg. Chem.* **24,** 51 (1978).
6. W. J. Evans, *in* "The Chemistry of the Metal–Carbon Bond" (F. R. Hartley and S. Patai, eds.), p. 489. Wiley, New York, 1982.
7. T. J. Marks and R. D. Ernst, *in* "Comprehensive Organometallic Chemistry" (G. Wilkinson, F. G. A. Stone, and E. W. Abel, eds.), Chapt. 21. Pergamon, Oxford, 1982.
8. A. J. Freeman and R. E. Watson, *Phys. Rev.* **127,** 2058 (1962).
9. A. P. Paszek and H. M. Crosswhite, personal communication, 1981.
10. T. Moeller, *in* "Comprehensive Inorganic Chemistry" (J. C. Bailar, Jr., H. J. Emeleus, R. Nyholm, and A. F. Trotman-Dickenson eds.), Vol. 4, p. 1. Pergamon, Oxford, 1973.
11. F. A. Cotton and G. Wilkinson, "Advanced Inorganic Chemistry," 4th Ed., p. 982. Wiley, New York, 1980.
12. L. R. Morss, *Chem. Rev.* **76,** 827 (1976).
13. J. E. Huheey, "Inorganic Chemistry," 3rd Ed., p. 912. Harper & Row, New York, 1983.
14. W. J. Evans, S. C. Engerer, P. A. Piliero, and A. L. Wayda, *in* "Fundamental Research in Homogeneous Catalysis" (M. Tsutsui, ed.) Vol. 3, p. 941. Plenum, New York, 1979.
15. A. L. Wayda, *Organometallics* **2,** 565 (1983).
16. S. A. Cotton, F. A. Hart, M. B. Hursthouse, and A. J. Welch, *J. Chem. Soc., Chem. Commun.,* p. 1225 (1972).
17. S. A. Cotton, *J. Organomet. Chem. Libr.* **3,** 189 (1977).

18. M. F. Lappert and R. Pearce, *J. Chem. Soc., Chem. Commun.*, p. 126 (1973).
19. J. L. Atwood, W. E. Hunter, R. D. Rogers, J. Holton, J. McMeeking, R. Pearce, and M. F. Lappert, *J. Chem. Soc., Chem. Commun.*, p. 140 (1978).
20. H. Schumann and J. Müller, *J. Organomet. Chem.* **146**, C5 (1978).
21. G. K. Barker and M. F. Lappert, *J. Organomet. Chem.* **76**, C45 (1974).
22. H. Schumann and J. Müller, *J. Organomet. Chem.* **169**, C1 (1979).
23. I. Sh. Guzman, N. N. Chigir, O. K. Sharaev, G. N. Bondarenko, E. I. Tinyakova, and B. A. Dolgoplosk, *Dokl. Akad. Nauk SSSR* **249**, 860 (1979); E. L. Vollerstein, V. A. Yakovlev, E. I. Tinyakova, and B. A. Dolgoplosk, *Dokl. Akad. Nauk SSSR* **250**, 365 (1980); B. A. Dolgoplosk, E. I. Tinyakova, I. Sh. Guzman, E. L. Vollerstein, N. N. Chigir, G. N. Bondarenko, O. K. Sharaev, and V. A. Yakovlev, *J. Organomet. Chem.* **201**, 249 (1980).
24. A. L. Wayda and W. J. Evans, *J. Am. Chem., Soc.* **100**, 7119 (1978).
25. W. J. Evans, A. L. Wayda, and D. L. Stanley, unpublished results, 1981.
26. W. J. Evans and A. L. Wayda, *J. Organomet. Chem.* **202**, C6 (1980).
27. H. Schumann and J. Müller, *Angew. Chem., Int. Ed. Engl.* **17**, 276 (1978).
28. H. Schumann, J. Pickardt, and N. Bruncks *Angew. Chem., Int. Ed. Engl.* **20**, 120 (1981).
29. H. Schumann, J. Müller, N. Bruncks, H. Lauke, J. Pickardt, H. Schwarz, and K. Eckart, *Organometallics* **3**, 69 (1984).
30. H. Schumann and S. Hohmann, *Chem.-Ztg.* **100**, 336 (1976).
31. J. Holton, M. F. Lappert, D. G. H. Ballard, R. Pearce, J. L. Atwood, and W. E. Hunter, *J. Chem. Soc., Dalton Trans.*, p. 54 (1979).
32. W. J. Evans, A. L. Wayda, W. E. Hunter, and J. L. Atwood, *J. Chem. Soc., Chem. Commun.*, p. 292 (1981).
33. H. Schumann, W. Genthe, N. Bruncks, and J. Pickardt, *Organometallics* **1**, 1194 (1982).
34. W. J. Evans, R. Dominguez, T. P. Hanusa, and R. J. Doedens, unpublished results, 1984.
35. J. L. Atwood, W. E. Hunter, A. L. Wayda, and W. J. Evans, *Inorg. Chem.* **20**, 4115 (1981).
36. W. J. Evans, I. Bloom, W. E. Hunter, and J. L. Atwood, *Organometallics* **2**, 709 (1983).
37. W. J. Evans, *J. Organomet. Chem.* **250**, 217 (1983).
38. W. J. Evans, A. L. Wayda, W. E. Hunter, and J. L. Atwood, *J. Chem. Soc., Chem. Commun.*, p. 706 (1981).
39. J. M. Manriquez, D. R. McAlister, R. D. Sanner, and J. E. Bercaw, *J. Am. Chem. Soc.* **100**, 2716 (1978); P. T. Wolczanski and J. E. Bercaw, *Acc. Chem. Res.* **13**, 121 (1980), and references therein.
40. P. J. Fagan, J. M. Manriquez, T. J. Marks, V. W. Day, S. H. Vollmer, and C. S. Day, *J. Am. Chem. Soc.* **102**, 5393 (1980).
41. W. J. Evans, J. H. Meadows, A. L. Wayda, W. E. Hunter, and J. L. Atwood, *J. Am. Chem. Soc.* **104**, 2008 (1982).
42. P. J. Brothers, *Prog. Inorg. Chem.* **28**, 1 (1981), and references therein.
43. H. A. Zinnen, J. J. Pluth, and W. J. Evans, *J. Chem. Soc., Chem. Commun.*, p. 810 (1980).
44. A. W. Langer, Jr., *Adv. Chem. Ser.* **130**, 1 (1974).
45. W. J. Evans, J. H. Meadows, W. E. Hunter, and J. L. Atwood, *J. Am. Chem. Soc.* **106**, 1291 (1984).
46. R. D. Fischer and G. Bielang, *J. Organomet. Chem.* **191**, 61 (1980).

47. D. G. H. Ballard, A. Courtis, J. Holton, J. McMeeking, and R. Pearce, *J. Chem. Soc., Chem. Commun.*, p. 994 (1978).
48. D. G. H. Ballard and R. Pearce, *J. Chem. Soc., Chem. Commun.*, p. 621 (1975).
49. W. M. Mueller, J. P. Blackledge, and G. G. Libowitz, "Metal Hydrides." Academic Press, New York, 1968.
50. W. J. Evans, S. C. Engerer, P. A. Piliero, and A. L. Wayda, *J. Chem. Soc., Chem. Commun.*, p. 1007 (1979); W. J. Evans, I. Bloom, and S. C. Engerer, *J. Catal.* **84,** 468 (1983).
51. W. J. Evans, S. C. Engerer, and K. M. Coleson, *J. Am. Chem. Soc.* **103,** 6672 (1981).
52. W. J. Evans, J. H. Meadows, A. G. Kostka, and G. L. Closs, *Organometallics* **4,** 324 (1985).
53. K. W. Bagnall and Li Xing-Fu, *J. Chem. Soc., Dalton. Trans.*, p. 1365 (1982).
54. W. J. Evans, J. H. Meadows, A. L. Wayda, W. E. Hunter, and J. L. Atwood, *J. Am. Chem. Soc.* **104,** 2015 (1982).
55. H. H. Wang and L. H. Pignolet, *Inorg. Chem.* **19,** 1470 (1980); D. F. Chodosh, R. H. Crabtree, H. Felkin, S. Morehouse, and G. E. Morris, *Inorg. Chem.* **21,** 1307 (1982).
56. W. J. Evans, J. H. Meadows, and T. P. Hanusa, *J. Am. Chem. Soc.* **106,** 4454 (1984).
57. H. Schumann and W. Genthe, *J. Organomet. Chem.* **213,** C7 (1981).
58. E. B. Lobkovsky, G. L. Soloveychik, A. B. Erofeev, B. M. Bulychev, and V. K. Bel'skii, *J. Organomet. Chem.* **235,** 151 (1982); E. B. Lobkovsky, G. L. Soloveychik, B. M. Bulychev, A. B. Erofeev, A. I. Gusev, and N. I. Kirillova, *J. Organomet. Chem.* **254,** 167 (1983); P. J. Fagan, G. W. Grynkewich, T. J. Marks, unpublished results, cited in ref. 7, p. 205.
59. W. J. Evans, J. H. Meadows, W. E. Hunter, and J. L. Atwood, *Organometallics* **2,** 1252 (1983).
60. W. J. Evans, *in* "The Rare Earths in Modern Science and Technology," (G. J. McCarthy, H. E. Silber, and J. J. Rhyne, eds.), Vol. 3, p. 61. Plenum, New York, 1982.
61. D. F. Evans, G. V. Fazakerley, and R. F. Phillips, *J. Chem. Soc. A*, p. 1931 (1971).
62. B. A. Dolgoplosk, E. I. Tinyakova, I. N. Markevich, T. V. Soboleva, G. M. Chernenko, O. K. Sharaev, and V. A. Yakovlev, *J. Organomet. Chem.* **255,** 71 (1983).
63. G. B. Deacon and D. G. Vince, *J. Organomet. Chem.* **112,** Cl (1976).
64. G. B. Deacon, W. D. Raverty, and D. G. Vince, *J. Organomet. Chem.* **135,** 103 (1977).
65. G. B. Deacon and A. J. Koplick, *J. Organomet. Chem.* **146,** C43 (1978).
66. G. B. Deacon, A. J. Koplick, W. D. Raverty, and D. G. Vince, *J. Organomet. Chem.* **182,** 121 (1979).
67. P. S. Skell and M. J. McGlinchey, *Angew. Chem., Int. Ed. Engl.* **14,** 195 (1975)
68. P. L. Timms and T. W. Turney, *Adv. Organomet. Chem.* **15,** 53 (1977).
69. K. J. Klabunde, "Chemistry of Free Atoms and Particles," Academic Press, New York, 1980.
70. J. R. Blackborow and D. Young, "Metal Vapor Synthesis in Organometallic Chemistry," Springer-Verlag, Berlin and New York, 1979.
71. M. Moskovits and G. A. Ozin, "Cryochemistry." Wiley, New York, 1976.
72. G. H. Dieke, *in* "Spectra and Energy Levels of Rare Earth Ions in Crystals," (H. M. Crosswhite and H. Crosswhite, eds.), p. 53. Wiley, New York, 1968.
73. J. L. Slater, T. C. DeVore, and V. Calder, *Inorg. Chem.* **12,** 1918 (1973).
74. J. L. Slater, T. C. DeVore, and V. Calder, *Inorg. Chem.* **13,** 1808 (1974).

75. Ref. 6, p. 501.
76. C. W. DeKock, S. R. Ely, T. E. Hopkins, and M. A. Brault, *Inorg. Chem.* **17,** 625 (1978).
77. K. O. Hodgson, F. Mares, D. F. Starks, and A. Streitwieser, Jr., *J. Am. Chem. Soc.* **95,** 8650 (1973).
78. W. J. Evans, S. C. Engerer, and A. C.Neville, *J. Am. Chem. Soc.* **100,** 331 (1978).
79. W. J. Evans, K. M. Coleson, and S. C. Engerer, *Inorg. Chem.* **20,** 4320 (1981).
80. E. O. Fischer and H. Fischer, *J. Organomet. Chem.* **3,** 181 (1965).
81. R. G. Hayes and J. L. Thomas, *Inorg. Chem.* **3,** 2421 (1969).
82. R. G. Hayes and J. L. Thomas, *J. Am. Chem. Soc.* **91,** 6876 (1969).
83. F. Calderazzo, R. Pappalardo, and S. Losi, *J. Inorg. Nucl. Chem.* **28,** 987 (1966).
84. G. W. Watt and E. W. Gillow, *J. Am. Chem. Soc.* **91,** 775 (1969).
85. M. F. Lappert, P. I. W. Yarrow, J. L. Atwood, R. Shakir, and J. Holton, *J. Chem. Soc., Chem. Commun.*, p. 987 (1980).
86. T. D. Tilley, R. A. Andersen, B. Spencer, H. Ruben, A. Zalkin, and D. H. Templeton, *Inorg. Chem.* **19,** 2999 (1980).
87. G. B. Deacon, P. I. MacKinnon, T. W. Hambley, and J. C. Taylor, *J. Organomet. Chem.* **259,** 91 (1983).
88. R. B. King, *Coord. Chem. Rev.* **20,** 155 (1976).
89. P. M. Maitlis, *Acc. Chem. Res.* **11,** 301 (1978).
90. P. T. Wolczanski and J. E. Bercaw, *Acc. Chem. Res.* **13,** 121 (1980), and references therein.
91. T. J. Marks, *Science* **217,** 989 (1982).
92. T. D. Tilley and R. A. Andersen, *Inorg. Chem.* **20,** 3267 (1981).
93. A. L. Wayda and W. J. Evans, *Inorg. Chem.* **19,** 2190 (1980).
94. P. L. Watson, J. F. Whitney, and R. L. Harlow, *Inorg. Chem.* **20,** 3271 (1981).
95. M. F. Lappert, A. Singh, J. L. Atwood, and W. E. Hunter, *J. Chem. Soc., Chem. Commun.*, p. 1190 (1981).
96. M. F. Lappert, A. Singh, J. L. Atwood and W. E. Hunter, *J. Chem. Soc., Chem. Commun.* 1191 (1981).
97. M. F. Lappert, A. Singh, J. L. Atwood, W. E. Hunter, and H.-M. Zhang, *J. Chem. Soc., Chem. Commun.*, p. 69 (1983).
98. T. D. Tilley, R. A. Andersen, A. Zalkin, and D. H. Templeton, *Inorg. Chem.* **21,** 2644 (1982).
99. P. L. Watson, *J. Chem. Soc., Chem. Commun.*, p. 652 (1980).
100. P. L. Watson, *J. Am. Chem. Soc.* **104,** 337 (1982).
101. P. L. Watson and T. Herskovitz, *ACS Symp. Ser.* **212,** 459 (1983).
102. A. Maercker and R. Stotzel, *J. Organomet. Chem.* **254,** 1 (1983); B. Klei, J. H. Teuben, and H. J. de Leifde Meijer, *J. Chem. Soc., Chem. Commun.*, p. 342 (1981); G. Fink and R. Rottler, *Angew. Makromol. Chem.* **94,** 25 (1981).
103. P. L. Watson and D. C. Roe, *J. Am. Chem. Soc.* **104,** 6471 (1982).
104. W. Pfohl, *Justus Liebigs Ann. Chem.* **629,** 207, 210 (1960).
105. W. J. Evans, I. Bloom, W. E. Hunter, and J. L. Atwood, *J. Am. Chem. Soc.* **105,** 1401 (1983).
106. P. L. Watson, *J. Am. Chem. Soc.* **105,** 6491 (1983).
107. P. L. Watson, *J. Chem. Soc., Chem. Commun.*, p. 276 (1983).
108. T. D. Tilley, R. A. Andersen, B. Spencer, and A. Zalkin, *Inorg. Chem.* **21,** 2647 (1982).
109. T. D. Tilley, R. A. Andersen, and A. Zalkin, *Inorg. Chem.* **22,** 856 (1983).
110. T. D. Tilley and R. A. Andersen, *J. Chem. Soc., Chem. Commun.*, p. 985 (1981).

111. T. D. Tilley and R. A. Andersen, *J. Am. Chem. Soc.* **104,** 1772 (1982).
112. J. M. Boncella and R. A. Andersen, *Inorg. Chem.* **23,** 432 (1984).
113. W. J. Evans and H. A. Zinnen, unpublished results, 1980.
114. W. J. Evans, I. Bloom, W. E. Hunter, and J. L. Atwood, *J. Am. Chem. Soc.* **103,** 6507 (1981).
115. W. J. Evans, I. Bloom, W. E. Hunter, and J. L. Atwood, *Organometallics* **4,** 112 (1985).
116. W. J. Evans, J. W. Grate, H. W. Choi, I. Bloom, W. E. Hunter, and J. L. Atwood, *J. Am. Chem. Soc.* **107,** 941 (1985).
117. P. Girard, J. L. Namy, and H. B. Kagan, *J. Am. Chem. Soc.* **102,** 2693 (1980).
118. W. J. Evans, J. W. Grate, I. Bloom, W. E. Hunter, and J. L. Atwood, *J. Am. Chem. Soc.* **107,** 405 (1985).
119. R. W. Broach, A. J. Schultz, J. M. Williams, G. M. Brown, J. M. Manriquez, P. J. Fagan, and T. J. Marks, *Science* **203,** 172 (1979).
120. J. W. Lauher and R. Hoffmann, *J. Am. Chem. Soc.* **98,** 1729 (1976).
121. W. J. Evans, L. A. Hughes, and T. P. Hanusa, *J. Am. Chem. Soc.* **106,** 4270 (1984).

ADVANCES IN ORGANOMETALLIC CHEMISTRY, VOL. 24

Silyl, Germyl, and Stannyl Derivatives of Azenes, N_nH_n Part II. Derivatives of Triazene N_3H_3, Tetrazene N_4H_4, and Pentazene N_5H_5 [1]

NILS WIBERG

Institut für Anorganische Chemie der Universität München
Munich, Federal Republic of Germany

I

INTRODUCTION

The purpose of this article is to summarize the preparation and properties of silyl, germyl, and stannyl derivatives of azenes N_nH_n ($n = 2$–5). This part II of the review deals with known Group IV derivatives of triazene, tetrazene, and pentazene ($n = 3$–5; excluding purely organic substituted triazenes and tetrazenes, cf. refs $2, 3$). Group IV derivatives of diazene were discussed in Part I of the review (1).

---

[1] For Part I ("Derivatives of Diazene") see Wiberg (1).

II

GROUP IV DERIVATIVES OF TRIAZENE, N_3H_3

A. Preparation

Triazenes can be prepared by the following three methods (Scheme 1): (1) *redox reactions* (reduction of azides, oxidation of triazanes), (2) *building or splitting of nitrogen chains*, and (3) *exchange of substituents* (mutual transformation of triazenes).

1. Redox Reactions

Oxidation of silylated hydrazines provides an easy approach to silylated diazenes (*1*). In an analogous manner, silylated triazenes should be formed by *oxidation of triazanes*. This process is still not applicable, however, as appropriate synthetic techniques for the preparation of silylated triazanes are not available. On the other hand, *reduction of azides*, as subsequently inferred, leads effectively to the synthesis of silyltriazenes. As observed by O. Dimroth at the beginning of the century (*4*), metal organyls MR' [M = MgHal (Hal = halogen), or, more recently Li and Na] react by reductive addition with organic azides to form triazenides of type **1** [path (a), Eq. (1)], which lead to diorganyltriazenes **2** (R, R' = organyl; R″ = H) through protonation, or to triorganyltriazenes **2** (R, R', R″ = organyl) through organylation [Eq. (1), path (b)].

$$R-N=N=N \quad \xrightarrow[\text{(a)}]{+MR'} \quad R-N=N-N\diagdown_{M}^{R'} \quad \xrightarrow[\text{(b)}]{+ R''Hal, \; -MHal} \quad R-N=N-N\diagdown_{R''}^{R'}$$
$$\underline{\underline{1}} \hspace{5.5cm} \underline{\underline{2}}$$

$$\tag{1}$$

A large number of organyltriazenes have been synthesized by the route Eq. (1) (*Preparation Method A*) (*5*). The reaction sequence of Eq. (1) has also been used for the preparation of *silyl-*, *germyl-*, and *stannylorganyltriazenes* (*6, 7*), using silyl, germyl, or stannyl halides in path (b) of Eq. (1).[2] Individual examples are listed in Table I. The partial reaction of path (b) in Eq. (1), involving exchange of substituents, is treated in Section I,A,3.

---

[2] Besides the reaction in Eq. (1), MeN_3 reacts with MeMgI and X_3SiCl: MeN_3 + MeMgI + 2 $X_3SiCl \rightarrow MeN(SiX_3)_2$ + MeI + N_2 + $MgCl_2$ (*7*). The share of side reaction increases with increasing temperature; it amounts to 0% at 0°C or below and 50% in boiling ether (cf. Section II,B,1).

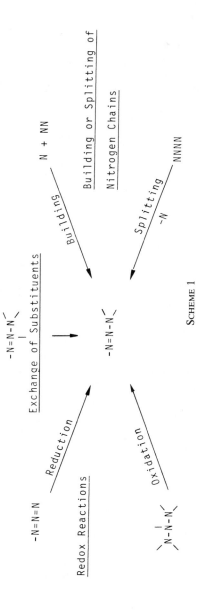

SCHEME 1

According to path (a) of Eq. (1), silyl azides (R = SiX$_3$) and metal organyls or metal silyls (R' = organyl or silyl) should form silyl-organyltriazenides[3] or disilyltriazenides **1**, whose subsequent silylation (R'' = SiX$_3$) according to path (b) of Eq. (1), must give *disilylorganyl-triazenes* or *trisilyltriazenes* (**2**)[4] (the synthesis of corresponding germyl and stannyl derivatives is analogous). In fact, the desired silyltriazenes are formed only if reactants with very bulky organyl and/or silyl groups are used. Thus, according to Eq. (2) tri-*tert*-butylsilylsodium adds to trimethyl-

$$
\begin{array}{c}
\text{X}_3\text{Si-N=N=N} \\[2mm]
\text{(a)}\ \ X=\text{Me} \quad \nearrow \quad \text{Me}_3\text{SiN=N-N}\begin{array}{l}\text{Si}^t\text{Bu}_3\\ \text{Na}\end{array} \underline{\underline{3a}} \xrightarrow[-\ \text{NaY}]{+\ \text{Me}_3\text{SiY}} {}^t\text{Bu}_3\text{SiN=N-N}\begin{array}{l}\text{SiMe}_3\\ \text{SiMe}_3\end{array} \underline{\underline{4a}} \\[4mm]
+ \\[2mm]
\text{NaSi}^t\text{Bu}_3 \\[2mm]
\text{(b)}\ \ X={}^t\text{Bu} \quad \searrow \quad {}^t\text{Bu}_3\text{SiN=N-N}\begin{array}{l}\text{Si}^t\text{Bu}_3\\ \text{Na}\end{array} \underline{\underline{3b}} \xrightarrow[-\ \text{NaOMe}]{+\ \text{MeOH}} {}^t\text{Bu}_3\text{SiN=N-N}\begin{array}{l}\text{Si}^t\text{Bu}_3\\ \text{H}\end{array} \underline{\underline{4b}}
\end{array}
\tag{2}
$$

silyl azide or tri-*tert*-butylsilyl azide at low temperatures to form silyltriaze-nides **3a** or **3b**, which are converted to the silyltriazenes **4a** or **4b** (*8, 9*) (silylation of **3b** is no longer possible because of steric hindrance).

In contrast, the reaction between sterically normal silyl azides and metal organyls (the same is true for germyl and stannyl azides) leads either to substitution of the silicon-bound azide group by an organyl group, or to cleavage of molecular nitrogen from the azide moiety (*10*) [Eq. (3)]. The preference for reaction in paths (a) or (b) of Eq. (3) depends on the kind

$$
\text{X}_3\text{Si-N=N=N} \xrightarrow{+\ \text{MR}'}
\begin{cases}
\text{(a)} & \text{X}_3\text{Si-R}' \ +\ \text{MN=N=N} \\[2mm]
\text{(b)} & \text{X}_3\text{Si-N}\begin{array}{l}\text{R}'\\ \text{M}\end{array} \ +\ \text{N}\equiv\text{N}
\end{cases}
\tag{3}
$$

---

[3] Compounds of type X$_3$Si—N=N—NMR' should also be accessible from organyl azides and metal silyls.

[4] Alkylation leads to silyldiorganyltriazenes or disilylorganyltriazenes protolysis to silylorganyltriazenes or disilyltriazenes.

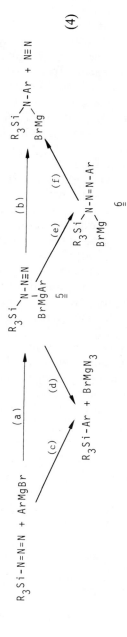

$$R_3Si-N=N=N + ArMgBr \xrightarrow{\quad (a) \quad} \underset{\underset{\textstyle 5}{\textstyle |}}{\underset{BrMgAr}{R_3Si}}N-N\equiv N \xrightarrow{\quad (b) \quad} \underset{BrMg}{R_3Si}N-Ar + N\equiv N$$

(4)

and concentration of metal organyls as well as the type of silylazide and the solvent (10). Thus, Ph_3SiN_3 reacts with LiPh or MeMgBr in diethyl ether only according to path (a) of Eq. (3), with $MgPh_2$ only according to path (b), and with ArMgBr (Ar = Ph, tolyl, or mesityl) according to both path (a) and path (b). In the latter case (ArMgBr, with Ar = Ph), the proportion of the substitution reaction [path (b)] increases with a decrease in concentration of metal organyl (e. g., 0.4 M: 27%; 0.1 M: 33%) and with increasing basicity of the solvent (e. g., toluene: 20%; ether: 30%; tetrahydrofuran (THF): 80%; monoglyme: 100%).[5] In addition, the proportion of substitution reaction increases to double or 100% if Ph_3SiN_3 is replaced by Me_3SiN_3 or $(MeO)_3SiN_3$ (Et_2O as solvent).

Kinetic and spectroscopic studies as well as experiments with ^{15}N-labeled azides (10–12) show that the reaction of silyl azides with Grignard compounds ArMgBr proceeds partly with nucleophilic substitution of the azide group by an aryl group [Eq. (4), path (c)] and partly with the formation of the silyl azide/Grignard adduct 5 [Eq. (4), path (a)]. The latter—also isolable—compound 5 decomposes (more slowly than it is formed) with migration of the aryl group either (1) from magnesium to silicon [Eq. (4), path (d)], (2) to the first nitrogen atom of the azide group [Eq. (4), path (b)] as well as (3) to the third nitrogen atom of the azide group [Eq. (4), path (e)]. In path (b) molecular nitrogen is cleaved simultaneously with rearrangement of the organyl group, whereas with path (e) nitrogen elimination follows the formation of an intermediate (6) [Eq. (4), path (f)]. Since the decomposition of 6 normally is faster than its formation, in most cases it could not be isolated.[6]

Besides the already mentioned "Dimroth method" [Eq. (1)], silyl-, germyl- or stannyltriazenes should also be obtainable by the direct silylation, germylation, or stannylation of azides, as shown in Eq. (5).

$$-N=N=N \xrightarrow[\text{(E = Si, Ge, Sn)}]{2EX_3} -N=N-N\begin{smallmatrix}EX_3\\EX_3\end{smallmatrix} \qquad (5)$$

Bis(silyl)diazenes ($X_3Si—N=N—SiX_3$), which transfer their silyl groups to double bond systems (1), may act as silylating agents. But neither organyl nor silyl azides could have, so far, been converted into silyltriazenes with bis(trimethylsilyl)diazene (X = Me). Either no reaction

---

[5] Silyl azide/Grignard reactions in toluene are preparatively interesting because of the possibility of transformation of halo- into aminoaromatics (10):

$$Ar\text{-}Hal \xrightarrow{+Mg} Ar\text{-}MgHal \xrightarrow[-N_2]{+R_3SiN_3} Ar\text{-}N(MgHal)(SiR_3) \xrightarrow{\text{hydrolysis}} Ar\text{-}NH_2$$

[6] The desired triazene 6 is isolable only in the presence of highly hindered silyl groups [cf. Eq. (2)].

occurs or amine and nitrogen are formed exclusively as reaction products as in the reactions with $(MeO)_3SiN_3$, p-$TolSO_2N_3$ (Tol = tolyl), or t-$BuOCON_3$ [Eq. (6)] (1).[7] Unstable intermediates of reaction [Eq. (6)] are

$$R-N=N=N \xrightarrow[\substack{R = MeO, p\text{-}TolSO_2, \\ {}^tBuOCO}]{Me_3Si-N=N-SiMe_3} \left\{ R-N=N-N \begin{matrix} {}^{SiMe_3} \\ {}_{SiMe_3} \end{matrix} \right\} \longrightarrow R-N \begin{matrix} {}^{SiMe_3} \\ {}_{SiMe_3} \end{matrix} \quad (6)$$

$$\underset{7}{} \qquad\qquad + \qquad N\equiv N$$

most probably silylated triazenes (7). (Attempts to germylate or stannylate azides are not known as yet.)

An additional possibility for preparing Group IV derivatives of triazene by azide reduction is the [2 + 3]-cycloaddition of azides and sila- or germaethenes (stannaethenes are still not known) with the formation of cyclic triazenes (8) (sila- or germatriazolines) [Eq. (7)] (*Preparation Method B*). Reaction of silaethene $Me_2Si=C(SiMe_3)_2$ (14) or germaethene $Me_2Ge=C(SiMe_3)_2$ (15)[8] with azides $R''N_3$ (e. g., $R'' = Me_3C$, Me_3Si) at low temperatures, smoothly provides the [2 + 3]-cycloadduct 8 with Me_2E- and $C(SiMe_3)_2$-ring members (14–16) (for individual compounds, cf. Table I).

$$\begin{matrix} {}^{\backslash}_{/}E=C^{/}_{\backslash} \\ + \\ R''-N=N=N \end{matrix} \xrightarrow{E = Si, Ge} \underset{8}{\begin{matrix} {\scriptstyle |}\;\;{\scriptstyle |} \\ -E-C- \\ R''-N \quad N \\ {}_{N}\nearrow \end{matrix}} \quad (7)$$

2. Building or Splitting of Nitrogen Chains

Under the triazene syntheses accomplished by *building of nitrogen chains*, the coupling of amides $MNR'R''$ (M = Li, Na) with aryldiazonium salts ArN_2Cl is worth special mention [Eq. (8)]. This reaction (*Preparation*

$$Ar-N\equiv N^+Cl^- \;+\; MN \begin{matrix} {}^{R'} \\ {}_{R''} \end{matrix} \xrightarrow{-MCl} Ar-N=N-N \begin{matrix} {}^{R'} \\ {}_{R''} \end{matrix} \quad (8)$$

$$\underset{9}{}$$

---

[7] Also, the action of silyl radicals (or of organyl radicals) on silyl azides leads finally to amines and nitrogen (13). As reaction intermediates, triazenyl radicals $-N\cdots\dot{N}\cdots N-$ (e. g., $Me_3SiN-N\cdots NSiMe_3$) have been identified by ESR spectroscopy.

[8] Generated by thermal decomposition of $Me_2EX-CLi(SiMe_3)_2$ (E = Si, Ge; X = halogen) at low temperatures (14, 15).

Method C), in the case of silylated amides (R′, R″ = SiX$_3$) in diethyl ether at −20°C, leads to *disilylaryltriazenes* (**9**) (*7*) in good yields (for individual compounds, cf. Table I). The latter compounds, formed according to Eq. (8), exist in the cis as well as the trans configuration. In fact, the formation of *cis*-triazene is favored with decreasing reaction temperature, whereas *trans*-triazene formation is favored with increasing reaction temperature.[9]

Other possible syntheses of silylated triazenes according to Eq. (8), based on the condensation principle —N=N— + —N< → —N=N—N<, have not been observed. Thus, bis(trimethylsilyl)diazene reacts with dimethylchloramine according to Eq. (9), with the elimination

$$\text{Me}_3\text{Si-N=N-SiMe}_3 + \text{Cl-NMe}_3 \xrightarrow[-\text{Me}_3\text{SiCl}]{} \text{Me}_3\text{Si-NMe}_2 + \text{N}\equiv\text{N} \qquad (9)$$

of trimethylchlorosilane to form trimethylsilyldiethylamine and nitrogen (*1*). However, the silylamine is formed by a route other than Me$_3$Si—N=N—SiMe$_3$ + Cl—NMe$_2$ → Me$_3$Si—N=N—NMe$_2$ + Me$_3$SiCl → Me$_3$Si—NMe$_2$ + N$_2$ + Me$_3$SiCl, involving silyldimethyltriazene (*1*).

On the other hand, silyltriazenes can also be synthesized by the condensation principle —N= + >N—N< → —N=N—N<. In this way, bis(trimethylsilyl)phenyltriazene, along with other side products, can be obtained by the reaction of nitrosobenzene with lithium tris(trimethylsilyl)-hydrazide [Eq. (10)] (*7*).[10]

$$\text{Ph-N=O} + \begin{array}{c} \text{Li} \quad \text{SiMe}_3 \\ \diagdown\text{N-N}\diagup \\ \text{Me}_3\text{Si}\diagup \quad \diagdown\text{SiMe}_3 \end{array} \xrightarrow[-\text{LiOSiMe}_3]{} \text{Ph-N=N-N}\begin{array}{c}\diagup\text{SiMe}_3\\\diagdown\text{SiMe}_3\end{array} \qquad (10)$$

An already reported example of triazene synthesis by *splitting of nitrogen chains* involves thermal decomposition of tetrazane (obtainable by the oxidation of hydrazine, e.g., with OH radicals) (*17*):

$$\begin{array}{c}\text{H}\quad\text{H}\\\diagdown\text{N-N-N-NH}_2\\\diagup\quad\mid\\\text{H}\quad\text{H}\end{array} \xrightarrow{\hspace{2cm}} \text{H-N=N-N}\begin{array}{c}\diagup\text{H}\\\diagdown\text{H}\end{array} + \text{NH}_3 \qquad (11)$$

In view of the lack of appropriate synthetic procedures for silylated, germylated, or stannylated tetrazanes, this mode of preparation cannot be applied for the synthesis of Group IV derivatives of triazene.

---

[9] For example, PhN$_2$Cl and NaN(SiMe$_3$)$_2$ form 100% *cis*-PhN$_3$(SiMe$_3$)$_2$ at −78°C and 3% *cis*- and 97% *trans*-PhN$_3$(SiMe$_3$)$_2$ at −20°C (cf. Section II,B,2).

[10] Triazenes have not so far been obtained from amides and *N*-nitroso compounds, e.g.,

(Me$_3$Si)$_2$NNa + O=N—NMe$_2$ ↛ Me$_3$Si—N=N—NMe$_2$ + NaOSiMe$_3$.

3. *Exchange of Substituents*

Exchange of a triazene-bound substituent by another substituent as in Eq. (12) [*Preparation Method D, Eq. (12)*] provides an important method

$$R-N=N-N\begin{matrix}R'\\R''\end{matrix} \xrightarrow{\quad R''\ \text{substitution by }R'''\quad} R-N=N-N\begin{matrix}R'\\R'''\end{matrix} \quad (12)$$

for the synthesis of Group IV derivatives of triazene. As described in path (b) of Eq. (1) (Preparation Method A), metal triazenides (**1**) will react with silyl, germyl, or stannyl halides to give replacement of $R'' = $ metal by EX_3. Metal triazenides (**1**) are synthesized not only from RN_3 and MR' according to path (a) of Eq. (1), but also from triazenes $RN=N-NHR'$ and strong bases. In this way, triazene **10** is obtained in good yields from diphenyltriazene (*18*):

$$Ph-N=N-N\begin{matrix}Ph\\H\end{matrix} \xrightarrow[-\ H_2]{+\ NaH} Ph-N=N-N\begin{matrix}Ph\\Na\end{matrix} \xrightarrow[-\ NaCl]{+\ Me_3SiCl}$$

$$Ph-N=N-N\begin{matrix}Ph\\SiMe_3\end{matrix} \quad (13)$$
$$\underline{\underline{10}}$$

Hydrogen in triazenes $RN=N-NHR'$ can also be substituted directly by EX_3 [replacement of $R'' = $ H by EX_3 in Eq. (12)] as shown in the reaction [Eq. (14)] of diphenyltriazene with trimethylstannyldiethylamine to form **11** (*19*) (cf. Table I). Also, a replacement of $R'' = $ silyl by EX_3 in Eq. (12)

$$Ph-N=N-N\begin{matrix}Ph\\H\end{matrix} \xrightarrow[-\ HNEt_2]{+\ Me_3SnNEt_2} Ph-N=N-N\begin{matrix}Ph\\SnMe_3\end{matrix} \quad (14)$$
$$\underline{\underline{11}}$$

is possible. Thus, triazene **12**, in the presence of silicon tetrachloride at room temperature, changes to triazene **13** (*7*) according to Eq. (15).

$$2Me-N=N-N\begin{matrix}Me\\SiMe_3\end{matrix} \xrightarrow[-\ 2Me_3SiCl]{+\ SiCl_4} Me-N=N-N\begin{matrix}Me\\Si\end{matrix}\begin{matrix}Me\\Cl_2\end{matrix}N-N=N-Me \quad (15)$$
$$\underline{\underline{12}} \qquad\qquad\qquad \underline{\underline{13}}$$

Triazene $MeN=N-NMe(SiCl_3)$ is formed as an intermediate but decomposes at once into **13** and $SiCl_4$. The same is observed in the reaction of **12** with $MeSiCl_3$ and Me_2SiCl_2.

Analogous to silyl-bound triazenes, germyl and stannyl groups also become substituted [replacement of $R'' = $ germyl or stannyl by EX_3 in Eq. (12)]. Even replacement of triazene-bound organyl groups R'' by EX_3

is possible, although only in a radical reaction. Thus, as shown in Eq. (16), action of triethylstannyl radicals on triorganyltriazene **14** provides stannyl-triazene **15** (*19*).[11] Accordingly, other organyl radicals, which are some-what stabilized, are substituted by stannyl radicals (*19*).

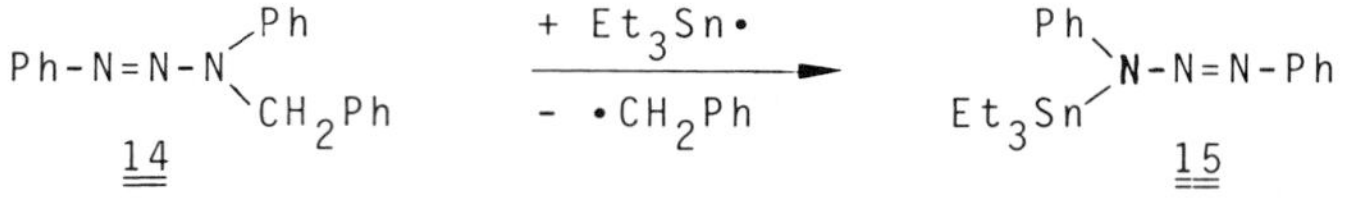

$$
\underset{\underline{14}}{Ph-N=N-N\overset{Ph}{\underset{CH_2Ph}{\diagdown}}}
\xrightarrow[-\ \cdot CH_2Ph]{+\ Et_3Sn\cdot}
\underset{\underline{15}}{Et_3Sn\overset{Ph}{\underset{}{\diagup}}N-N=N-Ph}
\qquad (16)
$$

Besides substitution on the triazene skeleton [cf. Eq. (12)], trans-formations of triazenes can also occur by substitution on the triazene substituents (*Preparation Method E*). For example, triazenes **16** react with LiOMe according to Eq. (17) to replace silicon-bound chloride by methyl-ate to form triazenes **17** (*7*). Since the silanes $(MeO)_nMe_{3-n}SiCl$ needed for direct preparation of **17** are difficult to obtain, the above-mentioned

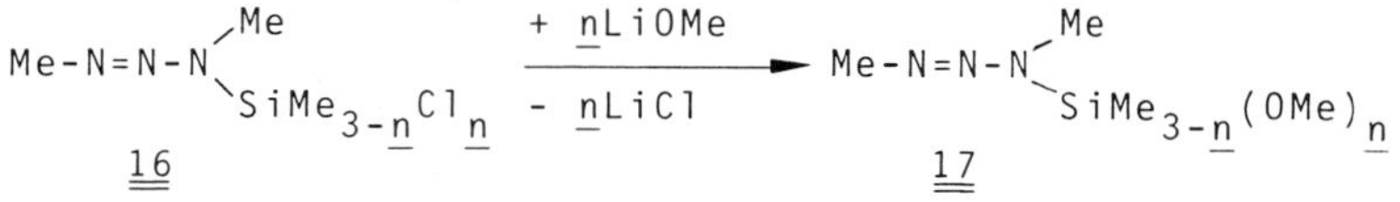

$$
\underset{\underline{16}}{Me-N=N-N\overset{Me}{\underset{SiMe_{3-n}Cl_n}{\diagdown}}}
\xrightarrow[-\ \underline{n}LiCl]{+\ \underline{n}LiOMe}
\underset{\underline{17}}{Me-N=N-N\overset{Me}{\underset{SiMe_{3-n}(OMe)_n}{\diagdown}}}
\qquad (17)
$$

approach is therefore favored.

B. *Properties*

1. *General Features, Structure, and Reactivity*

Some *physical properties* of silyl, germyl, and stannyl derivatives of triazene or organyltriazenes that have been isolated to date, along with their methods of preparation, are listed in Table I. In general, triazenes can be categorized into five types (**18a–18e**) of triazenes or organyl-

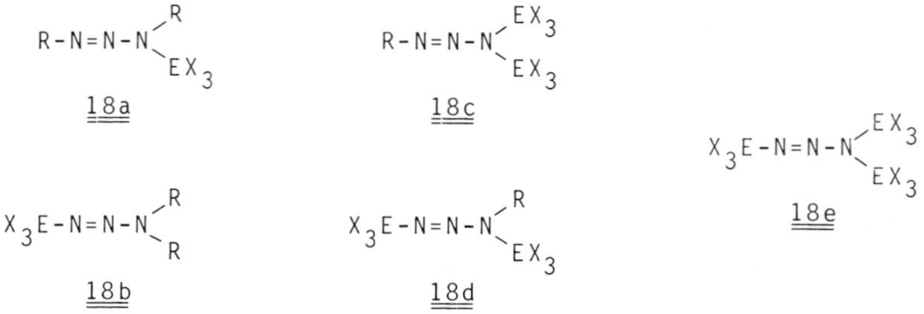

$$
\underset{\underline{18a}}{R-N=N-N\overset{R}{\underset{EX_3}{\diagdown}}}
\qquad\qquad
\underset{\underline{18c}}{R-N=N-N\overset{EX_3}{\underset{EX_3}{\diagdown}}}
$$

$$
\underset{\underline{18b}}{X_3E-N=N-N\overset{R}{\underset{R}{\diagdown}}}
\qquad\qquad
\underset{\underline{18d}}{X_3E-N=N-N\overset{R}{\underset{EX_3}{\diagdown}}}
$$

$$
\underset{\underline{18e}}{X_3E-N=N-N\overset{EX_3}{\underset{EX_3}{\diagdown}}}
$$

---

[11] Et₃Sn radicals are generated by the reaction of Et₃SnH with the existing CH₂Ph radicals. Accordingly, the formation of **15** involves a radical chain reaction. The overall equation runs as follows: **14** + Et₃SnH → **15** + CH₃Ph.

triazenes $[R_{3-n}(X_3E)_nN_3]$ containing one, two, or three EX_3 substituents (E = Si, Ge, Sn; R = H or organyl; n = 1, 2, 3) (cf. Section II,B,3). Example, for all of these except **18b** are listed in Table I. Acyclic triazenes, occur in the trans configuration (cf. Section III,B,2), whereas cyclic triazenes naturally display the cis configuration, as shown in the following structures:

Group IV derivatives of triazene and organyltriazenes are sometimes obtained colorless, sometimes yellow to red (Table I). The pure compounds are actually colorless except for triazenes of type $X_3E—N{=}N—N{<}(E$ = Si, Ge). The color originates from traces of impurities (azo compounds), which are difficult to remove. Triazenes studied by UV spectroscopy show an intense $\pi \to \pi^*$ absorption band at about $\bar{v}_{max}$ = 35000 cm^{-1}, which is also observed in organyltriazenes (Table I). In addition, t-Bu$_3$Si—N$=$N—NHSi(t-Bu)$_3$ shows a weak intensity band toward longer wavelength at $\bar{v}_{max}$(hexane) = 26050 cm^{-1}(ϵ = 60), which is assigned to an $n \to \pi^*$ transition and indicates a relatively high energy for the n_+ level (cf. Section III,B,2). This assignment is consistent with earlier results (1) which showed that the upper n level of an azo group is raised strongly on replacement of azo-bound alkyl by silyl groups.

Chemical properties of silyl-, germyl-, and stannyltriazenes have not been studied very systematically so far. Most of the known compounds are air stable and sensitive to hydrolysis. Thermostability of triazenes of type **18a** (R = organyl) is strikingly high, and these can often be heated above 150°C without decomposition. On the other hand, the known triazenes of type **18c** (R = aryl) decompose slowly at room temperature. Triazenes of type **18b**, **18d**, and **18e** are even more thermolabile and can be prepared only if triazene substituents are quite bulky [e.g., t-Bu$_3$ SiN$_3$H(Sit-Bu$_3$) is stable above 150°C]. For more details of triazene thermolysis, which occurs mainly with N_2 elimination, see Section II,C.

Some examples of substitution reactions of Group IV triazenes undergoing have already been discussed (Section II,A,3). In general, silyl, germyl, and stannyl groups are easily replaced by hydrogen, as already indicated by the hydrolytic sensitivity of these compounds. Further examples of substitution are the reactions of silyl- or stannyltriazenes with Mn(CO)$_5$Br, C$_5$H$_5$Mo(CO)$_3$Cl, or [Rh(CO)$_2$Cl]$_2$, which proceed accordingly to Eqs. (18), (19), and (20) (18). Radical substitution of stannyl

TABLE I.

$$R-N=N-N\diagdown_{R''}^{R'}$$

Type	R	R'	R"	Method	Equation	Melting point (°C)
		Acyclic triazenes				
18d	Si^{t}Bu$_3$	Si^{t}Bu$_3$	H	A	(1)	140
18e	Si^{t}Bu$_3$	SiMe$_3$	SiMe$_3$	A	(1)	ca. >35/dec
18e	Si^{t}Bu$_3$	Si^{t}Bu$_3$	SnMe$_3$	A	(1)	77–79
		Acyclic organyltriazenes				
18a	Me	Me	SiMe$_3$	A	(1)	<0
18a	Me	Me	SiMeEt$_2$	A	(1)	<0
18a	Me	Me	SiMe$_2$X^e	A	(1)	<0
18a	Me	Me	SiMe$_2$Cl	A	(1)	<0
18a	Me	Me	SiMe$_2$(OMe)	E	(17)	<0
18a	Me	Me	SiMeCl$_2$	A	(1)	<0
18a	Me	Me	SiMe(OMe)$_2$	E	(17)	<0
18a	Me	Me	Si(OMe)$_3$	A	(1)	<0
18a	Me	Me	SiCl$_2$X^e	A	(1)	99/dec
18a	Ph	Me	SiMe$_3$	A	(1)	<0
18a	Ph	Ph	SiMe$_3$^{g}	D	(13)^g	h
18d	^{t}Bu$_3$Si	Me	SiMe$_3$	A	(1)	<0
18d	^{t}Bu$_3$Si	Me	Si^{t}Bu$_3$	A	(1)	119
18c	Ph	SiMe$_3$	SiMe$_3$ cis	C	(8)	ca. −5
			trans	C	(8)	ca. −5
18c	Ph	SiEt$_3$	SiEt$_3$ i	C	(8)	−13
18c	Ph	SiMeEt$_2$	SiMeEt$_2$ cis	C	(8)	ca. −10
			trans	C	(8)	ca. −10
18c	p-Tolyl	SiMe$_3$	SiMe$_3$ cis	C	(8)	ca. −5
			trans	C	(8)	ca. −5
18c	o-Tolyl	SiMe$_3$	SiMe$_3$ cis	C	(8)	ca. −5
			trans	C	(8)	ca. −5
18c	Ph	Si(OMe)$_3$	Si(OMe)$_3$	C	(8)	ca. 0
18a	Me	Me	SnMe$_2$X^e	A	(1)	<0
18a	Me	Me	SnMe$_2$Br	A	(1)	<0
18a	Me	Me	SnMe$_2$I	A	(1)	<0
18a	Ph	Me	SnEt$_3$	D	(14)	<0
18a	Ph	Ph	SnMe$_3$	D	(14)	<0
18a	Ph	Ph	SnEt$_3$	D	(14)	<0
18a	Ph	CH$_2$Ph	SnEt$_3$	D	(14)	<0
18a	Ph	p-Tolyl	SnEt$_3$	D	(14)	<0
18a	p-Tolyl	p-Tolyl	SnEt$_3$	D	(14)	<0
		Cyclic triazenes^l				
18a	(Me$_3$Si)$_2$C—SiMe$_2$ (ring N–N–N, N—R")		p-Tolyl	B	(7)	67/dec
18a			CMe$_3$	B	(7)	76
18c			SiMe$_3$	B	(7)	15/dec
18c			SiMe$_2$^{t}Bu	B	(7)	12/dec
18c			SiMe^{t}Bu$_2$	B	(7)	4/dec
18c			Si^{t}Bu$_3$	B	(7)	n
18c			SiMe$_2$N(SiMe$_3$)$_2$	B	(7)	−10/dec

SILYL, GERMYL, AND STANNYL DERIVATIVES OF TRIAZENE (N_3H_3) AND ORGANYLTRIAZENES

Boiling point (sublimation point) (°C/torr)[c]	Color[a]	Decomposition (ca. °C)[b]	¹H-chemical shifts (TMS, ppm)[c]				References
			δ(R)	δ(R')	δ(R'')	Solvent	
110/HV	Light yellow[a]	>150	1.28	1.15	8.49	C_6D_6	8
Dec	Yellow[a]	ca. 70	1.23	0.299	0.299	C_6H_6	8, 9
	Light yellow	>150	1.25	1.25	0.487	C_6H_6	9
57–58/50	Colorless	>150	3.38	2.80	0.217	—	6, 7
56–57/5	Colorless	>150	3.38	2.80	0.233/0.9[d]	CH_2Cl_2	7
38/0.2	Colorless	>150	3.47	2.77	0.533	CH_2Cl_2[f]	6,7
49/14	Colorless	>150	3.47	2.80	0.583	C_6H_{12}[f]	7
50/4	Colorless	>150	3.45	2.78	0.233/3.52	CH_2Cl_2[f]	7
63–65/23	Colorless	>150	3.52	2.85	0.967	CCl_4[f]	7
50/4	Colorless	>150	3.50	2.83	0.317/3.52	CH_2Cl_2[f]	7
46–48/2	Colorless	>150	3.50	2.83	3.58	—	7
	Colorless	>150	3.40	3.17	—	CH_2Cl_2[f]	7
73–74/HV	Light yellow[a]	>150	m	3.02	0.267	—	7
h	Yellow	>150	m	h	h	—	18
Dec	Yellow[a]	ca. 70	1.27	2.73	0.215	C_6H_6	9
Dec	Yellow	>150	1.28	2.97	1.22	C_6H_6	9
Dec	Light yellow	70	m	0.083	0.083	C_5H_{12}	7
Dec	Orange[a]	70	m	0.317	0.317	C_5H_{12}	7
Dec	Orange	50	m	0.9–1.1	0.9–1.1	CCl_4	7
Dec	Yellow	60	m	0.050	0.050/0.94[d]	CH_2Cl_2	7
Dec	Orange	60	m	0.267	0.267/0.94[d]	CH_2Cl_2	7
Dec	Red	70	2.18[j]	0.067	0.067	Et_2O	7
Dec	Red	70	2.18[j]	0.267	0.267	Et_2O	7
Dec	Red	80	2.02[j]	0.067	0.067	CCl_4	7
Dec	Red	80	2.02[j]	0.267	0.267	CCl_4	7
Dec	Light red	60	m	3.52	3.52	CCl_4	7
50–52/0.2	Colorless	>150	3.31	3.31	0.72	C_6H_6	6
38/0.2	Colorless	>150	3.15	3.15	0.73	C_6H_6	6
50/0.2	Pale yellow	>150	3.18	3.18	0.96	C_6H_6	6
139–142/HV	Yellow	>150	m	3.42	1.1[d]	CCl_4	19
142–145/HV	Yellow	>150	m	m	0.35[d]	CCl_4	19
151–153/HV	Yellow	>150	m	m	1.13[d]	CCl_4	19
157–160/HV	Yellow	>150	m	5.27[k]	1.15[d]	CCl_4	19
155–158/HV	Yellow	>150	m	2.33[j]	1.15[d]	CCl_4	19
158–162/HV	Yellow	>150	2.33[j]	2.33[j]	1.15[d]	CCl_4	19
Dec	Colorless	45	0.163	0.487	2.3[j]	Et_2O	14
ca. 55/HV	Colorless	>50	0.212	0.403	1.405	Et_2O	16
Dec	Colorless	−14[m]	0.108	0.305	0.272	C_5H_{12}	14
Dec	Colorless	−17[m]	0.119	0.318	0.223/0.978	Et_2O	16
Dec	Colorless	−20[m]	0.144	0.357	0.188/1.04	Et_2O	16
Dec	Colorless	<−20	n	n	n		16
Dec	Colorless	<−20	0.135	0.388	0.366/0.244	Et_2O	16

(*continued*)

TABLE I (*continued*).

$R-N=N-N\begin{smallmatrix}R'\\R''\end{smallmatrix}$				Best preparation		Melting point (°C)
Type	R	R'	R"	Method	Equation	
18a	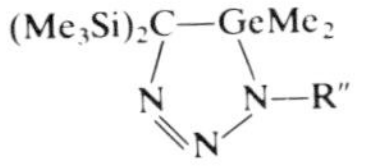 $(Me_3Si)_2C-GeMe_2$		CM_3	B	(7)	86
	$\underset{N\,N}{\overset{\displaystyle N-R''}{\underset{\displaystyle N}{\big|\quad\big|}}}$					
18c		SiMe₃		B	(7)	127
18c		SiMe₂'Bu		B	(7)	74
18c		SiMe'Bu₂		B	(7)	115
18c		SiMe₂N(SiMe₃)₂		B	(7)	86
18c		GeMe₂N(SiMe₃)₂		B	(7)	90

[a] UV absorption: PhN_3Me_2, $\bar{\nu}_{max} = 35250$ cm^{-1}; $PhN_3Me(SiMe_3)$, $\bar{\nu}_{max} = 35600$ cm^{-1}; $PhN_3(SiMe_3)_2$, $\bar{\nu}_{max} = 34050$ cm^{-1}; $^tBu_3SiN_3H(Si^tBu_3)$, $\bar{\nu}_{max} = 38610$ cm^{-1} ($\varepsilon = 15900$), 26040 cm^{-1} ($\varepsilon = 76$); $^tBu_3SiN_3(SiMe_3)_2$, $\bar{\nu}_{max} = 38760$ cm^{-1} ($\varepsilon = 2033$), 26040 cm^{-1} ($\varepsilon = 23$); $^tBu_3SiN_3Me(SiMe_3)$, $\bar{\nu}_{max} = 40000$ cm^{-1} ($\varepsilon = 11215$), 26880 cm^{-1} ($\varepsilon = 91$). With the exception of triazenes of type $X_3E-N=N-N<$ (E = Si, Ge), all other compounds are probably colorless.
[b] For half-life ca. 1 hour, and dilute solutions in benzene.
[c] HV, high vacuum; dec, decomposition; m, multiplet.
[d] Multiplet.
[e] X = $-MeN-N=NMe$.
[f] $< -40°C$.

groups by other stannyl groups, as shown in Eq. (21), must also be mentioned here (*19*).

$$Ar_2N_3EMe_3 + Mn(CO)_5Br \longrightarrow Ar_2N_3Mn(CO)_4 + Me_3EBr + CO \quad (18)$$

$$Ar_2N_3EMe_3 + C_5H_5Mn(CO)_3Cl \longrightarrow Ar_2N_3Mn(CO)_2(C_5H_5) + Me_3ECl + CO \quad (19)$$

$$2\ Ar_2N_3EMe_3 + [Rh(CO)_2Cl]_2 \longrightarrow [Ar_2N_3Rh(CO)_2]_2 + 2\ Me_3ECl \quad (20)$$

$$ArN=N-NAr(SnEt_3) + \cdot SnMe_3 \longrightarrow (Me_3Sn)ArN-N=NAr + \cdot SnEt_3 \quad (21)$$

Substitutions of the types shown in Eqs. (18)–(20) are probably acid/base reactions between substrates acting as Lewis acids, and the triazenes behaving as Lewis bases. Also, in the reaction between trimethyl-silyldimethyltriazene and trimethylchlorosilane in the presence of MgClI [$MeN=N-NMe(SiMe_3) + Me_3SiCl + MgClI \rightarrow MeN(SiMe_3)_2 + MeI + N_2 + MgCl_2$], a Lewis acid/base adduct of MgClI and $(Me_3Si)N_3Me_2$

SILYL, GERMYL, AND STANNYL DERIVATIVES OF TRIAZENE (N_3H_3) AND ORGANYLTRIAZENES

Boiling point (sublimation point) (°C/torr)[c]	Color[a]	Decomposition (ca. °C)[b]	[1]H-chemical shifts (TMS, ppm)[c]				
			δ(R)	δ(R')	δ(R'')	Solvent	References
60/HV	Colorless	90	0.092	0.640	1.388	Et_2O	15
60/HV	Colorless	80	0.087	0.268	0.564	Et_2O	15
Dec	Colorless	50	0.111	0.412	0.040/1.08	C_6H_6	15
Dec	Colorless	60	0.136	0.450	0.084/1.13	C_6H_6	15
Dec	Colorless	40	0.103	0.593	0.084/1.13	Et_2O	15
Dec	Colorless		0.092	0.585	0.360/0.238	Et_2O	15
Dec	Colorless				0.664/0.211	Et_2O	15

[g] In an analogous manner, triazenes with R = R' =p-tolyl, and p-chlorophenyl are prepared (18); properties not reported.
[h] Not reported.
[i] Probably mainly trans (cf. Section II,B,2).
[j] MeC$_6$H$_4$ protons; MeC$_6$H$_4$: multiplet.
[k] CH$_2$C$_6$H$_4$ protons; CH$_2$C$_6$H$_4$: multiplet.
[l] Sila- and germatriazolines. Not included are triazolines and triazoles, which have only silyl, germyl, or stannyl ring substituents (cf. ref. 19a).
[m] In Et_2O.
[n] Not isolable, because of decomposition during preparation.

most probably plays a role as an intermediate (7). Finally, preliminary results (9) on the possibility of protonation of (t-Bu$_3$Si)$_2$N$_3$H, according to path (a) of Eq. (22), indicate Brönsted base character for the triazene.[12] In contrast, triazenes are also able to act as Brönsted acids (2, 8) as is shown by the easy deprotonation of (t-Bu$_3$Si)$_2$N$_3$H and other disubstituted triazenes with bases [path (b) of Eq. (22)].

$$\left[\underset{H}{\overset{\diagdown}{}}N \cdot\cdot N \cdot\cdot N \underset{H}{\overset{\diagup}{}} \right]^{+} \xrightarrow[(a)]{+H^+} \ -N=N-N \underset{H}{\overset{\diagup}{\diagdown}} \ \xrightarrow[(b)]{-H^+} \ \left[-N \cdot\cdot N \cdot\cdot N - \right]^{-} \quad (22)$$

As an intermediate in Eq. (21), an adduct [19 (E = Sn)] of a trimethyl-stannyl radical with triethylstannyldimethyltriazene may be formed according to path (a) of Eq. (23) (19). Radicals of type 19 provide possible

$$X_3E \overset{\diagdown}{}N-\dot{N}-N \overset{\diagup}{} EX_3 \xleftarrow[(a)]{+EX_3} \ -N=N-N \overset{\diagup}{\underset{EX_3}{}} \xrightarrow[(b)]{-EX_3} \ -N \cdot\cdot \dot{N} \cdot\cdot N - \quad (23)$$
$$\underline{\underline{19}} \qquad\qquad\qquad\qquad\qquad\qquad\qquad\qquad\qquad \underline{\underline{20}}$$

[12] Certainly, the proton adducts are unstable. (t-Bu$_3$Si)$_2$N$_3$H, for example, reacts with methanol with splitting of the nitrogen chain (9).

intermediate steps leading to triazane derivatives by redox reactions. In an analogous manner, radicals of type **20** could be the intermediates of redox reactions of triazenes leading to azide derivatives. Whether such radicals are formed according to path (b) of Eq. (23) is not known as yet. Many cases of radicals of type **20** can be produced by radical addition to azides (*13*) and may perhaps be formed by the oxidation of $(t\text{-}Bu_3Si)_2N_3^-$ with atmospheric oxygen (*8, 9*).

2. Cis/Trans *Isomerizations*

As a rule, the cis–trans equilibrium of acyclic azo compounds (—N=N—) lies extensively on the trans side. Thus, the reaction enthalpy for cis → trans transformation of azobenzene amounts to almost 40 kJ/mol (*20*). The same is probably true for triazene and its derivatives (—N=N—N⌐).[13] The disilylaryltriazenes are exceptions (cf. Table I), wherein the cis forms are only slightly higher in energy than the trans forms and a moderate concentration of the cis isomer is present even at room temperature [Eq. (24)] (*22*). Thus, the free reaction enthalpy (ΔG) of the

$$
\underset{\text{trans}}{\overset{Ar}{\underset{}{}}\hspace{-0.5em}\diagdown N{=}N\diagup N(SiX_3)_2} \;\;\rightleftarrows\;\; \underset{\text{cis}}{\overset{Ar}{}\hspace{-0.5em}\diagup \; N{=}N\diagup N(SiX_3)_2} \tag{24}
$$

weakly endothermic and exotropic trans → cis transformation of Ph—N=N—N(SiMe₃)₂ in pentane at 25°C amounts to only 4.9 kJ/mol [K (25°C) = 0.14 in pentane; ΔH = 14.1 kJ/mol; ΔS = −63.6 eu]. The equilibrium shifts more strongly toward the right with an increasing polarity of the solvent [K (19.5°C) = 0.32 in ether; K (10°C) ≃ 3 in THF]. A similar relationship of the trans → cis transformation equilibrium is observed in the case of *p*-TolN=N—N(SiMe₃)₂, although the proportion of cis-Ph—N=N—(SiMeEt₂)₂ in equilibrium with *trans*-Ph—N= N—N(SiMeEt₂)₂ is small because of steric factors (K (10°C) ≃ 0.01 in THF). The reason for the surprisingly "balanced" equilibrium of Eq. (24) must be due to an unexplained special stability of the cis compounds, because from steric and mesomeric relationships, the trans compounds should be favored energetically (*22*).

The configuration change of disilylaryltriazene, analogous to that of azobenzene, proceeds by inversion of the azo-nitrogen atom bound to the

---

[13] In the case of BrC₆H₄—N=N—NHC₆H₄Br, the trans configuration has been established by X-ray analysis (*21*).

aryl group, as shown by the arrow in Eq. (24).[14] Consequently, cis → trans isomerization of PhN=NPh and PhN=N—N(SiMe$_3$)$_2$ is comparatively fast. The rate of isomerization of the latter compound ($\tau_{1/2} \simeq 1$ hour at room temperature) depends little on the polarity of the reaction medium. If the configuration change occurred with rotation about the N=N double bond, a large increase in the rate of isomerization would have been expected with increasing solvent polarity because of the polarity of the rotational transition intermediate [Ar—N$^-$—N=N$^+$(SiMe$_3$)$_2$] (22).

3. 1,3-Migrations

As expected, the ^{1}H-NMR spectrum of MeN=N—NMe(SiMe$_3$) consists of three signals with relative areas 1:1:3. Signals due to both N-methyl groups are relatively broad. With increasing probe temperature, these signals broaden further and coalesce at 50.5°C. This phenomenon, which is also observed in other triazenes of type RN=N—NR(SiX$_3$) (cf. Table II), indicates rapid exchange of silyl groups between nitrogen atoms 1 and 3 of the trans triazene chain (6, 24), as shown in Eq. (25). The 1,3-silyl group

$$\text{(25)}$$

rearrangement must occur intramolecularly because the coalescence temperatures are independent of concentration of the compound. The transition state of the rearrangement involves an N$_3$Si four-membered ring with penta-coordinated silicon.

As inferred from values in Table II, the tendency of silyl groups to migrate increases with silicon-bound substituents in the order Et < Me < OMe < Cl. These results can be explained as follows: In the direction Et < Me < OMe < Cl, the substituents raise the acidity of silicon either through decreasing bulkiness or by withdrawing electrons, and thus the bonding capacity of silicon increases with the result that silyl migration can occur more easily.

In contrast to triazenes RN=N—NR(SiX$_3$) with identical organyl groups, the molecules PhN=N—NMe(SiMe$_3$) and ArN=N—N(SiX$_3$)$_2$ exist only in the form **21a**. Evidently, the difference in the free enthalpies of tautomers **21a** and **21b** is so large that the energetically unstable isomer

---

[14] The rate of nitrogen inversion in systems R$_2$C=N—Z increases in the series Z = R$_2$N < alkly < aryl (23). Accordingly, for triazenes ArN=N—NR$_2$ an inversion at the amino-nitrogen atom is ruled out.

TABLE II

COALESCENCE OF ^{1}H-NMR SIGNALS OF THE METHYL GROUPS OF MeN=N—NMe(SiX$_3$)

	SiX$_3$								
	SiMe$_3$	SiMeEt$_2$	SiMe$_2$X^a	SiMe$_2$Cl	SiMe$_2$(OMe)	SiMeCl$_2$	SiMe(OMe)$_2$	Si(OMe)$_3$	SiCl$_2$X^a
T_c^{Me} (°C)	50.5	54.0	33.0	8.0	33.0	3.0	27.0	26.0	18.0^b
$\Delta G_c^{\ddagger}$ (kJ/mol)	67.4	68.2	63.2	57.8	63.2	56.9	62.4	62.0	63.2
Solvent	—	C$_5$H$_{12}$	C$_5$H$_{12}$	C$_5$H$_{12}$	C$_5$H$_{12}$	C$_5$H$_{12}$	C$_5$H$_{12}$	C$_5$H$_{12}$	Et$_2$O

^{a} X = —MeN—N=NMe.
^{b} The two signals for the NMe protons coalesce at 18°C and each splits into two at temperatures < -20°C (δ = 3.41 and 3.39; 3.18 and 3.16).

21b, with an SiX$_3$ group on the less basic aryl bound nitrogen atom, is not present in amounts detectable by ^{1}H-NMR (24). An indication that **21b** is in equilibrium with **21a** follows from the thermolysis of the compound (Section II,B,5).

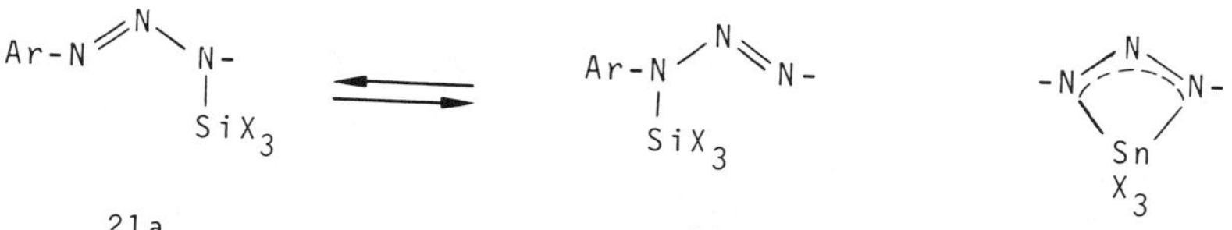

By analogy to silyl groups, hydrogen in triazenes of type RN=N—NHR can migrate between positions 1 and 3 (2). Thus, for example, coalescence of both the ^{1}H-NMR signals of the t-Bu$_3$Si groups in t-Bu$_3$Si—N=N—NHSi(t-Bu)$_3$ is 55°C in THF ($\Delta G_c^{\ddagger}$ = 78 kJ/mol) (8). Further, 1,3-rearrangements are also expected for germyl groups in triazenes of type RN=N—NR(GeX$_3$). On the other hand, stannyl groups in triazenes of type RN=N—NR(SnX$_3$) are believed to bind simultaneously to both nitrogen atoms 1 and 3 of the triazene chain as shown in formulation **22**, so that the transition state of 1,3-silyl group migration [Eq. (25)] becomes the molecular ground state.

4. Hindered Rotations

The ^{1}H-NMR spectrum of *trans*-PhN=N—N(SiMe$_3$)$_2$ is temperature dependent; with decreasing temperature of the NMR probe below the coalescence temperature T_c = −70.5°C (in pentane), the proton resonance signal of the Me$_3$Si groups broadens to split into two signals of equal intensity. These findings indicate an increasingly "frozen" rotation around the N—N single bond with decreasing temperature [Eq. (26)] (25).

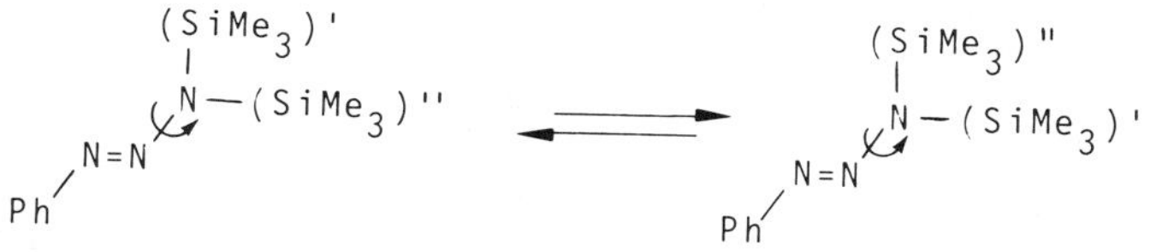

$$(26)$$

The free activation enthalpy of rotation ($\Delta G_c^{\ddagger}$) is calculated to be 46 kJ/mol at the coalescence temperature; this is significantly higher than the rotation barrier for a normal single bond, which is ~10 kJ/mol (26). The increased barrier in triazenes is consistent with resonance (**23a** ↔ **23b**) (27), providing partial double bond character between nitrogen atoms 2 and 3, as suggested in the following:

$$\left[-\overset{..}{N}=\overset{..}{N}-\overset{..}{N}\diagdown \quad \longleftrightarrow \quad -\overset{-}{\overset{..}{N}}-\overset{..}{N}=\overset{+}{N}\diagdown \right]$$

$$\underset{23a}{} \qquad\qquad\qquad \underset{23b}{}$$

In rotation to the transition state, the contribution of **23b** falls off to zero. The rotation barrier of *trans*-PhN=N—NMe$_2$ (ΔG_c = 53 kJ/mol at T_c = −23.5°C in HCCl$_3$) is higher than that of *trans*-PhN=N—N(SiMe$_3$)$_2$ (*28*) because carbon, in contrast to silicon, has no ability to stabilize by $d_\pi p_\pi$-backbonding the free electron pair of the amino nitrogen in rotation transition state **23a**. If the $d_\pi p_\pi$-backbonding strength is reduced by replacing trimethylsilyl groups in *trans*-Ph—N=N—N(SiMe$_3$)$_2$ by ethyldimethylsilyl groups, the rotation barrier ($\Delta G^\ddagger$) increases as expected and amounts to 47 kJ/mol at T_c = −65.0°C in pentane.

The ^{1}H-NMR spectra of *cis*-PhN=N—N(SiMe$_2$R)$_2$ (R = Me, Et) do not change with decreasing temperature (*25*). These findings agree with the structure of the compounds; steric factors hinder the rotation of N(SiMe$_2$R)$_2$ groups around the N—N bond and force the N(SiMe$_2$R) groups to a position with symmetrically equivalent silyl groups (*22*).

Similarly, in the case of *trans*-PhN=N—NMe(SiMe$_3$) and *trans-t*-Bu$_3$SiN=N—NH[Si(*t*-Bu)$_3$], ^{1}H-NMR signals do not split as a result of decreasing probe temperatures. In these compounds, one rotamer is much more stable than the other due to steric and other factors, therefore one of the two isomeric forms is not detectable by NMR spectroscopy because of its small equilibrium concentration.

C. Thermolysis

1. Acyclic Triazenes

Tri- and diorganyltriazenes (RN=N—NR$_2'$ and RN=N—NHR') thermolyze, in general above 100°C, according to Eq. (27) (*Thermolysis*

$$R-N=N-N\diagdown \quad \xrightarrow{\quad \text{Thermolysis path I} \quad} \quad R\cdot \; + \; N\equiv N \; + \; \cdot N\diagdown \qquad (27)$$

Pathway I), with the evolution of nitrogen, into radicals R and NR$_2'$, which combine to form amines (RNR$_2'$) or abstract hydrogen to form RH and HNR$_2'$ (*5, 19*). Silyl-, germyl-, and stannyltriazenes of type RN=N—NR'(EX$_3$) are more stable than the above-mentioned triazenes but decompose by the same radical Thermolysis Pathway I (*19, 29*). In the best studied case of PhN=N—NMe(SiMe$_3$) (**24**), thermolysis proceeds at 180°C in cyclopentane (12 hours) according to Eq. (28), with the elimination of N$_2$ to form PhNMe(SiMe$_3$) [path (a), Eq. (28)] or PhNH(SiMe$_3$)

$$Ph-N\overset{Me}{\underset{SiMe_3}{\diagdown}} + N_2$$

$$Ph-N=N-N\overset{Me}{\underset{SiMe_3}{\diagdown}} \quad \overset{(a)}{\underset{(81\%)}{\nearrow}} \quad \overset{+2H\ (19\%)}{\underset{(b)}{\searrow}} \quad Ph-N\overset{H}{\underset{SiMe_3}{\diagdown}} + HMe + N_2 \tag{28}$$

$$\underset{24}{}$$

and methane [path (b), Eq. (28)]. The actual cleavage of triazene **24** into radicals probably proceeds with 3,1-silyl group migration to form first an energy-rich triazene **25**; this compound decomposes more rapidly to give the resonance-stabilized PhNSiMe$_3$ radical [Eq. (29)]. Equilibrium **24** ⇄ **25** lies extensively to the left (**25** could not be detected by ^{1}H-NMR spectrum). This factor probably accounts for the high metastability of **24**.

$$\underset{24}{} \rightleftharpoons \quad \overset{Ph}{\underset{Me_3Si}{\diagdown}}N-N=N-Me \longrightarrow \overset{Ph}{\underset{Me_3Si}{\diagdown}}N\cdot + N\equiv N + \cdot Me \tag{29}$$

$$\underset{25}{} \qquad\qquad\qquad cf.\ Eq.\ (24)$$

In contrast to the comparatively more stable silyldiorganyltriazenes RN=N—NR′(SiX$_3$) mentioned above, the disilylorganyltriazenes RN=N—N(SiX$_3$)$_2$ are quite thermolabile. For example, PhN=N—N(SiMe$_3$)$_2$ decomposes near room temperature, very slowly in nonpolar solvents or rapidly in polar solvents, with the cleavage of nitrogen to form mainly PhN(SiMe$_3$)$_2$ [Eq. (30)] (29). Analagous reactions occur for other disilyl-

$$Ph-N=N-N\overset{SiMe_3}{\underset{SiMe_3}{\diagdown}} \longrightarrow Ph-N\overset{SiMe_3}{\underset{SiMe_3}{\diagdown}} + N_2 \tag{30}$$

triazenes. In addition, Me$_3$SiC$_6$H$_4$NHSiMe$_3$ is formed (19) in increasing amounts with rising solvent polarity.[15]

In contrast to triazenes RN=N—NR′(SiX$_3$), the thermolysis of di-silylaryltriazenes (**26**) occurs with a nonradical mechanism, as shown in Eq. (31). In this reaction, a silyl group at nitrogen atom 3 first migrates

$$\overset{Ar}{\underset{\underset{26}{|}}{\diagdown}}\overset{\displaystyle\ddot{N}=\ddot{N}}{\underset{N-SiX_3}{}} \rightleftarrows \overset{Ar}{X_3Si}\overset{\diagdown\ddot{N}-\ddot{N}}{\underset{\ddot{N}-SiX_3}{\diagup}} \overset{}{\underset{-N_2}{\longrightarrow}} \overset{Ar}{X_3Si}\overset{\diagdown N-SiX_3}{\underset{}{\diagup}} \tag{31}$$

$$\underset{26}{SiX_3} \qquad\qquad \underset{27}{}$$

---

[15] For example, in pentane (THF) the yield is 80% (67%) PhN(SiMe$_3$)$_2$, and 20% (30%) Me$_3$SiC$_6$H$_4$NHSiMe$_3$.

to nitrogen atom 1 with the formation of triazene **27** (not detectable by ^{1}H NMR), which, on transfer of the second silyl group, evolves nitrogen (29). The initial silyl group migration occurs from the trans-triazenes **26**; cis-triazenes, therefore, must first isomerize to trans-configurated triazenes **26** before thermolysis. The rate of nitrogen elimination [Eq. (31)] is influenced by electronic as well as steric effects of substituents bound to silicon and to aryl groups. The rate increases in the sequence X = MeO < Me < Et that is with decreasing electronegativity of substituents X and in the sequence Ar = o-Tol < Ph < p-Tol.[16]

The decomposition of triazenes into amine and nitrogen according to Eq. (32) (*Thermolysis Pathway II*), occurs fast and nonradically when the

$$\begin{array}{c} \text{Thermolysis path II} \\ \searrow N-N=N- \xrightarrow{\hspace{3cm}} N\equiv N \ + \ -N\hspace{-0.2em}\diagdown \end{array} \qquad (32)$$

substituents bound to the azo nitrogen are able to migrate, as in the case of the silyl groups in **27**. Because the azo group of disilylorganyltriazenes of type **18c** [RN=N—N(EX$_3$)$_2$ (E = Si and probably Ge and Sn also)] is bound to an organyl group R (slow in migration), the triazene decomposition can therefore follow Pathway II after rearrangement to R(X$_3$E)N—N=N—EX$_3$, which possesses an azo-bound EX$_3$ group capable of migration. The rate of triazene decomposition depends on the substituents R and EX$_3$, which determine the amount of energy required for triazene rearrangement. For disilylaryltriazenes ArN=N—N(SiX$_3$)$_2$, the energy of rearrangement is so high that the triazenes are just isolable under normal conditions. Higher thermolability is expected for disilylorganyltriazenes of type **18d** [X$_3$E—N=N—NR(EX$_3$)], unless X$_3$E migration is strongly hindered on steric grounds. The same is true for silyldiorganyltriazenes of type **18b** (X$_3$E—N=N—NR$_2$) or **18e** [X$_3$E—N=N—N(EX$_3$)$_2$], or for triazenes in which EX$_3$ groups are partly or fully replaced by hydrogen, which is equally capable of migration. Thus, according to our investigations, a triazene of the type Me$_3$Si—N=N—N (SiMe$_3$)$_2$ is unstable under normal conditions (cf. Section II,A). Moreover, the bulky triazene t-Bu$_3$Si—N=N—NH[Si(t-Bu)$_3$] decomposes, into nitrogen and the amine t-Bu$_3$SiNH[Si(t-Bu)$_3$] with migration of an azo-bound silyl group, but only at higher temperatures (>150°C) (8, 9) [Eq. (32)]. In the case of silyldiorganyltriazenes type **18a** [RN=N—NR(EX$_3$)], the azo nitrogen carries an organyl group in the unrearranged form as well as in the rearranged triazene

---

[16] Values of $\tau_{1/2}$ (in minutes, at 80°C, mesitylene) for various Ar/SiX$_3$ are as follows: Ph/Si(OMe)$_3$, 96; Ph/SiMe$_3$ 16; Ph/SiEt$_3$, 2; o-Tol/SiMe$_3$, 26; p-Tol/SiMe$_3$, 13.

$[(X_3E)RN—N≡N—R]$. Such triazenes can thus decompose only by the free radical path of Thermolysis Pathway I and are, therefore, thermally as stable as triorganyltriazenes $(RN≡N—NR_2)$.

2. Cyclic Triazenes

Sila- and germatriazolines of type **28** (cf. Table I) decompose thermally neither according to Eq. (27) nor Eq. (32). 3-*tert*-Butyl-4,4-dimethyl-5,5-bis(trimethylsilyl)-4-silatriazoline and its germa analogue (**28**; E = Si, Ge; R″ = Me_3C) decompose quantitatively in benzene at 100°C according to path (a) of Eq. (33) to form the diazomethane derivates **29** (*15, 16*).

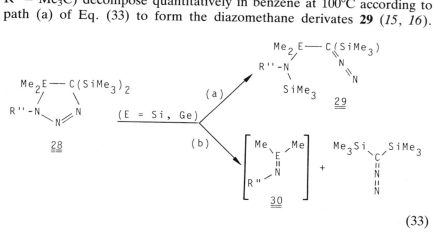

(33)

Accordingly, the ring in **28** opens between nitrogen atoms 2 and 3 of the triazene chain with simultaneous migration of the carbon-bound silyl group to the amino nitrogen of the triazene chain. Molecular nitrogen is not set free; it remains as a diazo group in the molecule. 3,5,5-Tris(trimethylsilyl) -4,4-dimethyl-4-silatriazoline (**28**; E = Si; R″ = Me_3Si) decomposes even below 0°C ($\tau_{1/2}$ at −17°C = 161 minutes in Et_2O), according to path (a) of Eq. (33). The lower thermolysis temperature is probably due to limited steric shielding of the amino nitrogen on the triazene chain. In fact thermolysis occurs both by path (a) [Eq. (33)] (57%) and by [2 + 3]-cycloreversion (~43% yield) according to path (b) of Eq. (33), leading to bis(trimethylsilyl)diazomethane as well as the unsaturated silicon compound N-trimethylsilyldimethylsilaketimine (**30**). The latter dimerizes or polymerizes promptly to form $[—Me_2Si—N(SiMe_3)—]_n$ (**14**). The germanium analogue **28** (E = Ge; R″ = Me_3Si) decomposes even more slowly than the silicon compound exclusively according to path (b) (*15*). Analogous behaviour is observed for sila- and germatriazolines with bulky groups R″, such as t-Bu$_n$Me$_{3-n}$Si, which sterically block silyl group rearrangement according to path (a) of Eq. (33), cf. ref. (14a).[17]

[2 + 3]-Cycloreversion of triazolines of type **28** provides an especially useful preparative route to unsaturated silicon and germanium compounds of type **30** (cf. Section III,C,2). These sila- and germaketimines may be intercepted in the presence of an appropriate trapping agent. Thus, the intermediates **30** react (1) with silicon, germanium, and tin halides with insertion in the Si—X, Ge—X, and Sn—X bonds, (2) with acetone or isobutene to form an ene reaction product, and (3) with dimethylbutadiene, covalent azides, or ketimines in a [2 + 4]-, [2 + 3]-, or [2 + 2]-cycloaddition (*14a*).

III

GROUP IV DERIVATIVES OF TETRAZENE, N₄H₄

A. Preparation

Tetrazenes can be prepared by the two methods shown in Scheme 2: (1) *building of nitrogen chains*, and (2) *exchange of substituents* (mutual transformation of tetrazenes).

1. Building of Nitrogen Chains

Organic tetrazenes can be formed in high yield by the oxidation of 1,1-bis(organyl)hydrazines R$_2$N—NH$_2$ (*31*). On the other hand, bis(silyl)hydrazine (**31**) is converted only indirectly [Eq. (34)] to tetrakis(silyl)tetrazene (**32**) (*1, 32, 33*).[18] The last synthetic step, involving dimerization of azosilane, shows low yield thermally, but high yield (90%) in the presence of the Lewis acid SiF$_4$ as catalyst.

---

[17] Ratio (%) of products [a/b, Eq. (33)] is as follows:

					R″			
E		Me$_3$Si	t-BuMe$_2$Si	t-Bu$_2$MeSi	t-Bu$_3$Si	Ph$_3$Si	Me$_3$C	p-Tol
Si		57/43	44/56	0/100	0/100	70/30	100/0	34/66
Ge		0/100	0/100	0/100	—	—	—	—

---

[18] Direct oxidation of **31** leads to molecular nitrogen.

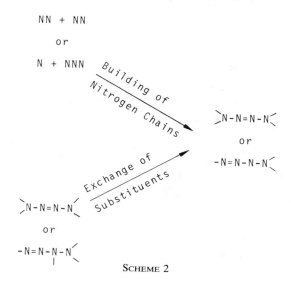

SCHEME 2

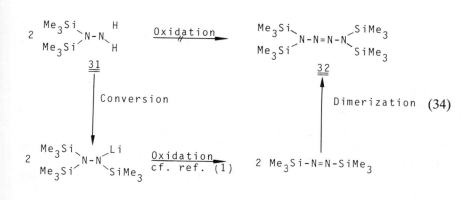

Tetrazene formation from inorganic diazenes, as in Eq. (35), is a

$$-N=N- + -N=N- \longrightarrow \text{>}N-N=N-N\text{<} \qquad (35)$$

general reaction (*Preparation Method F*). Thus, besides Me₃Si—N≡N—SiMe₃, other silyldiazenes such as Me₃Si—N≡N—H (*34*), Me₃Si—N≡N—SiF₃ (*1, 33*), and Me₂Si(N≡N—SiMe₃)₂ (*1, 35*) also react by dimerization of the azo system,[19] e.g., Eq. (36). Furthermore, germyldiazene

---

[19] Silyl- and germylorganyldiazenes (X₃E—N=N—R) (*1*) could not be dimerized so far.

$$
\begin{array}{c}
\underset{\underset{Me_3Si}{N}}{\overset{\overset{Me_2}{\underset{|}{Si}}}{N}} \quad \overset{\overset{N}{||}}{\underset{SiMe_3}{N}}
\end{array}
\quad \xrightarrow{\quad cf.\ ref.\ (1)\quad} \quad
\begin{array}{c}
Me_3Si-N \overset{\overset{Me_2}{\underset{}{Si}}}{\diagdown} N-SiMe_3 \\
N=N
\end{array}
\qquad (36)
$$

Me_3Ge—N≡N—$GeMe_3$ yields germyltetrazene $(Me_3Ge)_2N$—N≡ N—$N(GeMe_3)_2$ (36), along with other products. Finally, diazene N_2H_2 itself can also be transformed by acid catalysis into tetrazene N_4H_4 (37). The observed catalysis of dimerization [Eq. (35)] by an acid A probably depends on the formation of an adduct —AN≡N—, which can further add to free diazene (cf. refs 1 and 38).

Another method for synthesizing Group IV derivatives of tetrazenes by constructing nitrogen chains depends on coupling of hydrazides with aryldiazonium salts (*Preparation Method G*). Thus, the reaction of ArN_2Cl (Ar = Ph, *p*-Tol) with lithium tris(trimethylsilyl)hydrazide in diethyl ether leads to 1-tetrazene (33) in ~50% yield [path (a), Eq. (37)]. At the same

$$
\begin{array}{c}
\overset{+}{Ar}-N≡N\ Cl^{-} \\[6pt]
+ \\[6pt]
\underset{Me_3Si}{\overset{Li}{\diagup}} N-N \overset{SiMe_3}{\underset{SiMe_3}{\diagdown}}
\end{array}
\quad
\xrightarrow[\underset{(b)}{-LiCl}\ +2H]{\overset{(a)}{\underset{Ar=Ph,\ p-Tol}{}}}
\quad
\begin{array}{c}
Ar-N=N-N-N \overset{SiMe_3}{\underset{\underset{SiMe_3}{|}}{\diagdown}} SiMe_3 \\
\underline{\underline{33}} \\[10pt]
Ar-H\ +\ N≡N\ +\ \underset{Me_3Si}{\overset{H}{\diagup}} N-N \overset{SiMe_3}{\underset{SiMe_3}{\diagdown}}
\end{array}
\qquad (37)
$$

time, an unwanted decomposition develops according to path (b) of Eq. (37). The intermediates in the latter reaction are evidently Ar and $(Me_3Si)_3N_2$ radicals (39).

The tetrazene syntheses [Eq. (35) and path (a), Eq. (37)] develop according to the general scheme NN + NN → NNNN. One possible synthesis of Group IV derivatives of tetrazene according to scheme N + NNN → NNNN consists of [2 + 3]-cycloaddition of azides to sila- or germaketimines (stannaketimines are still not known) to form cyclic tetrazenes 34 (sila- or germatetrazolines) [Eq. (38)] (*Preparation*

$$
\begin{array}{c}
\overset{\diagdown}{\underset{\diagup}{E}}=N \diagdown R' \\[6pt]
+ \\[6pt]
R-N=N=N
\end{array}
\quad \xrightarrow{\quad E\ =\ Si,\ Ge\quad} \quad
\begin{array}{c}
R-N \overset{\overset{\diagdown E \diagup}{}}{\diagdown} N-R' \\
N=N \\[4pt]
\underline{\underline{34}}
\end{array}
\qquad (38)
$$

Method H). Thus, the reaction of sila- or germaketimines Me$_2$E=NR′ (R′ = SiMe$_n$t-Bu$_{3-n}$, SiPh$_3$; prepared by the thermolysis of silatriazolines, Section II,C,2) with azides RN$_3$ (R = Me$_3$C, Me$_n$t-Bu$_{3-n}$Si) proceeds smoothly to [2 + 3]-cycloadducts **34** with Me$_2$E ring members (*15, 16, 40*) (for individual compounds see Table III).

2. *Exchange of Substituents*

An exchange of a tetrazene-bound substituent by another substituent, as in Eq. (39) (*Preparation Method I*), is an especially important and

$$\underset{R'}{\overset{R}{\diagdown}}N-N=N-N\underset{R'''}{\overset{R''}{\diagup}} \quad \xrightarrow{\text{R''' substitution by R''''}} \quad \underset{R'}{\overset{R}{\diagdown}}N-N=N-N\underset{R''''}{\overset{R''}{\diagup}} \quad (39)$$

versatile technique for synthesis of Group IV derivatives of tetrazene (cf. Section III,D,2). Above all, we mention the *protolysis* of tetrazenes (Me$_3$E)$_n$N$_4$H$_{4-n}$ ($n \leqslant 4$, E = only one or mixed Si, Ge, Sn), which in general follows the path from top left to bottom right in Scheme 3 (*41, 42*). In this way, the reaction of **35** (E = Si) with CF$_3$COOH in methylene chloride at $-78°$C leads to tetrazene (N$_4$H$_4$) (*43*) via **36**, **37b**, and **38** (E = Si). Corresponding to the molar ratio of reactants, mixtures containing **36**, **37b**, or **38** as chief products are formed, and these can be separated into individual components (*41*). Tetrazene **37a** (E = Si) does not arise from the protolysis of **35** with CF$_3$COOH but can be obtained by the protolysis of **36** (E = Si) with MeOH in pentane at 5°C (*41*).

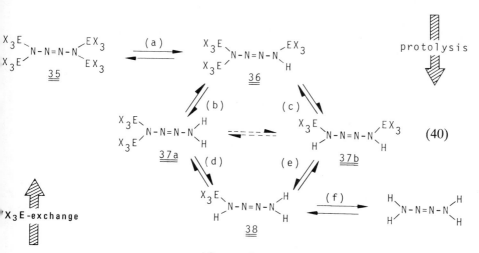

SCHEME 3

Germyl groups on the tetrazene skeleton are more easily replaced by hydrogen than are silyl groups. Thus, **35** (E = Ge) reacts even with MeOH in benzene at room temperature to form a mixture of partially germylated tetrazenes [**36** and **37** (E = Ge)] (*41*). In an analogous manner, the methanolysis of $(Me_3Si)_2N$—N=N—$N(GeMe_3)_2$ or (Me_3Si) $(Me_3Ge)N$—N=N—$N(GeMe_3)(SiMe_3)$ leads to **37a** or **37b** (E = Si) via monogermylated tetrazenes.

Partially substituted or unsubstituted tetrazenes $(Me_3E)_{4-n}N_4H_n$ ($n < 4$; E = Si, Ge, Sn), obtained by the protolysis of **35** can be changed further into new tetrazenes by *silylation, germylation,* or *stannylation* (Scheme 3), along the route from bottom right to top left. To introduce X_3E groups, amines of type X_3ENR_2 (E = Si, Ge, Sn; R = Me, Et) are the best suited and react with tetrazenes (*42*) according to Eq. (41a). However, it is advantageous in many cases to transform tetrazenes $(Me_3E)_{4-n}N_4H_n$ to tetrazenides $(X_3E)_{4-n}N_4H_{n-m}Li_m$ (*44*), which react, in general at low temperatures, with dimethyl sulfate Me_2SO_4 [Eq. (41b)] or with halides X_3EHal (E = Si, Ge, Sn) [Eq. (41c)] (*42, 44–46*). Instead of hydrogen or lithium, other tetrazene-bound groups such as Me_3Sn [Eq. (41d)] or CHO [Eq. (41e)] can also be exchanged with X_3E groups. Finally, the tetrazene bound substituents can also, in a few cases, undergo intramolecular exchange. This technique, discussed in more detail in

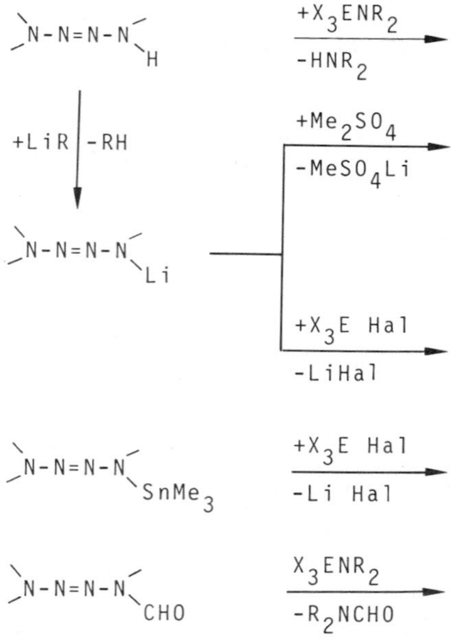

$$\diagdown N\text{-}N\text{=}N\text{-}N\diagdown_H \quad \xrightarrow[-HNR_2]{+X_3ENR_2} \quad \diagdown N\text{-}N\text{=}N\text{-}N\diagdown_{EX_3} \qquad (41a)$$

$$+LiR\ \Big|\ -RH$$

$$\xrightarrow[-MeSO_4Li]{+Me_2SO_4} \quad \diagdown N\text{-}N\text{=}N\text{-}N\diagdown_{Me} \qquad (41b)$$

$$\diagdown N\text{-}N\text{=}N\text{-}N\diagdown_{Li}$$

$$\xrightarrow[-LiHal]{+X_3E\ Hal} \quad \diagdown N\text{-}N\text{=}N\text{-}N\diagdown_{EX_3} \qquad (41c)$$

$$\diagdown N\text{-}N\text{=}N\text{-}N\diagdown_{SnMe_3} \quad \xrightarrow[-Li\ Hal]{+X_3E\ Hal} \quad \diagdown N\text{-}N\text{=}N\text{-}N\diagdown_{EX_3} \qquad (41d)$$

$$\diagdown N\text{-}N\text{=}N\text{-}N\diagdown_{CHO} \quad \xrightarrow[-R_2NCHO]{X_3ENR_2} \quad \diagdown N\text{-}N\text{=}N\text{-}N\diagdown_{EX_3} \qquad (41e)$$

Section III,B,2,d, has only limited utility for the preparation of Group IV derivatives of tetrazene.

B. *Properties*

1. *General Features*

Some *physical properties* of liquid or crystalline silyl, germyl, and stannyl derivatives of tetrazene and organyltetrazenes, along with methods of their preparation, are given in Table III. Except for the light yellow tetrazene derivatives of Sn(II) [$ClSnN_4(SiMe_3)_3$ and $(Me_3Si)_3N_4SnN_4(SiMe_3)_3$], all other compounds are colorless (cf. Section III,B,2). Silylated aryltetrazenes, $ArN_4(SiMe_3)_3$, are red colored when prepared and even after several recrystallizations retain a yellowish tinge, which is attributed to almost inseparable impurities (isomeric azo compounds).

Among the *chemical properties* of compounds listed in Table III, the thermal stability of tetrasubstituted tetrazenes is strikingly high; some can be heated above 150°C without decomposition. The stability of $(Me_3E)_nN_4H_{4-n}$ ($n = 1$–4) decreases approximately in the order E = Si > Ge >Sn and with decreasing number n of Me_3E groups (see Table III). Apart from completely substituted silylated tetrazenes, which react only very slowly with water, other compounds in Table III are all very sensitive to hydrolysis. Stannylated tetrazenes, in contrast to silylated and germylated tetrazenes, are air sensitive also. Silyl-, germyl-, and stannyltetrazenes decompose via irradiation. (For details of thermolysis and photolysis, cf. Section III,B,3 and Section III,C; and for redox and substitution reactions such as oxidation and hydrolysis, cf. Section III,D,1 and Section III,D,2.)

2. *Structure*

a. Geometric Structure. The following four structural possibilities for tetrazenes need discussion:

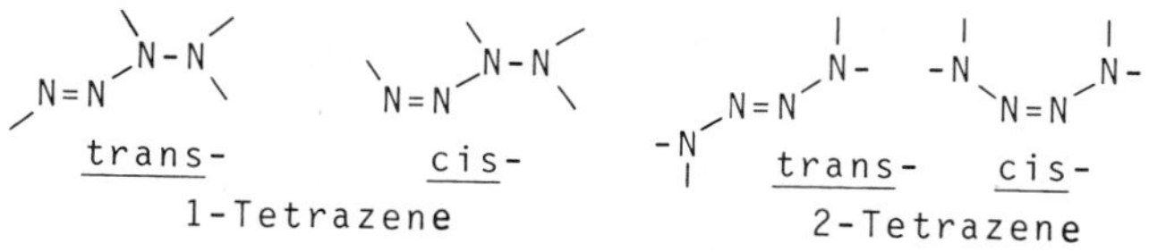

The known acyclic silylated, germylated, and stannylated tetrazenes and alkyltetrazenes generally exhibit 2-tetrazene constitution. These can be

TABLE III.

	R	R'	R"	R"'	Best preparation		Melting Point (°C)
Type	R	R'	R"	R"'	Method	Equation	
Acyclic silyltetrazenes							
39a	$SiMe_3$	$SiMe_3$	$SiMe_3$	$SiMe_3$	F	(34)	46
39a	$SiMe_3$	$SiMe_3$	$SiMe_3$	$SiMe_2Ph$	I	d	
39a	$SiMe_3$	$SiMe_3$	$SiMe_3$	$SiMe_2Cl$	I[e]	(41c)	48
39a	$SiMe_3$	$SiMe_3$	$SiMe_3$	SiF_3	I[e]	(41c)	Liquid
39a	$SiMe_3$	$SiMe_3$	$SiMe_3$	SiF_2X^g	I[e]	(41c)	Liquid
39a	$SiMe_3$	SiF_3	$SiMe_3$	SiF_3	I[e]	(41c)	Liquid
39a	$SiMe_3$	$SiMe_3$	$SiMe_3$	$SiCl_3$	I[e]	(41c)	56
39a	$SiMe_3$	$SiMe_3$	$SiMe_3$	$SiBr_3$	I[e]	(41c)	54
39b	$SiMe_3$	$SiMe_3$	$SiMe_3$	H	I	(40a)	Liquid
39c	$SiMe_3$	$SiMe_3$	H	H	I	(40b)	Liquid
39d	$SiMe_3$	H	$SiMe_3$	H	I	(40a,b)	Liquid
39e	$SiMe_3$	H	H	H	I	(40e)	i
Acyclic germyltetrazenes							
39a	$GeMe_3$	$GeMe_3$	$GeMe_3$	$GeMe_3$	I	(41a)	71
39a	$GeMe_3$	$GeMe_3$	$GeMe_3$	$SiMe_3$	I	(41d)	41
39a	$GeMe_3$	$GeMe_3$	$SiMe_3$	$SiMe_3$	I[e]	(41a)[j]	45
39a	$GeMe_3$	$SiMe_3$	$GeMe_3$	$SiMe_3$	I	(41d)[j]	57
39a	$GeMe_3$	$SiMe_3$	$SiMe_3$	$SiMe_3$	I	(41d)[j]	41
39a	$GeMe_3Cl$	$SiMe_3$	$SiMe_3$	$SiMe_3$	I[e]	(41c)	62
39a	$GeCl_3$	$SiMe_3$	$SiMe_3$	$SiMe_3$	I[e]	(41c)	76
39b	$GeMe_3$	$GeMe_3$	$GeMe_3$	H	I	(40a)	k
39c	$GeMe_3$	$GeMe_3$	H	H	I	(40a,b)	k
39d	$GeMe_3$	H	$GeMe_3$	H	I	(40a,c)	k
39b	$GeMe_3$	$SiMe_3$	$SiMe_3$	H	I[e]	(41a)	Liquid
39b	$GeMe_3$	H	$SiMe_3$	$SiMe_3$	I[e]	(41c)[j]	Liquid
Acyclic Stannyltetrazenes							
39a	$SnMe_3$	$SnMe_3$	$SnMe_3$	$SnMe_3$	I	(41a)	88
39a	$SnMe_3$	$SnMe_3$	$SnMe_3$	$SiMe_3$	I	(41d)	84
39a	$SnMe_3$	$SnMe_3$	$SiMe_3$	$SiMe_3$	I	(41d)	45
39a	$SnMe_3$	$SiMe_3$	$SnMe_3$	$SiMe_3$	I	(41d)	47
39a	$SnMe_3$	$SiMe_3$	$SiMe_3$	$SiMe_3$	I[e]	(41a)	37
39a	$SnMe_2Cl$	$SiMe_3$	$SnMe_2Cl$	$SiMe_3$	I[e]	(41c)	160
39a	$SnMe_2Cl$	$SiMe_3$	$SiMe_3$	$SiMe_3$	I[e]	(41c)	75
39a	$SnMe_2X^g$	$SiMe_3$	$SiMe_3$	$SiMe_3$	I[e]	(41c)	Solid
39a	$SnCl_2X^g$	$SiMe_3$	$SiMe_3$	$SiMe_3$	I[e]	(41c)	95
39a	$SnCl$	$SiMe_3$	$SiMe_3$	$SiMe_3$	I[e]	(41c)	Solid
39a	SnX^g	$SiMe_3$	$SiMe_3$	$SiMe_3$	I[e]	(41c)	85
39a	$SnMe_3$	$SnMe_3$	$SnMe_3$	$GeMe_3$	I	(41d)	Solid[l]
39b	$SnMe_3$	$SiMe_3$	$SiMe_3$	H	I[e]	(41a)	Solid[m]
39b	$SnMe_3$	H	$SiMe_3$	$SiMe_3$	I[e]	(41a)[j]	Solid[m]
Acyclic organyltetrazenes							
39b	$SiMe_3$	$SiMe_3$	$SiMe_3$	Me	I	(41b)	19
39c	$SiMe_3$	$SiMe_3$	Me	Me	I	(41b)	Liquid

ILYL, GERMYL, AND STANNYL DERIVATIVES OF TETRAZENE (N_4H_4) AND ORGANYLTETRAZENES

Boiling point (sublimation point) (°C/torr)[c]	Decomposition (ca. °C)[a]	^{1}H-chemical shifts (C_6H_6, internal TMS)[c]				UV, AP[b]	References
		$\delta(R)$	$\delta(R')$	$\delta(R'')$	$\delta(R''')$		
40/HV	190	0.278	0.278	0.278	0.278	b	33, 41
70/HV	190	0.258	0.258	0.258	0.358 + m		53
40/HV	120	0.245	0.245	0.356	0.588		45
50/HV	120	0.236	0.236	0.225[f]	—		45
80/HV	130	0.287	0.287	0.375[f]	—		45
60/HV		0.17[f]	—	0.172[f]	—		45
40/HV	85	0.248	0.248	0.315	—		45
50/HV	60	0.273	0.273	0.363	—		45
31–33/HV	140	0.311	0.311	0.100	6.2	b	41, 42
15–20/HV	70[h]	0.267	0.267	6.3	6.3		41, 42
10/HV	>150[h]	0.142	6.0	0.142	6.0		41, 42
i	>0	0.202[i]	—	—	—		41
45/HV	180	0.442	0.442	0.442	0.442	b	33, 41
40–45/HV		0.428	0.428	0.438	0.310		41
40–45/HV	165	0.423	0.423	0.308	0.308		41
40–45/HV	165	0.445	0.298	0.445	0.298		41
40–45/HV	<190	0.410	0.253	0.260	0.260		41
50/HV	130	0.883	0.433	0.183	0.183		45
55/HV	100		0.317	0.250	0.250		45
		0.467	0.467	0.317			41, 42
	<100	0.428	0.428				41, 42
		0.317		0.317			41, 42
40/HV	<140	0.463	0.325	0.125			42
35/HV	<100	0.263		0.328	0.328		42
60/HV	150	0.290	0.290	0.290	0.290	b	33, 41
	>140	0.290	0.290	0.308	0.273		41
	>140	0.290	0.290	0.280	0.280		41
	>150	0.333	0.267	0.333	0.267		41
50–60/HV	<160	0.297	0.258	0.258	0.258		41
100/HV	150	0.690	0.290	0.690	0.290		45
60/HV	160	0.740	0.473	0.127	0.127		45
90–100/HV		0.637	0.357	0.297	0.297		45
	140		0.483	0.280	0.280		45
	130		0.492	0.283	0.283		45
			0.343	0.287	0.287		45
		0.295	0.295	0.313	0.395		41
	m	0.303	0.325	0.112			42
	m	0.183		0.308	0.308		42
25/HV	>180	0.337	0.337	0.172	2.77		46
40/HV	>170	0.228	0.228	2.88	2.88		45

(*continued*)

TABLE III (*cont.*)

$$\begin{array}{c}R\\ \diagdown\\ \diagupN{-}N{=}N{-}N\diagup\\ R'\diagdown\\ R'''\end{array}\quad\begin{array}{c}R''\\ \diagup\end{array}$$

Type	R	R'	R''	R'''	Best preparation Method	Best preparation Equation	Melting Point (°C)
39e	SiMe$_3$	Me	Me	Me	I	(41e)	Liquid
39e	GeMe$_3$	Me	Me	Me	I	(41d)	Liquid
39e	SnMe$_3$	Me	Me	Me	I	(41e)	−10
39f	R—N=NNN(SiMe$_3$)SiMe$_3$, SiMe$_3$ R = Ph				G	(37)	−10
39f	R = *p*-tolyl				G	(37)	38

Cyclic tetrazenes[n]

$$\begin{array}{c}Me_2\\ E\\ \diagup\diagdown\\ R{-}NN{-}R'\\ \diagdown\diagup\\ N{=}N\end{array}$$

Type	E	R	R'			
39b	Si	SiMe$_3$	CMe$_3$	H	(38)	68
39b		SiPh$_3$	CMe$_3$	H	(38)	*o*
39a		SiMe$_3$	SiMe$_3$[p]	F	(34)[q]	89
39a		SiMe$_3$	SiMe^{t}Bu$_2$	H	(38)	70
39a		SiMe$_2^t$Bu	SiMe$_2^t$Bu	H	(38)	72
39a		SiMe^{t}Bu$_2$	SiMe$_2^t$Bu	H	(38)	106
39a		SiMe^{t}Bu$_2$	SiMe^{t}Bu$_2$	H	(38)	162
39a		Si^{t}Bu$_3$	Si^{t}Bu$_3$	H	(38)	133
39a		SiPh$_3$	SiPh$_3$	H	(38)	224
39b	Ge	SiMe$_3$	CMe$_3$	H	(38)	71
39a		SiMe$_3$	SiMe$_3$	H	(38)[r]	112
39a		SiMe$_2^t$Bu	SiMe$_2^t$Bu	H	(38)	103
39a		SiMe^{t}Bu$_2$	SiMe^{t}Bu$_2$	H	(38)	153
39a		Si^{t}Bu$_3$	Si^{t}Bu$_3$	H	(38)	132
39a		SiPh$_3$	SiPh$_3$	H	(38)	157

[a] For half life ca. 1 hour, and dilute solutions in benzene.

[b] (Me$_3$Si)$_4$N$_4$ (**33**): AP (appearance potential) = 7.37 V (adiabatic), $\bar{\nu}_{max}$ = 39370 cm^{-1} (ε = 9700), 41840 cm^{-1} (ε = 5500); (Me$_3$Ge)$_4$N$_4$ (**33**): AP = 7.10 V (adiabatic), $\bar{\nu}_{max}$ = 33370 cm^{-1} (ε = 3120), 39370 cm^{-1} (ε = 1800); (Me$_3$Sn)$_4$N$_4$ (**33**): $\bar{\nu}_{max}$ = 31850 cm^{-1} (ε = 17400), 38170 cm^{-1} (ε = 3300); (Me$_3$Si)$_3$N$_4$H (**34**): $\bar{\nu}_{max}$ = 35460 cm^{-1} (ε = 1000); Me$_2$SiN$_4$ (SiMe$_3$)$_2$ (**35**): AP = 7.56 V (vertical), $\bar{\nu}_{max}$ = 40720 cm^{-1} (ε = 5900).

[c] HV, High vacuum; m, multiplet.

[d] Prepared by silylation of (Me$_3$Si)$_3$N$_4$H with CF$_3$COOSiMe$_2$Ph + NEt$_3$ in CH$_2$Cl$_2$. (PhMe$_2$Si)(Me$_3$Si)N$_4$(SiMe$_3$)(SiMe$_3$Ph) was prepared (not pure) in an analogous manner.

[e] Reactants (Me$_3$Si)$_3$N$_4$Li, (Me$_3$Si)$_2$N$_4$Li$_2$, and (Me$_3$Si)$_2$N$_4$HLi.

[f] Quartet; J ca. 0.7 Hz.

[g] X = —(Me$_3$Si)N—N=N—N(SiMe$_3$)$_2$.

[h] Pure (Me$_3$Si)$_2$N$_4$H$_2$ and (Me$_3$Si)HN$_4$H(SiMe$_3$) decompose at room temperature ($\tau_{1/2}$ = 1 hour, and some days, respectively), and must be stored at low temperatures.

[i] So far, only in CH$_2$Cl$_2$ solution.

SILYL, GERMYL, AND STANNYL DERIVATIVES OF TETRAZENE (N_4H_4) AND ORGANYLTETRAZENES

Boiling point (sublimation point) (°C/torr)[c]	Decomposition (ca. °C)[a]	¹H-chemical shifts (C_6H_6, internal TMS)[c]				UV, AP[b]	References
		δ(R)	δ(R')	δ(R'')	δ(R''')		
70/30	>170	0.183	2.80	2.73	2.73		39
25/2	>160	0.305	2.92	2.75	2.75		39
−8/HV	<50	0.260	3.18	2.67	2.67		39
100–100/HV	>150	m	0.255	0.240	0.240		39
110–120/HV	>150	2.10 + m	0.268	0.253	0.253		39

	δ(R)	δ(EMe₂)	δ(R')		
>160	0.257	0.173	1.29		16
>160	m	−0.032	1.30		16
140	0.244	0.140	0.244	b	16, 35
>150	0.247	0.198	0.036/1.09		16
>150	0.196/0.998	0.203	0.196/0.998		16
>160	0.040/1.09	0.201	0.223/1.00		16
>160	0.051/1.08	0.236	0.051/1.08		16
>140	1.24	0.556	1.24		16
>200	m	0.150	m		16
>80	0.266	0.344	1.29		15
130	0.250	0.293	0.250[s]		15
>130	0.243/1.06	0.422	0.243/1.06		15
>140	−0.005/1.11	0.410	−0.005/1.11		15
>120	1.234	0.708	1.234		15
	m	−0.146	m		15

[j] Also preparable according to I [Eq. (41c)] from $(Me_3Si)_3N_4Li$, $(Me_3Si)_2N_4Li_2$, or $(Me_3Si)_2N_4HLi$ and Me_3ECl (44).

[k] So far isolated only as a mixture: $(Me_3Ge)_3N_4H/(Me_3Ge)_2N_4H_2$.

[l] So far isolated only in mixture with 20% $(Me_3Sn)_2(Me_3Ge)_2N_4$.

[m] Stable only in solution at low temperatures. Dismutation on trying to isolate the compound.

[n] Sila- and germatetrazolines. Not included are tetrazolines and tetrazoles, which have only silyl, germyl, or stannyl ring substituents.

[o] In mixture with $(Me_3Si)(Ph_3Si)N—SiMe_2—(Me_3Si)C=N=N$.

[p] Probably cyclic tetrazenes with F or Cl instead of Me (R = R' = Me_3Si) are formed in low yields by thermolysis of $(Me_3Si)_3N_4SiHal_3$ (16); ¹H NMR($C_6H)_6$, Hal = F: δ($SiMe_3$) = 0.250 (triplet, J = 0.7 Hz); Hal = Cl: δ ($SiMe_3$) = 0.292.

[q] Preparable also from $Me_2Si=NSiMe_3$ + Me_3SiN_3 [H, Eq. (38)], and from $(Me_3Si)_3(Me_2ClSi)N_4$ by heating (cf. Section III,C).

[r] Preparable also by thermolysis of $(Me_3Si)_3(Me_2ClGe)N_4$ in 20% yield.

[s] In Et_2O.

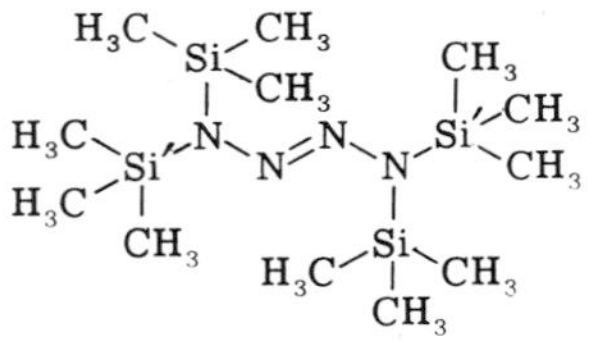

FIG. 1. Structure of tetrakis(trimethylsilyl)tetrazene. Bond lengths (Å) are for N—N, 1.394; N=N, 1.268; Si—N, 1.770; and Si'—N, 1.789. Bond angles (in degrees) are Si—N—N, 118.2; Si'—N—N, 113.3; Si—N—Si', 128.5; and N—N—N, 112.4.

categorized into five types (**39a–39e**) of 2-tetrazenes or 2-organyltetrazenes [$R_{4-n}(X_3E)_nN_4$ (E = Si, Ge, Sn; R = H or organyl; n = 1–4)]. Examples of each type of compound are known and are given in Table III. On the other hand, preparation of only one type of the conceivable possibilities for 1-tetrazenes and their organyl derivatives could be accomplished: fully silylated aryltetrazenes exhibit the constitution **39f**.

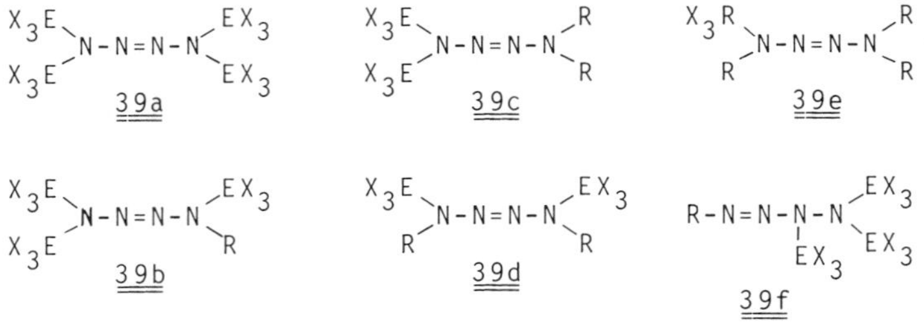

An X-ray structural analysis by Veith (*47*) showed (Me$_3$Si)$_2$N—N=N—N(SiMe$_3$)$_2$ to have the *trans configuration* with a planar Si$_4$N$_4$ framework (Fig. 1). The remarkable feature is a comparatively short N—N single bond (1.40 Å) and a long N=N double bond (1.27 Å).[20] The bond relationships indicate a π delocalization of the nonbonding amino electron pairs and of the π azo electron pair over the four-membered nitrogen chain. The Si—N bond distances (average 1.78 Å) are longer than in any of the compounds with (Me$_3$Si)$_2$N groups investigated so far, and they agree closely with the calculated Si—N single bond distance of 1.81 Å (*47*). The $d_\pi p_\pi$-backbonding discussed to explain the planarity of nitrogen with silyl substituents, does not lead to a significant shortening of the Si—N distances in tetrasilyltetrazene. On the other hand, the SiNSi bond angle (128°) is comparable with the SiNSi angle of other (Me$_3$Si)$_2$N-containing

[20] The N—N distance amounts to 1.45 Å in H$_2$N—NH$_2$ and 1.48 Å in (H$_3$Si)$_2$N—N(SiH$_3$)$_2$ (*47*), and the N=N distance is 1.23–1.27 Å in RN=NR and 1.17 Å in Me$_3$SiN=NSiMe$_3$ (*48*)

compounds. Similarly, the NNN bond angle (112.4°) lies in the usual angle range for azo-nitrogen atoms (106–113°) (*48*).

X-ray structures of other acyclic silyl-, germyl-, and stannyltetrazenes are not yet known. Spectroscopic investigations of these tetrazenes (*33, 42*), as well as the possibility of their conversion into and their formation from trans-configurated tetrazene (H_2N—N=N—NH_2) (*49*) by exchange of substituents, suggest the trans configuration. Of course, the amino-nitrogen atoms in N_4H_4 are not planar as in $(Me_3Si)_4N_4$ but are pyramidal, thereby showing sp^3 instead of sp^2 hybridization. In addition, relatively long N—N single bond distance (1.429 Å) and short double bond distance (1.205 Å) has been observed. This, however, according to other points discussed below, does not argue against π delocalization in N_4H_4.[21] The *conformation* determined for amino groups in N_4H_4 may also belong to unsubstituted amino groups in tetrazenes of type $(X_3E)_2N$—N=N—NH_2, whereas $(X_3E)_2N$ groups may, in general, be of planar conformation, as in $(Me_3Si)_4N_4$.

Cis-silylated, germylated, or stannylated acyclic tetrazenes have not yet been identified (*33*). These are, probably in all cases, substantially higher in energy than the corresponding trans compounds. The cyclic tetrazenes (sila- and germatetrazolines) listed in Table III, however, are in the cis configuration.

b. Electronic Structure and Absorption Spectra. A typical *energy level diagram* of the inner π- and n-molecular orbitals of *trans*-2-tetrazene, along with the corresponding energy level diagram of *trans*-triazene, is shown in Fig. 2 (*50*). Molecular orbitals n_- and n_+, each occupied by two electrons, arise from interaction of sp^2-hybrid orbitals of the electron pairs on the azo-nitrogen atoms. On the other hand, π-molecular orbitals arise from the interaction of the p_z atomic orbitals of the azo nitrogen atoms with orbitals of the electron pairs of sp^2- or sp^3-hybridized amino-nitrogen atoms. Two out of the three π-molecular orbitals of triazene and three out of the four π-molecular orbitals of tetrazene possess two electrons each. The π conjugation along the nitrogen chain of tetrazene does not necessarily lead to a marked shortening of the N—N single bond and lengthening of the N=N double bond, because the delocalized electron pair in the π_3-orbital counters this change in distances.

---

[21] The N—N and N=N bond distances are also influenced by the hybridization of nitrogen atoms, by the repulsion of free electron pairs of nitrogen atoms, and by the electronegativity of substituents. Higher electronegativity of NH_2 groups, as compared with that of $N(SiMe_3)_2$ groups, may also lead to the decrease of the NNN angle from 112° in $(Me_3Si)_4N$ to 109° in N_4H_4.

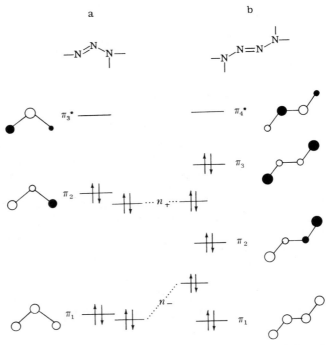

FIG. 2. Energy level diagram of the inner molecular orbitals of the triazene group —N=N—N< (a) as well as the 2-tetrazene group >N—N=N—N< (b).

In unsubstituted tetrazene (N_4H_4), the photoelectron (PE) spectrum shows the following energy states of inner molecular orbitals (50): 14.7 (π_1), 13.31 (n_-), 11.62 (π_2), 10.04 (n_+) and 8.99 eV (π_3). CNDO/S-calculations (51) show that substitution should lead to an increase or decrease of individual energy states, but not to any change in the sequence of inner molecular orbitals of tetrazene. Therefore, the highest occupied molecular orbitals of the π type have different energy depending upon the tetrazene substituents (e.g., Me_4N_4, 7.56 eV; $(Me_3Si)_4N_4$, 7.37 eV; $(Me_3Ge)_4N_4$, 7.10 eV; cf. Table III).

The energy states of unsubstituted triazene are not experimentally accessible because the hydride has not been isolated so far (cf. ref. 17). According to CNDO/S-calculations, the values are (50, 51): 14.41 (n_-), 14.09 (π_1), 10.12 (n_+), and 9.63 eV (π_2). Because of the comparatively small energy difference between the n_+- and π_2-molecular orbitals, the energy sequence of both these orbitals can reverse with appropriate choice of substituents on triazene (cf. Section II,B,1).

The UV spectra of silylated, germylated and stannylated tetrazenes and organyltetrazenes are analogous to tetrazene and its organyl derivatives and consist of a strong *absorption* around 35000 cm^{-1} (Table III); e.g., $(Me_eSi)_4N_4$, $\bar{\nu}_{max}$ =39370 cm^{-1} (*33*); N_4H_4, 38000 cm^{-1} (*50*); R_4N_4, 35000 cm^{-1} (*52*); $(Me_3Ge)_4N_4$, 33560 cm^{-1} (*33*); $(Me_3Sn)_4N_4$, 31850 cm^{-1} (*33*). This is attributed to a $\pi_3 \rightarrow \pi_4^*$ electronic transition (Fig. 2) (*50*). Furthermore, a medium intensity absorption is usually observed around 40,000 cm^{-1} (Table III).

3. Isomerizations

Just like triazenes (Sections II,B,2–II,B,4), tetrazenes can isomerize with change of *conformation* (rotational isomerization),[22] *configuration* (cis–trans isomerization), and *constitution* (1,3- and 1,4-rearrangements).

a. Hindered Rotations. The activation barrier for rotation of azo-bound amino groups about the N—N single bond in $(Me_3E)_4N_4$ (E = Si, Ge, Sn) and other tetrazenes, as shown in Eq. (42), is very small

$$ (42) $$

(<40 kJ/mol), since the ^{1}H-NMR spectra show no signal splitting even at $-120°C$ (*33*). In this regard, tetrazenes behave quite differently from the corresponding triazenes (Section II,B,4). For example, $ArN=N-N(SiX_3)_2$, exhibit higher barriers observable by ^{1}H-NMR spectroscopy. The probable reason for this difference is that although rotation of amino groups disturbs the π conjugation, the tetrazenes continue to possess good π conjugation over the three-membered triazene chain. Tetrazenes do however exhibit certain barriers for the rotation of amino groups, as shown from vibrational spectroscopic investigations of partially substituted tetrazenes $(Me_3E)_nN_4H_{4-n}$ (*42*).

b. Trans/Cis Isomerizations. Photolytic transformation of *trans-* to *cis-2*-tetrazenes [Eq. (43)] might be expected, but has not been observed. A benzene solution of *trans-*$(Me_3E)_4N_4$ decomposes under UV radiation, without any evidence for the formation of the cis isomer, in a few minutes

---

[22] In tetrazenes with pyramidal amino-nitrogen atoms, the conformation can also change by inversion of these nitrogen atoms.

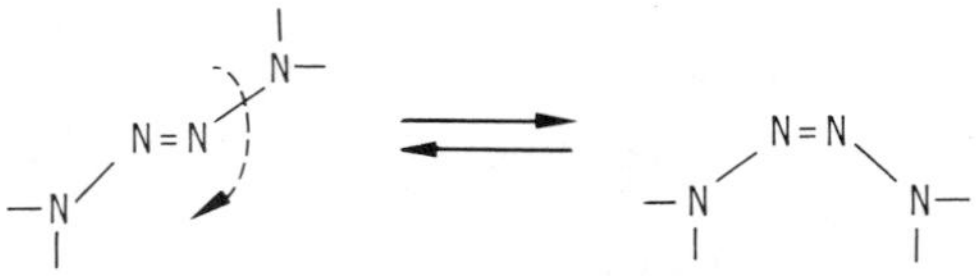

$$\text{(structure)} \qquad (43)$$

(E = Si, Ge) or hours (E = Sn) with the formation of $(Me_3E)_3N$ and Me_3EN_3 as well as $(Me_3E)_2NH$ and $N_2(E = Si, Ge)$, or with the formation of $(Me_3E)_3N$, N_2 and $(Me_3E)_2$ (E = Sn) (33). The same is true in the photolysis of SnX_2 [X = $(Me_3Si)_3N_4$], which leads immediately to $(Me_3Si)_3N_2H$ and N_2 (45) with the precipitation of Sn. Evidently, the interaction with photons mainly induces a radical decomposition of the compound as in Eq. (44). The radicals themselves then abstract protons or react further with the initial reactants.

$$\text{\textbackslash}N-N=N-N\diagup \xrightarrow[-N_2]{h\nu} \quad 2 \;\diagdown N\cdot \quad or \quad \diagdown N-N\cdot \;\; + \;\; \text{—}\cdot \qquad (44)$$

Thermal trans/cis rearrangement of the tetrazene skeleton has been demonstrated for tetrazenes of type $(Me_3Si)_3N_4EX_3$ (EX_3, e.g., $SiMe_2Cl$, $GeMe_2Cl$, SiF_3, $SiCl_3$, $GeCl_3$) [Eq. (45)] (45). However, because of the

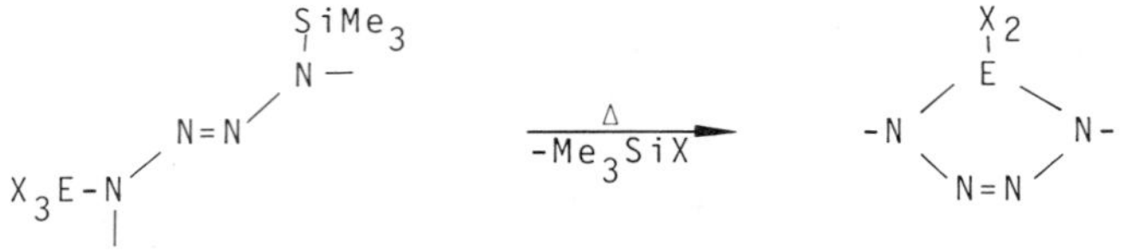

$$X_3E-N\diagdown \underset{N=N}{\overset{\underset{N-}{SiMe_3}}{}} \quad \xrightarrow[-Me_3SiX]{\Delta} \quad -N\diagdown \underset{N=N}{\overset{\underset{E}{X_2}}{}} N- \qquad (45)$$

cleavage of Me_3SiX, Eq. (45) does not exemplify isomerization in the true sense. Perhaps cis-tetrazenes appear as intermediate products in the thermolysis of some trans-tetrazenes (for details, cf. Section III,C,1).

c. 1,3-Migrations. In view of the migratory tendencies of X_3E and H groups in the tetrazenes listed in Table III, kinetics suggest the likelihood of 1,3-rearrangement as in Eq. (46). However, of the possible isomers

$$\text{(structure)} \qquad (46)$$

only the 2-isomer has ever been found in the ^{1}H-NMR spectra. The difference in free enthalpy between the isomers is so large that the energetically unstable isomer does not exist in amounts detectable by ^{1}H-NMR spectroscopy.

The equilibrium position of Eq. (46) depends on the steric and electronic effects of the tetrazene substituents. Naturally, the 1-tetrazene form is favored with 1-substituents that are either very bulky, but these have not been studied. Stabilization of the 1-isomer should also be favored by 1-substituents which interact both with n and π electron pairs of the azo group [e.g., the aryl group in $ArN_4(SiMe_3)_3$]. These effects are apparently weaker for silylated, germylated, and stannylated tetrazenes and alkyl-tetrazenes, which assume the 2-tetrazene structure exclusively. Isomerization into the higher energy 1-tetrazene [the corresponding argument is true for the transition of $ArN_4(SiMe_3)_3$ into 2-tetrazene] occurs during the thermolysis reaction, in which the 1-tetrazene appears, in many cases, as a reactive intermediate (cf. Section III,C,1). Its ease of formation according to

$$(Me_3Si)(X_3E)N—N=N—N(SiMe_3)_2 \rightleftharpoons (X_3E)N=N—N(SiMe_3)—N(SiMe_3)_2$$

depends on the EX_3 group in the sequence (with increasing ease) $SiMe_3 <$ $SnMe_2Cl < GeMe_2Cl < SiMe_2Cl < SiF_3 < SiCl_3 < SiBr_3 < SiI_3$ (decreasing thermolysis temperature, cf. Table III).

d. 1,4-Migrations. Besides constitution isomerization [represented in Eq. (46)] by 1,3-rearrangement of tetrazene substituents, silyl-, germyl-, and stannyl derivatives of acyclic *trans*-2-tetrazenes can still change constitution by intramolecular 1,4-exchange of substituents according to Scheme 4. Thus, dilute $(Me_3Si)_2N—N=N—NH_2$ in benzene at 100°C changes rapidly and quantitatively to $(Me_3Si)HN—N=N—NH(SiMe_3)$ (*42*). The half-life of this first-order isomerization at 94°C is 11 minutes ($E_a = 97.4$ kJ/mol; $\Delta H^{\ddagger} = 94.2$ kJ/mol; $\Delta G^{\ddagger} = 111.0$ kJ/mol; $\Delta S^{\ddagger} = -45.6$ eu). In an analogous manner, $(Me_3Ge)_2N—N=N—NH_2$ isomerizes into $(Me_3Ge)HN—N=N—NH(GeMe_3)$ (*42*), $(Me_3Si)_2-N—N=N—NH(GeMe_3)$ into $(Me_3Si)(Me_3Ge)N—N=N—NH(SiMe_3)$ (*42*), and $(Me_3Si)_2N—N=N—N(GeMe_3)_2$ into $(Me_3Si)(Me_3Ge)-N—N=N—N(GeMe_3)(SiMe_3)$ (*33*). In the latter case in pentane at 140°C, isomerization occurs at only 70% yield. The half-life of this first-order isomerization is 53 minutes ($E_a \sim 170$ kJ/mol) (*53*).[23]

---

[23] Since the thermolysis of $(Me_3Si)_2(Me_3Ge)_2N_4$ in benzene proceeds faster than in pentane, isomerization of the tetrazene in C_6H_6 does not occur independent of its thermolysis. Thermolysis of $(Me_3Si)_2(Me_3Sn)_2N_4$ in pentane and benzene proceeds even faster than isomerization, so that the latter is not observed at all. That isomerization takes place to some extent is indicated by the thermolysis products (Section III,C,1). Also, the isomerization of $(Me_3Si)_2N—N=N—NH(SnMe_3)$ into $(Me_3Si)(Me_3Sn)N—N=N—NH(SiMe_3)$ has not been observed because this compound dismutates immediately into $(Me_3Si)_2N—N=N—NH_2$ and $(Me_3Si)_2N—N=N—N (SnMe_3)_2$.

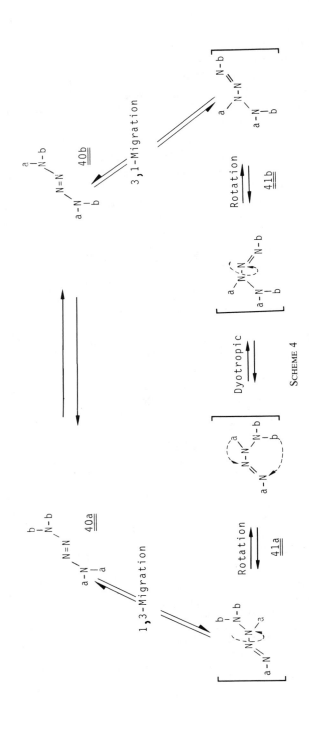

SCHEME 4

Isomerization of *trans*-2-tetrazenes (generally **40a**) may take place by 1,3-rearrangement (see Section III,B,3,c), first to *trans*-1-tetrazenes (**41a**) (single-step 1,4-rearrangement of substituents, starting from *trans*-2-tetrazenes, are sterically impossible), which then—most probably dyotropically (54)—isomerize into *trans*-1-tetrazenes (**41b**). The latter then reorganizes internally by 3,1-rearrangement of substituents to form *trans*-2-tetrazenes (**40b**) (Scheme 4).

Which group migrates first is not yet known. Thus, 1,3-rearrangement of $(Me_3Si)_2N \rightarrow N=N-NH_2$ leads to $(Me_3Si)N=N-N(SiMe_3)-NH_2$ if a = Me_3Si in Scheme 4 and to $(Me_3Si)_2N-NH-N=NH$ if a = H. It appears that the 1-tetrazene **41a** with less energy content is formed (cf. 1,3-rearrangement, Section III,B,3,c); this is because the energy content of **41a**, influenced by electronic and steric effects, goes essentially into activation energy of the endothermic rearrangement **40a** $\rightleftarrows$ **40b**.

The position of the equilibrium **40a** $\rightleftarrows$ **40b** naturally depends on the substituents. In the case of silylated compounds, tetrazenes with amino groups containing at least one silyl group are preferred (see above). The rate of isomerization **40a** $\rightleftarrows$ **40b** increases in these cases in the sequence $(Me_3Si)_2N_4(GeMe_3)_2 < (Me_3Si)_2N_4H(GeMe_3) < (Me_3Si)_2N_4H_2 < (Me_3Ge)_2N_4H_2$ (42). Hence, it follows that even in tetrazene $(Me_3Ge)_4N_4$, rearrangement of Me_3Ge groups occurs.

For discussion of constitution isomerization of cyclic tetrazenes by 1,4-rearrangement see Section III,C,2.

C. Thermolysis

1. Acyclic Tetrazenes

Investigations on thermolysis of Group IV derivatives of *trans*-2-tetrazene indicate that these decompose by a free radical mechanism according to Eqs. (48) and (49) (*Thermolysis Pathways I and II*) as well as by non-free radical pathways [Eqs. (50) and (51)] (*Thermolysis Path-*

$$\text{\textbackslash N-N=N-N\textbackslash} \xrightarrow{\text{Thermolysis Path I}} \text{\textbackslash N}\cdot \;+\; N\equiv N \;+\; \cdot N\textbackslash \qquad (48)$$

$$\text{\textbackslash N-N=N-N\textbackslash} \xrightarrow{\text{Thermolysis Path II}} \cdot + \; N\equiv N \;+\; \overset{\bullet}{N}\text{-N\textbackslash} \qquad (49)$$

$$\text{\textbackslash N-N=N-N\textbackslash} \xrightarrow{\text{Thermolysis Path III}} N\equiv N \;+\; \text{\textbackslash N-N\textbackslash} \qquad (50)$$

$$\text{\textbackslash N-N=N-N\textbackslash} \xrightarrow{\text{Thermolysis Path IV}} -N=N=N \;+\; -N\textbackslash \qquad (51)$$

ways III and *IV*). The radicals formed initially abstract hydrogen from the chemical environment or react with reactant tetrazenes or their thermolysis products. The isomerizations in Eqs. (45) and (47) can proceed before or run parallel to the thermolyses of Eqs. (48)–(51). For many compounds, the activation energies for these thermolysis reactions are of a comparable order of magnitude, so that the reactions proceed side by side, and the percentage share of the competing decompositions on the total thermolysis can be changed significantly by changes in reaction conditions.

   a. Completely Substituted Tetrazenes $(Me_3E)_n(Me_3E')_{4-n}$. Tetraorganyltetrazenes $(R_2N{-}N{=}N{-}NR_2)$ thermolyze generally above 100°C according to Thermolysis Pathway I with cleavage of nitrogen to form $R_2N\cdot$ radicals, which can dimerize to hydrazine, react to form amines (R_2NH), or react in other ways (*31, 55*). Still more stable than these tetrazenes—with some exceptions (cf. Table III)—are silylated, germylated, and stannylated tetrazenes of type $(Me_3E)_n(Me_3E')_{4-n}N_4$ (E, E' = Si, Ge, Sn), which also thermolyze mostly by radical pathways (I or II). Thus, tetrasilyltetrazene $[(Me_3Si)_4N_4$ (*42a*)$]$ decomposes in benzene at 150°C in a first-order reaction with a half-life of 53.2 hours $[E_a = 186$ kJ/mol; by comparison, $\tau_{1/2}$ (150°C) for $Me_4N_4 = 2.4$ hours; $E_a = 151$ kJ/mol]. Products of thermolysis are mainly (>80%) bis-(trimethylsilyl)amine and nitrogen [Eq. (52), path (a)], minor amounts of nitrogen and tetrakis(trimethylsilyl)hydrazine [Eq. (52), path (b)], trimethylsilylazide and tris(trimethylsilyl)amine [Eq. (52), path (c)], as well as some unidentified N- and Me_3Si-containing substitution products of benzene (*33*).

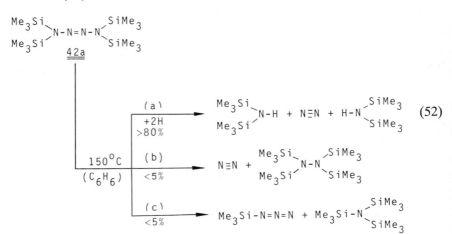

Mechanistically, the decomposition takes place exclusively (or mainly) according to Thermolysis Pathway I; cleavage of molecular nitrogen from positions 2 and 3 of the tetrazene chain forms radicals $(Me_3Si)_2N$, which become saturated mainly by abstraction of hydrogen from the chemical environment [Scheme 5, Eq. (54), paths (a) and (b)], and to a slight extent by reaction with benzene.[24] The findings that **42a** thermolyzes much slower than Me_4N_4 suggest that $(Me_3Si)_2N$ radicals have higher energy than Me_2N radicals. Products N_2 and $(Me_3Si)_4N_2$ from thermolysis of **42a** are possibly formed by the dimerization of $(Me_3Si)_2N\cdot$ or, more probably, by Thermolysis Pathway II [cf. Scheme 5, Eq. (54), paths (c), (d), and (e)], and products Me_3SiN_3 and $(Me_3Si)_3N$ by Thermolysis Pathway IV [cf. Scheme 5, Eq. (54), paths (c) and (f)].

Whereas tetrasilyltetrazene (**42a**) decomposes essentially into nitrogen and $(Me_3Si)_2NH$, tetragermyltetrazene $(Me_3Ge)_4N_4$ (**42b**) thermolyzes mainly (in benzene $\simeq$ 80%) with the formation of nitrogen and $(Me_3Ge)_4N_2$ [Eq. (53), path (a)] (33). In addition, $(Me_3Ge)_3N_2H$,

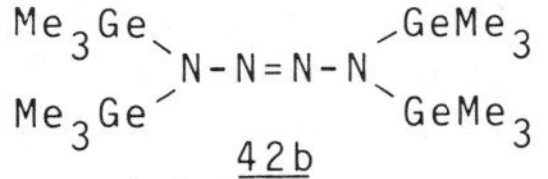

$$
\begin{array}{c}
Me_3Ge \\
\diagdown \\
N-N=N-N \\
Me_3Ge \diagup \qquad\qquad\quad\diagdown \\
\underline{42b}
\end{array}
\quad
\begin{array}{c}
GeMe_3 \\
\\
\\
GeMe_3
\end{array}
$$

$$140°C \;\Big|\; (C_6H_6)$$

(a)
+H
ca. 80%
$\longrightarrow\; N_2 \; + \; $ Me_3Ge, Me_3Ge N–N $GeMe_3$, $GeMe_3$

(b)
+H
ca. 10%
$\longrightarrow\; N_2 \; + \; \tfrac{1}{2}$ Me_3Ge, H N–N $GeMe_3$, $GeMe_3$ $+$

$\tfrac{1}{2}$ Me_3Ge, Me_3Ge NH $+ \tfrac{1}{2}$ Me_3Ge, Me_3Ge N–$GeMe_3$

(c)
+H
ca. 10%
$\longrightarrow\; N_2 \; + \;$ Me_3Ge, H N–N $GeMe_3$, $GeMe_3$ $+ \tfrac{1}{2} Me_3Ge-GeMe_3$

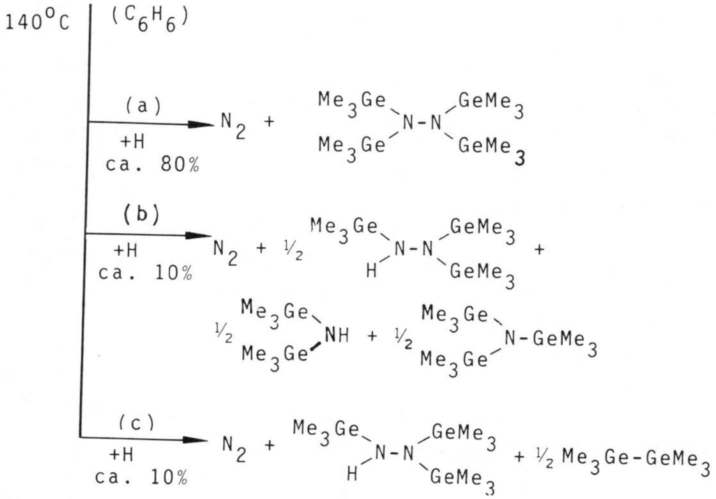

(53)

---

[24] Cleavage according to Eq. (48) (Thermolysis Pathway I) is supported by (1) the formation of $(Me_3Si)_2NH$, (2) thermolysis of ^{15}N-labeled **42a**, (3) formation of biphenyl and $(Me_3Si)_2N$-containing aromatic compounds in the case of thermolysis of **42a** in benzene, (4) the fact that the reaction is first-order, (5) the large positive reaction entropy (+78 eu), and (6) direct mass spectrometric proof of $(Me_3Si)_2N\cdot$ as a product of pyrolysis of **42a** in the gas phase (33).

$(Me_3Ge)_2NH$, $(Me_3Ge)_3N$, and $(Me_3Ge)_2$ are also formed in small amounts [Eq. (53), paths (b) and (c)]. On the other hand, Me_3GeN_3 is not produced.

The first-order decomposition of **42b** ($\tau_{1/2}$ in benzene at 150°C = 15.4 hours; E_a = 165 kJ/mol) takes place predominantly (or exclusively) by Thermolysis Pathway II. The reaction involves the formation of an intermediate isomer of 1-tetrazene **43** (E = Ge). After cleavage of molecular nitrogen from positions 1 and 2 of the tetrazene chain, **43** gives radicals $Me_3Ge\cdot$ and $\cdot N_2(GeMe_3)_3$. These radicals recombine mainly to form hydrazine $(Me_3Ge)_4N_2$ [Scheme 5, Eq. (54), paths (c), (d), and (e)].[25] In addition, the hydrazyl radical abstracts hydrogen from the chemical environment and the germyl radical abstracts $(Me_3Ge)_2N$— or Me_3Ge— from **42b** [Scheme 5, Eq. (54), path (i), and paths (g) and (h)]. To a certain extent, radicals $(Me_3Ge)_2N\cdot$ and $(Me_3Ge)_3N_2\cdot$ probably react also with **42b** by abstraction of Me_3Ge [formation of $(Me_3Ge)_3N$ and $(Me_3Ge)_4N_2$ as well as N_2 and $(Me_3Ge)_3N_2\cdot$]. Since the radical pair $Me_3Ge/N_2(GeMe_3)_3$ is also the primary product of the decomposition of Me_3Ge—$N{=}N$—$GeMe_3$ (*1*), it is not astonishing that thermolysis of **42b** and of the latter compound lead to the same products.

The isomerization [Eq. (54), path (c)] connected with the actual thermolysis of **42b** is also supported by the reversible monomolecular rearrangement of $(Me_3Si)_2N$—$N{=}N$—$N(GeMe_3)_2$ into $(Me_3Si)(Me_3Ge)$ N—$N{=}N$—$N(GeMe_3)(SiMe_3)$, which also occurs via 1-tetrazene (cf. Section III,B,3,d). Since this rearrangement occurs faster than the thermolysis of **42b** ($\tau_{1/2}$ (150°C) < $\frac{1}{2}$ hour), the exchange of germyl groups via **43** (E = Ge) happens before the thermolysis of **42b**. Thermolysis intermediate **43** explains the observed drastic increase in the yield of hydrazine $(Me_3E)_4N_2$ from **42a** versus **42b**. The isomerization in Eq. (54), path (c), is sterically more hindered for **42a** than **42b** and is therefore considerably slower. [On the other hand, the thermolysis reaction in path (a) of Eq. (54) is comparably fast for **42a** and **42b**.]

Tetrazenes $(Me_3E)(Me_3Si)N$—$N{=}N$—$N(SiMe_3)_2$ (E = Ge, Sn) evidently decompose according to Thermolysis Pathways I and II. According to the reactions in Scheme 5 nitrogen is eliminated with the formation of amines $(Me_3E)_m(Me_3Si)_nNH_p$ ($m + n + p = 3$) and the hydrazines

---

[25] Cleavage according to Eq. (49) (Thermolysis Pathway II) is supported [apart from the formation of $(Me_3Ge)_3N_2H$] by (1) the thermolysis of ^{15}N-labeled **42b**, (2) the first-order reaction rate, (3) the positive activation entropy (+39 eu), and (4) a strong increase in the yield of $(Me_3Ge)_3N_2H$ in the case of thermolysis of **42b** in toluene, a better hydrogen donor, at the cost of the yield of $(Me_3Ge)_4N_2$ (*33*). It is not clear whether $(Me_3Ge)_4N_2$ is formed solely by radical recombination or also by a nonradical route (Thermolysis Pathway III) through β-elimination from **43** (E = Ge).

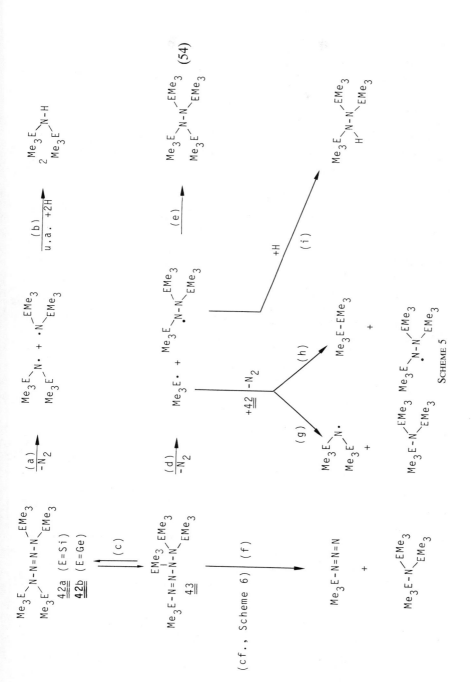

SCHEME 5

$(Me_3E)(Me_3Si)_3N_2$ and $(Me_3Si)_3N_2H$ (53, 57). The rate of tetrazene decomposition (Table III) and the yield of thermolysis products (Me_3E)-$(Me_3Si)_3N_2$ and $(Me_3Si)_3N_2H$ depend on E, increasing in the order Si < Ge < Sn for tetrazenes $(Me_3E)(Me_3Si)_3N_4$. This can be interpreted as follows: The activation energies E_a for Thermolysis Pathway II for the tetrazenes involved, consist of the activation energies E'_a for rearrangement [Eq. (54), path (c)] and E''_a for l-tetrazene decomposition [Eq. (54), path (d)]. The E—N bond energy (BE) diminishes considerably according to E in the order Si > Ge > Sn (58) ($BE_{Si-N} \simeq 315$ kJ/mol; $BE_{Ge-N} \simeq 230$ kJ/mol; $BE_{Sn-N} \simeq 170$ kJ/mol). Accordingly, E''_a decreases in the same direction. On the basis of the discussion in Section III,B,3,c, E'_a for the formation of l-tetrazenes Me_3E—N=N—$N_2(SiMe_3)_3$ may slightly increase in the same order. In any case, the resulting activation energy of thermolysis $E_a = E'_a + E''_a$ decreases in order given. On the other hand, the activation energy for pathway I should remain constant [$DE_{N-N} \approx 250$ kJ/mol (59)[26]].

In principle, symmetric tetrazenes $(Me_3E)(Me_3Si)N$—N=N—N-$(SiMe_3)(EMe_3)$ (E = Ge, Sn) thermolyze analogously to (Me_3E)-$(Me_3Si)_3N_4$ (53). The asymmetric tetrazene $(Me_3Si)_2N$—N=N—N-$(GeMe_3)_2$ rearranges into the symmetric forms (Section III,B,3,d). However, tetrazene $(Me_3Si)_2N$—N=N—$N(SnMe_3)_2$ decomposes in benzene mainly into Me_3SiN_3 and $Me_3SiN(SnMe_3)_2$. The products are formed according to path (c) of Eq. (54) via l-tetrazene $(Me_3Si)N$=N—N-$(SiMe_3)$—$N(SnMe_3)_2$, whose decomposition according to path (f) of Eq. (54) is apparently faster than the conceivable free radical secondary reaction.[27] Minor products (20%) of thermolysis of asymmetric disilyldistannyltetrazene are again the typical thermolysis products of symmetric disilyldistannyltetrazene, which indicates a possible preisomerization of the tetrazene (Section III,B,3,d).

In the thermolysis of $(Me_3Sn)_n(Me_3E)_{4-n}N_4$, with increasing number n of stannyl groups, there is an increase in production of distannane $(Me_3Sn$—$SnMe_3)$ and trisubstituted amines $[(Me_3Sn)_n(Me_3E)_{3-n}N]$, and a decrease in disubstituted amines $[(Me_3Sn)_n(Me_3E)_{2-n}NH]$ and hydrazines $[(Me_3Sn)_n(Me_3E)_{4-n}N_2]$. Thus, $(Me_3Sn)(Me_3Si)N$—N=N—$N(SnMe_3)_2$ decomposes in benzene at 150°C with the elimination of nitrogen to form mainly $(Me_3Sn)_2$, $(Me_3Sn)_3N$, and $(Me_3Sn)_2N(SiMe_3)$, and to a slight

---

[26] Evidently the total activation energy of thermolysis in each case is also smaller than that of cleavage of tetrazene into azide and amine as in Eq. (54), paths (c) and (f) (Thermolysis Pathway IV).

[27] Obviously the formation of $Me_3SiN(SnMe_3)_2$ is required sterically and/or electronically by both stannyl groups bound to nitrogen.

extent $(Me_3Sn)(Me_3Si)NH$ (*45*). Tetrastannyltetrazene $(Me_3Sn)_4N_4$ (**42c**) thermolyzes almost exclusively into $(Me_3Sn)_2$ and $(Me_3Sn)_3N$ [Eq. (55)] (*33*).[28] The extent of reaction of path (a) [Eq. (55)] rises at

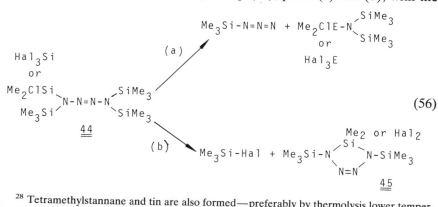

the expense of the reaction in path (b) with decreasing temperature of thermolysis (from ~60% at 190°C to ~70% at 90°C), and with high concentration of reactants (>0.4 *M*). The mechanism of this first-order thermolytic reaction [Eq. (55)] [τ (150°C) ~1 hour; E_a = 129 kJ/mol] is not clear (*53*).[29]

   b. Completely Substituted Tetrazenes $(Me_nX_{3-n}E)(Me_3Si)_3N_4$. The replacement of an Me_3Si group in $(Me_3Si)_4N_4$ (**42a**) by a sterically comparable Me_2ClSi group leads to a thermolabile tetrazene (Table III), whose mode of thermolysis is drastically changed: Instead of cleavage of nitrogen to form amine radicals according to Thermolysis Pathway I, dimethylchlorosilyltris(trimethylsilyl)tetrazene (**44**) decomposes exclusively by a non-free radical mechanism according to Eq. (56), paths (a) and (b), with the

$$(56)$$

---

[28] Tetramethylstannane and tin are also formed—preferably by thermolysis lower temperature —as a result of catalytic decomposition of Me_3Sn—$SnMe_3$ (*33*).
[29] Upon thermolysis of ^{15}N-labeled **42c** (0.1 *M* solution in benzene, 130°C), ~80% of the eliminated molecular nitrogen comes from positions 1 and 2 (or 3 and 4) and ~20% from positions 2 and 3 of the tetrazene chain.

TABLE IV

THERMOLYSIS OF $(Me_nX_{3-n}E)(Me_3Si)_3N_4$

	$Me_nX_{3-n}E$						
	Me_2ClSi	Me_2ClGe	Me_2ClSn	F_3Si	Cl_3Si	Br_3Si[a]	Cl_3Ge[b]
Δ (°C)	120	120	160	120	80	80	100
$\tau_{1/2}$ (minutes)	42	113	88	42	87	12	59
Equation (56), path (a) (%)	25	80	100	>90	95	100	80
Equation (56), path (b) (%)	75	20	0	<10	5	0	20

[a] $(I_3Si)(Me_3Si)_3N_4$ decomposes rapidly. Accordingly, thermolysis of $(Me_3Si)_3N_4Li$ with SiI_4 in benzene at 6°C does not yield tetrazene, but instead gives mainly Me_3SiN_3 and $(I_3Si)(Me_3Si)_2N$.
[b] $(Cl_3Sn)(Me_3Si)_3N_4$ was not isolable. In the reaction of $(Me_3Si)_3N_4Li$ with $SnCl_4$, $Cl_2Sn[N_4(SiMe_3)_3]_2$ was found.

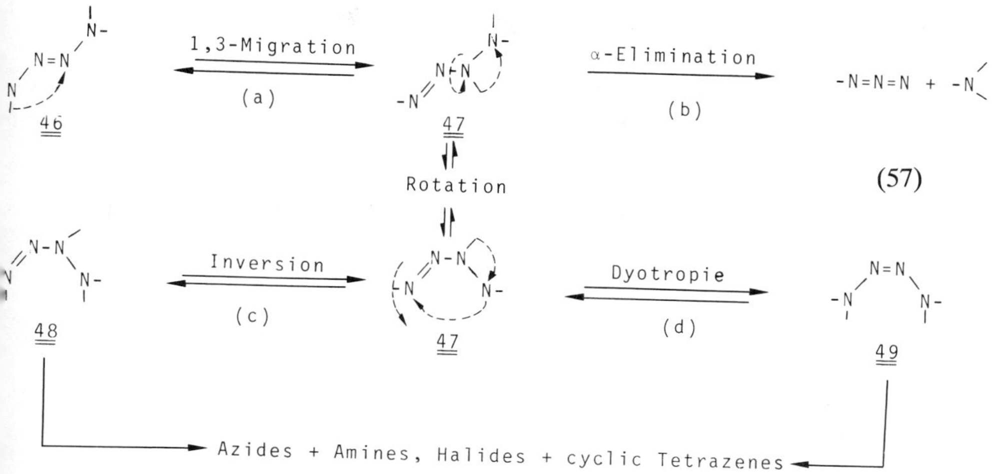

Scheme 6

(57)

formation of trimethylsilyl azide and dimethylchlorosilylbis(trimethyl-silyl)amine (Thermolysis Pathway III), as well as additional trimethylchlorosilane and the cyclic tetrazene **45**. Other tetrazenes of type $(Me_nX_{3-n}E)(Me_3Si)_3N_4$, with $Me_nX_{3-n}E = Me_2ClE$ and Hal_3E (E = Si, Ge, Sn) (**45**), decompose correspondingly. The rate of the first-order thermolytic reaction increases for the sequence of compounds $(Me_nX_{3-n}E)(Me_3Si)_3N_4$ with E = Sn < Ge < Si, X = F < Cl < Br < I, and $Me_nX_{3-n}E = Me_2ClE < Cl_3E$ (Table IV). On the other hand, the percentage share of the thermolytic reaction in path (b) [Eq. (56)] rises at the cost of the thermolytic reaction in path (a) in the direction of compounds with $Me_nX_{3-n}E = Me_2ClSn$, $Br_3Si < Cl_3Si < F_3Si < Cl_3Ge < Me_2ClGe << Me_2ClSi$ (Table IV). Evidently, the cyclization in path (b) [Eq. (56)] is slow and is, therefore, observed as a competitive reaction only when cleavage is very slow [Eq. (56), path (a)].

The initiating reaction step of decomposition in path (a) of Eq. (56) consists of an isomerization of *trans*-2-tetrazene into *trans*-1-tetrazene, which, with α-elimination of amine, produces azides [Scheme 6, Eq. (57), paths (a) and (b)]. The isomerization in path (a) [Eq. (57)] may also be the first reaction step toward transformation [Eq. (56), path (b)]. The next reaction step of *trans*-1-tetrazene (**47**), necessary for the formation of products according to Eq (56b), probably consists of an inversion according to path (c) of Eq. (57),[30] or a dyotropic rearrangement [cf. ref. *54*] as in

---

[30] Inversion of azo–nitrogen atoms R—N= (R = H or organyl) generally requires high activation energy (*23*). Small activation energies are expected for the inversion of azo-nitrogen atoms R—N= (R = silyl or germyl) [cf. planar $(H_3Si)_3N$].

path (d), leading to *cis*-tetrazenes (**48** or **49**). The latter compounds in turn may react to form the cyclic *cis*-2-tetrazenes **45** after elimination of silyl or germyl halides, and possibly also to form azides upon γ-elimination of amines (cf. Scheme 6).

Tetrazenes $F_2Si[N_4(SiMe_3)_3]_2$ and $Cl_2Sn[N_4(SiMe_3)_3]_2$ (*45*) also decompose analogously to **44** with the formation of Me_3SiN_3 and Me_3SiF or Me_3SiCl as well as an uncharacterized product [thermolysis goes according to a first-order reaction: $\tau_{\frac{1}{2}}$ (120°C) = 123 minutes and 1666 minutes, respectively].

   c. Partially Substituted Tetrazenes $(Me_3E)_nN_4H_{4-n}$. In contrast to the tetrasubstituted tetrazene $(Me_3Si)_4N_4$, the *trisubstituted* tetrazene **50** decomposes mainly (>84%) via a non-free radical pathway (Thermolysis Pathway IV) to form trimethylsilyl azide and bis(trimethylsilyl)amine [Eq. (58)] (*42*).[31] In a corresponding manner, the tetrazene $(Me_3Si)_2N$—

$$
\begin{array}{c}
Me_3Si \diagdown \\
N\text{-}N\text{=}N\text{-}N \diagup^{SiMe_3}_{\diagdown H} \\
Me_3Si \diagup \\
\underline{50}
\end{array}
\longrightarrow
Me_3Si\text{-}N\text{=}N\text{=}N \;+\; Me_3Si\text{-}N\diagup^{SiMe_3}_{\diagdown H}
$$

(58)

N=N—NH(GeMe₃) thermolyzes via $(Me_3Si)(Me_3Ge)N$—N=N—NH(SiMe₃) [cf. Section III,B,3,c] to Me_3GeN_3 and $(Me_3Si)_2NH$ (74%) and to Me_3SiN_3 and $(Me_3Si)(Me_3Ge)NH$ (26%). On the other hand, stannyltetrazenes $(Me_3Si)_2N$—N=N—NH(SnMe₃) and (Me_3Si)-$(Me_3Sn)N$—N=N—NH(SiMe₃) dismutate into $(Me_3Si)_2(Me_3Sn)_2N_4$ and $(Me_3Si)_2N_4H_2$ (*42*). The same is true for germyltetrazene $(Me_3Ge)_2N$—N=N—NH(GeMe₃) (*42*).

The thermolysis in Eq. (58) is first order, and must, therefore, proceed via intramolecular rearrangement of substituents. The half-life for thermolysis at 140°C for **50** as well as for *N*-deuterated **50** is 44 minutes, and the activation energy is 139.8 kJ/mol ($\Delta H^{\ddagger}$ = 136.3 kJ/mol; $\Delta G^{\ddagger}$ = 130.5 kJ/mol; $\Delta S^{\ddagger}$ = 14.0 eu). The thermolytic course probably follows Eq. (57), paths (a) and (b), and the absence of an isotope effect supports the conclusion that the first reaction step consists of rearrangement of 1,3-silyl groups, and not 4,2-hydrogen rearrangement.

Thermolysis of *disubstituted* tetrazenes $(Me_3Si)_2N$—N=N—NH₂ (**51**) and $(Me_3Si)HN$—N=N—NH(SiMe₃) (**52**) is quite complicated. Thus, the decomposition of **51** runs parallel to isomerization of **52** (cf. Section III,B,3,c) and follows Thermolysis Pathway IV to give trimethylsilyl azide

---

[31] Of the remaining 16%, 14% thermolyzes radically as in pathway I, and 2% thermolyzes according to pathways II and III.

and trimethylsilylamine, which in turn react further with each other to form bis(trimethylsilyl)amine and ammonium azide [Eq. (59)] (42). In

$$\begin{array}{c} Me_3Si \\ Me_3Si \end{array}\!\!N-N=N-N\!\!\begin{array}{c} H \\ H \end{array} \quad \longrightarrow \quad Me_3Si-N=N=N \;+\; Me_3Si-N\!\!\begin{array}{c} H \\ H \end{array} \longrightarrow$$

$$\underset{\underline{\underline{51}}}{}$$

$$\tfrac{2}{3}\,Me_3SiN_3 \;+\; \tfrac{2}{3}\,(Me_3Si)_2NH \;+\; \tfrac{1}{3}\,NH_4N_3 \quad (59)$$

addition, the thermolysis products of **51** react with **51** to give **50**. In contrast to the rate of isomerization, the rate of decomposition of **51** increases with increasing concentration of tetrazene (cf. Table III). Accordingly, thermolysis of pure **51** at room temperature runs almost entirely according to Eq. (59) with the formation of Me_3SiN_3, $(Me_3Si)_2NH$, and NH_4N_3, as well as **50**, but thermolysis of highly dilute **51** results almost exclusively in isomerization to **52** (42).

The rate of decomposition [Eq. (59)] increases not only with increasing concentration but also with the addition of base, such as Me_3SiNMe_2. The amine formed from the thermolysis of **51** probably also catalyzes the decomposition of **51**. Moreover, it is quite conceivable that **51** acting as a base catalyzes its own decomposition. This may explain the increase in the rate of thermolysis with increasing concentration of **51**. Accordingly, the "uncatalyzed thermolysis" of **51** consists only of rearrangement to isomer **52** and thermal decomposition of the latter compound.

Pure tetrazene **52** is substantially more thermostable than pure tetrazene **51** (Table III). Also, the decomposition of **52** occurs according to Thermolysis Pathway IV to give hydrazoic acid and bis(trimethysilyl)amine, which in turn react further with each other to form trimethylsilylamine and ammonium azide (42),[32] as shown in Eq. (60).

$$\begin{array}{c} Me_3Si \\ H \end{array}\!\!N-N=N-N\!\!\begin{array}{c} SiMe_3 \\ H \end{array} \quad \longrightarrow \quad HN=N=N \;+\; Me_3Si-N\!\!\begin{array}{c} SiMe_3 \\ H \end{array} \longrightarrow$$

$$\underset{\underline{\underline{52}}}{}$$

$$\tfrac{2}{3}\,Me_3SiN_3 \;+\; \tfrac{2}{3}\,(Me_3Si)_2NH \;+\; \tfrac{1}{3}\,NH_4N_3 \quad (60)$$

In solution, but not as a pure substance, **52** decomposes at room temperature additionally according to Thermolysis Pathway III into nitrogen and 1,1-bis(trimethylsilyl)hydrazide,[33] which, under thermolytic conditions,

---

[32] Silylation of **52** (formation of **50**) does not occur due to reasons discussed in Section III,D,2,b.

[33] Strikingly, symmetric tetrazene **52** changes into an asymmetric hydrazine. Possibly, it is not **52** that decomposes into the products given in Eq. (61), but tetrazene **51**, which exists in a small equilibrium concentration with **52**.

rearranges slowly into 1,2-bis(trimethysilyl)hydrazine [Eq. (61)] (*42*). The percentage share of thermolysis in Eq. (61) increases with decreasing

$$N_2 + (Me_3Si)HN-NH(SiMe_3) \quad (61)$$

concentration of **52** (*42*), as does the *rate* of thermolysis (Table III). As in the case of tetrazene **51**, these results indicate a (base−) self-catalyzed decomposition of **52**. Actually, the base Me_3SiNMe_2 accelerates the decomposition in Eq. (60), so that dilute solutions of **52** in the presence of Me_3SiNMe_2 practically do not decompose according to Eq. (61). Therefore, "uncatalyzed thermolysis" of **52** leads to formation of N_2 and $(Me_3Si)_2N_2H_2$ (Thermolysis Pathway III). With that, it distinguishes itself from the "uncatalyzed thermolysis" of tetrazene **50** (Thermolysis Pathway IV). Thermodynamically, thermolysis III is definitely favored over thermolysis IV. Transformation of **50** into N_2 and $(Me_3Si)_3N_2H$ is obviously restrained due to steric considerations.

Monosubstituted tetrazene $(Me_3Si)HN-N{=}N-NH_2$ is very thermolabile (*41, 42*). In methylene chloride it decomposes above ca. −40°C with the formation of trimethylsilyl azide and bis(trimethylsilyl)amine, as well as ammonium azide. Thus, the (catalyzed?) decomposition of the compound follows Thermolysis Pathway IV.

d. Completely Substituted Organyltetranzenes $(Me_3E)_nR_{4-n}N_4$. The hitherto reported silylated, germylated, and stannylated organyltetrazenes $(Me_3E)_nR_{4-n}N_4$ (Table III) decompose via a free radical as do tetrazenes $(Me_3E)_n(Me_3E')_{4-n}N_4$, but at higher temperatures. Thus, $(Me_3Si)_3N_4Me$ thermolyzes in benzene at 165°C according to a first-order reaction $(\tau_{1/2} = 7.3$ hours) to give the main product $(Me_3Si)_2NH$ (*53*) and a number of other unidentified products. Evidently, the compound decomposes according to thermolysis Pathway I with initial evolution of nitrogen to form radicals $(Me_3Si)_2N\cdot$ and $\cdot NMe(SiMe_3)$, which react further in many ways. That $(Me_3Si)_3N_4Me$ [unlike $(Me_3Si)_3N_4H$] does not decompose into Me_3SiN_3 and $(Me_3Si)_2NMe$ is not due to steric factors but depends on the fact that the rearrangement into a 1-tetrazene requires a higher energy barrier, and cleavage into nitrogen and amine radicals a lower energy barrier.

Corresponding to $(Me_3Si)_3N_4Me$, the tetrazenes $(Me_3Si)_2N_4Me_2$ (first-order reaction, $\tau_{1/2} = 10.5$ hours) $(Me_3Si)N_4Me_3$, and $(Me_3Ge)N_4Me_3$

decompose at 160°C with the formation of a large number of products [e.g., $(Me_3Si)_2NH$ in the case of decomposition of $(Me_3Si)_2N_4Me_2$], which can arise only via a radical intermediate (Thermolysis Pathway I) (39, 45). In contrast, tetrazene $Me_3SnN_4Me_3$ decomposes, even at 50°C, mainly (ca. 85%) according to Thermolysis Pathway IV with the formation of methyl azide and trimethylstannyldimethylamine [(Eq. (62)] (39). A large number

$$Me_3Sn\diagdown_{Me}N-N=N-N\diagup^{Me}_{\diagdown Me} \longrightarrow Me-N=N=N + Me_3Sn-N\diagup^{Me}_{\diagdown Me} \quad (62)$$

of other products are formed, of which Me_3SnN_3 and Me_3N could be identified. Consequently, thermolysis of $(Me_3Sn)N_4Me_3$ may also be based on a radical mechanism.[34]

The silylated 1-aryltetrazenes compared with the organylated 1-aryltetrazenes known to date are surprisingly stable and thermolyze only slowly and only at temperatures above 150°C into nitrogen trimethylsilyl-arylamine, and bis(trimethylsilyl)amine [Eq. (63)] (39). Once again, a

$$Ar-N=N-N-N\diagdown_{SiMe_3}^{SiMe_3} \xrightarrow[\substack{Ar = Ph, \\ p-Tol}]{+ 2H} Ar\diagdown_{Me_3Si}N-H + N\equiv N + H-N\diagup^{SiMe_3}_{\diagdown SiMe_3} \quad (63)$$

$$\underline{53a}$$

free radical decomposition is involved. The actual decomposition of tetrazene **53a** into radicals follows a 3,1-silyl-group migration to form a higher energy tetrazene **53b**, which, being the only source of resonance-stabilized radicals $PhNSiMe_3$, thermolyzes more easily than **53a** [cf. thermolysis of $PhN=N—NMe(SiMe_3)$, Section II,C,1] [Eq. (64)].

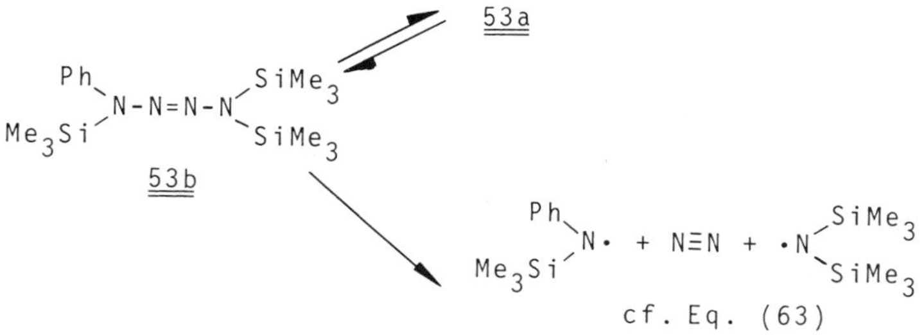

$$\underline{53a}$$
$$Ph\diagdown_{Me_3Si}N-N=N-N\diagup^{SiMe_3}_{\diagdown SiMe_3}$$
$$\underline{53b}$$
$$(64)$$
$$Ph\diagdown_{Me_3Si}N\cdot + N\equiv N + \cdot N\diagup^{SiMe_3}_{\diagdown SiMe_3}$$
$$cf. \; Eq. \; (63)$$

[34] For example, the decomposition in Eq. (62) may develop by a radical chain reaction:

$$Me_3Sn\cdot + (Me_3Sn)MeN—N=N—NMe_2 \rightarrow (Me_3Sn)MeN—N=N\cdot + Me_3SnNMe_2;$$
$$(Me_3Sn)MeN—N=N \rightarrow Me_3Sn\cdot + MeN=N=N.$$

2. *Cyclic Tetrazenes*

The hitherto synthesized sila- and germatetrazolines of type **54** (cf. Table III) decompose according to Thermolysis Pathway IV. Thus, 5,5-dimethyl-1,4-bis(trimethylsilyl)-5-sila-2-tetrazoline (**54**; E = Si, R = Me$_3$Si in benzene undergoes quantitative first-order decomposition [$\tau_{1/2}$ (150°C) = 22.4 minutes; $\Delta H^{\ddagger}$ = 138 kJ/mol, $\Delta S^{\ddagger}$ = 20 eu] into the azidosilylamine **55** (35), according to Eq. (65).

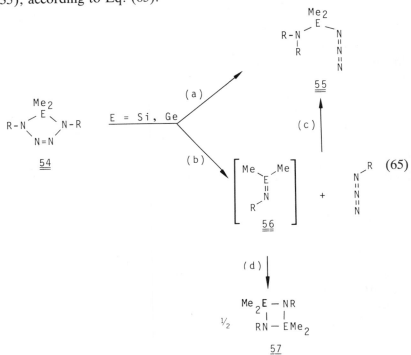

In an analogous manner, but faster (cf. Table III), silatetrazolines **54** (R = ^{t}BuMe$_2$Si and ^{t}Bu$_2$MeSi) decompose to form azidosilylamines **55** in low yield. The main product in this reaction is, besides ^{t}Bu$_2$MeSiN$_3$, the disilylazetidine **57** (E = Si; R = ^{t}Bu$_2$MeSi). Silatetrazoline **54**, with R = ^{t}Bu$_3$Si, thermolyzes exclusively into the four-membered-ring compound **57** (*14a, 16*). The same is true for germatetrazolines **54** (E = Ge) with R = ^{t}Bu$_n$Me$_{3-n}$Si (*14a, 15*). On the other hand, 5,5-dimethyl-1,4-bis(trimethylsilyl)-5-germa-2-tetrazoline (**54**; E = Ge, R = Me$_3$Si) undergoes first-order decomposition in benzene into azidogermylamine **55** (27%) and into Me$_3$SiN$_3$ and [—Me$_2$GeNSiMe$_3$—]$_n$ (73%).

The thermolysis products of sila- and germatetrazolines **54** (E = Si, Ge; R = ^{t}Bu$_n$Me$_{3-n}$Si) are formed by a [2 + 3]-cycloreversion via sila- and germaketimines **56** [Eq. path (b)],[35] which either insert into the Si—N bond of simultaneously formed silyl azides RN_3 [Eq. (65) path (c)] or dimerize (polymerize) [Eq. (65) path (d).[36] Evidently insertion [path (c)] is inhibited by the increasing bulkiness of the silyl groups R, so that its share decreases, while dimerization increases [path (d)] (*14a*).

The [2 + 3]-cycloreversion of tetrazolines of type **54** like the more rapid [2 + 3]-cycloreversion of triazolines of type **28** provides a good preparative approach to unsaturated silicon and germanium compounds of type **56**. In the presence of appropriate trapping agents for **56**, the thermolysis of **54** leads accordingly to products of sila- and germaketimines with these trapping agents (cf. Section II,C,2).

D. Reactivity

1. *Redox Reactions*

Oxidation of Group IV derivatives of tetrazenes $(Me_3E)_4N_4$ with halogens occurs faster with increasing oxidizing potential and depends on E in the sequence Si < Ge < Sn (*41*). Thus, $(Me_3Si)_4N_4$ in methylene chloride reacts with chlorine at $-80°C$ and with bromine at room temperature according to Eq. (66) to form nitrogen and trimethylsilyl chloride or bromide. Iodine does not react with tetrasilyltetrazene even at 80°C.

$$(Me_3Si)_4N_4 + 2 \ Hal_2 \xrightarrow{\text{(Hal = Cl, Br)}} 2 \ N_2 + Me_3SiHal \quad (66)$$

p-Benzoquinone oxidizes $(Me_3Si)_4N_4$ slowly in boiling benzene and $(Me_3Sn)_4N_4$ rapidly even at room temperature [Eq. (67)]. Under normal conditions, $(Me_3Si)_4N_4$ and $(Me_3Ge)_4N_4$ are stable toward oxygen, whereas $(Me_3Sn)_4N_4$ is attacked by oxygen at room temperature [Eq. (68)].

[35] In the case of **54** (E = Si; R = Me_3Si) the product **55** possibly forms directly [path (a), Eq. (65)]. The same may be true for the silatetrazolines **54** (E = Si; R/R = Me_3C/Me_3Si or Me_3C/Ph_3Si), which thermolytically change into $(Me_3C)(Me_3Si)NSiMe_2N_3$ or $(Me_3C)(Ph_3Si)NSiMe_2N_3$ (*16*).

[36] For mixed, substituted silatetrazolines **54** ($R_{left} \neq R_{right}$), there are naturally two possible [2 + 3]-cycloreversions; accordingly, **56** (R_{left}) + $R_{right}N_3$ or **56**(R_{right}) + $R_{left}N_3$ can be formed. Actually a mixture of all four products results.

	R_{left}/R_{right}	
	$Me_3Si/^tBu_2MeSi$	$^tBuMe_2Si/^tBu_2MeSi$
%**56**(R_{left})	23	46
%**56**(R_{right})	77	54

Evidently, the silaketimine (**56**), which has a larger number of ^tBu groups, is preferred.

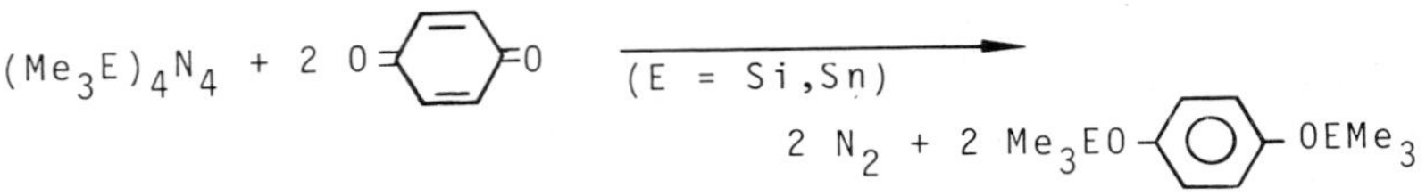

$$(Me_3E)_4N_4 + 2 \ O = \!\!\!\langle \ \rangle \!\!\!= O \quad \overline{(E = Si, Sn)}$$

$$2 \ N_2 + 2 \ Me_3EO - \!\!\langle O \rangle - OEMe_3 \quad (67)$$

$$(Me_3Sn)_4N_4 + O_2 \quad \longrightarrow \quad 2 \ N_2 + 2(Me_3Sn)_2O$$

$$(68)$$

Naturally, oxidation occurs in many steps. Tetrazadienes $(Me_3E)_2$-N—N=N=N or $(Me_3E)N$=N—N=N(EMe_3) have not been detected as intermediates, probably because their formation is slower than their further reaction. The only identified oxidation intermediate, generated by single-electron oxidation of $(Me_3Si)_4N_4$ with $AlCl_3/CH_2Cl_2$, and stable in solution at room temperature, is the trans acyclic radical-cation **58**, whose electronic structure according to its ESR spectrum is similar to that of the radical-cation $Me_4N_4^{\cdot \ +}$ *(61)*. In contrast to **58**, the cis-configurated cyclic radical cation **59** is unstable in CH_2Cl_2 solution. It is probably formed as a primary product in the oxidation of the silatetrazoline $Me_2SiN_4(SiMe_3)_2$ with $AlCl_3/CH_2Cl_2$ but changes instantly even at $-80°C$ into the radical-cation **60** *(60)*.

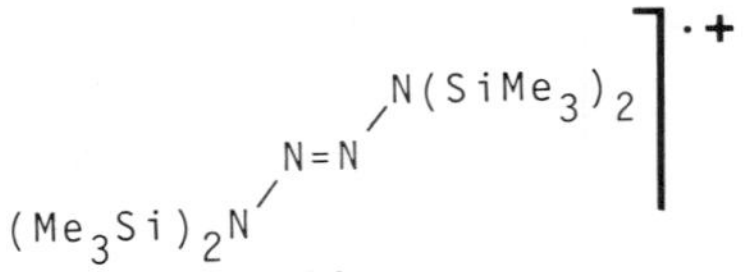

$$\left[\begin{array}{c} N(SiMe_3)_2 \\ N=N \\ (Me_3Si)_2N \end{array} \right]^{\cdot +}$$

$$\underline{58}$$

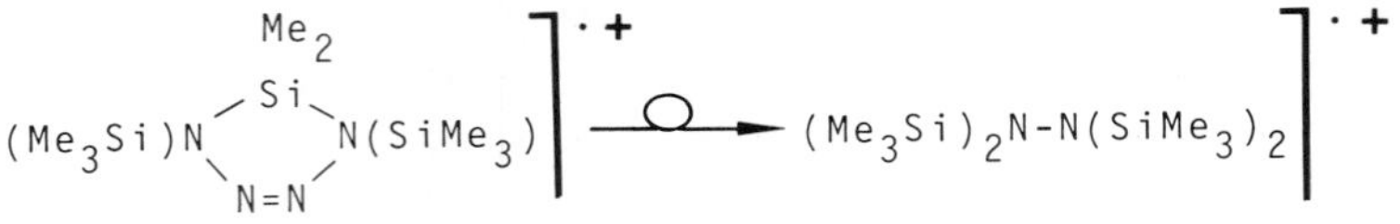

$$\left[\begin{array}{c} Me_2 \\ Si \\ (Me_3Si)N \quad N(SiMe_3) \\ N=N \end{array} \right]^{\cdot +} \xrightarrow{\ -O\ } \left[(Me_3Si)_2N-N(SiMe_3)_2 \right]^{\cdot +}$$

$$\underline{59} \hspace{5cm} \underline{60}$$

Reduction of the silyltetrazene $(Me_3Si)_4N_4$ leads, with the evolution of nitrogen, to bis(trimethylsilyl)amine *(41)*. With alkali metals it occurs faster with increasing reduction potential. Thus, an ether solution of $(Me_3Si)_4N_4$ reacts with lithium, sodium, or potassium at about 35°C, 0°C, or $-25°C$, respectively, according to Eq. (69a). Hydrogen reacts with $(Me_3Si)_4N_4$ in inpentane only under pressure (200 bar) and only in the presence of a hydrogenation catalyst (Pd) [Eq. 69b)].

$$(Me_3Si)_4N_4 + 2 \ M \quad \overline{(M = Li, Na, K)} \quad N_2 + 2(Me_3Si)_2NM \quad (69a)$$

$$(Me_3Si)_4N_4 + H_2 \quad \overline{(Pd, 200 \ bar)} \quad N_2 + 2(Me_3Si)_2NH \quad (69b)$$

The reduction processes take place in many steps. Tetrazanes
(Me$_3$Si)$_2$N—N—N—N(SiMe$_3$)$_3$ were not identified as intermediates. Also, (Me$_3$Si)$_4$N$_4$·$^-$, as a product of one-electron reduction, could not be isolated (62).

2. Substitution Reactions

a. Substitution of EX$_3$ Groups. As explained in Section III,A,2, tetrazene-bound EX$_3$ groups can be protolytically replaced by hydrogen. The rate of protolysis rises with the acidic strength of HX (e.g., MeOH < CF$_3$CO$_2$H) and depends on E in the sequence Si < Ge < Sn (41). Protolysis of tetrazenes with Me$_3$E groups provides a good route to hydrogen-bearing tetrazenes (42). Thus, protolysis of (Me$_3$Si)$_4$N$_4$ with CF$_3$CO$_2$H at −78°C (Fig. 3) or of (Me$_3$Ge)$_4$N$_4$ with MeOH at room temperature (Fig. 4) leads to partially substituted tetrazenes (Me$_3$E)$_n$-N$_4$H$_{4-n}$ (cf. Scheme 3 and Table III). As an end product of (Me$_3$Si)$_4$N$_4$/CF$_3$CO$_2$H protolysis, trans-2-tetrazene was synthesized for the first time [Eq. (70a)] as a colorless, crystalline substance, sublimable at −35°C in high vacuum, which decomposes at ca. 0°C to form N$_2$ + N$_2$H$_4$ as well as HN$_3$ + NH$_3$ (43, 49). On the other hand, total protolysis of the sila-triazoline Me$_2$SiN$_4$(SiMe$_3$)$_2$ [Eq. (70b)] with trifluoroacetic acid in meth-

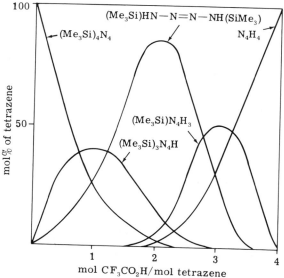

FIG. 3. Protolysis of (Me$_3$Si)$_4$N$_4$ in CH$_2$Cl$_2$ at −78°C.

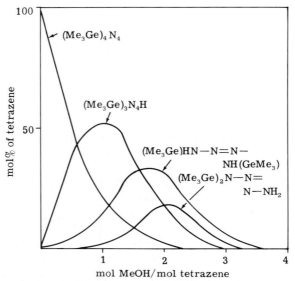

FIG. 4. Protolysis of $(Me_3Ge)_4N_4$ in C_6H_6 at 20°C; $(Me_3Ge)N_4H_3$ and N_4H_4 are unstable at 20°C, and therefore, undetectable.

ylene chloride at −78°C does not lead to isolable *cis*-2-tetrazene, but instead to ammonium azide. Evidently, *cis*-2-tetrazene, which probably is formed as an intermediate, is more thermolabile than *trans*-2-tetrazene and hence decomposes even during its synthesis (*35*).

$$(Me_3Si)_4N_4 \xrightarrow[- \ 4 \ CF_3CO_2SiMe_3]{+ \ 4 \ CF_3CO_2H} \quad \begin{matrix} & NH_2 \\ & N=N \\ H_2N & \end{matrix} \xrightarrow[>0°C]{} \begin{matrix} 25\% & HN_3 + NH_3 \\ & \\ 75\% & N_2 + N_2H_4 \end{matrix}$$

$$(70a)$$

$$(Me_3Si)N \overset{\overset{\displaystyle Me_2}{Si}}{\underset{N=N}{\diagup}} N(SiMe_3) \xrightarrow[\substack{-2 \ CF_3CO_2SiMe_3 \\ - \ (CF_3CO_2)_2SiMe_2}]{+ \ 4 \ CF_3CO_2H}$$

$$\begin{matrix} & N=N \\ H_2N & NH_2 \end{matrix} \quad\longrightarrow\quad HN_3 + NH_3 \quad (70b)$$

Other examples of protolysis include (1) methanolysis of *s*- or *a*-$(Me_3Si)_2(Me_3Ge)_2N_4$ at 20°C to form *s*- or *a*-$(Me_3Si)_2N_4H_2$ via *s*- or *a*-$(Me_3Si)_2(Me_3Ge)N_4H$ (*42*), (2) methanolysis of $(Me_3Si)_3N_4H$ at 20°C to form $a\text{-}(Me_3Si)_2N_4H_2$ (42), and (3) protolysis of $(Me_3E)Me_3N_4$ (E = Si, Ge, Sn) with CF_3CO_2H at $-78°C$, leading to Me_3N_4H (39). Trimethyltetrazene, first prepared in this fashion, is a colorless solid at $-40°C$ and sublimes under high vacuum. It decomposes above $-30°C$ to form MeN_3 and $HNMe_2$.[37]

As with hydrogen, tetrazene-bound Me_3E groups can also be exchanged with Me_3E' groups (cf. Section III, A, 2). Accordingly, a benzene solution of $(Me_3Sn)_4N_4$ reacts with Me_3ECl (E = Si, Ge) to form a mixture of tetrazenes according to Eq. (71). The predominant tetrazene product is

$$(\text{in each case} +Me_3ECl, -Me_3SnCl)$$

$$(Me_3Sn)_4N_4 \longrightarrow \begin{array}{c} Me_3E \\ Me_3Sn \end{array}\!\!>\!N\text{-}N\text{=}N\text{-}N\!<\!\!\begin{array}{c} SnMe_3 \\ SnMe_3 \end{array} \longrightarrow \begin{array}{c} Me_3E \\ Me_3Sn \end{array}\!\!>\!N\text{-}N\text{=}N\text{-}N\!<\!\!\begin{array}{c} EMe_3 \\ SnMe_3 \end{array} \longrightarrow$$

$$\begin{array}{c} Me_3E \\ Me_3E \end{array}\!\!>\!N\text{-}N\text{=}N\text{-}N\!<\!\!\begin{array}{c} EMe_3 \\ SnMe_3 \end{array} \longrightarrow (Me_3E)_4N_4 \qquad (71)$$

that for which the number of Me_3E groups equals the chosen reactant ratio of $Me_3ECl:(Me_3Sn)_4N_4$ (41).

Reaction of Me_3SiCl with $(Me_3Ge)_4N_4$ is much slower than that with $(Me_3Sn)_4N_4$. The formation of the primary substitution product $(Me_3Si)\text{-}(Me_3Ge)_3N_4$ is, therefore, associated with its thermolysis (Section III,C,1). Mixed silylgermyltetrazenes are easily formed by the reaction of silylstannyltetrazenes with Me_3GeCl [Eq. (71)] (41).

Thermolytic decomposition of a substitution product, as a consequence of its slower formation, is observed in other cases also. Thus, reaction of $(Me_3Sn)_4N_4$ with trimethylstannane at $60°C$, which leads primarily to an Me_3Sn/H exchange ($>\!N\text{—}SnMe_3 + Me_3SnH \rightarrow >\!N\text{—}H + Me_3Sn\text{—}SnMe_3$; cf. ref. 63), yields ammonium azide as well as hydrazine and nitrogen as typical decomposition products of thermolabile tetrazene N_4H_4 (43). Other examples include reactions of $(Me_3Si)_4N_4$ with aluminum chloride or nitrosyl chloride, as well as reactions of $(Me_3Sn)_4N_4$ with acetone or nitrobenzene [Eqs. (72)–(75)] (43). The products of these reactions are consistent with a substitution reaction involving formation of $(Me_3Si)_3N_4AlCl_2$, $(Me_3Si)_3N_4NO$, $(Me_3Sn)_2N_4CMe_2$, and $(Me_3Sn)_2N_4NPh$, respectively. These substitution products then decompose rapidly with cleavage into azide and amine [formation of $Me_3SiN_3/(Me_3Si)_2NAlCl_2$ as well as $Me_3SiN_3/(Me_3Si)_2NNO$] or into hydrazine and nitrogen [formation of $(Me_3Sn)_2N\text{—}N\text{=}CMe_2/N_2$ as well

---

[37] H-NMR chemical shift (CH_2Cl_2, TMS): $\delta = 2.80$ (Me_2N), 2.82 ($NHMe$).

$$(Me_3Si)_4N_4 + AlCl_3 \xrightarrow[-78°C]{(CH_2Cl_2)}$$

$$(Me_3Si)_2N-AlClN_3 + 2\ Me_3SiCl \quad (72)$$

$$(Me_3Si)_4N_4 + NOCl \xrightarrow[-78°C]{(Et_2O)}$$

$$Me_3SiN_3 + N_2 + (Me_3Si)_2O + Me_3SiCl \quad (73)$$

$$(Me_3Sn)_4N_4 + 2\ Me_2CO \xrightarrow[25°C]{(C_6H_6)}$$

$$Me_2C=N-N=CMe_2 + N_2 + 2(Me_3Sn)_2O \quad (74)$$

$$(Me_3Sn)_4N_4 + 2\ PhNO \xrightarrow[25°C]{(C_6H_6)}$$

$$PhN=NPh + 2\ N_2 + 2(Me_3Sn)_2O \quad (75)$$

as $(Me_3Sn)_2N—N{=}NPh/N_2$], which in turn proceed to reaction products [chloride/azide exchange in the case of Eq. (72)), nitrosamine thermolysis in case of Eq. (73), $(Me_3Sn)_2NNCMe_2/Me_2CO$ condensation in the case of Eq. (74), and triazene thermolysis and $(Me_3Sn)_2NPh/PhNO$ condensation in the case of Eq. (75)].

An example of substitution followed by cyclization of the substituted tetrazene has been observed in the reaction of $(Me_3Si)_4N_4$ with phenylsulfonyl isocyanate [Eq. (76)] (43), leading to cyclic tetrazene **61**, which also is obtainable by the action of CO_2 on $Me_3Si—N{=}N—SiMe_3$, cf. ref. 56.]

$$(Me_3Si)_4N_4 + PhSO_2-N=C=O \xrightarrow[25°C]{(Et_2O)}$$

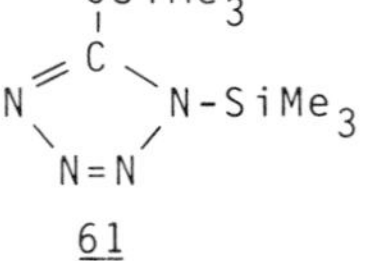

$$\begin{array}{c} OSiMe_3 \\ | \\ C \\ N \diagup \quad \diagdown N{-}SiMe_3 \\ \diagdown \ N{=}N \ \diagup \end{array} + PhSO_2N(SiMe_3)_2 \quad (76)$$

$$\underline{\underline{61}}$$

b. Substitution of Hydrogen. Action of amines Me_3ENR_2 (E = Si, Ge, Sn; R = Me, Et) on partially substituted tetrazenes can lead to substitution of tetrazene-bound hydrogen by Me_3E groups, as discussed in Section III,A,2. The tendency of H/EMe_3 exchange increases for amines in the order $Me_3SiNR_2 < Me_3GeNR_2 < Me_3SnNR_2$ and for tetrazenes in the sequence $>N\!\!-\!\!N\!\!=\!\!N\!\!-\!\!NH(SiMe_3) < >N\!\!-\!\!N\!\!=\!\!N\!\!-\!\!NH(GeMe_3) <$ $>N\!\!-\!\!N\!\!=\!\!N\!\!-\!\!NH(SnMe_3) < >N\!\!-\!\!N\!\!=\!\!N\!\!-\!\!NH_2$. Thus, $(Me_3Si)_2$-$N\!\!-\!\!N\!\!=\!\!N\!\!-\!\!NH(SiMe_3)$ and $(Me_3Si)HN\!\!-\!\!N\!\!=\!\!N\!\!-\!\!NH(SiMe_3)$ can only be stannylated, as described, whereas $(Me_3Ge)_2N\!\!-\!\!N\!\!=\!\!N\!\!-\!\!NH(GeMe_3)$ and $(Me_3Si)_2N\!\!-\!\!N\!\!=\!\!N\!\!-\!\!NH_2$ can be stannylated or germylated (42). On the other hand, the amine Me_3SiNMe_2 does not act as a silylating agent toward $(Me_3Si)_2N_4H_2$ but catalyzes the thermolysis of the tetrazene (cf. Section II,C,1,c).

Due to its acidic character, tetrazene-bound hydrogen can be replaced by alkali metals. Accordingly, the action of BuLi with $(Me_3Si)_3N_4H$ or $(Me_3Si)_2N_4H_2$ in organic solvents such as ether, pentane, or benzene leads to tetrazenides **62–64** (44) which, on removal of the solvent, can be isolated in the form of colorless to light yellow solids, which are extremely sensitive to hydrolysis and easily flammable in air (Table V).[38] The most unstable is **62**, while **64** is surprisingly the most stable of the isolated tetrazenides. Thermal decomposition of tetrazenides, the rate and course of which depend upon the nature of the reaction medium, occurs partly with the formation of nitrogen and hydrazides and partly with the formation of azides and amides [Eqs. (77)–(88)].

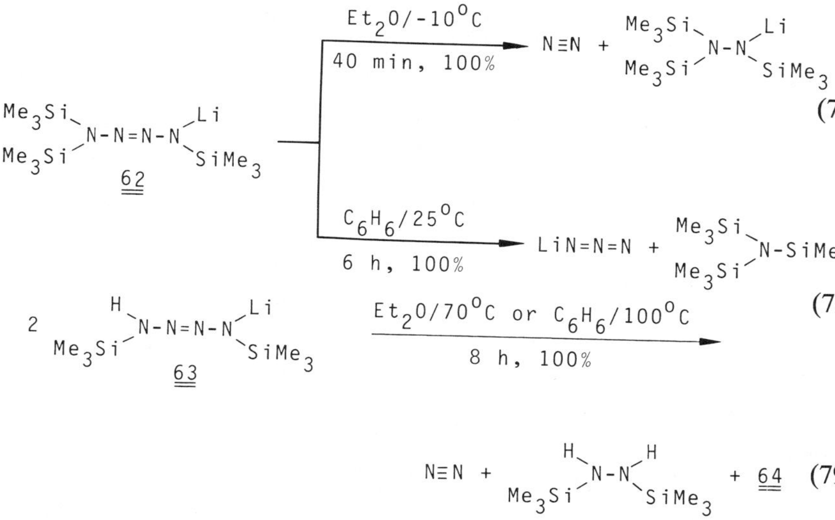

---

[38]Action of BuLi on N_4H_4 in Et_2O in the molar ratio 4:1 at $-78°C$ leads to the hitherto unknown lithium nitride (Li_4N_4) as a yellowish solid, insoluble in pentane and ether, which decomposes above $-30°C$ (45).

$$\underset{\substack{\text{Me}_3\text{Si} \\ \underset{\underline{\underline{64}}}{}}}{\overset{\text{Li}}{\diagdown}}\text{N-N=N-N}\overset{\text{Li}}{\underset{\text{SiMe}_3}{\diagup}}$$

$$\xrightarrow[\text{15 h, 100\%}]{\text{Et}_2\text{O}/120^\circ\text{C}} \quad \text{LiN=N=N} \; + \; \underset{\text{Me}_3\text{Si}}{\overset{\text{Me}_3\text{Si}}{\diagdown}}\text{N-Li} \tag{80}$$

$$\xrightarrow[\text{40 h, 90\%}]{\text{C}_6\text{H}_6/120^\circ\text{C}} \quad \text{N}\equiv\text{N} \; + \; \underset{\text{Me}_3\text{Si}}{\overset{\text{Li}}{\diagdown}}\text{N-N}\overset{\text{Li}}{\underset{\text{SiMe}_3}{\diagup}} \tag{81}$$

c. *Substitution of Lithium.* As compared with Me_3E or hydrogen, lithium in tetrazenides (62–64) undergoes easier replacement by other groups, so that 62–64 behave as valuable agents for the synthesis of new inorganic tetrazenides.

Thus, 62 in benzene/ether at 0°C reacts promptly with Me_2SO_4, Me_3ECl, Me_2ECl_2, EHal_4, or SnCl_2 (E = Si, Ge, Sn) to form tetrazenes $(\text{Me}_3\text{Si})_2\text{N—N}{=}\text{N—N}(\text{SiMe}_3)\text{X}$ (X = Me, EMe_3, EClMe_2, EHal_3, SnCl), and with Me_2SnCl_2, SiF_4, SnCl_4, and SnCl_2 to form tetrazenes $[(\text{Me}_3\text{Si})_2\text{N—N}{=}\text{N—N}(\text{SiMe}_3)]_2\text{Y}$ (Y = SnMe_2, SiF_2, SnCl_2, Sn). In an analogous manner, 63 reacts with Me_3ECl to form tetrazenes $(\text{Me}_3\text{Si})\text{HN—N}{=}\text{N—N}(\text{EMe}_3)(\text{SiMe}_3)$, and 64 combines with Me_2SO_4, Me_3ECl, or Me_2ECl_2 to form tetrazenes $(\text{Me}_3\text{Si})_2\text{N—N}{=}\text{N—NX}_2$ (X = Me) and $(\text{Me}_3\text{Si})\text{XN—N}{=}\text{N—NX}(\text{SiMe}_3)$ (X = EMe_3, EClMe_2). Regarding characterization of the tetrazene derivatives refer to Table III, and for their thermolyses refer to Section III,C,1.

Analogous to the mentioned halides of silicon, germanium, and tin, *beryllium chloride,* (as its adduct with THF) react with 62 in ether at low temperature to form tetrazenide 65, as in Eq. (82). Compound 65

$$62 + \text{BeCl}_2 \xrightarrow{(\text{Et}_2\text{O})} \underset{\text{Me}_3\text{Si}}{\overset{\text{Me}_3\text{Si}}{\diagdown}}\text{N-N=N-N}\underset{\text{Be}}{\overset{\text{SiMe}_3}{\diagup}}\underset{\underset{\text{Me}_3\text{Si}}{}}{\overset{}{\diagdown}}\text{N-N=N-N}\overset{\text{SiMe}_3}{\underset{\text{SiMe}_3}{\diagup}} \tag{82}$$

$$\underline{\underline{65}}$$

decomposes slowly at room temperature (Table V); in ether it yields exclusively $(\text{Me}_3\text{Si})_3\text{N}_2\text{H}$, Be, and N_2 whereas in benzene it forms mainly (90%) N_2 and $(\text{Me}_3\text{Si})_3\text{N}_2\text{—Be—N}_2(\text{SiMe}_3)_3$, as well as small amounts (10%) of Me_3SiN_3 and $(\text{Me}_3\text{Si})_2\text{N—Be—N}(\text{SiMe}_3)_2$ (45).

Boron tribromide (BBr_3) also reacts with 62 at low temperatures to give tetrazenide 66 [Eq. (83)] (45), which under reaction conditions decom-

TABLE V
Derivatives of Silyltetrazenes

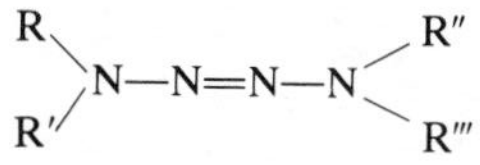

$$\begin{array}{c} R \\ {\textstyle \diagdown} \\ R' \end{array} N\!-\!N\!=\!N\!-\!N \begin{array}{c} R'' \\ {\textstyle \diagup} \\ R''' \end{array}$$

Compound	R	R'	R"	R'''	Color	State (room temperature)	Decomposition temperature ($\tau_{1/2}$, minutes)[a]	δ(R')	δ(R")	δ(R''')	Reference
								\multicolumn{3}{c}{¹H-Chemical shifts (Et₂O, TMS)}	
62	Li	SiMe₃	SiMe₃	SiMe₃	Colorless	Solid	ca. 25°C	0.083	0.183	0.183	44
63	Li	SiMe₃	H	SiMe₃	Colorless	b	>50°C	0.110[c]		0.110[c]	44
64	Li	SiMe₃	Li	SiMe₃	Colorless	Solid	>100°C	0.078		0.115	44
65	BeX[d]	SiMe₃	SiMe₃	SiMe₃	Colorless	Liquid	25°C (996)	0.135	0.182	0.182	45
67	PPh₂	SiMe₃	SiMe₃	SiMe₃	Colorless	Liquid	50°C (62)	0.323[e]	−0.023	−0.023	45
68	PCl₂	SiMe₃	SiMe₃	SiMe₃	Colorless	b	ca. 25°C	0.325[f]	0.235	0.235	45
69	PBr₂	SiMe₃	SiMe₃	SiMe₃	Colorless	b	ca. 25°C	0.322[f]	0.253	0.253	45
70	AsClX[d]	SiMe₃	SiMe₃	SiMe₃	Colorless	Liquid	100°C (200)	0.357	0.225	0.225	45
71	SC₆F₅	SiMe₃	SiMe₃	SiMe₃	Colorless	Liquid	40°C (210)	0.355[g]	0.166	0.166	45
72	SX[d]	SiMe₃	SiMe₃	SiMe₃	Pale yellow	Liquid	25°C (1053)	0.288	0.240	0.240	45

[a] In benzene.
[b] In solution, not isolated.
[c] Broad signal.
[d] $X = N_4(SiMe_3)_3$.
[e] Doublet, $J(^{31}P) = 1.6$ Hz.
[f] Doublet, $J(^{31}P) = 2.0$ Hz.
[g] Multiplet.

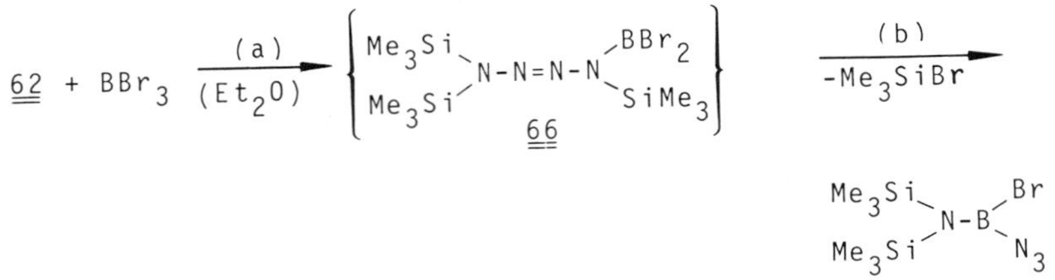

$$\underline{62} + BBr_3 \xrightarrow[(Et_2O)]{(a)} \left\{ \begin{matrix} Me_3Si \\ Me_3Si \end{matrix} N-N=N-N \begin{matrix} BBr_2 \\ SiMe_3 \end{matrix} \right\} \xrightarrow[-Me_3SiBr]{(b)}$$

$$\begin{matrix} Me_3Si \\ Me_3Si \end{matrix} N-B \begin{matrix} Br \\ N_3 \end{matrix} \quad (83)$$

poses into possibly polymeric aminoboronazide bromide (melting point and sublimation point under high vacuum >150°C).

Carbon halides, CCl_4 and CBr_4 react with **62** in ether according to Eq. (84) with the formation of nitrogen and hydrazones $(Me_3Si)_2N—N= CX_2$, as well as tetrakis(trimethylsilyl)tetrazene, trimethylsilyl halide, and

$$\underline{62} + CX_4 \xrightarrow{(Et_2O; \ X = Cl, \ Br)}$$
$$N_2 + (Me_3Si)_2N-N=CX_2 + Me_3SiX + LiX \quad (84a)$$

$$\underline{62} + Me_3SiX \xrightarrow{}$$
$$(Me_3Si)_2N-N=N-N(SiMe_3)_2 + LiX \quad (84b)$$

$$\underline{62} + CX_4 \xrightarrow[-\underline{a}N_2, \ -\underline{a}Me_3SiX, \ -\underline{a}LiX]{(\underline{a} + \underline{b} = 1)}$$
$$\underline{a}(Me_3Si)_2N-N=CX_2 + \underline{b}(Me_3Si)_4N_4 \quad (84)$$

lithium halide (45) (cf. the reaction of $Me_3Si—N=N—SiMe_3$ with CX_4, ref. *1*). Compound **62** and CX_4 probably undergo an initial radical substitution reaction to form an intermediate tetrazene $(Me_3Si)_2N— N=N—N(CX_3)(SiMe_3)$, which, with the cleavage of Me_3SiX as well as N_2, converts to the products shown in Eq. (84a). Subsequently, Me_3SiX partially interacts further with **62** according to Eq. (84b). The resulting bromine-containing hydrazone formed then reacts to some extent with **62** according to Eq. (85) to form isocyanide $(Me_3Si)_2N—N≡C$, which is therefore a side product of the reaction of **62** and CBr_4. Obviously, $(Me_3Si)_2N—N=CBr_2$ can transfer bromide to **62** to form $(Me_3Si)_2N—N=N—NBr(SiMe_3)$, $(Me_3Si)_2N—N≡C$, and LiBr. The bromine-containing tetrazene then decomposes with elimination of Me_3SiBr and N_2 into diazene $Me_3Si—N=N—SiMe_3$, which in turn brings about transformation of hydrazone into isocyanide (*1*). Therefore, toward **62**

$$\underline{\underline{62}} + 2(Me_3Si)_2N-N=CBr_2 \longrightarrow$$

$$2(Me_3Si)_2N-N\equiv C + 2 N_2 + 3 Me_3SiBr + LiBr \quad (85)$$

$$\underline{\underline{62}} + \quad 2 X_2 \longrightarrow 2 N_2 + 3 Me_3SiX + LiX \quad (86)$$

$(Me_3Si)_2N_2CBr_2$ acts as a supplier of bromine, which like other *halogens* reacts with **62** (*45*) as in Eq. (86).

Phosphorus and arsenic halides (Ph_2PCl, PCl_3, PBr_3, and $AsCl_3$) react with **62** in ether at low temperatures to form tetrazenides **67–70** (Table V)[39] [Eqs. (87) and (88)]. Whereas **67** decomposes at higher temperatures (Table V) according to Eq. (89), **68** and **69** thermolyze even at room

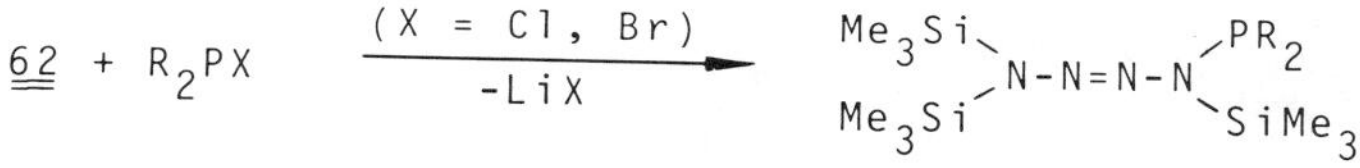

$$\underline{\underline{62}} + R_2PX \xrightarrow[\text{$-LiX$}]{\text{(X = Cl, Br)}} \begin{array}{c} Me_3Si \\ {}^{Me_3Si} \end{array}\!\!N-N=N-N\!\!\begin{array}{c} PR_2 \\ SiMe_3 \end{array} \quad (87)$$

$$R = \underline{Ph} \quad \underline{Cl} \quad \underline{Br}$$
$$\underline{\underline{67}} \quad \underline{\underline{68}} \quad \underline{\underline{69}}$$

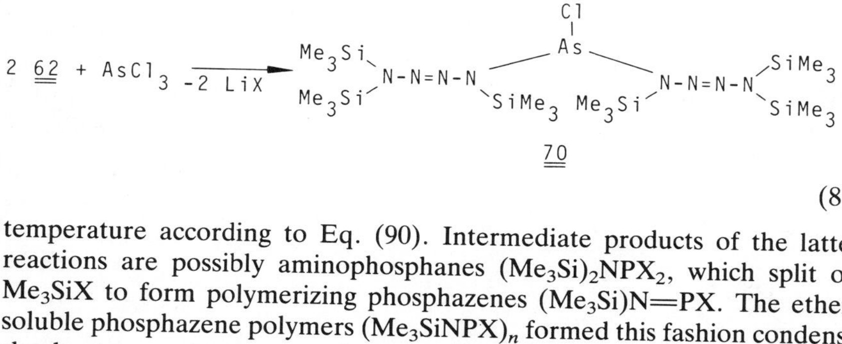

$$2\ \underline{\underline{62}} + AsCl_3 \xrightarrow{\text{$-2\ LiX$}} \begin{array}{c} Me_3Si \\ Me_3Si \end{array}\!\!N-N=N-N\!\!\begin{array}{c} \\ SiMe_3 \end{array}\!\!\overset{\overset{\displaystyle Cl}{|}}{As}\!\!\begin{array}{c} \\ Me_3Si \end{array}\!\!N-N=N-N\!\!\begin{array}{c} SiMe_3 \\ SiMe_3 \end{array}$$
$$\underline{\underline{70}}$$

$$(88)$$

temperature according to Eq. (90). Intermediate products of the latter reactions are possibly aminophosphanes $(Me_3Si)_2NPX_2$, which split off Me_3SiX to form polymerizing phosphazenes $(Me_3Si)N=PX$. The ether-soluble phosphazene polymers $(Me_3SiNPX)_n$ formed this fashion condense slowly at room temperature into ether-insoluble polymers $Me_3Si(NP)_nX$.

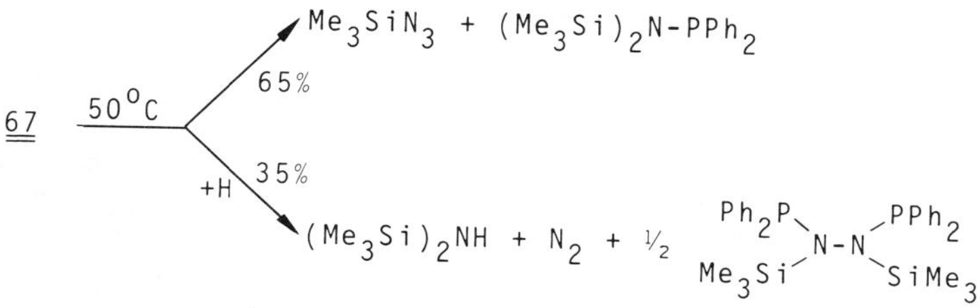

$$\underline{\underline{67}} \xrightarrow[\text{$+H$}]{50°C} \begin{array}{c} \nearrow Me_3SiN_3 + (Me_3Si)_2N-PPh_2 \quad 65\% \\ \\ \searrow (Me_3Si)_2NH + N_2 + \tfrac{1}{2}\ \begin{array}{c} Ph_2P \\ Me_3Si \end{array}\!\!N-N\!\!\begin{array}{c} PPh_2 \\ SiMe_3 \end{array} \quad 35\% \end{array} \quad (89)$$

[39] $SbCl_3$ also reacts with **62** in a molar ratio of 1:2 (*45*).

$$\underline{68}, \ \underline{69} \quad \xrightarrow[\text{(X = Cl, Br)}]{25^\circ C} \quad \left| Me_3SiN_3 + (Me_3Si)_2NPX_2 \right|$$

$$\xrightarrow[-Me_3SiN_3]{} \quad \frac{1}{n} \ (Me_3SiNPX)_n \quad (90)$$

Compound **70** is more stable than **68** and **69**, and like **67** decomposes only at higher temperatures to give mixtures of products that have not been characterized completely. Like **62**, tetrazenide **64** reacts—but faster—with PX_3 (X = Cl, Br). Even at $-78°C$, no monomer, but only polymer [dimer: $(Me_3SiNPX)_2$] is formed.

Sulfur halides, C_6F_5SCl and SCl_2, react with **62** to form tetrazenides **71** and **72** (Table V) according to Eqs. (91) and (92). Both compounds

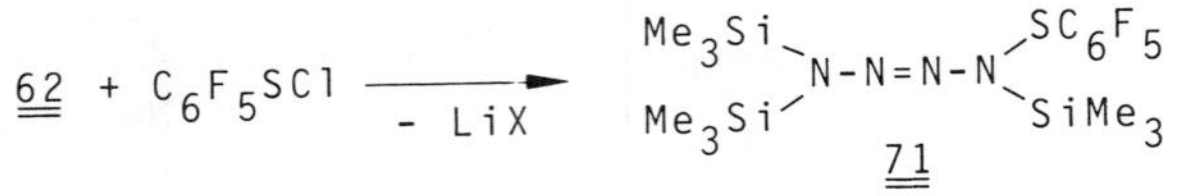

$$\underline{62} + C_6F_5SCl \xrightarrow[-\ LiX]{} \quad \begin{array}{c} Me_3Si \\ \diagdown \\ Me_3Si \diagup \end{array} N{-}N{=}N{-}N \begin{array}{c} \diagup SC_6F_5 \\ \diagdown SiMe_3 \end{array} \quad (91)$$
$$\underline{71}$$

$$2 \ \underline{62} + SCl_2 \xrightarrow[-\ 2\ LiX]{} \quad \begin{array}{c} Me_3Si \\ \diagdown \\ Me_3Si \diagup \end{array} N{-}N{=}N{-}N \underset{\underset{SiMe_3}{|}}{} {-}S{-} N \underset{\underset{SiMe_3}{|}}{} {-}N{=}N{-}N \begin{array}{c} \diagup SiMe_3 \\ \diagdown SiMe_3 \end{array} \quad (92)$$
$$\underline{72}$$

thermolyze slowly at room temperature (Table V) according to Eqs. (93) and (94). Decomposition of **71** in ether or benzene [Eq. (93)] is similar to that of **67** [Eq. (89)].[40]

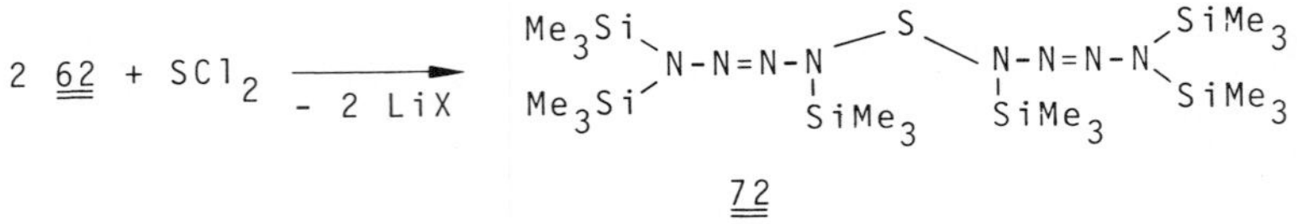

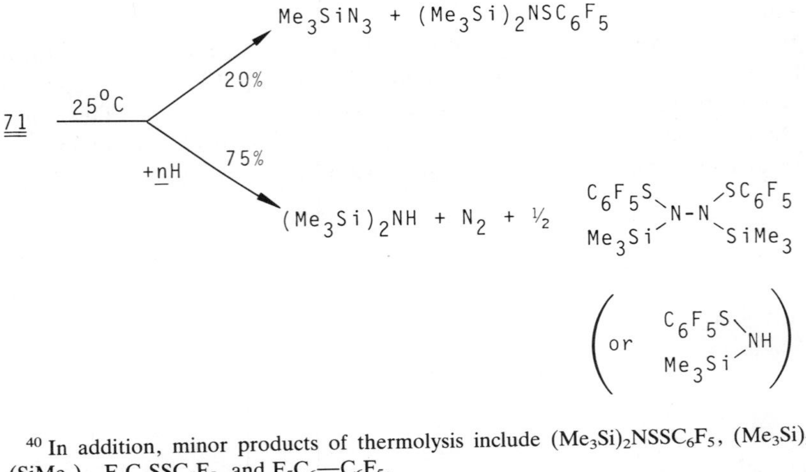

$$\begin{array}{c} \underline{71} \xrightarrow{25^\circ C} \\ +nH \end{array} \begin{cases} \xrightarrow{20\%} Me_3SiN_3 + (Me_3Si)_2NSC_6F_5 \\[2em] \xrightarrow{75\%} (Me_3Si)_2NH + N_2 + \frac{1}{2} \begin{array}{c} C_6F_5S \diagdown N{-}N \diagup SC_6F_5 \\ Me_3Si \diagup \diagdown SiMe_3 \end{array} \\[2em] \left(or \ \begin{array}{c} C_6F_5S \diagdown NH \\ Me_3Si \diagup \end{array} \right) \end{cases} \quad (93)$$

[40] In addition, minor products of thermolysis include $(Me_3Si)_2NSSC_6F_5$, $(Me_3Si)_2NSN(SiMe_3)_2$, $F_5C_6SSC_6F_5$, and $F_5C_6{-}C_6F_5$.

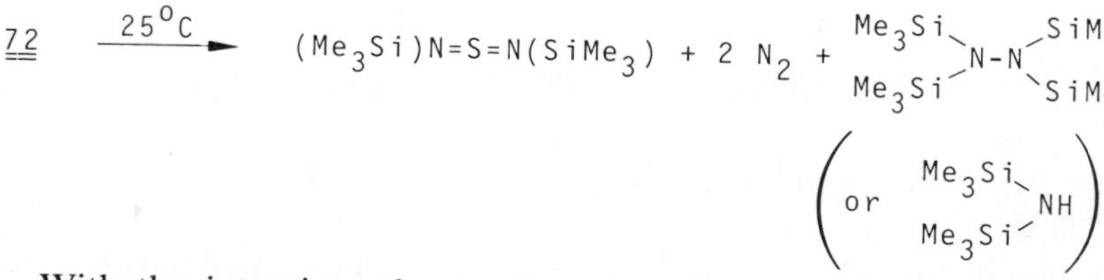

$$\underline{\underline{72}} \xrightarrow{25^\circ C} (Me_3Si)N=S=N(SiMe_3) + 2\ N_2 + \begin{matrix} Me_3Si \\ \\ Me_3Si \end{matrix} N-N \begin{matrix} SiMe_3 \\ \\ SiMe_3 \end{matrix}$$

$$\left(or\ \begin{matrix} Me_3Si \\ \\ Me_3Si \end{matrix} NH \right) \quad (94)$$

With the intention of extending the nitrogen chain of tetrazene, tetrazenides **62** and **64** were allowed to react with *nitrogen halides* as well as their derivatives. The desired objective could not be reached in any case because of decomposition of the expected compounds. Thus, reactions of **62** with nitrosyl chloride (NOX; X = Cl) and **64** with isoamyl nitrite [NOX; X = O(*i*-C$_5$H$_{11}$)] at low temperature in ether occur according to Eqs. (95) and (96) via **73** and **74** (45), respectively, whereas reactions of **62** with phenyldiazonium chloride (PhN≡N$^+$Cl$^-$) or tosyl azide (TsN$_3$; Ts = *p*-TolSO$_2$) at −78°C develop according to Eqs. (97) and (98), probably via **75** and **76**, respectively.

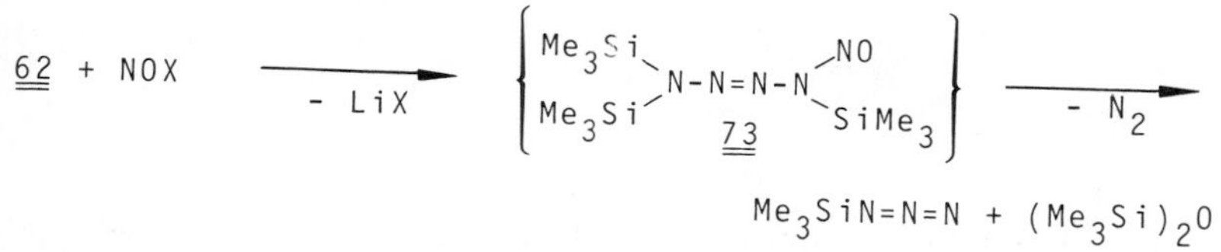

$$\underline{\underline{62}} + NOX \xrightarrow[-LiX]{} \left\{ \begin{matrix} Me_3Si \\ \\ Me_3Si \end{matrix} N-N=N-N \begin{matrix} NO \\ \\ SiMe_3 \end{matrix} \right\} \xrightarrow{-N_2}$$
$$\underline{73}$$

$$Me_3SiN=N=N + (Me_3Si)_2O \quad (95)$$

$$\underline{\underline{64}} + NOX \xrightarrow[-LiX]{} \left\{ \begin{matrix} Li \\ \\ Me_3Si \end{matrix} N-N=N-N \begin{matrix} NO \\ \\ SiMe_3 \end{matrix} \right\} \xrightarrow{-N_2}$$
$$\underline{74}$$

$$LiN=N=N + (Me_3Si)_2O \quad (96)$$

$$\underline{\underline{62}} + PhN_2Cl \xrightarrow[-LiCl]{} \left\{ \begin{matrix} Me_3Si \\ \\ Me_3Si \end{matrix} N-N=N-N \begin{matrix} N_2Ph \\ \\ SiMe_3 \end{matrix} \right\} \xrightarrow{-N_2}$$
$$\underline{75}$$

$$N_2 + \begin{matrix} Me_3Si \\ \\ Me_3Si \end{matrix} N-N-N=NPh \quad (97)$$
$$\underset{\underset{SiMe_3}{|}}{}$$

$$\underline{\underline{62}} + TosN_3 \xrightarrow{} \left\{ \begin{matrix} Me_3Si \\ \\ Me_3Si \end{matrix} N-N=N-N \begin{matrix} LiN_3Tos \\ \\ SiMe_3 \end{matrix} \right\} \xrightarrow{-2N_2}$$
$$\underline{76}$$

$$Me_3SiN=NSiMe_3 + (Me_3Si)LiNTos \quad (98)$$

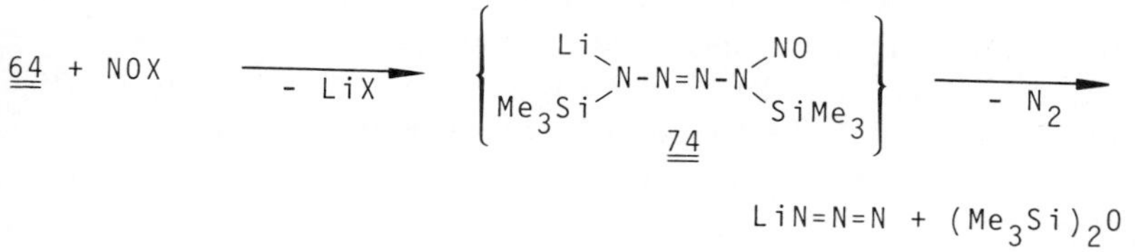

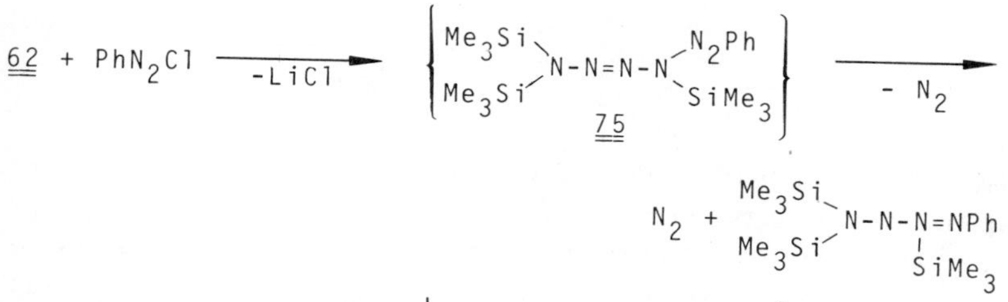

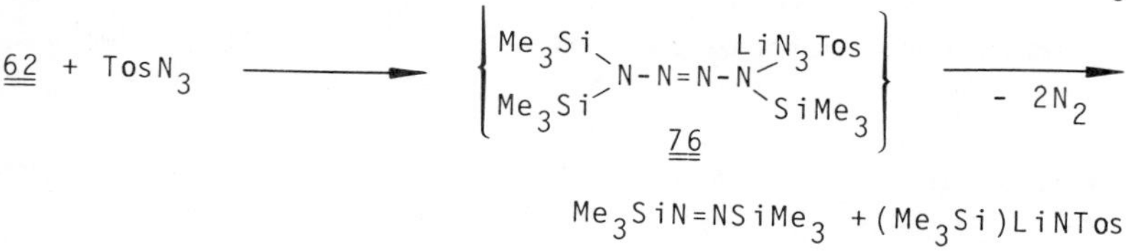

IV

GROUP IV DERIVATIVES OF PENTAZENE, N_5H_5

Apart from organic derivatives of pentazene, N_5H_5, the cyclic compound **78** is, so far, the only other Group IV derivative of this hydride (*35*). It is formed by the action of ethyl nitrite on the silylated lithium hydrazide **77** in diethyl ether at $-78°C$ [Eq. (99)]. It is a colorless crystalline

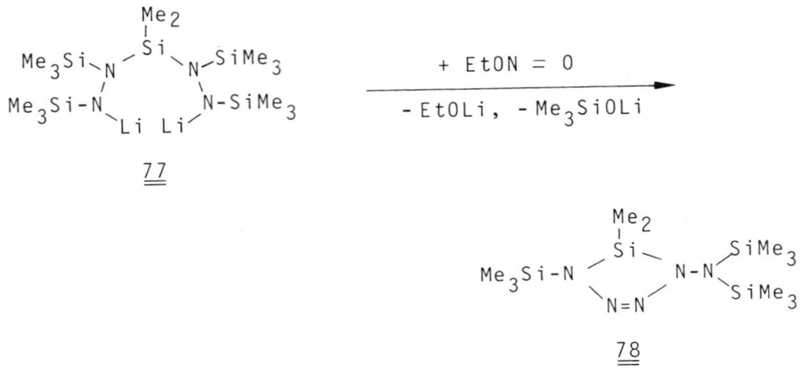

$$ (99) $$

77 → **78**

compound with a melting point of 39–41°C and it sublimes at 64–66°C in high vacuum.[41] Air- and somewhat water-resistant pentazene derivative **78** decomposes thermally by a first-order reaction [$\tau_{1/2}$ (150°C) = 4.5 minutes; $\Delta H^{\ddagger}$ = 124 kJ/mol; $\Delta S^{\ddagger}$ = 2.1 eu] with N_2 evolution into the ring compound **79** [Eq. (100)]. Protolysis of **78** with trifluoroacetic acid at $-78°C$

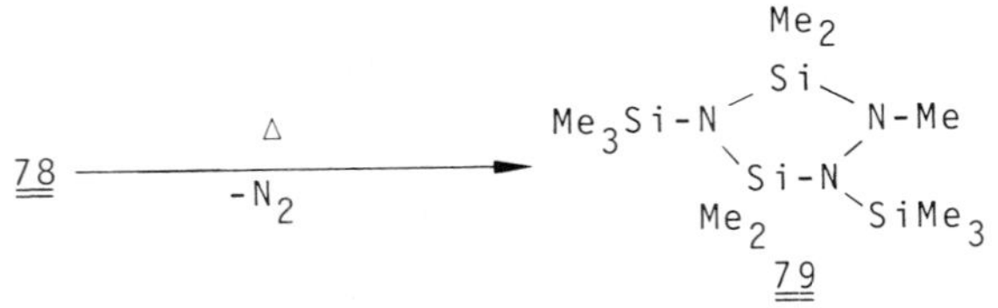

$$ (100) $$

in CH_2Cl_2 does not lead to the parent hydride N_5H_5, but rather to its decomposition products. Formation of hydrolysis products N_2, NH_3, N_2H_4, and HN_3 (part of the nitrogen is liberated at $-78°C$ and the rest at $-60°C$) can possibly be explained by Eq. (101) (*35*) by analogy with the observed thermolysis routes of 2-tetrazene [cf. Eqs. (70a and 70b)].

---

[41] Proton NMR (C_6H_6, TMS): δ = 0.205, 0.217, and 0.237 with relative areas in the ratio 6:2:3. UV (hexane): $\bar{\nu}_{max}$ = 38910 cm^{-1} (ϵ = 4300). Ionization energy: IE = 7.65 eV.

$$
\underline{\underline{78}} \xrightarrow[\substack{- 3\ CF_3CO_2SiMe_3 \\ - (CF_3CO_2)_2SiMe_2}]{+ 5\ CF_3CO_2H} \left\{ H_2N-N=N-NH-NH_2 \right\} \xrightarrow{<-78^\circ C}
$$

$$
HN_3 + N_2H_4 \quad 17\%
$$

$$
N_2 + \left\{ N_3H_5 \right\} \quad 83\%
$$

$$
0.5N_2 + NH_3 + 0.5N_2H_4 \longleftarrow
$$

$$(101)$$

ACKNOWLEDGMENTS

I wish to express my appreciation to my co-workers, who have carried out with enthusiasm all the experiments described above: Dr. H. Bayer, Dr. G. Fischer, Dr. H. -W. Häring, Prof. Dr. W. -Ch. Joo, Dipl. -Chem. P. Karampatses, Dipl. -Chem. Ch. K. Kim, Dipl. -Chem. H. Köpf, Dr. F. Monzel, Dr. R. Meyers, Dr. H. J. Pracht, Dr. G. Preiner, Dr. W. Uhlenbrock, Dr. S. K. Vasisht, and Dr. G. Ziegleder.

REFERENCES

1. N. Wiberg, *Adv. Organomet. Chem* **23**, 131 (1984).
2. P. A. S. Smith, "The Chemistry of Open-Chain Organic Nitrogen Compounds," pp. 336–340. Benjamin, New York, 1966.
3. P. A. S. Smith, "The Chemistry of Open-Chain Organic Nitrogen Compounds," pp. 345–348. Benjamin, New York, 1966.
4. O. Dimroth, *Ber. Dtsch. Chem. Ges.* **36**, 909 (1903).
5. "Houben-Weyl's Methoden der Organischen Chemie," 4th ed., Vol. 10, Part 3, p. 699. Thieme, Stuttgart, 1965.
6. F. E. Brinkman, H. S. Haiss, and R. A. Robb, *Inorg. Chem.* **4**, 936 (1965).
7. N. Wiberg and H. J. Pracht, *Chem. Ber.* **105**, 1377 (1972).
8. N. Wiberg, G. Fischer, and P. Karampatses, *Angew. Chem., Int. Ed. Engl.* **23**, 59 (1984).
9. N. Wiberg, G. Fischer, P. Karampatses, and E. Kühnel, unpublished results.
10. N. Wiberg and W.-C. Joo, *J. Organomet. Chem.* **22**, 333 (1970).
11. N. Wiberg, W.-C. Joo, and P. Olbert, *J. Organomet. Chem.* **22**, 341 (1970).
12. N. Wiberg and W.-C. Joo, *J. Organomet. Chem.* **22**, 349 (1970).
13. B. P. Roberts and J. N. Winter, *J. Chem. Soc., Perkin Trans. 2*, p. 1353 (1979).
14. N. Wiberg, G. Preiner, and O. Schieda, *Chem. Ber.* **114**, 3518 (1981).
14a. N. Wiberg, *J. Organomet. Chem.* **273**, 141 (1984).
15. N. Wiberg and C. K. Kim, unpublished results.
16. N. Wiberg and P. Karampatses, unpublished results.
17. E. Hayon and M. Simic, *J. Am. Chem. Soc.* **94**, 42 (1972).
18. E. W. Abel and I. D. H. Towle, *J. Organomet. Chem.* **155**, 299 (1978).
19. J. Hollaender, W. P. Neumann, and G. Alester, *Chem. Ber.* **105**, 1540 (1972).
19a. R. M. Pike and N. Sobinski, *J. Organomet. Chem.* **253**, 183 (1983).
20. R. J. Corruccini and E. C. Gilbert, *J. Am. Chem. Soc.* **61**, 2925 (1939).
21. M. H. Akhtar, R. S. McDaniel, M. Feser, and A. C. Oehlschlager, *Tetrahedron* **24**, 3901 (1968).

22. N. Wiberg and H. J. Pracht, *Chem. Ber.* **105**, 1392 (1972).
23. H. Kessler, *Angew. Chem., Int. Ed. Engl.* **10**, 219 (1970).
24. N. Wiberg and H. J. Pracht, *Chem. Ber.* **105**, 1388 (1972).
25. N. Wiberg and H. J. Pracht, *Chem. Ber.* **105**, 1399 (1972).
26. D. G. Lister, J. N. MacDonald, and N. L. Owen, "Internal Rotation and Inversion," Academic Press, London, 1978.
27. Cf. H. Kessler, *Angew. Chem., Int. Ed. Engl.* **10**, 219 (1970).
28. M. H. Akhtar, R. S. McDaniel, M. Freser, and A. C. Oehlschlager, *Tetrahedron* **24**, 3899 (1968).
29. N. Wiberg and H. J. Pracht, *J. Organomet. Chem.* **40**, 289 (1972).
30. H. Musso, *Chem. Ber.* **92**, 2881 (1959).
31. "Houben-Weyl's Methoden der Organischen Chemie," 4th Ed., Vol. 10, Part 2, p. 834. Thieme, Stuttgart, 1967.
32. N. Wiberg and W. Uhlenbrock, *J. Organomet. Chem.* **70**, 239 (1974).
33. N. Wiberg, S. K. Vasisht, H. Bayer, and R. Meyers, *Chem. Ber.* **112**, 2718 (1979).
34. N. Wiberg and W. Uhlenbrock, *Chem. Ber.* **105**, 63 (1972).
35. N. Wiberg and G. Ziegeleder, *Chem. Ber.* **111**, 2123 (1978).
36. N. Wiberg, S. K. Vasisht, and G. Fischer, *Angew. Chem., Int. Ed. Engl.* **15**, 236 (1976).
37. N. Wiberg, H.-W. Häring, and S. K. Vasisht, *Z. Naturforsch., B: Anorg. Chem., Org. Chem.* **34B**, 356 (1979).
38. W. R. McBride and H. W. Kruse, *J. Am. Chem. Soc.* **79**, 572 (1957).
39. N. Wiberg, H. J. Pracht, and F. Monzel, unpublished results.
40. N. Wiberg and G. Preiner, *Angew. Chem., Int. Ed. Engl.* **17**, 362 (1978).
41. N. Wiberg, H. Bayer, S. K. Vasisht, and R. Meyers, *Chem. Ber.* **113**, 2916 (1980).
42. N. Wiberg, R. Meyers, S. K. Vasisht, and H. Bayer, *Chem. Ber.* **117**, 2886 (1984).
43. N. Wiberg, H. Bayer, and H. Bachhuber, *Angew. Chem., Int. Ed. Engl.* **14**, 177 (1975).
44. N. Wiberg, H. Bayer, and S. K. Vasisht, *Chem. Ber.* **114**, 2658 (1981).
45. N. Wiberg and S. K. Vasisht, unpublished results.
46. N. Wiberg and H. Bayer, unpublished results.
47. M. Veith, *Acta Crystallogr., Sect. B* **B31**, 678 (1975).
48. M. Veith and H. Bärnighausen, *Acta Crystallogr., Sect. B* **B30**, 1806 (1974).
49. M. Veith and G. Schlemmer, *Z. Anorg. Allg. Chem.* **494**, 7 (1982).
50. J. Kroner, N. Wiberg, and H. Bayer, *Angew. Chem., Int. Ed. Engl.* **14**, 178 (1975).
51. J. Kroner, unpublished results.
52. W. R. McBride and E. M. Bens, *J. Am. Chem. Soc.* **81**, 5546 (1959).
53. N. Wiberg, H. Bayer, H. Köpf, and G. Rauth, unpublished results.
54. N. T. Reetz, *Adv. Organomet. Chem.* **16**, 33 (1977).
55. B. R. Cowley and W. A. Waters, *J. Chem. Soc.*, p. 1228 (1961).
56. N. Wiberg and G. Schwenk, *Chem. Ber.* **104**, 3986 (1971).
57. N. Wiberg and R. Meyers, unpublished results.
58. J. C. Baldwin, M. F. Lappert, J. B. Pedley, and J. S. Paland, *J. Chem. Soc., Dalton Trans.* p. 1943 (1972).
59. N. Wiberg, G. Fischer, and H. Bachhuber, *Z. Naturforsch., B: Anorg. Chem., Org. Chem.* **34B**, 1385 (1979).
60. H. Bock, W. Kaim, N. Wiberg, and G. Ziegleder, *Chem. Ber.* **111**, 3150 (1978).
61. W. M. Tolles, D. W. Moore, and W. E. Thun, *J. Am. Chem. Soc.* **88**, 3476 (1966).
62. H. Bock, W. Kaim, N. Wiberg, and H. Bayer, unpublished results.
63. W. P. Neumann, "The Organic Chemistry of Tin," 1st ed., p. 78, Wiley, London, 1970.

ADVANCES IN ORGANOMETALLIC CHEMISTRY, VOL. 24

Photochemistry of Alkyl, Alkylidene, and Alkylidyne Complexes of the Transition Metals

DANIEL B. POURREAU and GREGORY L. GEOFFROY

Department of Chemistry
The Pennsylvania State University
University Park, Pennsylvania

I
INTRODUCTION

Transition metal alkyl (**1**), alkylidene (**2**), and alkylidyne (**3**) complexes

$$L_nM\text{—}R \qquad L_nM\text{=}C\begin{matrix}R\\R'\end{matrix} \qquad L_nM\equiv CR$$

(**1**) (**2**) (**3**)

constitute an extremely important class of organometallic compounds because of their many roles in catalytic and synthetic chemistry. Alkyl complexes are key intermediates in such diverse catalytic reactions as hydrogenation (*1*), hydroformylation (*2*), hydrocyanation (*3*), and the

ISBN 0-12-031124-0

synthesis of acetic acid (4), to name a few. Likewise, alkylidene complexes are involved in olefin metathesis (5) and cyclopropanation (6) chemistry and may be intermediates in olefin polymerization (7). Recent results promise the development of alkyne metathesis catalysts which involve alkylidyne complexes (8).

The photochemistry of this class of compounds was reviewed in the 1979 book *Organometallic Photochemistry* (9) which reviewed the literature comprehensively through 1977. In the years since that review, a number of significant advances have been made. Indeed, the general conclusions that can now be drawn concerning the photochemistry of alkyl complexes are quite different from those given in that earlier review. Then, few alkylidene and alkylidyne complexes had been studied, but that too has changed. A comprehensive review on this subject is thus timely.

Throughout this article the terms alkylidene and carbene will be used interchangeably, although arguably one could restrict the definition of alkylidene ligands to those with only alkyl substituents. Likewise, alkylidyne and carbyne will be understood to signify the same type of ligand. Alkyl complexes will be discussed first, followed by alkylidene and alkylidyne compounds. Each section will begin with the transition metals farthest to the left and then proceed systematically to the right. Literature coverage is through the end of 1983.

II

PHOTOCHEMISTRY OF ALKYL COMPLEXES

A. Th, U, and Yb Complexes

1. $ThCp_3R$

Irradiation of these thorium complexes gives $ThCp_3$ ($Cp = \eta^5-C_5H_5$)[1] in essentially quantitative yield, as well as almost equimolar amounts of alkane and alkene [Eq. (1)] (10, 10a). The authors suggested that the

$$2 \ ThCp_3\{CHRCH_2R'\} \xrightarrow[\substack{\phi_{313} = 1.7 \\ \text{toluene}}]{h\nu} 2 \ ThCp_3 + RCH_2CH_2R' + RCH=CHR'$$
$$\text{(4)}$$

$$(R = CH_3, \ R'=H)$$
$$(R=H, \ R'=C_2H_5) \qquad\qquad (> 90\%)$$

(1)

reaction proceeds via photoinduced β-hydride elimination (Scheme 1). The starting complexes **4** are coordinatively saturated, and a coordination

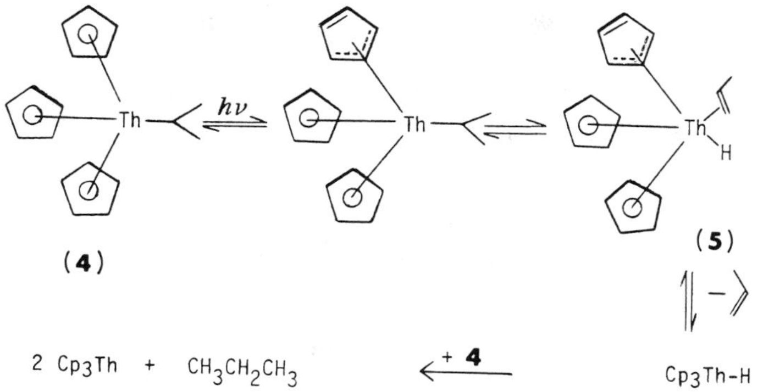

SCHEME 1

site must open before β-hydride elimination can occur. Irradiation into a Cp $\rightarrow$ M charge-transfer (CT) transition was suggested to cause conversion of one of the η^5-C_5H_5 ligands into an η^3-C_5H_5 ligand which opens a coordination site and permits this reaction. The 313 nm quantum yield for disappearance of 4 is 1.7 ± 0.1, consistent with the mechanism in which one photon causes loss of two molecules of 4. The decrease in the relative yield of alkane with respect to alkene upon photolysis in a rigid toluene glass also argues for the proposed bimolecular alkane formation. A mechanism involving photoinduced Th—R bond homolysis to give free radicals is inconsistent with the finding that >96% of the hydrogens incorporated into the product alkane come from the original alkyl ligands.

Photolysis of the related Th(idenyl)$_3$(n-C_4H_9) complex gave a 96.5% yield of butane and butene but in vastly different relative proportions

[1] Abbreviations: acac, Acetyl acetonate; AIBN, 2,2-azobis(isobutyro) nitrile ; bae, dianion of N,N'-ethylenebis(acetylacetoneimine); bdhc, 1,19-dimethyl-AD-bisdehydrocorrinato; bdm 1,3 pn, bis(1,3-diacetylmonoximeimino)propane; bipy, bipyridyl; Bz, Benzyl; CIDNP, chemically induced dynamic nuclear polarization; cod, cyclooctadienyl; COD, cyclooctadiene; Cp, η^5-C_5H_5; Cp*, η-C_5Me_5; CT, charge transfer; diphos, P(Ph)$_2$CH$_2$CH$_2$P(Ph)$_2$; dmgH, dimethylglyoxime; DMPO, 5,5-dimethyl-1-pyrroline-N-oxide; DMSO, dimethyl sulfoxide; dppm, Ph$_2$PCH$_2$PPh$_2$; [14]aneN$_4$, 1,3,8,11-tetraazacyclotetradecane; Hacac, acetylacetone; LF, ligand field; LMCT, ligand-to-metal charge transfer; MLCT, metal-to-ligand charge transfer; MMA, methyl methacrylate; ND, nitrosodurene; neo, neopentyl; nor, norbornyl; oep, octaethylporphyrin; PBN, phenyl-N-t-butylnitrone; phen, phenanthroline; PVC, poly-(vinyl chloride); py, pyridyl; thf, tetrahydrofuranyl; THF, tetrahydrofuran; TMPNO, 2,2,6,6-tetramethylpiperidine N-oxide; p-tol, p-$C_6H_4CH_3$.

[Eq. (2)] (*10*). The only organometallic product was $Th(\eta^5\text{-}C_9H_7)_3$. The authors suggested that the steric bulk of the indenyl ligands made formation of an alkene–hydride intermediate analogous to **5** unfavorable

$$Th(\eta^5\text{-}C_9H_7)_3(n\text{-}C_4H_9) \xrightarrow[\text{toluene}]{h\nu} Th(\eta^5\text{-}C_9H_7)_3 + C_4H_{10} + C_4H_8 \qquad (2)$$
$$\qquad\qquad\qquad\qquad\qquad\qquad (96\%) \quad (4\%)$$

and thus suppressed the β-elimination pathway. The origin of the hydrogen incorporated into the product alkane was not clearly established nor was the mechanism of photoinduced cleavage of the Th—R bond elucidated.

The photochemistry of the methyl complex $ThCp_3CH_3$ differed markedly from that of thorium alkyls containing β hydrogens [Eq. (3)]. The reaction was appreciably slower and was only $\sim 50\%$ complete after

$$ThCp_3CH_3 \xrightarrow[\text{toluene}]{h\nu} \text{decomposition} + CH_4 + C_2H_6 + H_2 \qquad (3)$$

20 hours (*10, 10a*). The principal organic product was methane and labeling studies showed the Cp rings and the solvent, in that order, to be the major sources of hydrogen uptake in its production.

The above-mentioned results suggest photoinduced homolysis of the Th—CH$_3$ bond to produce free methyl radicals which abstract hydrogen from solvent and Cp ligands. However, it is important to point out for this study and for many of the other studies discussed later in this article that Cp and solvent hydrogen abstraction can also occur by processes that do not involve the intermediacy of free radicals. Thus, the observation of ligand and solvent hydrogen abstraction is *not* proof of a radical mechanism. Marks (*11*) has demonstrated that *thermolysis* of these thorium alkyls results in an *intramolecular abstraction of a ring hydrogen*, apparently via an intermediate such as **6**, to yield the corresponding hydrocarbon and the

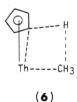

(6)

organometallic product $[Cp_2Th(\eta^1,\eta^5\text{-}C_5H_4)]_2$ in essentially quantitative yield. Furthermore, these authors noted that hydrogen abstraction from *solvent* molecules can occur via a *nonradical hydrocarbon metathesis* [Eq. (4)], as has been shown (*11*) to occur in the thermolysis of

$$M\text{—}R + C_6D_6 \rightarrow M\text{—}C_6D_5 + RD \qquad (4)$$

the related complex $ThCp_2^*(CH_3)_2$ [$Cp^* = \eta^5\text{-}C_5(CH_3)_5$]. Likewise, a photogenerated reactive organometallic product can oxidatively add solvent C—H bonds, particularly those of aromatic solvents, which if followed by coupling of residual alkyl ligands with the resultant hydride ligand would give "hydrogen abstraction from solvent "[Eq. (5)]. These

$$\underset{\overset{|}{M}}{\overset{R}{|}} + C_6D_6 \rightarrow \underset{\overset{|}{M}}{\overset{R}{\underset{C_6D_5}{\overset{|}{\diagdown}}}} \overset{D}{\diagup} \rightarrow RD + M—C_6D_5 \qquad (5)$$

points illustrate the danger of assuming photoproduction of *free* radicals solely upon the basis of hydrogen abstraction from ligands and solvent.

2. UCp_3R

Marks and co-workers (*10, 10a*) also studied the photochemistry of uranium alkyls in toluene solution at ambient (10°C) and low (−196°C) temperatures. As with the thorium alkyls, mixtures of the corresponding alkane and alkene were produced in near quantitative yield [e.g., Eq. (6)].

$$Cp_3U(i\text{-}C_3H_7) \xrightarrow{\text{toluene}} Cp_3U + C_3H_8 + C_3H_6 \qquad (6)$$
$$(83\%) \quad (17\%)$$

However, the relative yield of alkane versus alkene was much larger than in the thorium case and decreased with increasing steric bulk of the starting alkyl ligand. These results are summarized in Table I. Labeling studies showed that the hydrogen incorporated into the alkane and alkene originated mostly from the Cp rings but also from the solvent (toluene). It appears that both β-hydride elimination and Cp-ring hydrogen abstraction are photoinduced.

TABLE I

Yields of Photolysis Products for UCp_3R Compounds in Toluene

Compound	Yield of gaseous products (%)	Organic products	Relative yield (%)		Isolated yield of UCp_3 (%)
			10°C	−196°C	
$Cp_3U(i\text{-}C_3H_7)$	95	Propane	83	—	27
		Propene	17	—	
$Cp_3U(n\text{-}C_4H_9)$	91	n-Butane	91	58	12
		1-Butene	9	42	
$Cp_3U(i\text{-}C_4H9)$	96	n-Butane	71	53	41
		Butenes	29	47	

Significantly different results and conclusions were obtained by Gian-notti *et al.* (*12*, *12a*) who irradiated the same complexes in tetrahydrofuran (THF) solution and invoked photoinduced homolysis of the U—alkyl bonds. They observed that rapid photochemical reactions occur only at elevated temperatures and most of their studies were conducted at 60°C. Irradiation of the complexes $Cp_3U(i\text{-}C_3H_7)$, $Cp_3U(i\text{-}C_4H_9)$, and $Cp_3U(n\text{-}C_4H_9)$ in THF produced more alkene than in toluene solution, although the yield of alkane still predominated. As Marks *et al.* had noted (*10*, *10a*), the yield of UCp_3 increased to 60–70% when the photolyses were conducted in THF in which the complex was isolated as the THF adduct.

No free radicals were detected upon photolysis of $Cp_3U(n\text{-}C_4H_9)$ in toluene solution at ambient temperature in the presence of various spin traps (*12*, *12a*). On the other hand, photolysis in THF solution in the presence of phenyl-*N*-*t*-butylnitrone (PBN) gave clear ESR signals attrib-uted to an *n*-butyl spin adduct (*12*). Similarly, photolysis of $Cp_3U(i\text{-}C_4H_9)$ and $Cp_3U(i\text{-}C_3H_7)$ at ambient temperature using nitrosodurene (ND) as a spin trap gave rise to ESR signals attributed to the corresponding ND–alkyl spin adducts (*12a*). Mechanistic information derived solely from spin-trapping experiments should be viewed with caution since ESR spectroscopy is so sensitive that it can detect minute quantities of radicals produced from minor reaction pathways (*13*). However, it is apparent that the photochemistry of uranium alkyls in THF differs from that in toluene solution. Giannotti *et al.* (*12*) suggested that the role of THF as a solvent may be to decrease the stability of the uranium–carbon σ-bond, and to stabilize photogenerated radicals and radical pairs.

Marks *et al.* (*10*, *10a*) noted that the photoinduced decomposition of UCp_3CH_3 gave mostly methane (98%) and some ethane (2%) but the overall reaction rate was much slower than that of alkyl complexes with β hydrogens. Giannotti *et al.* (*12*) also observed methane formation upon irradiation of UCp_3CH_3 at elevated temperature (60°C). Deuterium labeling studies identified the solvent as the major source of the hydrogen incorporated into methane in contrast with the Cp rings being the major source of hydrogen when the reaction was run under strictly thermal conditions (*12*). Spin-trapping experiments using PBN or ND gave ESR signals attributed to a methyl spin adduct only when the reactions were run in THF and under photolytic conditions (*12*).

As the above results indicate, the photochemistry of these uranium–alkyl complexes is not fully resolved. The olefin formation during Marks's study of UCp_3R in toluene solution suggests that photoinduced β-hydride elimination occurs to some extent, presumably via a mechanism similar to that of Scheme 1. However, the other data appear to indicate a predomi-nant radical mechanism via photoinduced M—R bond homolysis.

3. $[Yb(MeC_5H_4)_2Me]_2$

The title complex afforded a 25% yield of $(MeC_5H_4)_2Yb$ upon irradiation in toluene solution at 5°C overnight, but no further details were given (14).

B. Ti, Zr, and Hf Complexes

1. MR_4 (M = Ti or Zr)

Both $Ti(CH_2Ph)_4$ and $Zr(CH_2Ph)_4$ have been shown to be sensitive to light (15). Irradiation of toluene solutions of $Ti(CH_2Ph)_4$ at −78°C gave nearly 50% reduction to Ti, but no mechanistic details were given. The zirconium derivative is more light sensitive than $Ti(CH_2Ph)_4$, and its solutions gave a rapid color change from yellow to brown when irradiated. Reaction of the resulting solutions with CH_3OD gave formation of HD, and this suggested the presence of a Zr—H bond in the photoproduct. A brown solid which decomposes at 0°C was isolated by precipitation from the irradiated solutions. Analytical data were consistent with the formulation 7, and the overall photoreaction shown in Eq. (7) was suggested.

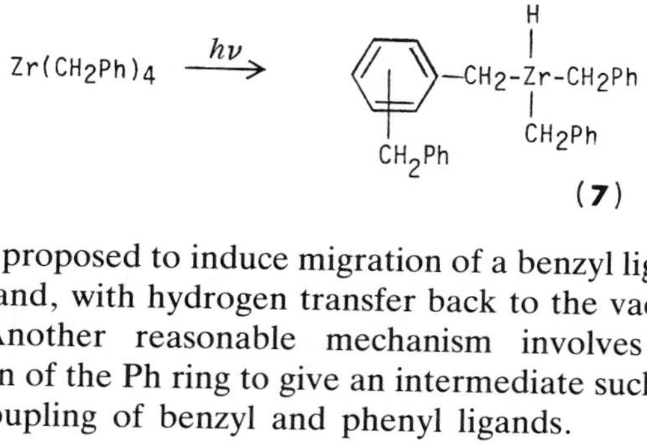

$$Zr(CH_2Ph)_4 \xrightarrow{h\nu}$$

(7)

(**7**)

Irradiation was proposed to induce migration of a benzyl ligand to another coordinated ligand, with hydrogen transfer back to the vacant site on the metal (15). Another reasonable mechanism involves photoinduced orthometallation of the Ph ring to give an intermediate such as 8, followed by reductive coupling of benzyl and phenyl ligands.

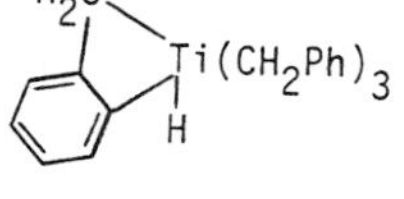

(**8**)

Ballard and Van Lienden (16) showed that irradiation of $Zr(CH_2Ph)_4$ solutions in the presence of vinyl monomer increased the rate of thermal polymerization by a factor of four. Kinetic rate data suggested that the mechanism of styrene polymerization was wavelength dependent. For wavelengths between 450 and 600 nm, the photopolymerization rate was completely independent of the radiation intensity, but below 450 nm the rate was directly proportional to $I^{1/2}$. On the basis of these results the authors discounted a free radical mechanism for long-wavelength irradiation and suggested that the reaction proceeded via an intermediate **9**

$$(PhCH_2)_3Zr \overset{\displaystyle (CHRCH_2)_nCH_2Ph}{\underset{\displaystyle \overset{\|}{CHR}}{\diagdown\; CH_2}}$$

(9)

similar to that postulated for the thermal polymerization. Irradiation with near-UV light presumably initiated a free radical process.

Recent photochemical studies have focused on complexes with alkyl ligands either devoid of β hydrogens, such as tetraneopentyltitanium (*17*), or with bulky bridgehead carbons, such as tetranorbornyltitanium (*18*). Such ligands are effective in increasing the stability of tetraalkyl complexes by suppressing the β-elimination decomposition pathway and reducing their sensitivity to oxygen and protic reagents.

The electronic absorption spectrum of Ti(norbornyl)$_4$ (**10**) is shown in Fig. 1. The intense band at 245 nm ($\epsilon = 29,200 \text{ mol}^{-1} \text{ cm}^{-1}$ in hexane) was assigned to a fully allowed ligand-to-metal charge-transfer (LMCT) transition (*18*). The weaker band at 367 nm ($\epsilon = 253 \text{ mol}^{-1} \text{ cm}^{-1}$ in hexane) and shoulders at 312 and 412 nm were attributed to other spin or orbitally forbidden LMCT transitions. Near-UV irradiation of yellow hexane solu-

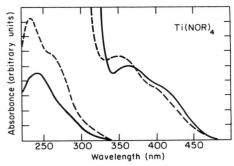

FIG. 1. Electronic absorption spectrum of Ti(nor)$_4$ in EPA (5:5:2 diethyl ether:2-methylbutane:ethanol) solution at 298 K (———) and 77 K (———), not corrected for solvent contraction. The portion on the right shows the spectrum of a more concentrated solution (*18*).

tions of **10** produce norbornane and 1,1′-binorbornyl [Eq. (8)] with low quantum yield. The relative or absolute yields of organic products were not

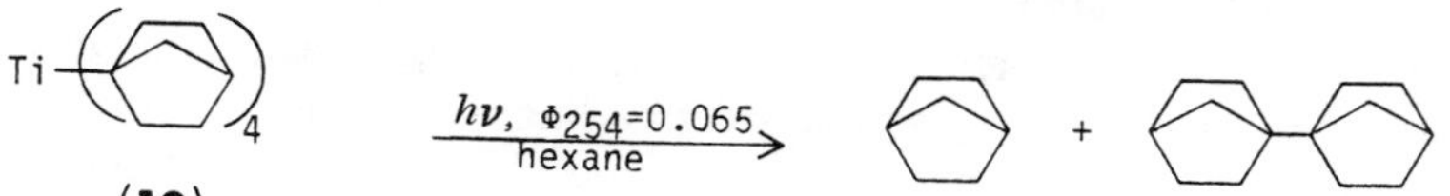

$$\text{Ti}\left(\text{Nor}\right)_4 \quad \xrightarrow[\text{hexane}]{h\nu,\ \Phi_{254}=0.065} \quad \bigcirc\!\!\!\triangle \quad + \quad \bigcirc\!\!\!-\!\!\!\bigcirc \quad (8)$$

(10)

reported, nor were the organometallic products identified. However, magnetic susceptibility measurements of the dark, air, and thermally sensitive solid produced upon photolysis showed the presence of no unpaired electrons, implying that the expected Ti(III) primary photoproducts had dimerized or diproportionated to yield a spin-paired system. Photolysis of **10** (nor = norbornyl) in the presence of carbon tetrachloride gave 1-chloronorbornane and hexachloroethane as the major organic products but also some norbornane and binorbornyl [Eq. (9)]. These

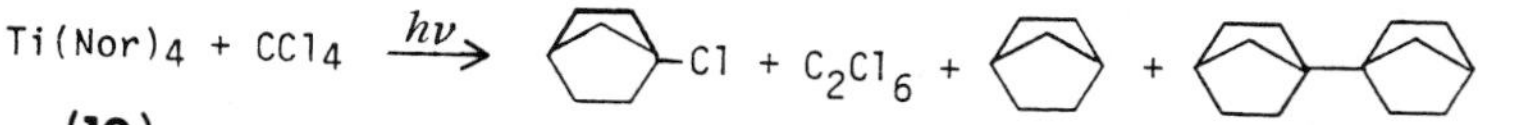

$$\text{Ti(Nor)}_4 + \text{CCl}_4 \xrightarrow{h\nu} \bigcirc\!\!\!-\!\!\text{Cl} + \text{C}_2\text{Cl}_6 + \bigcirc + \bigcirc\!\!\!-\!\!\!\bigcirc \quad (9)$$

(10)

results are consistent with the formation of norbornyl radicals by photoinduced homolytic cleavage of titanium–norbornyl bonds.

Irradiation of tetraneopentyltitanium (*11*) in benzene/toluene or hexafluorobenzene solutions gave neopentane as the major organic product along with some 2,2,5,5-tetramethylhexane [Eq. (10)] and a black precipi-

$$\text{Ti}\left(\text{neo}\right)_4 \xrightarrow{h\nu} \times \quad + \quad \times\!\!/\!\!\times \quad (10)$$

(11)

tate which was not characterized (*17*). The reaction was followed by ESR spectroscopy which revealed the presence of four Ti(III) species when irradiation was conducted at room temperature. Cooling the solutions to −77 k led to the appearance of only two of the four signals but no assignments were made and the organometallic products were not identified.

Addition of styrene or methyl methacrylate to the irradiated solutions produced a mixture of low- and high-molecular-weight (lmw and hmw) homopolymers [Eq. (11)]. This distribution was the same as when

$$\text{Ti(neo)}_4 + \text{H}_2\text{C}=\text{CRR}' \xrightarrow{h\nu} \text{lmw polymer} + \text{hmw polymer} \quad (11)$$

(11) (R = H or CH$_3$)   (72–88%)   →   (12–28%) →
    (R′ = Ph or OCH$_3$)
(neo = neopentyl)

polymerization was induced by the thermal radical initiator 2,2′-azobis-(isobutyro)nitrile (AIBN) and when the monomers were added in the dark to previously irradiated solutions of **11**. Furthermore, irradiation of **11** in the presence of monomer gave rise to ESR signals attributed to Ti(III) species which had different g values depending on the monomer used. The authors thus proposed a mechanism for the photoinduced polymerization of styrene and methyl methacrylate involving a caged radical species such as **12** (*17*). The authors concluded that photolysis of Ti(neo)$_4$ gave

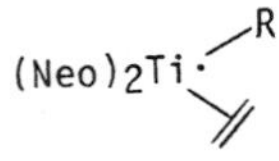

$$(Neo)_2Ti\cdot\overset{R}{\diagup}$$

(12)

homolytic cleavage of titanium–neopentyl bonds to generate neopentyl radicals and presumably Ti(neo)$_3$ (*17*).

2. Ti(CH₃)Cl₃

De Vries (*19*) reported that solutions of Ti(CH$_3$)Cl$_3$ can be photodecomposed, but obtained no evidence for production of methyl radicals. Deuterium labeling studies indicated that the hydrogen incorporated into the methane produced originated from the methyl ligands. No hydrogen abstraction from the solvent was detected and no further details were presented.

3. Zr[CH₂C(CH₃)₃]₂(PMe₃)₂Cl₂

Schrock and co-workers (*20*) noted that irradiation of the title complex **13** in benzene solution gave a green complex analyzing for ZrCl$_3$(PMe$_3$)$_2$, along with neopentane and 2,2,5,5-tetramethylhexane [Eq. (12)]. The

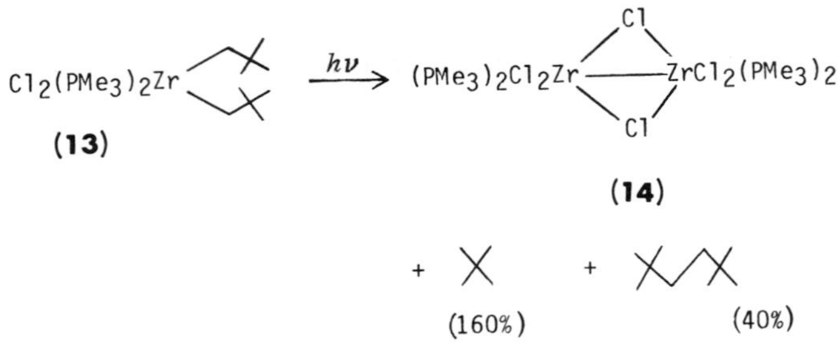

$$Cl_2(PMe_3)_2Zr\diagup\diagdown \quad \overset{h\nu}{\longrightarrow} \quad (PMe_3)_2Cl_2Zr\overset{Cl}{\underset{Cl}{\diagup\diagdown}}ZrCl_2(PMe_3)_2$$

(13)   **(14)**   (12)

$$+ \; \diagup\!\!\diagdown \quad + \quad \diagup\!\!\diagdown\!\!\diagup$$

(160%)   (40%)

authors postulated that the organometallic product had the dimeric structure (**14**). The authors also suggested that 2,2,5,5-tetramethylhexane was produced by the coupling of two photogenerated radicals.

4. $MCp_2R_2(M = Ti, Zr, or Hf)$

The photochemistry of this class of compounds has been extensively investigated (*21–33*) and it is well established that irradiation leads to loss of one or both of the R ligands. Irradiation of $TiCp_2(CH_3)_2$ in $CHCl_3$ solution yields $TiCp_2(CH_3)Cl$, $TiCp_2Cl_2$, and $TiCpCl_3$ (*21*), and photolysis of the series $MCp_2(CH_3)_2$ (M = Ti, Zr, or Hf) in hydrocarbon solution gives mainly production of methane along with metallocene-type products [Eq. (13)] (*22*). The latter are not simple Cp_2M complexes but are

$$MCp_2(CH_3)_2 \xrightarrow[\text{hexane}]{hv} \text{``}MCp_2\text{''} + CH_4 + C_2H_6 + C_2H_4 \qquad (13)$$

$$(>99\%) \quad (\text{trace}) \quad (\text{trace})$$

probably oligomeric and may contain η^1,η^5-C_5H_4-bridging ligands. However, the *mechanism* by which these reactions occur has been very difficult to resolve and consequently differing opinions have been offered. As detailed below, the situation has been clarified somewhat by recent studies (*24*) which show that irradiation does lead to homolysis of the metal–alkyl bonds to give radical pairs that rapidly recombine and that can be trapped only by very efficient scavengers [Eq. (14)].

$$Cp_2M\overset{\displaystyle CH_3}{\underset{\displaystyle CH_3}{\diagup\diagdown}} \underset{\text{fast}}{\overset{hv}{\rightleftharpoons}} \text{``}Cp_2\overset{\cdot}{M}\!-\!CH_3 + \cdot CH_3\text{''} \rightarrow \text{further decay} \qquad (14)$$

Rausch, Alt, and co-workers (*22, 23, 27, 28, 31*) conducted several investigations of this class of compounds and reported that irradiation of $TiCp_2(CH_3)_2$ in hydrocarbon solvent gave methane and black titanocene [Eq. (13)]. Deuterium labeling studies (*22, 23, 25*) showed that the hydrogen incorporated into the methane originated predominantly from the Cp rings. Little, if any, hydrogen abstraction from solvent occurred, even with toluene, which would give relatively stable benzyl radicals. These results tend to suggest that free radicals are not produced upon photolysis.

Puddephatt *et al.* (*26*) noted that irradiation of $TiCp_2(CH_3)_2$ in the presence of methyl methacrylate (MMA) or styrene gave polymerization of these monomers, a result that can be interpreted as evidence for free radical formation. However, these workers ruled out photoinduced homolysis of the Ti—CH_3 bonds mainly on the basis of their observed lack of hydrogen abstraction from solvent. Instead they preferred the pathway

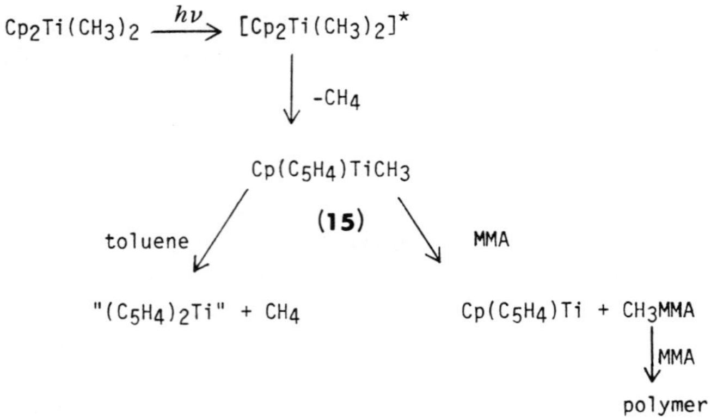

$$Cp_2Ti(CH_3)_2 \xrightarrow{h\nu} [Cp_2Ti(CH_3)_2]^*$$

$$\downarrow -CH_4$$

$$Cp(C_5H_4)TiCH_3$$

toluene ╱   **(15)**   ╲ MMA

$$"(C_5H_4)_2Ti" + CH_4 \qquad\qquad Cp(C_5H_4)Ti + CH_3MMA$$

$$\downarrow MMA$$

polymer

SCHEME 2

of Scheme 2, in which irradiation was implied to lead to *intramolecular* hydrogen abstraction from a Cp ring by a methyl ligand, probably through an intermediate similar to **6**. Polymerization was suggested to occur through $CH_3\cdot$ abstraction by MMA from the photogenerated $Cp(C_5H_4)TiCH_3$.

However, Samuel and Giannotti (*25*) showed that irradiation of $TiCp_2(CH_3)_2$ and $ZrCp_2(CH_3)_2$ in the presence of the radical scavengers nitrosodurene (ND) and 5,5-dimethyl-1-pyrroline *N*-oxide (DMPO) gave ESR signals characteristic of the adducts formed by combination of these scavengers with methyl radicals. Likewise, Lappert and co-workers (*29*) showed that low-temperature photolysis ($-90°C$) of the related zirconium alkyls $ZrCp_2(CH_2Ph)_2$ (**16**) and $ZrCp_2(CH_2SiMe_3)_2$ gave ESR signals attributed to the benzyl and CH_2SiMe_3 radicals as well as signals attributed to an unidentified zirconium hydride species. Also, irradiation of these complexes in the presence of added tertiary phosphine gave ESR signals attributed to the complexes $ZrCp_2(CH_2Ph)(PPh_3)$ and $ZrCp_2(CH_2SiMe_3)(PEt_3)$. These results suggest that homolysis of titanium– and zirconium–carbon bonds *does* occur in the primary photoprocess for these classes of compounds [Eq. (15)], in contrast to the non-free-radical conclusions drawn above.

$$ZrCp_2(CH_2R)_2 \xrightarrow{h\nu} \cdot ZrCp_2(CH_2R) + \cdot CH_2R \qquad (15)$$

(16)

Recent results by Van Leeuwen and co-workers (*24*) appear to reconcile these differences. These workers demonstrated that irradiation of $MCp'_2(CH_3)_2$ (M = Ti or Zr; $Cp' = \eta^5\text{-}CH_3C_5H_4$) does indeed give homolysis of the metal–methyl bonds but that the resultant radical pair recombines unusually rapidly. The methyl radicals so produced can be scavenged only by very efficient scavengers. Otherwise, they recombine with the metal or abstract hydrogen from a Cp ligand or the other CH_3 ligand. Irradiation of the dimethyl complexes $MCp'_2(CH_3)_2$ with traces of oxygen present in solution resulted in a three- to fourfold enhancement of the ^1H-NMR signal due to the titanium-bonded methyl groups. This phenomenon is known as chemically induced dynamic nuclear polarization (CIDNP)(34) and requires the formation of a radical pair involving radicals with different g values. These the authors attributed to titanium(III) and methyl radicals. No CIDNP effect was observed when oxygen and water were scrupulously excluded from the irradiated solutions. The authors interpreted these results in terms of a mechanism involving the initial photoinduced homolysis of the metal–methyl bond followed by a very fast recombination of the radical pair or the occasional formation of methane and other products in the absence of an efficient radical scavenger. The role of oxygen in the system is presumably to provide an "escape route" for the radicals or radical pair, which is necessary for CIDNP to occur. Of the common alkyl radical scavengers used, only thiophenol and 2,2,6,6-tetramethylpiperidine N-oxide (TMPNO) were shown to be reactive enough to intercept the radical pairs formed upon irradiation of the dimethyl complexes. From Samuel and Giannotti's results (25), ND and DMPO must also be sufficiently efficient but Puddephatt's results (26) imply that MMA and styrene are not. Van Leeuwen et al. estimated the rate of recombination of the photogenerated radical pair to be close to diffusion controlled (24). In later work, Tyler and co-workers (35) assigned the electronic absorption spectra of $TiCp_2(CH_3)_2$ and attributed the Ti—R homolysis to a methyl-to-metal CT excited state.

These $MCp_2(CH_3)_2$ complexes have also been irradiated in the presence of other substrates to give interesting transformations. Rausch and co-workers (22) showed that photolysis of the $MCp_2(CH_3)_2$ complexes in the presence of alkynes gave both metallacycles (17) and insertion products (18) [Eq. (16)], as well as numerous organic products. Irradiation of the

$$(16)$$

zirconium derivative in the presence of PhC≡CPh gave a metallacyclic species analogous to 17 as well as methyl adducts of diphenylacetylene, but no insertion products analogous to 18 were reported (24).

Dimethyltitanocene did not react with ethylene under photolytic conditions (24). However, irradiation of the zirconium analogue in benzene solution in the presence of ethylene gave methane, ethane, and polyethylene [Eq. (17)], but the organometallic products were not identified. However, a triplet at $\delta 1.01$ was observed in the ^{1}H-NMR spectrum

$$ZrCp_2(CH_3)_2 + H_2C{=}CH_2 \xrightarrow[\text{benzene}]{h\nu} CH_4 + C_2H_6 + (-CH_2CH_2-)_n \qquad (17)$$

of the irradiated solutions and was attributed to a Zr–propyl group. The signal disappeared when irradiation was stopped. The authors interpreted these results in terms of a mechanism involving the coordination of ethylene to a photogenerated zirconium–methyl radical species (19) followed by insertion of the coordinated ethylene into the zirconium–methyl bond [Eq. (18–21)]. However, none of the postulated intermediates 19–21

$$Cp_2Zr\!\!\begin{array}{c} {\scriptstyle CH_3} \\ {} \\ {\scriptstyle CH_3} \end{array} \xrightarrow[\text{benzene}]{h\nu} Cp_2Zr\cdot\!\!\begin{array}{c} {\scriptstyle CH_3} \\ {} \end{array} + \;\cdot CH_3 \qquad (18)$$

$$(19)$$

$$\mathbf{19} + C_2H_4 \rightarrow Cp_2\dot{Z}r\!\!\begin{array}{c} {\scriptstyle CH_3} \\ {} \\ {\scriptstyle CH_2} \\ \Vert \\ CH_2 \end{array} \qquad (19)$$

$$(20)$$

$$\mathbf{20} \rightarrow Cp_2Zr\cdot\!\!\begin{array}{c} {\scriptstyle CH_2CH_2CH_3} \\ {} \end{array} \qquad (20)$$

$$(21)$$

$$\mathbf{21} + \;\cdot CH_3 \rightarrow Cp_2Zr\!\!\begin{array}{c} {\scriptstyle CH_2CH_2CH_3} \\ {} \\ {\scriptstyle CH_3} \end{array} \qquad (21)$$

$$(22)$$

was conclusively identified, and the authors did not suggest a route for the formation of polyethylene, although 21 could be the polyethylene chain carrier.

Samuel (30) recently reported that irradiation of complexes of the type $TiCp_2R_2$ (R = CH_3 or CH_2Ph) under H_2 generates species capable of effecting the rapid and photocatalytic hydrogenation of linear and cyclic olefins under mild conditions [Eq. (22)]. Irradiation of dimethyltitanocene

$$H_2 + HRC{=}CHR' \xrightarrow[TiCp_2(CH_3)_2]{h\nu} CH_4 + H_2RC{-}CH_2R' \qquad (22)$$

and H_2 in hexane or benzene solution caused a rapid color change and the production of methane. Introduction of olefin into the resultant solution led to complete hydrogenation within ~30 minutes. The nature of the active catalytic species was not established, but ESR studies revealed the presence of signals attributed to a titanium(III) species. The author postulated that homolytic cleavage of the Ti—alkyl bond was the primary photoprocess followed by formation of carbene species **23** and the corresponding hydrocarbon [Eq. (23)]. It was further suggested that the carbene

$$\text{Cp}_2\text{Ti}\Big\langle\begin{matrix}\text{CH}_3\\\text{CH}_3\end{matrix}\quad\xrightarrow{h\nu}\quad \text{Cp}_2\text{Ti}\cdot\overset{\text{CH}_3}{} + \cdot\text{CH}_3 \rightarrow \text{Cp}_2\text{Ti}=\text{CH}_2 + \text{CH}_4 \qquad (23)$$

(23)

then reacts with hydrogen to give an alkyl- or a methylene-bridged hydride dimer which is the active catalyst under thermal conditions.

Photolysis of the dimethylmetallocenes of titanium (*22*) and zirconium (*24*) and their fluorenyl (*31*) and indenyl (*22*) derivatives under CO atmospheres gave high yields of $\text{MCp}_2(\text{CO})_2$, $\text{M}(\eta^5\text{-C}_9\text{H}_7)(\text{CO})_2$, and $\text{M}(\eta^5\text{-C}_{13}\text{H}_9)(\text{CO})_2$ [e.g., Eq. (24)]. Substitution on the Cp rings did not

$$\text{Cp}_2\text{Ti}(\text{CH}_3)_2 + \text{CO} \xrightarrow{h\nu} \text{Cp}_2\text{Ti}(\text{CO})_2 \qquad (24)$$

seem to alter the photochemistry significantly. However, the mechanism of formation of the dicarbonyl products was not established.

Samuel and Giannotti (*32*) have shown that photolysis of TiCp_2R_2 ($\text{R} = \text{CH}_3$ or CH_2Ph) in the presence of elemental sulfur gave **24** [Eq. (25)]. Photolysis of $\text{ZrCp}_2(\text{CH}_3)_2$ under identical conditions did not produce zirconocene pentasulfide.

$$\text{TiCp}_2\text{R}_2 + \text{S}_8 \xrightarrow[\text{C}_6\text{H}_6]{h\nu} \text{Cp}_2\text{Ti}\Big\langle\begin{matrix}\text{S}\diagdown\text{S}\\\text{S}\diagdown\text{S}\diagup\text{S}\end{matrix} + \text{RS}_{\underline{n}}\text{R} \qquad (25)$$

(24)

Photolysis of $\text{Cp}_2\text{Ti}(\text{CH}_3)_2$ and $\text{Cp}_2\text{Ti}(\text{CH}_3)\text{Cl}$ in the presence of another transition metal complex has been shown to result in methyl transfer from titanium to the second transition metal and also to other ligand transfer reactions (*32a*). Photolysis in the presence of certain transition metal halides gave simple metathetical exchange of halide for methyl [e.g., Eq. (26)]. The authors postulated titanium–carbon bond

$$\text{TiCp}_2(\text{CH}_3)_2 + \text{ReBr}(\text{CO})_5 \xrightarrow{h\nu} \text{TiCp}_2(\text{CH}_3)\text{Br} + \text{ReCH}_3(\text{CO})_5 \qquad (26)$$

homolysis as the primary photochemical step followed by attack of the photogenerated methyl radical on the transition metal halide. However, it seems unlikely that a transition metal complex would be such a good radical scavenger so as to compete effectively with back recombination.

Recall that Van Leeuwen *et al.* (*24*) had shown that only the most efficient radical scavengers would intercept the photogenerated CH_3 radicals. A more probable mechanism involves bimolecular reaction between the transition metal complex and a photogenerated Ti species such as **15** in Scheme 2. This could lead to CH_3/Br exchange via a bridged intermediate such as **25**. Yields of these reactions were not reported and thus their

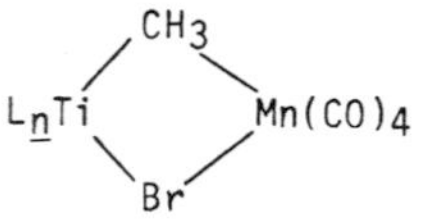

$$L_{\underline{n}}Ti \underset{Br}{\overset{CH_3}{\diamond}} Mn(CO)_4$$

(25)

synthetic utility cannot be determined. Photolysis of $TiCp_2(CH_3)_2$ in the presence of $Mn_2(CO)_{10}$ gave the mononuclear complexes $Mn(CO)_5CH_3$ and $CpMn(CO)_3$ [Eq. 27)]. This result was interpreted in terms of transfer

$$TiCp_2(CH_3)_2 + Mn_2(CO)_{10} \xrightarrow{h\nu} Mn(CO)_5CH_3 + CpMn(CO)_3 \qquad (27)$$

of Cp and CH_3 radicals to $Mn(CO)_5 \cdot$, photogenerated via cleavage of the Mn—Mn bond in $Mn_2(CO)_{10}$.

5. $M(\eta^5-C_5R_5)(Ar)_2$ $(M = Ti$ or $Zr)$

Photolysis of $TiCp_2Ph_2$ in benzene solution has been reported to give production of biphenyl, benzene, and an oligomeric material formulated as $(TiCp_2H)_n$ [Eq. (28)] although yields were not given (*36*). Photolysis of

$$TiCp_2Ph_2 \xrightarrow[benzene]{h\nu} (TiCp_2H)_n + PhH + Ph\text{—}Ph \qquad (28)$$

diphenyltitanocene in the presence of carbon monoxide and diphenyl acetylene was also reported to give $Cp_2Ti(CO)_2$ and complex **17**. It was suggested that these reactions occur through photoinduced homolysis of Ti—Ph bonds.

However, Rausch *et al.* (*37*) showed that *both* titanium–aryl bond homolysis and reductive coupling of the two aryl ligands occur upon photolysis of $TiCp_2(Ar)_2$. For example, irradiation of $TiCp_2(Ph)_2$ in benzene-d_6 solution produced biphenyl-d_0 and biphenyl-d_5 in equal amounts. Photolysis of $TiCp_2(p\text{-tol})_2$ (p-tol $= p\text{-}C_6H_4CH_3$) produced a mixture of toluene, 4-methylbiphenyl, and 4,4'-dimethylbiphenyl. Irradiation of $TiCp_2Ph_2$ in the presence of CO gave moderate yields of $TiCp_2(CO)_2$ (*37*).

In contrast to the chemistry observed with $TiCp_2Ph_2$, Erker (*38*) suggested that photolysis of bis(aryl)zirconocene complexes leads only to

$$\text{Cp}_2\text{Zr}\left(\!\!-\!\!\left\langle\!\!\bigcirc\!\!\right\rangle\!\!-\text{CH}_3\right)_2 \xrightarrow{h\nu} \text{H}_3\text{C}-\left\langle\!\!\bigcirc\!\!\right\rangle\!\!-\!\!\left\langle\!\!\bigcirc\!\!\right\rangle\!\!-\text{CH}_3 + \text{"ZrCp}_2\text{"} \quad (29)$$

coupling of the aryl ligands and formation of substituted biphenyls. For example, photolysis of bis(p-tolyl)zirconocene in benzene solution gave almost quantitative yield of 4,4′-dimethylbiphenyl [Eq. (29)]. The solution turned deep blue-brown upon photolysis, and it was suggested that the color arose from production of "zirconocene." Irradiation of the corresponding diphenyl complex gave biphenyl, and 3,3′-dimethylbiphenyl resulted from photolysis of bis(m-tolyl)zirconocene. The biphenyl products probably arise through concerted reductive elimination and coupling of the aryl ligands. The near-quantitative yield of 4,4′-dimethylbiphenyl from photolysis of $ZrCp_2(p\text{-tol})_2$ in benzene solution appears to rule out Zr—tolyl bond homolysis to produce tolyl radicals. Such a radical process in benzene solution should produce substantial amounts of C_6H_5—$C_6H_4CH_3$, which were apparently not detected.

Lappert and co-workers (29) showed that photolysis of $ZrCp_2Ph_2$ in toluene solution gave rise to an ESR signal attributed to the same zirconium hydride species detected during photolysis of $ZrCp_2(CH_2Ph)_2$ (*vide supra*). Irradiation of the diphenylzirconocene complex in the presence of tertiary phosphines also gave ESR signals attributed to the zirconium hydride species and to $ZrCp_2(Ph)(PR_3)$. These results imply that zirconium–aryl bond homolysis does occur to some extent, but given the ultrasensitivity of the ESR technique, it is not clear how important a process this is.

Substitution of the Cp ring hydrogens has a pronounced effect on the photochemistry of these diphenyl complexes. For example, when $TiCp^*_2Ph_2$ ($Cp^* = C_5Me_5$) was irradiated in benzene-d_6 solution, biphenyl-d_0 and biphenyl-d_5 were formed in a ratio of 36:1 [Eq. (30)] (39).

$$\text{Cp}^*_2\text{TiPh}_2 \xrightarrow[\text{C}_6\text{D}_6]{h\nu} C_6H_5\text{—}C_6H_5 + C_6H_5\text{—}C_6D_5 \quad (30)$$

$$ 36 1$$

This suggested that reductive coupling of the two phenyl ligands predominated over Ti—aryl bond homolysis in the primary photochemical event. However, Tung and Brubaker (39) showed that Ti—phenyl bond homolysis does occur to a certain extent, as indicated by a spin-trapping experiment. Irradiation of $TiCp^*_2Ph_2$ in toluene solution in the presence of nitrosodurene gave a triplet ESR signal attributed to the 2,3,4,5-tetramethylphenyl(phenyl)nitroxide spin adduct. Even without added spin trap, a persistent ESR signal was observed that was attributed to species **26**. On the basis of these results, the authors proposed the mechanisms of

Scheme 3 for formation of **26**, with reductive coupling path B being the more important (*39*).

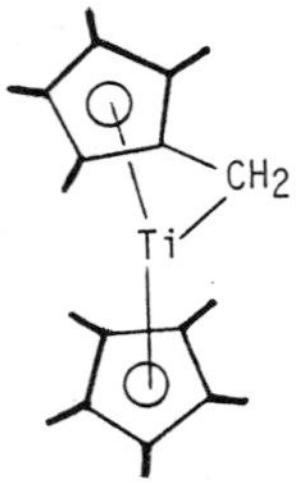

(26)

When $Cp^*_2ZrPh_2$ was irradiated under similar conditions, biphenyl-d_0 and biphenyl-d_5 were formed in a 1:3 ratio indicating that the stepwise homolysis of the Zr—phenyl bond (path A of the Scheme 3) was now predominant (*39*). Benzene-d_1 was also detected in the irradiated solutions and probably formed by deuterium abstraction from solvent molecules by photogenerated phenyl radicals (path A) although it could have arisen by H/D exchange catalyzed by the "zirconocene" intermediate. Furthermore, irradiation of $ZrCp^*_2Ph_2$ in benzene-d_6 solution gave a complex ESR spectrum due to several photogenerated paramagnetic species. These thermally decomposed to a single compound which the authors identified as $ZrCp^*_2Ph$ on the basis of its ESR spectrum. These results all point to stepwise zirconium–phenyl bond homolysis as the major photodegradation pathway. This is in direct contrast to the results obtained for the unsubstituted cyclopentadienyl (Cp) derivatives of this metal as summarized above (*38*).

These results show that both metal–aryl bond homolysis and reductive coupling of the aryl ligands can occur for these derivatives as a consequence of irradiation (Scheme 3). However, which path predominates is

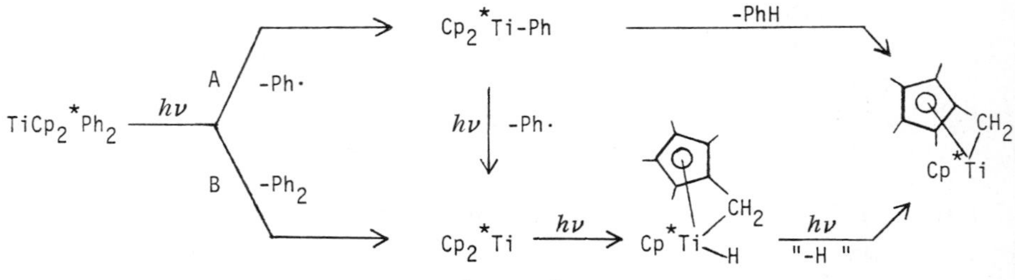

SCHEME 3

TABLE II

M-ARYL BOND HOMOLYSIS VERSUS REDUCTIVE COUPLING OF ARYL LIGANDS

Complex[a]	Path A[b] M—aryl homolysis	Path B[b] Reductive coupling	Reference
TiCp$_2$Ph$_2$	Equal	Equal	37
TiCp*$_2$Ph$_2$	Trace	Predominant	39
ZrCp$_2$(p-tol)$_2$[c]	None	All	38
ZrCp*$_2$Ph$_2$	Major	Minor	39

[a] Cp* = C$_5$Me$_5$.
[b] Scheme 3.
[c] p-Tol, p-C$_6$H$_4$CH$_3$.

dramatically dependent upon the metal and Cp ring substitution, as indicated by the data summarized in Table II. It is easy to rationalize the Ti data by assuming that the increased steric bulk of the 10 methyl groups on the Cp rings in moving from TiCp$_2$Ph$_2$ to TiCp*$_2$Ph$_2$ would promote reductive coupling, as this would better release steric strain. However, the Zr chemistry shows the opposite effect, although the larger size of Zr compared to Ti would greatly lower the steric sensitivity of the reaction and may permit control by electronic effects. However, it should be noted that the photochemistry of these bis(aryl) derivatives could exactly parallel that of the corresponding dimethyl complexes discussed earlier (Section II, B, 4), with M—R homolysis to generate a very short-lived radical pair being the dominant photochemical reaction. Escape of the aryl radical from the solvent cage would correspond to path A of Scheme 3 whereas "reductive coupling" (path B) could arise by coupling of the second aryl ligand with the initially generated radical before it escaped the solvent cage.

Irradiation of the complexes TiCp$_2$Ph$_2$ and ZrCp$_2$Ph$_2$ in the presence of other substrates has led to several interesting transformations, some of which closely resemble TiCp$_2$(alkyl)$_2$ photoreactions. For example, irradiation of TiCp$_2$Ph$_2$ under H$_2$ in the presence of ethylene or 1-hexene gave rapid, photocatalytic hydrogenation of olefin as was reported for TiCp$_2$(CH$_3$)$_2$ (30). In contrast, no hydrogenation was observed under similar conditions when TiCp*$_2$(Ph)$_2$ was used (39). Photolysis of TiCp$_2$Ph$_2$ in the presence of styrene was also reported to yield small amounts of polystyrene (39).

Erker and others (40–40d) have used the efficient reductive coupling reaction observed upon the photolysis of ZrCp$_2$Ph$_2$ to generate and transfer the reactive ZrCp$_2$ species to a variety of substrates. For example, if photolysis of ZrCp$_2$Ph$_2$ is carried out in the presence of suitable conjugated dienes, the corresponding monomeric η^2-dienezirconocene complexes are formed in good yield [Eq. (31)] (40a, b).

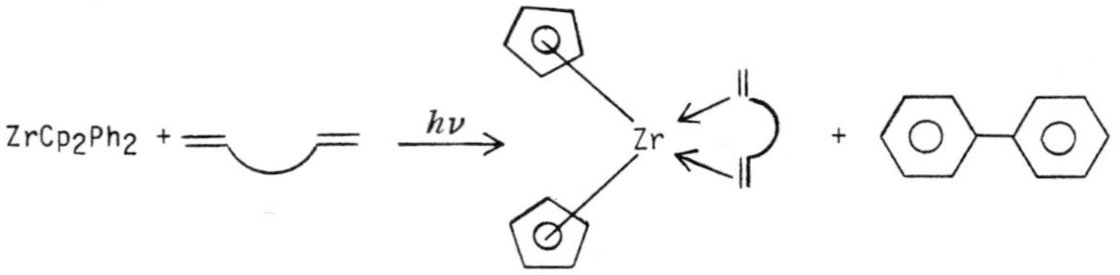

However, when a 1:1 mixture of cyclohexadiene and cyclooctadiene was irradiated in the presence of $ZrCp_2Ph_2$, a catalytic hydrogen transfer reaction occurred to give benzene and cyclooctene [Eq. (32)].

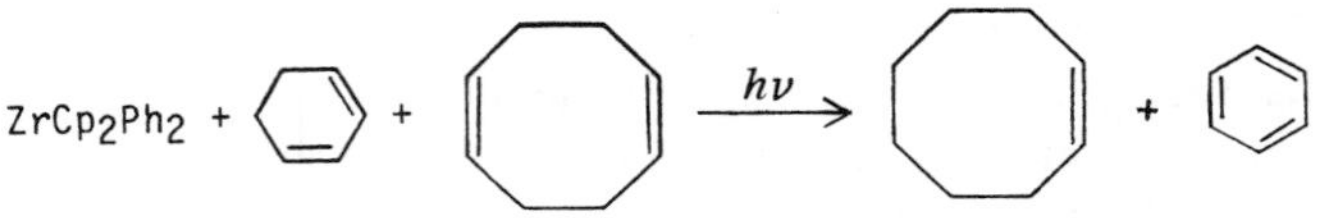

Photochemically generated zirconocene also reacts with various terminal alkynes to give metallacycles (40) [Eq. (33)]. The location of the R

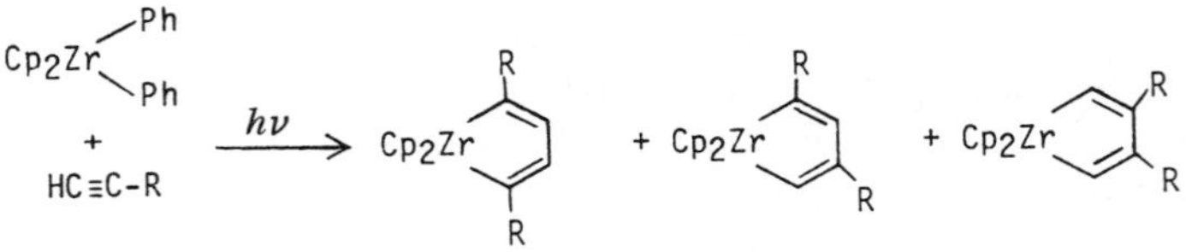

substituents in the metallacycle seemed independent of the steric bulk of the substituent. Tainturier and co-workers (40c) have shown that irradiation of $ZrCp'_2Ph_2$ in heptane solution in the presence of diselenides gave the products of Eq. (34).

$$ZrCp'_2Ph_2 + PhSeSePh \xrightarrow{h\nu} ZrCp'_2(SePh)_2 + PhPh \qquad (34)$$

$$(Cp' = C_5H_5 \text{ or } C_5H_4Bu^t)$$

6. $MCp_2(R)(X)$ $(M = Ti, Zr)$

Little is known about the photochemistry of these compounds but, from the few reports published (29, 32a, 33, 40c), it appears that metal–methyl bond homolysis occurs upon irradiation. For example, Lappert and co-workers (29) reported that irradiation of $ZrCp_2(CH_3)(Cl)$ in toluene solution gave an ESR signal that the authors attributed to $ZrCp_2Cl$. The same signal was observed when $ZrCp_2(Ar)(Cl)$ and $ZrCp_2(H)(Cl)$ were irradiated, suggesting that zirconium–methyl, –aryl, and –hydride bond homolysis occurs upon irradiation [Eq. (35)]. However, the relative importance of this homolysis reaction was not determined, nor were the organic products identified.

$$ZrCp_2(Cl)(CH_3) \xrightarrow{h\nu} ZrCp_2Cl + \cdot CH_3 \tag{35}$$

Similarly, irradiation of the complexes $ZrCp_2(EPh)(CH_3)$ (E = S or Se) in the presence of added selenoether resulted in the substitution of the methyl ligand by selenide [Eq. (36)] (*40c*). The authors suggested that this reaction proceeded via zirconium–methyl bond homolysis.

$$ZrCp_2(EPh)(CH_3) + PhECH_3 \xrightarrow{h\nu} ZrCp_2(EPh)_2 \tag{36}$$

Irradiation of $TiCp_2(CH_3)Cl$ (**27**) in the presence of Group VIII transition metal complexes resulted in transfer of both the methyl and chloride ligands from the titanium to the Group VIII metal [Eqs. (37, 38)] (*32a*). The authors suggested that these reactions proceeded via homolytic

$$TiCp_2(CH_3)Cl + Fe(CO)_3L_2 \xrightarrow{h\nu} Fe(CH_3)Cl(CO)_2L_2 \tag{37}$$
$$(\mathbf{27})$$

$$TiCp_2(CH_3)Cl + IrCl(CO)(PPh_3)_2 \xrightarrow{h\nu} Ir(CH_3)Cl_2(CO)(PPh_3)_2 \tag{38}$$
$$(\mathbf{27})$$

cleavage of the Ti—CH_3 bond to generate methyl radicals [Eq. (39)], which then reacted thermally with coordinatively unsaturated transition metal

$$Cp_2Ti(CH_3)Cl \xrightarrow{h\nu} Cp_2TiCl + \cdot CH_3 \tag{39}$$
$$(\mathbf{27})$$

$$\cdot CH_3 + [M] \rightarrow [M]CH_3 \xrightarrow{Cp_2TiCl} \text{``}Cp_3Ti\text{''} + [M]Cl(CH_3) \tag{40}$$

fragments [M], namely $IrCl(CO)(PPh_3)_2$ and photogenerated $Fe(CO)_2L_2$, to generate a 17-electron [M]CH_3 species. This, they assumed, reacted further with Cp_2TiCl to form titanocene and the observed chloromethyl complexes [Eq. (40)]. When the analogous ethyl complex ($Cp_2TiEtCl$) was irradiated in the absence of added substrate the main organic products were ethylene and ethane along with traces of butane (*33*). These results were also interpreted in terms of initial photoinduced Ti—C bond homolysis followed by a major secondary thermal reaction of the photogenerated ethyl radicals.

7. *ZrCp₂(X)(CH=CHPh)*

Erker and co-workers (*41, 41a*) have studied the photochemistry of several zirconium alkenyl complexes and observed both isomerization of the alkenyl ligand and reductive coupling of the alkenyl ligand with X.

The extent to which each reaction occurred was found to relate closely to the nature of the ligand X. For example, irradiation of (E)-ZrCp$_2$(Cl)(CH=CHPh) gave, after 2.5 hours, a photostationary mixture of E and Z isomers of the zirconium alkenyl complex [Eq. (41)], in a 30:70 ratio.

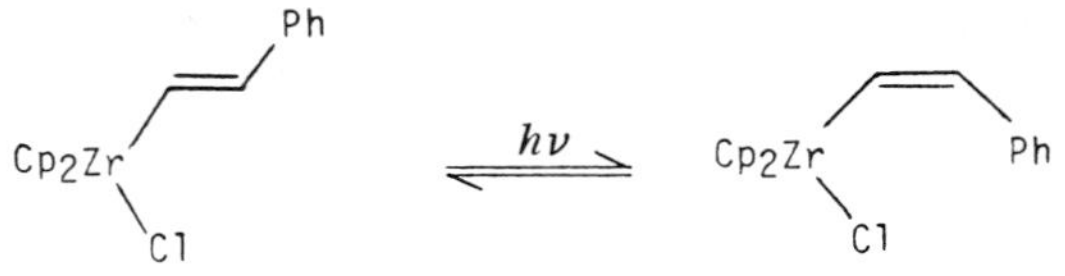

$$
\underset{(E)\text{-}(\mathbf{28})}{\text{Cp}_2\text{Zr}\backslash_{\text{Cl}}^{\diagup\overset{\diagup\text{Ph}}{=}}}
\quad\underset{}{\overset{h\nu}{\rightleftharpoons}}\quad
\underset{(Z)\text{-}(\mathbf{28})}{\text{Cp}_2\text{Zr}\backslash_{\text{Cl}}^{\diagup\overset{=}{\diagdown}\text{Ph}}}
\tag{41}
$$

Prolonged irradiation led to slow decomposition of both isomers and the products of Eq. (42). The mechanism of formation of complex **29** was not

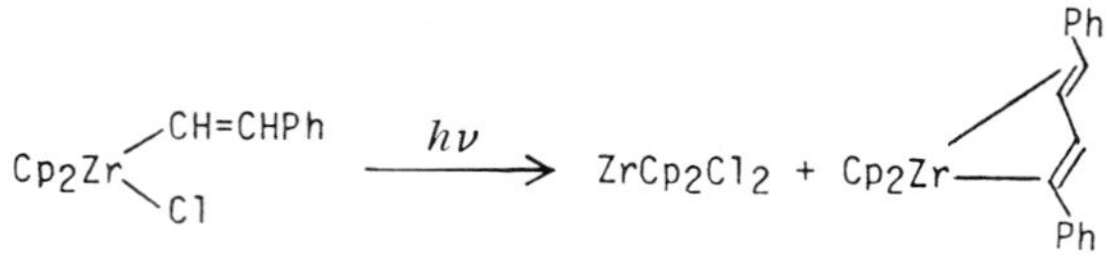

$$
\text{Cp}_2\text{Zr}\overset{\diagup\text{CH=CHPh}}{\diagdown_{\text{Cl}}}\quad\overset{h\nu}{\longrightarrow}\quad \text{ZrCp}_2\text{Cl}_2 + \underset{(\mathbf{29})}{\text{Cp}_2\text{Zr}-\!\!-\!\!\overset{\text{Ph}}{\underset{\text{Ph}}{\diagup\diagdown}}}
\tag{42}
$$

elucidated but it is reasonable to postulate a bimolecular ligand-exchange reaction via an intermediate similar to **25**.

Photostationary equilibria between E and Z isomers were also observed with the three zirconium alkenyl complexes shown in Eq. (43), but these

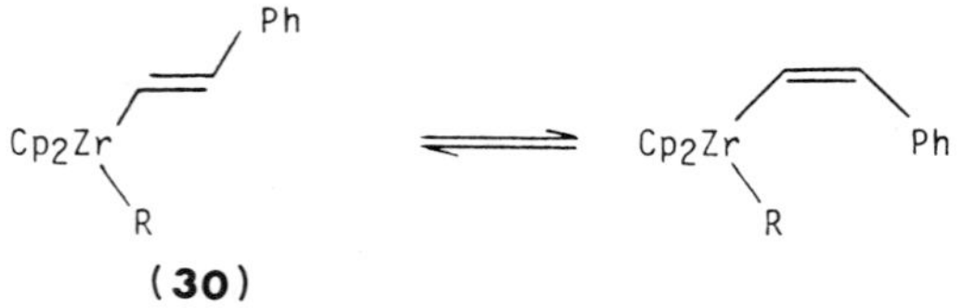

$$
\underset{(\mathbf{30})}{\text{Cp}_2\text{Zr}\backslash_{\text{R}}^{\diagup\overset{\diagup\text{Ph}}{=}}}
\quad\rightleftharpoons\quad
\text{Cp}_2\text{Zr}\backslash_{\text{R}}^{\diagup\overset{=}{\diagdown}\text{Ph}}
\tag{43}
$$

$$
\text{R= -CH}_3,\ \text{-CH}_2\text{Ph},\ \text{-Ph}
$$

reactions occurred much more rapidly than with the chloroalkenyl complex **28** (*41, 41a*). In each case, however, reductive coupling of the alkenyl and R ligands was significant, especially when R was phenyl. This competing reaction led to the decomposition of both isomers and the formation of binuclear, alkenyl-bridged zirconocene complexes **31** [Eq. (*44*)]. Irradia-

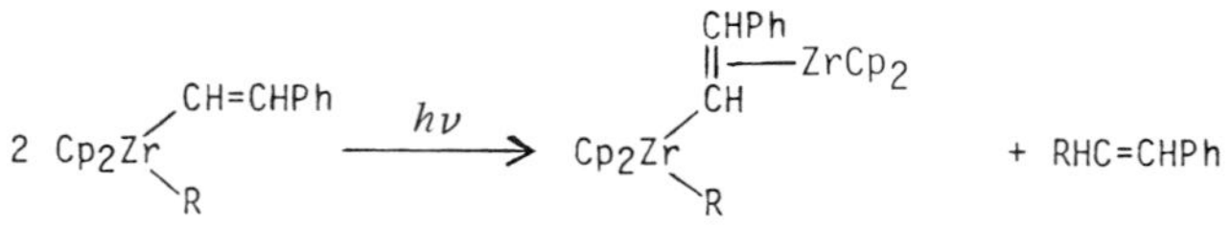

$$
2\ \underset{(\mathbf{30})}{\text{Cp}_2\text{Zr}\backslash_{\text{R}}^{\diagup\text{CH=CHPh}}}\quad\overset{h\nu}{\longrightarrow}\quad \underset{(\mathbf{31})}{\text{Cp}_2\text{Zr}\backslash_{\text{R}}^{\diagup\overset{\text{CHPh}}{\underset{\text{CH}}{\|\!\!-\!\!\text{ZrCp}_2}}}}\quad + \text{ RHC=CHPh}
\tag{44}
$$

tion of a solution containing a mixture of **28** and **30** (R = Ph) gave the reaction shown in Eq. (45) (*41a*). The authors suggested that irradiation of

$$Cp_2Zr\begin{smallmatrix}CH=CHPh\\Ph\end{smallmatrix} + Cp_2Zr\begin{smallmatrix}CH=CHPh\\Cl\end{smallmatrix} \xrightarrow{h\nu} Cp_2Zr\begin{smallmatrix}CH=CH\overset{Ph}{}\\Cl\end{smallmatrix}ZrCp_2 + trans\text{-}PhHC{=}CHPh \quad (45)$$

(30)   **(28)**   **(32)**

                            (>80%)   (100%)

30 resulted in reductive coupling of the phenyl and alkenyl ligands to give *trans*-stilbene and zirconocene, and the latter then reacted thermally with unreacted **30** or added **28** to give the dimeric complexes **31** and **32**, respectively.

The above results show that alkenylzirconocene complexes can undergo two distinct photoinduced transformations depending on the other ligand present. When this ligand is chloride, reductive coupling with the alkenyl ligand does not readily occur, and complex **28** undergoes reversible isomerization when irradiated. However, if X is an alkyl or aryl ligand, reductive elimination occurs upon irradiation and competes efficiently with the isomerization reaction.

C. V, Nb, and Ta Complexes

1. $MCp_2(CH_3)_2$ (M = V, Nb, or Ta)

Rausch and co-workers (*42, 42a*) investigated the photochemistry of the dimethylmetallocenes of Group V and found the nature of the primary photochemical event to depend markedly on the metal. For example, irradiation of $TaCp_2(CH_3)_2$ in pentane solution gave methane and traces of ethane and ethylene [Eq. (46)] (*42, 42a*). The only two organometallic

$$TaCp_2(CH_3)_2 \xrightarrow{h\nu} (TaC_{10}H_{10})_x + TaCp_2(CH_3)_3 + CH_4 + C_2H_4 + C_2H_6 \quad (46)$$

                                    (>99%) (trace) (trace)

products detected after prolonged photolysis were $TaCp_2(CH_3)_3$ and an insoluble compound analyzing for $TaC_{10}H_{10}$ and which the authors formulated as a polymeric form of tantalocene. Cyclopentadiene was also detected when the irradiation was conducted in benzene solution. Deuterium labeling experiments showed that the hydrogen incorporated into methane came from all three possible sources: solvent, Cp rings, and methyl ligands. Irradiation of an equimolar mixture of $Ta(\eta^5\text{-}C_5D_5)_2(CH_3)_2$ and $Ta(\eta^5\text{-}C_5H_5)_2(CD_3)_2$ gave methane with a labeling pattern that indicated that *intermolecular* hydrogen abstraction from the

Cp rings of another molecule also occurred to some extent. The authors concluded from these results that homolysis of the tantalum–methyl bond. was the major photochemical reaction. The product $TaCp_2(CH_3)_3$ was assumed to form by the reaction of the photogenerated 16-electron species $TaCp_2CH_3$ (33) with starting $TaCp_2(CH_3)_2$ [Eqs. (47) and (48)]. Evidence

$$TaCp_2(CH_3)_2 \xrightarrow{h\nu} [TaCp_2CH_3] + \cdot CH_3 \tag{47}$$

(33)

$$[TaCp_2CH_3] + TaCp_2(CH_3)_2 \rightarrow TaCp_2(CH_3)_3 + (1/n)(TaC_{10}H_{10})_n \tag{48}$$

33

for the coordinatively unsaturated intermediate 33 was the formation of $TaCp_2(CO)CH_3$ in low yield (3%) from the photolysis of $TaCp_2(CH_3)_2$ under CO.

Surprisingly, irradiation of $TaCp_2(CH_3)_2$ in the presence of styrene or methyl methacrylate did not give polymerization (42). In fact, more polymer was produced when $TaCp_2(CH_3)_2$ was not present in the monomer solution. This result would seem to argue against a free radical process as the primary mode of photodecomposition of $TaCp_2(CH_3)_2$. However, the situation could be similar to that found for $TiCp_2(CH_3)_2$ (24). Recall (Section II,B,4) that irradiation of this complex gave homolysis of the Ti—CH_3 bonds but the radical pair generated was so short-lived that monomer polymerization could not be initiated but yet hydrogen abstraction products did form.

Irradiation of $NbCp_2(CH_3)_2$ also gave methane as the principal organic product [Eq. (49)], along with traces of ethane and ethylene (42, 42a). The

$$NbCp_2(CH_3)_2 \xrightarrow{h\nu} CH_4 + C_2H_4 + C_2H_6 \tag{49}$$

(>99%) (trace) (trace)

organometallic products were not reported. Deuterium labeling studies showed that the hydrogen incorporated into the photogenerated methane came from the solvent, methyl groups, and Cp rings. Irradiation of $NbCp_2(CH_3)_2$ in the presence of CO also gave low yields (5%) of $NbCp_2(CH_3)(CO)$, suggesting that niobium–methyl bond homolysis occurs upon irradiation.

However, the photochemistry of the vanadium analogue $VCp_2(CH_3)_2$ differs in detail from that of Nb or Ta. For example, irradiation of $VCp_2(CH_3)_2$ in hexane solution gave substantial quantities of ethane in addition to methane, along with traces of ethylene [Eq. (50)] (42a). No

$$VCp_2(CH_3)_2 \xrightarrow{h\nu} CH_4 + C_2H_6 + C_2H_4 \tag{50}$$

(64%) (36%) (trace)

volatile organometallic products were recovered from the irradiated solutions. Deuterium labeling studies showed that the hydrogen incorporated into methane originated from all three possible sources: solvent, Cp rings, and CH_3 ligands. Irradiation of an equimolar mixture of $VCp_2(CH_3)_2$ and $VCp_2(CD_3)_2$ gave C_2D_6, C_2H_6, and $C_2D_3H_3$, indicating that ethane was produced both by dimerization of methyl groups from the same and from different molecules. However, intermolecular hydrogen abstraction from the rings did *not* occur upon irradiation of $VCp_2(CH_3)_2$. Irradiation of $VCp_2(CH_3)_2$ in the presence of CO gave no $VCp_2(CH_3)(CO)$ but instead $CpV(CO)_4$ in low (3%) yield (*40a*).

The experimental results clearly show that radical-derived chemistry is more predominant for $VCp_2(CH_3)_2$ than for its Ta and Nb analogues. However, it is not clear why this is the case and whether or not it represents a fundamental difference in the primary photochemistry of these complexes. It may be that photolysis of all three complexes induces homolysis of the metal–alkyl bonds and that the $(MCp_2CH_3 \cdot + CH_3 \cdot)$ radical pairs for Nb and Ta recombine much faster than the analogous radical pair from vanadium. Thus, the vanadium chemistry shows more "free" radical character. The *intermolecular* Cp-hydrogen abstraction probably arises from metallation of a Cp ring of $MCp_2(CH_3)_2$ by a photogenerated metal fragment, either MCp_2CH_3 or MCp_2. This may occur readily for M = Nb and Ta, but not for V, and thus the latter would not show intermolecular Cp-hydrogen abstraction.

The dimethylmetallocenes of Nb and Ta were also shown to react photochemically with S_8 to give disulfido methyl complexes [Eq. (51)] in

$$Cp_2M\underset{CH_3}{\overset{CH_3}{<}} \quad + \quad S_8 \quad \xrightarrow[C_6H_6]{h\nu} \quad Cp_2M\overset{CH_3}{\underset{S}{\diagup}}\diagdown_{S}\diagup \qquad (51)$$

moderate yields (50% and 15%, respectively) (*43*). Irradiation of $NbCp_2(CH_3)_2$ in the presence of red selenium also gave the diselenido complex **34** (*43*).

$$Cp_2Nb\overset{CH_3}{\underset{Se}{<}}\diagdown_{Se}\diagup Se$$

(**34**)

2. $MCp_2(X)R$ (M = Nb or Ta)

Schrock and co-workers (*44*) studied the photodecomposition of $TaCp_2(C_2H_4)CH_3$ (**35**) in the presence of various substrates and found that

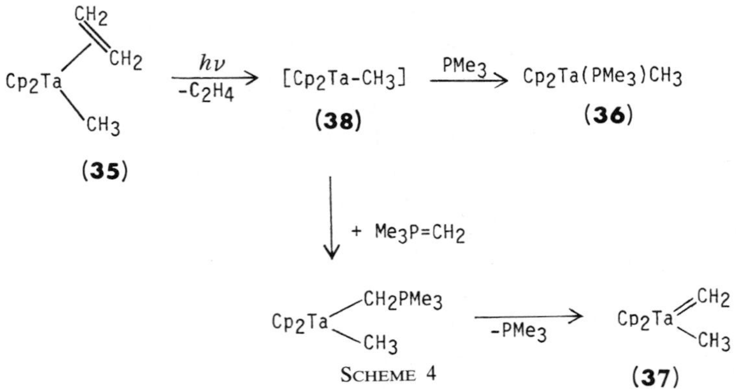

SCHEME 4

dissociative loss of C_2H_4 occurs upon irradiation. For example, irradiation of $TaCp_2(C_2H_4)CH_3$ in the presence of PMe_3 gave $TaCp_2(PMe_3)Me$ (36) in high yield [Eq. (52)]. Irradiation of complex 35 in the presence of the

$$TaCp_2(C_2H_4)CH_3 + PMe_3 \xrightarrow[C_6D_6]{h\nu} TaCp_2(PMe_3)CH_3 \qquad (52)$$
$$\quad (35) \qquad\qquad\qquad\qquad\qquad\quad (36)$$

Wittig reagent $Me_3P{=}CH_2$ gave the methylene–methyl complex 37 [Eq. (53)]. This complex decomposed thermally to $TaCp_2(C_2H_4)CH_3$ and

$$35 + Me_3P{=}CH_2 \xrightarrow{h\nu} Cp_2Ta\!\!\begin{array}{c}{}^{CH_2}\\{}_{CH_3}\end{array} + PMe_3 + C_2H_4 \qquad (53)$$
$$(37)$$

"TaCpCH$_3$" under these conditions and was not isolated although it was observed spectroscopically. The authors postulated that both complexes 36 and 37 were formed via the common intermediate 38 by the mechanism shown in Scheme 4.

Guerchais and co-workers (43) recently found that irradiation of $NbCp_2(\eta^2\text{-}S_2)CH_3$ does not result in the expulsion of the disulfido ligand but rather in the low-yield formation of the monosulfido methyl complex 39 [Eq. (54)]. No further details were provided.

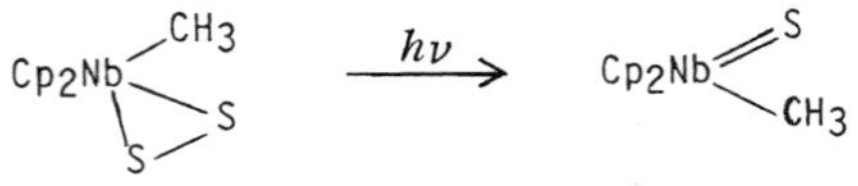

$$(39) \qquad\qquad\qquad (54)$$
$$(20\%)$$

3. $TaCp^*(CH_2Ph)(Cl)_2$

Schrock and co-workers (*44a*) have found that irradiation of TaCp*-$(CH_2Ph)Cl_2$ with UV light in benzene or toluene solution produced $TaCp^*(CHPh)Cl_2$ in low yield along with toluene and 2,2-diphenylethane [Eq. (55)], but no further details were given.

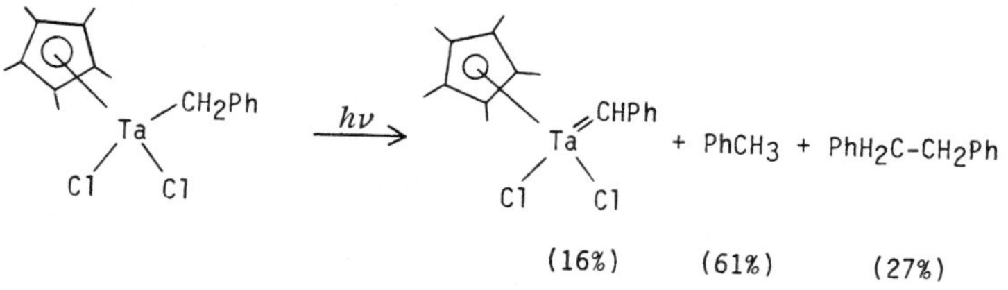

$$\xrightarrow{h\nu} \quad + \ PhCH_3 \ + \ PhH_2C\text{-}CH_2Ph \quad (55)$$

$$(16\%) \qquad (61\%) \qquad (27\%)$$

D. Cr, Mo, and W Complexes

1. $Cr(R)_4$

UV irradiation of the homoleptic alkyl complexes of chromium was found to give homolysis of the Cr—alkyl bonds and the generation of free radicals (*17, 45, 46*). When $Cr(nor)_4$ was irradiated with near-UV light in CCl_4/hexane mixtures, clean changes in the UV–visible absorption spectrum resulted and norbornane, 1-chloronorbornane, and hexachloroethane were produced [Eq. (56)] (*45*). An orange, chromium-containing solid

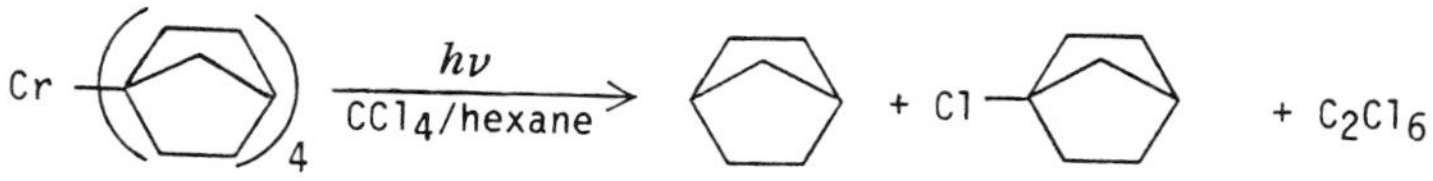

$$Cr \xrightarrow[CCl_4/hexane]{h\nu} \quad + \ Cl \qquad + \ C_2Cl_6 \quad (56)$$

was also produced and tentatively identified as $Cr_2(nor)_6$. The authors concluded that irradiation into the ligand-to-metal charge-transfer (LMCT) band of the complex resulted in homolytic cleavage of the Cr—norbornyl bond and the formation of free norbornyl radicals.

Similar results were obtained with $Cr(neo)_4$ and $Cr(mes)_3(thf)$ (mes = mesityl; thf = tetrahydrofuranyl) (*46*). Rausch and co-workers showed that irradiation of $Cr(neo)_4$ in solution gave neopentane and 2,2,5,5-tetramethyl hexane [Eq. (57)]. Chromium metal and an

$$Cr(neo)_4 \xrightarrow{h\nu} C(CH_3)_4 + CH_3\overset{\overset{\displaystyle CH_3}{|}}{\underset{\underset{\displaystyle CH_3}{|}}{C}}CH_2CH_2\overset{\overset{\displaystyle CH_3}{|}}{\underset{\underset{\displaystyle CH_3}{|}}{C}}CH_3 \quad (57)$$

unidentified black pyrophoric solid were also produced. The origin of the hydrogen incorporated into neopentane was not determined nor were the relative yields of organic products. Evidence for homolytic cleavage of the Cr—alkyl bonds to form free radicals came from irradiation of Cr(neo)$_4$ and Cr(mes)$_3$(thf) in the presence of vinyl monomers. Neither complex is an efficient polymerization catalyst in the dark. However, irradiation of both in solution in the presence of styrene, methyl acrylate, methyl methacrylate, or vinyl acetate resulted in the rapid formation of polymer. The authors noted that both the temperature dependence of the polymerization and the tacticity of the polymer produced were consistent with a free-radical mechanism.

These two complexes were also found to be active photocatalysts for the polymerization of ethylene (17, 47, 47a). Irradiation of solutions of either complex under 1 atm ethylene resulted in the rapid formation of high-molecular-weight polyethylene. The rate of ethylene polymerization was increased by the addition of various metal halides prior to photolysis. The mechanism of this reaction was not investigated but the authors postulated that the active species were photogenerated Ziegler–Natta-type catalysts (17).

2. [Cr(CO)₃Ph]₂Hg

Ph$_2$Hg, benzenechromium tricarbonyl, and diphenylbis(chromium tricarbonyl) were identified in the products of photolysis of the title complex by chromatographic analysis (47). No further details were given.

3. M₂(CH₂SiMe₃)₆ (M = Mo or W)

Chisholm et al. (47a) noted that these compounds are photosensitive and that 24 hours UV irradiation of an equimolar mixture of the title complexes led to considerable decomposition. The photoproducts, however, were not characterized and no further details were given.

4. MCp₂R₂ (M = Mo or W)

Dias and co-workers (48, 48a) investigated the photochemistry of the complexes MoCp$_2$(CH$_3$)$_2$ (48) and [MCp$_2$(CH$_3$)$_2$]PF$_6$ (M = Mo or W) (48, 48a) and found that only the cationic complexes were photosensitive. Irradiation of MoCp$_2$(CH$_3$)$_2$ with pyrex-filtered light ($\lambda > 310$ nm) led to no detectable reaction. However, irradiation of the complexes [MCp$_2$(CH$_3$)$_2$]$^+$ in the presence of added donor ligands led to photoproducts which depended markedly on the metal and the nature of the added ligand. For example, irradiation of $[MoCp_2(CH_3)_2]PF_6$ in the presence of pyridine gave only the monosubstituted product **40** [Eq. (58)]. On the

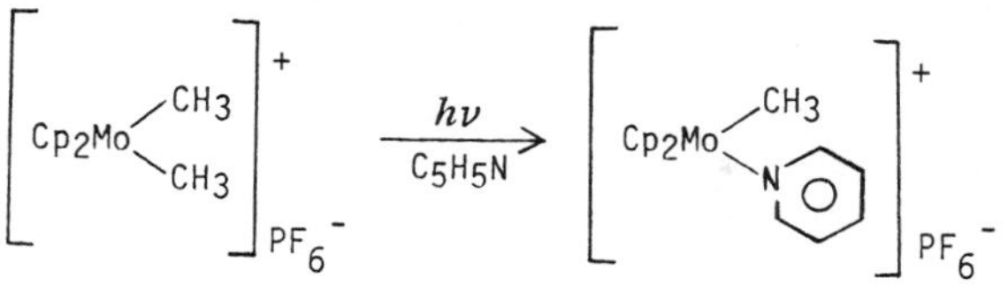

$$
\left[Cp_2Mo\underset{CH_3}{\overset{CH_3}{\diagup}} \right]^{+} PF_6^{-} \xrightarrow[\text{C}_5\text{H}_5\text{N}]{h\nu} \left[Cp_2Mo\underset{N\bigcirc}{\overset{CH_3}{\diagup}} \right]^{+} PF_6^{-} \qquad (58)
$$

(40)

other hand, irradiation of the analogous W complex **41** gave a mixture of the monosubstituted derivative **42** and $[WCp_2(H)(py)][PF_6]$ (*48a*) [Eq. (59)]. Furthermore, irradiation of $[WCp_2(CH_3)_2]^{+}$ in the presence of

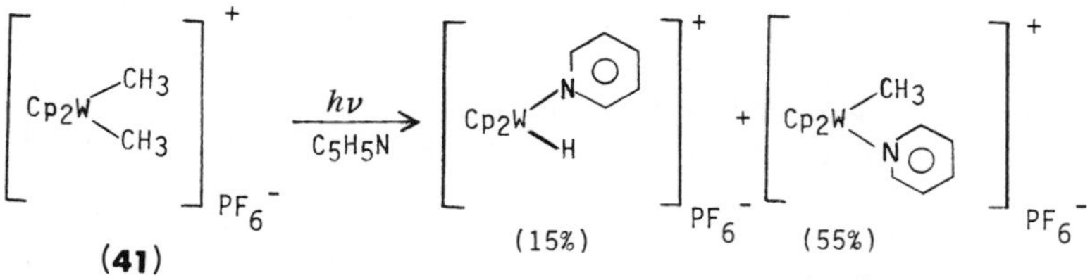

$$
\left[Cp_2W\underset{CH_3}{\overset{CH_3}{\diagup}} \right]^{+} PF_6^{-} \xrightarrow[\text{C}_5\text{H}_5\text{N}]{h\nu} \left[Cp_2W\underset{H}{\overset{N\bigcirc}{\diagup}} \right]^{+} PF_6^{-} + \left[Cp_2W\underset{N\bigcirc}{\overset{CH_3}{\diagup}} \right]^{+} PF_6^{-} \qquad (59)
$$

(41)　　　　　　　　　　　　(15%)　　　　(55%)

(42)

PMe_2Ph gave the complex $[WCp_2(H)CH_2PMe_2Ph][PF_6]$ in high yield, as well as the methylphosphine complex **43** [Eq. (60)]. The fate of the lost

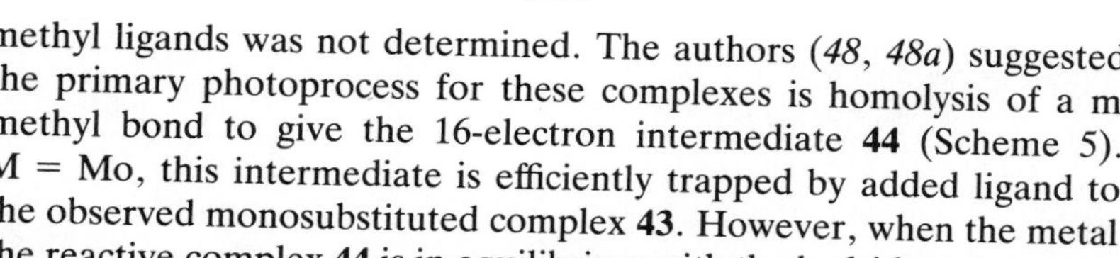

$$
\left[Cp_2W\underset{CH_3}{\overset{CH_3}{\diagup}} \right]^{+} PF_6^{-} \xrightarrow[PMe_2Ph]{h\nu} \left[Cp_2W\underset{PMe_2Ph}{\overset{CH_3}{\diagup}} \right]^{+} PF_6^{-} + \left[Cp_2W\underset{H}{\overset{CH_2PMe_2Ph}{\diagup}} \right]^{+} PF_6^{-} \qquad (60)
$$

(10%)　　　　　　(60%)

(43)

methyl ligands was not determined. The authors (*48, 48a*) suggested that the primary photoprocess for these complexes is homolysis of a metal–methyl bond to give the 16-electron intermediate **44** (Scheme 5). For M = Mo, this intermediate is efficiently trapped by added ligand to give the observed monosubstituted complex **43**. However, when the metal is W the reactive complex **44** is in equilibrium with the hydridocarbene complex **45** via a reversible α-abstraction process. Small and good σ-donor ligands such as PMe_2Ph and MeCN can trap complex **45** by nucleophilic attack at the carbene carbon to give **46** which can be isolated in 60% yield when

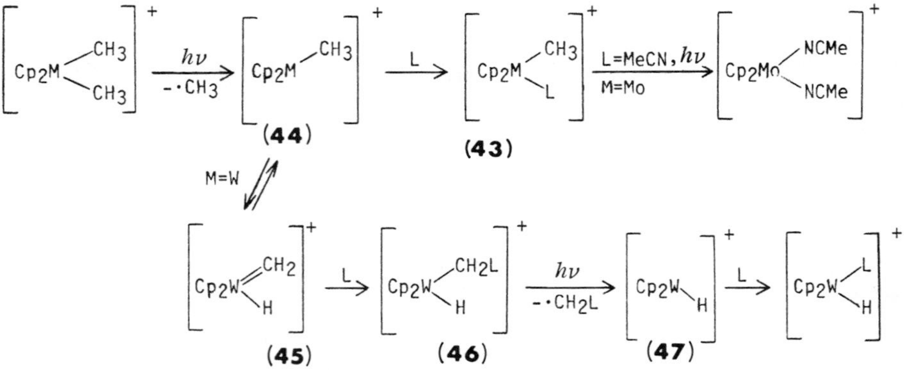

SCHEME 5

$L = PMe_2Ph$. The hydrido complexes $[WCp_2(H)(L)]^+$ ($L = PMe_2Ph$ or MeCN) are presumably formed by thermal or photoinduced cleavage of the substituted alkyl ligands ($-CH_2L$) from species **46** followed by reaction of excess ligand with the coordinatively unsaturated hydride complex **47**.

Green and co-workers (*49, 49a*) studied the photochemistry of several metallacyclobutane derivatives of tungstenocene (**48–51**) and found that

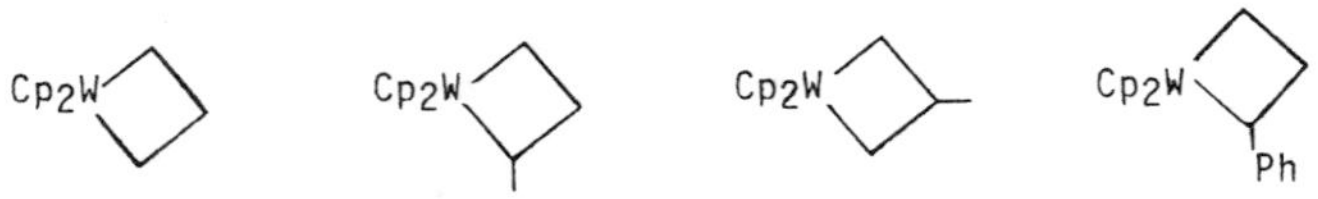

irradiation of these compounds gave a complex mixture of organic products. For example, irradiation of the unsubstituted metallacyclobutane complex **48** gave mostly propene and ethylene but also traces of propane, methane, cyclopropane, and ethane [Eq. (61)]. The nature and

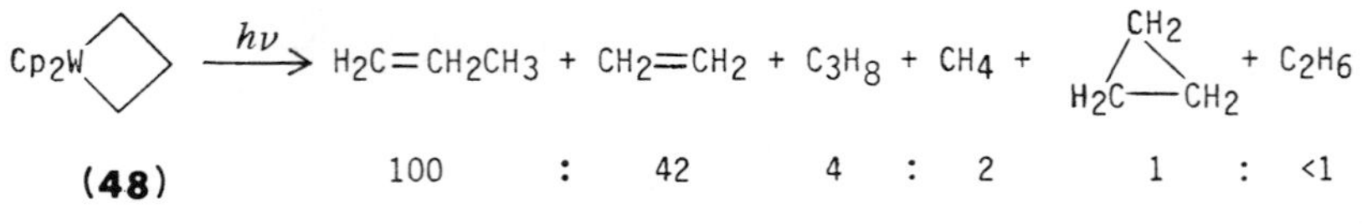

$$
\begin{array}{ccccccc}
100 & : & 42 & 4 & : & 2 & 1 & : & <1
\end{array}
$$

$$(61)$$

relative yields of organic products from complexes **49** and **50** were similar. However, irradiation of complex **51** gave mostly cyclopropylbenzene and propylbenzene along with small amounts of other organic products

[Eq. (62)] (49a). The organometallic products were not identified but the

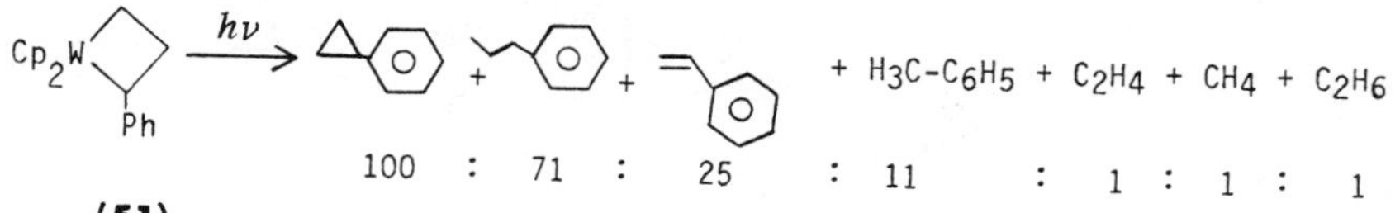

$$Cp_2W \underset{Ph}{\overset{}{\triangleleft}} \quad \xrightarrow{h\nu} \quad \triangle\!\!-\!\!\bigcirc \quad + \quad \sim\!\!\bigcirc \quad + \quad =\!\!\bigcirc \quad + \quad H_3C-C_6H_5 \quad + \quad C_2H_4 \quad + \quad CH_4 \quad + \quad C_2H_6$$

(51)

$$100 \quad : \quad 71 \quad : \quad 25 \quad : \quad 11 \quad : \quad 1 \quad : \quad 1 \quad : \quad 1$$

(62)

overall yields of organic products were low (22–39%), indicating that these photodegradation reactions are inefficient. The nature of the primary photochemical process is not clear. In fact, based on the nature of the organic products, several types of initial photoprocess could occur. Photolysis of each metallacyclobutane complex yielded substantial quantities of olefin which had fewer carbon atoms than were present in the original metallacycle. The authors suggested that these olefins were produced via mechanism involving a photoinduced $\eta^5-\eta^3$ Cp-ring shift in the initial photochemical step to produce the alkene–carbene complex **52** (Scheme 6). Decomposition of this complex would produce olefin and the reactive carbene complex **53**. However, this mechanism cannot account for the major olefins produced upon irradiation of complexes **48–50**, namely propene, *trans*-2-butene and 2-methylpropene, respectively. The formation of these olefins is consistent with a mechanism involving homolysis of

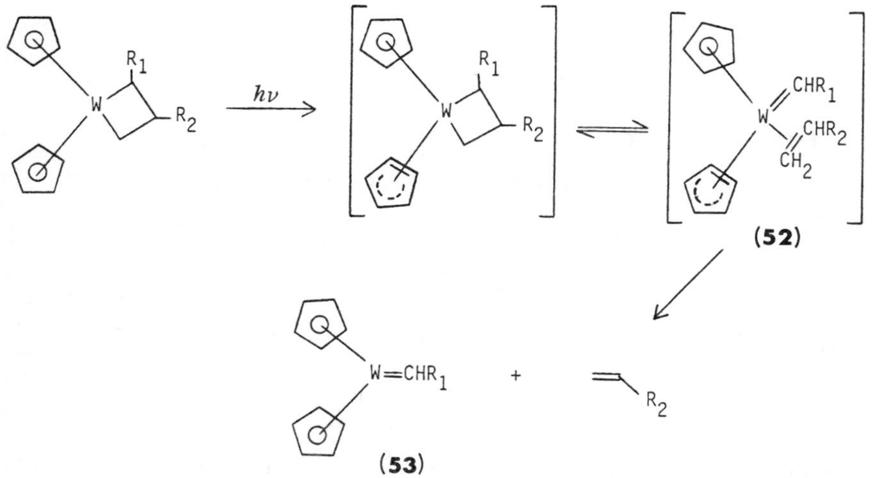

SCHEME 6

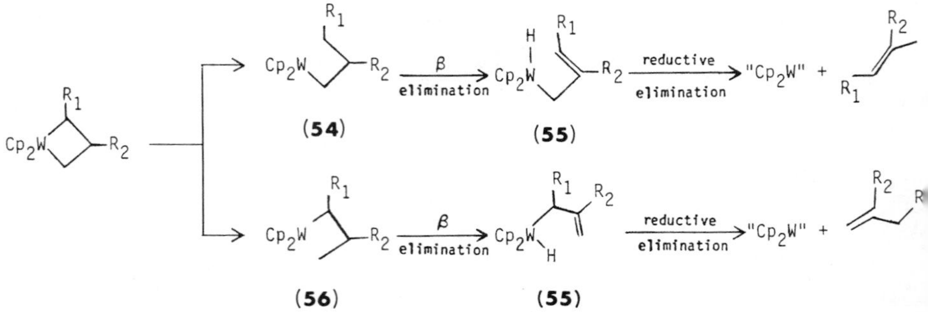

SCHEME 7

one of the metal–carbon bonds (Scheme 7) to give initially species **54** and **56**. Thermal β-hydrogen elimination to give the alkyl–hydride complexes **55** followed by reductive elimination would yield the observed olefins.

The above results suggest that the metallacyclobutane complexes **48–51** have at least two photodecomposition pathways available to them: (1) $\eta^5 - \eta^3$ shift of the Cp rings with homolysis of one of the metal–carbon bonds, and (2) possibly a concerted, photoinduced reductive-elimination reaction for the formation of cyclopropylbenzene from **51** [Eq. (63)].

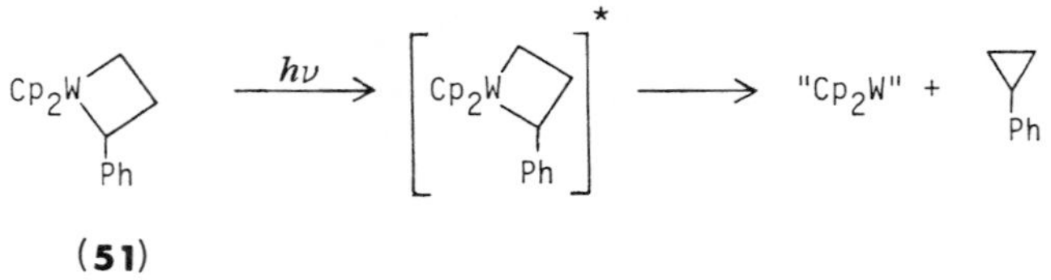

5. CpM(CO)₃R (M = Cr, Mo, or W)

The photochemistry of these complexes has been studied extensively (*50–63a*), and the nature of the primary photoprocess has been a controversial subject. Both carbon monoxide dissociation [Eq. (64)] (*50, 51a,b, 52–56, 59, 60, 61–63a*) and metal–alkyl bond homolysis [Eq. (65)] (*51, 51c, 57, 58, 60a*) have been proposed on the basis of different experimental results. However, more recent work has shown that CO

$$\text{CpM(CO)}_3\text{R} \xrightarrow{h\nu} \text{``CpM(CO)}_2\text{R''} + \text{CO} \tag{64}$$

$$\text{CpM(CO)}_3\text{R} \xrightarrow{h\nu} \text{``CpM(CO)}_3\text{''} + \cdot\text{R} \tag{65}$$

dissociation occurs as a *direct consequence* of photoexcitation and that products which implicate metal–alkyl homolysis form in subsequent thermal reactions of the photogenerated CpMo(CO)₂R intermediate

(52–52c, 56). Furthermore, it has been found that the nature of the subsequent thermal reactions is determined by the presence or absence of β hydrogens on the alkyl ligand and also, but to a lesser extent, by the nature of the metal.

Initial studies had shown that substitution of other ligands for CO could be photoinduced in these complexes. For example, Barnett and Treichel (50) demonstrated that reaction of $CpM(Co)_3CH_3$ (M = Mo or W) with PPh_3 proceeded much faster under UV irradiation [Eq. (66)] than under

$$CpMo(CO)_3CH_3 + PPh_3 \xrightarrow{h\nu} CpMo(CO)_2CH_3(PPh_3) \qquad (66)$$

(57)

(14%)

thermal conditions. Donor ligands which have been photosubstituted for CO include the tertiary phosphines PMe_3 (51b, c), $P(OMe)_3$ (51), and PPh_3 (50–53), various alkynes (54), CO (52–52c, 55), and even N_2 (55). Both mono- (58) and disubstituted (59) derivatives can be obtained, depending upon experimental conditions [Eq. (67)] (51, 51b, 53).

$$CpM(CO)_3R + L \xrightarrow{h\nu} CpM(CO)_2L(R) \xrightarrow[-CO, +L]{} CpM(CO)L_2(R) \qquad (67)$$

(58) (59)

However, these photosubstitution reactions do not appear to be very synthetically useful since the yields of the substituted products 58 and 59 are usually low with a maximum of ~60%, and other products are also formed. This suggested that competing thermal and/or photochemical processes occur. For example, Alt and Schwärzle (51b) showed that irradiation of $CpM(CO)_3CH_3$ (M = Cr, Mo, or W) in the presence of PMe_3 gave not only the mono- and disubstituted derivatives 58 and 59 but also the binuclear, asymmetrically substituted complexes $Cp(CO)(L)_2M—M(CO)_3Cp$ (60 and 61). Note that the CH_3 ligand has

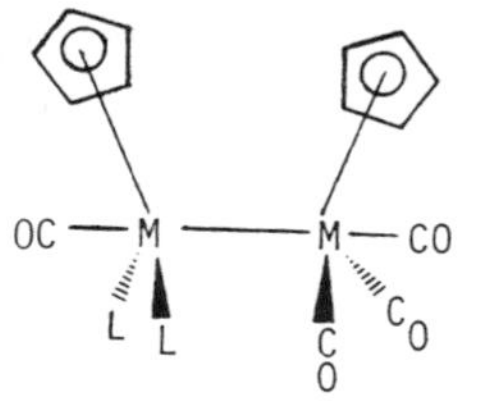

(60) 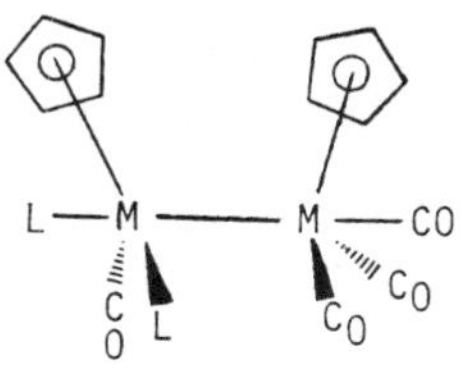   (61)

been lost in forming 60 and 61. The relative quantities of complexes 58–61 depend on the amount of donor ligand present as well as the length of irradiation time, with the dinuclear complexes being obtained in higher yields with longer irradiation. Irradiation of $Cp(CO)_3MoCH_3$ in CH_2Cl_2 solution in the presence of carboxylic acid azides and azidocarboxylic acid esters also led to dimeric, nitrene-bridged complexes $[Cp(O)M]_2(\mu\text{-}O)(\mu\text{-}NCOR)$ and $[Cp(O)M]_2(\mu\text{-}O)$ (μ-NCOOR) although in very low ($\leqslant 2\%$) yields (51d).

Irradiation of $CpM(CO)_3R$ (M = Cr, Mo, or W; R = CH_3 or Et) in the absence of added ligand has been reported to give loss of the R ligand and formation of the binuclear complexes $[CpM(CO)_3]_2$ and $[CpM(CO)_2]_2$ (51, 51c, 53). These results initially suggested (51) photoinduced metal–alkyl bond homolysis in these $CpM(Co)_3R$ complexes [Eq. (65)], with the binuclear organometallic complexes arising from coupling of photo-generated $CpM(CO)_3\cdot$ radicals [Eq. (68)]. Methane was the organic

$$CpM(CO)_3R \xrightarrow{h\nu} R\cdot + CpM(CO)_3\cdot \longrightarrow 1/2Cp_2M_2(CO)_6$$

$$\Big\downarrow {-CO} \tag{68}$$

$$CpM(CO)_2\cdot \longleftarrow 1/2Cp_2M_2(CO)_4$$

product obtained upon irradiation of the corresponding methyl complexes and labeling studies (56) showed that it formed by hydrogen abstraction from the Cp rings.

Tyler and co-workers (56) demonstrated that methane and the binuclear product $Cp_2W_2(CO)_6$ form from $CpW(CO)_3CH_3$ *by a process which involves initial photoinduced CO dissociation* and not W—CH_3 bond homolysis. They showed that irradiation of $CpW(CO)_3CH_3$ in THF solution at $-78°C$ led cleanly to formation of $CpW(CO)_2(thf)CH_3$ (**62**) which was spectroscopically characterized. When these solutions were allowed to warm, this species disappeared as $Cp_2W_2(CO)_6$ formed, and the mechanism of Scheme 8 was proposed. While this demonstrates the formation of $Cp_2W_2(CO)_6$ by a process involving initial CO dissociation, the mechanism by which $CpW(CO)_2(thf)CH_3$ yields $Cp_2W_2(CO)_6$ remains obscure. Labeling studies showed that the solvent was one of the sources of

$$CpW(CO)_3CH_3 \xrightarrow[-CO]{h\nu} CpW(CO)_2CH_3 \xrightarrow{+THF} CpW(CO)_2(THF)CH_3$$

$$\textbf{(62)}$$

$$\Big\downarrow {+CpW(CO)_3CH_3}$$

$$Cp_2W_2(CO)_6 \xleftarrow{+CO} Cp_2W_2(CO)_5 + 2\text{"}CH_3\cdot\text{"}$$

Scheme 8

the hydrogen incorporated into methane; no abstraction of Cp hydrogens occurred within the time scale of the photochemical experiments.

Wrighton and co-workers (52–52c) have also investigated the photochemistry of these complexes (M = Mo or W) in hydrocarbon matrices at 77 K and found no evidence for metal–methyl bond homolysis. Irradiation of $CpM(CO)_3CH_3$ under these conditions led to CO loss and reversible formation of $CpM(CO)_2CH_3$ (63) which was spectroscopically characterized [Eq. (69)]. The coordinatively unsaturated complex 63 was stable

$$CpM(CO)_3CH_3 \overset{h\nu}{\rightleftharpoons} CpM(CO)_2CH_3 + CO \qquad (69)$$

$$(M = Mo \text{ or } W) \qquad (63)$$

only at low temperatures and upon warming the matrix it reacted *quantitatively* with the photoejected CO or with added ligand L to regenerate starting material or form $CpM(CO)_2L(CH_3)$, respectively. Similar results were obtained for photolysis of the surface-confined species $CpW(CO)_3(CH_3)$ (52b). However, unlike with the corresponding complexes in solution, prolonged photolysis did not yield W—W bonded products, presumably because the immobilized centers remained anchored sufficiently far apart that W—W bonds could not form.

Irradiation of $CpCr(CO)_3CH_3$ in CH_4 matrices at 12K has been reported to give the methylene hydride species 64, identified by IR spectroscopy (60). When ^{13}CO-doped CH_4 matrices were used, initial $^{12}C/^{13}C$ exchange occurred followed by formation of both the methylene hydride species 64 and the complexes $CpCr(^{12}CO)_{2-n}(^{13}CO)_n(CH_2)H$. The authors interpreted these results in terms of the mechanism outlined in Scheme 9, which involves reversible photoinduced CO loss followed by reversible α-hydrogen elimination to give the methylene–hydride species 64. No metal–methyl bond homolysis was detected under these experimental conditions. Similarly, irradiation of $CpMo(Co)_3CH_3$ in frozen gas matrices at 12 K led to clean CO dissociation and to the spectroscopic observation of $MoCp(CO)_2CH_3$ (55). In an N_2 matrix, irradiation gave the N_2 complex $MoCp(CO)_2(N_2)(CH_3)$ (Scheme 10). However, no α-hydrogen abstraction process was observed, unlike the Cr analogue discussed above.

In contrast, irradiation of $CpMo(CO)_3CH_3$ in poly(vinyl chloride) (PVC) matrices at 12–293 K led to both CO dissociation and apparent Mo—CH_3 bond homolysis (60a). The evidence for the latter process came from the disappearance of IR bands due to $CpMo(CO)_3CH_3$ and the growth of bands due to $CpMo(CO)_3Cl$ [Eq. (70)]. The latter product was presumed to form by Cl abstraction from the matrix material by photo-

$$CpMo(CO)_3CH_3 \xrightarrow{h\nu, \text{ PVC}} CpMo(CO)_3Cl \qquad (70)$$

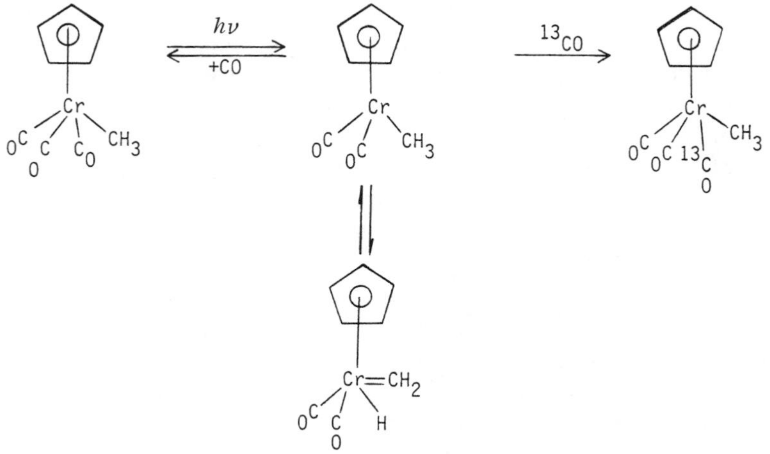

(64)

SCHEME 9

$$CpMo(CO)_3CH_3 \; \underset{h\nu}{\rightleftharpoons} \; CpMo(CO)_2CH_3 + CO \; \xrightarrow{N_2} \; CpMo(CO)_2(N_2)CH_3$$

SCHEME 10

generated $CpMo(CO)_3\cdot$ radicals. The authors suggested competing CO dissociation and Mo—CH_3 bond homolysis processes. Also, they suggested that the reason why the latter process was not detected in inert gas matrices is because the latter matrix molecules can pack tightly around a substrate molecule to form a "tight cage" in which the radicals produced on photolysis are unable to diffuse apart, and hence rapidly recombine. In the PVC matrice, Cl abstraction was proposed to compete effectively with back recombination. It should be noted, however, that the $CpMo(CO)_3Cl$ product could also form by oxidative addition of a C—Cl bond of the PVC matrix to photogenerated $CpMo(CO)_2CH_3$, followed by rapid reductive coupling of the alkyl ligands.

ESR studies on this class of compounds have not been particularly informative. Rausch and Samuel (58) found that irradiation of the complexes $Cp(CO)_3MCH_3$ (M = Cr or Mo) in toluene solution at room temperature gave ESR signals due to paramagnetic species in which the methyl–metal bond was preserved. However, no such signal was detected for the W analogue. Irradiation in the presence of the spin-trapping reagent PBN gave rise to signals which were identified as the methyl radical–PBN spin adduct, but only in the Mo case. In contrast, Lappert *et al.* (*59*) reported that irradiation of $MoCp(CO)_3R$ (R = Me, Et, or CH_2Ph) in the presence of various spin-trapping agents did not generate ·R radicals since they were unable to trap these under their experimental conditions.

The above results suggest that the principal mode of photo-decomposition in $Cp(CO)_3MCH_3$ (M = Cr, Mo or W) complexes is CO dissociation. However, the nature of the subsequent thermal processes depends markedly on the metal and the experimental conditions. It is apparent that metal–methyl bond homolysis does not readily occur from irradiation of the starting alkyls but it is possible that in the presence of suitable radical-trapping agents this process may become significant from the photogenerated 16-electron species **63**, as was suggested for the related $Cp(CO)_2W(\eta^1\text{-}CH_2C_6H_5)$ complex (*53*).

The photochemistry of $Cp(CO)_3MR$ complexes in which R has β hydrogens has been investigated by Wrighton (*52–52c*) and Alt (*61, 61a*) and their co-workers. These studies established that the primary photo-process is CO dissociation to give a 16-electron species analogous to **63**. However, this coordinatively unsaturated species (**65**) undergoes rapid β-hydride elimination to yield a *trans*-alkene–hydride complex as the final product. The overall reaction profile of Scheme 11 was proposed by Wrighton and co-workers to explain their results. Carbon monoxide loss from the excited complex first gave $Cp(CO)_2$ MEt (**65**) which accumulated in a rigid 77 K matrix and was spectroscopically characterized. However, in fluid solution this species equilibrated with another $Cp(CO)_2MEt$ complex (**66**) which is presumed to have a developing metal–hydride interaction. The latter complex undergoes β-hydride elimination to give a cis-hydridoethylene complex that rearranges to the trans product in the overall rate-determining step. The trans product can subsequently lose ethylene to give $CpM(CO)_2H$.

When these $CpM(CO)_3R$ complexes were irradiated at room tempera-ture in the presence of excess phosphine, the monosubstituted complexes **58** were the only products detected [Eq. (71)]. This presumably occurs

$$CpM(CO)_3R + PR_3 \xrightarrow{h\nu} CpM(CO)_2(PR_3)R + CO \qquad (71)$$

$$(58)$$

because the added phosphine efficiently traps all of **65** before it can yield **66**. Deuterium-labeling studies (*52a, 61a*) showed that all of the $CpM(CO)_3H$ observed in the absence of added phosphines formed only by the β-hydrogen elimination process [Eq. (72)]. Prolonged photolysis in the

$$CpW(CO)_3CD_2CH_3 \xrightarrow{h\nu} CpW(CO)_3H \xrightarrow[\text{or } \mathbf{66}]{\mathbf{67}} Cp_2W_2(CO)_6 \qquad (72)$$

$$(67)$$

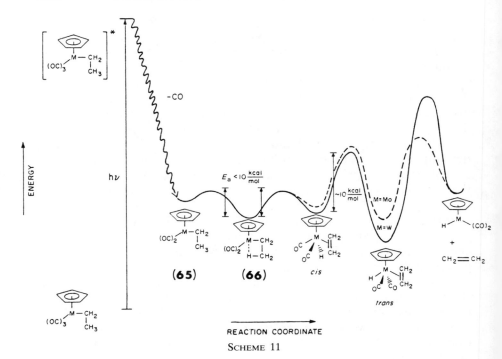

REACTION COORDINATE

Scheme 11

absence of added ligand resulted in the quantitative formation of the dinuclear complexes $Cp_2M(CO)_6$ [Eq. (72)], presumably via secondary photolysis of $CpM(CO)_3H$.

Further evidence against metal–alkyl bond homolysis in the $CpW(CO)_3R$ complexes was provided by Wojcicki's group (62) who showed that irradiation of *threo*-$CpW(CO)_3CHDCHDC_6H_5$ in cyclohexane solution did not change the diastereomeric purity of the recovered starting material [Eq. (73)]. Prolonged irradiation led to the formation of

$$\text{threo-}(\textbf{67}) \quad \xrightarrow[\text{45 min}]{h\nu} \quad \text{threo-}(\textbf{67}) \quad + \quad (\textbf{68}) \tag{73}$$

the η^3-allyl complex **68**, presumably via CO dissociation followed by β-hydrogen elimination. The authors suggested that W—alkyl bond homolysis followed by fast recombination of the radical pair could *not* have occurred since this would have resulted in racemization of the starting alkyl complex [Eq. (74)] by rapid epimerization of the alkyl radical **69**.

threo-(**67**) (**69**) threo-(**67**) erythro-(**67**) (74)

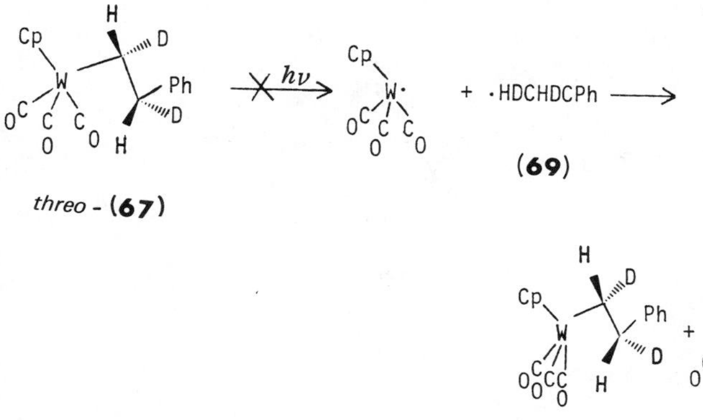

In summary, the photochemistry of these $CpM(CO)_3R$ complexes now appears to involve dissociation of CO in the primary photochemical process. The photogenerated $CpM(CO)_2R$ intermediates can then add a donor ligand L, undergo α- and β-hydrogen elimination reactions depending on the nature of R, or can decay to yield binuclear $Cp_2M_2(CO)_x$ ($x = 4$ or 6) products. The major remaining uncertainty concerns details of the $CpM(CO)_2R \rightarrow Cp_2M_2(CO)_x$ conversion since it is not at all clear how this occurs.

6. *WCp₂(CH₃)(H)*

Photolysis of the title complex in argon and N_2 matrices at 10 K led to tungstenocene and methane [Eq. (75)]. No intermediates were detected

$$Cp_2W\begin{matrix}CH_3\\H\end{matrix} \xrightarrow[10\ K]{h\nu} Cp_2W + CH_4 \qquad (75)$$

even in the N_2 matrix in which any methyl radicals produced would have yielded a sharp IR signal at 611 cm^{-1}. The authors suggested that methane was formed via concerted "reductive-elimination" directly from the starting alkyl–hydride complex.

7. *WMe₆(PMe₃)*

Wilkinson *et al.* (*64*) found that irradiation of this complex in the presence of PMe_3 at $-20°C$ gave CH_4 and the tungsten–carbyne complex

70 in high yield [Eq. (76)]. The authors suggested that complex **70** formed by successive α-hydrogen abstraction reactions from a coordinated methyl ligand (Scheme 12).

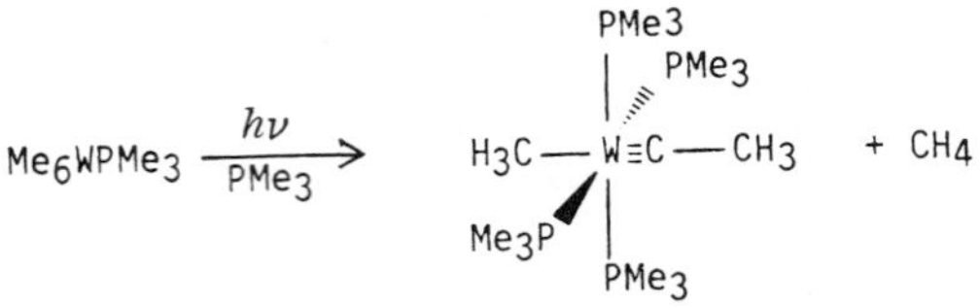

$$Me_6WPMe_3 \xrightarrow[PMe_3]{h\nu} \quad \underset{(\mathbf{70})}{\overset{\displaystyle \begin{array}{c} PMe_3 \\ | \quad PMe_3 \\ H_3C-W\equiv C-CH_3 \\ Me_3P \quad | \\ PMe_3 \end{array}}{}} \quad + \quad CH_4 \tag{76}$$

E. Mn and Re Complexes

1. $Mn(CO)_5R$ ($R = CH_3$, $CH_2C_6H_5$, or CH_2SiMe_3)

The photochemistry of these complexes has been studied in solution (65–68) and in solid gas matrices (69, 70), and both CO dissociation (66–70) and Mn—R bond homolysis (65) have been proposed to be primary photoprocesses. The evidence for Mn—R bond homolysis [Eq. (77)], was provided by Lappert and co-workers (59, 65) who showed

$$Mn(CO)_5R \xrightarrow[CHCl_3]{h\nu} R\cdot + \cdot Mn(CO)_5 \tag{77}$$

that irradiation of the complexes $Mn(CO)_5R$ ($R = CH_3$, CH_2SiMe_3, or CH_2Ph) in $CHCl_3$ solution and in the presence of the spin-trapping agent

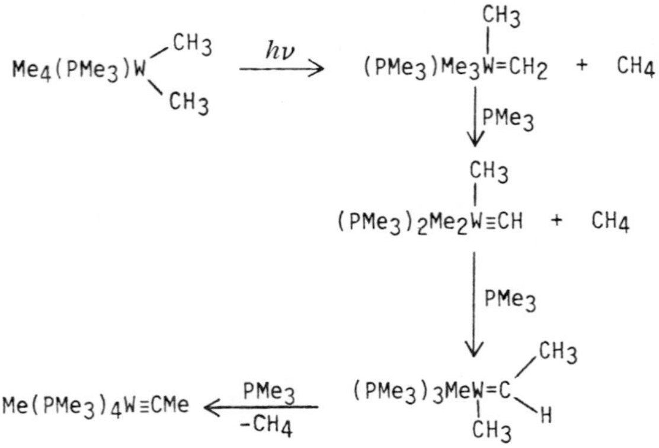

$$Me_4(PMe_3)W\overset{\displaystyle CH_3}{\underset{\displaystyle CH_3}{\big\backslash}} \xrightarrow{h\nu} (PMe_3)Me_3\overset{\displaystyle CH_3}{\overset{|}{W}}=CH_2 \quad + \quad CH_4$$

$$\downarrow PMe_3$$

$$(PMe_3)_2Me_2\overset{\displaystyle CH_3}{\overset{|}{W}}\equiv CH \quad + \quad CH_4$$

$$\downarrow PMe_3$$

$$Me(PMe_3)_4W\equiv CMe \xleftarrow[-CH_4]{PMe_3} (PMe_3)_3Me\overset{\displaystyle CH_3}{W}=C\overset{\displaystyle \diagup}{\underset{\displaystyle CH_3}{\diagdown}}H$$

$$(\mathbf{70})$$

SCHEME 12

nitrosodurene gave rise to ESR signals attributed to radical spin adducts **71** and **72**.

$$Ar - N - Mn(CO)_5 \qquad\qquad Ar - N - R$$
$$| \qquad\qquad\qquad\qquad |$$
$$\cdot O \qquad\qquad\qquad\qquad \cdot O$$
$$\textbf{(71)} \qquad\qquad\qquad \textbf{(72)}$$

Bamford and Mullik (*66*) showed that irradiation of $Mn(CO)_5CH_3$ in methyl methacrylate brought about free radical-induced polymerization of the monomer. The ketone methyl 2-methyl-4-oxopentanoate (**73**) was also a major product of this reaction [Eq. (78)]. Polymerization was strongly inhibited by CO which led the authors to suggest that the primary photochemical step was CO dissociation to give the coordinatively unsaturated complex **74**, [Eq. (79)]. This, they suggested, decomposed *thermally*

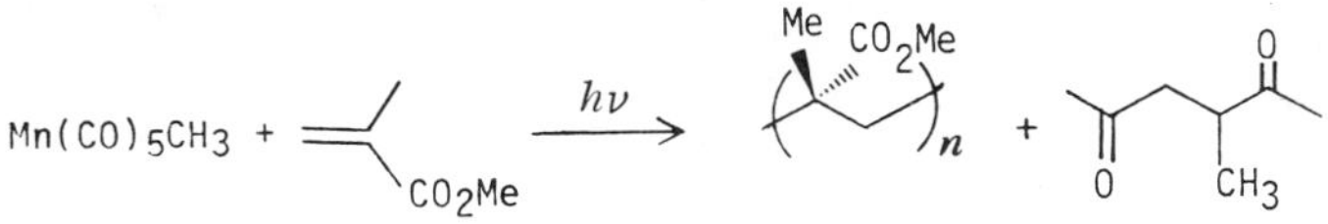

$$Mn(CO)_5CH_3 + \text{[diene with } CO_2Me\text{]} \xrightarrow{h\nu} \text{[polymer]}_n + \text{[ketone]} \qquad (78)$$

(73)

$$Mn(CO)_5CH_3 \xrightarrow[-CO]{h\nu} Mn(CO)_4CH_3 \rightarrow \cdot Mn(CO)_4 + \cdot CH_3 \qquad (79)$$
$$\textbf{(74)}$$

under their conditions to generate methyl radicals. Ketone **73** was proposed to form via photolysis of intermediate **75** generated by CO insertion into the $Mn-CH_3$ bond of the photoexcited complex **76** [Eq. (80)].

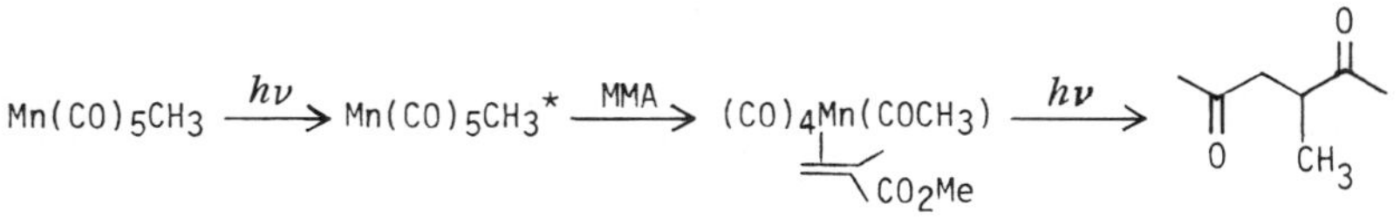

$$Mn(CO)_5CH_3 \xrightarrow{h\nu} Mn(CO)_5CH_3^* \xrightarrow{MMA} (CO)_4Mn(COCH_3) \xrightarrow{h\nu} \text{[ketone]} \qquad (80)$$

(76)   **(75)**   **(73)**

However, complex **75** would probably be present in very low concentration at any given time, and it is not clear how the relatively high yield of **73** could be produced by this process.

Irradiation of the related complexes $Mn(CO)_5R$ ($R = Ph$ or CH_2Ph) in the presence of 1,3-dienes also led to CO dissociation and formation of η^3-allyl complexes such as **78** [Eq. (81)] (*67*). When $Mn(CO)_5(CH_2Ph)$ was

$$(CO)_5MnR + \text{[diene]} \xrightarrow[\underline{n}\text{-hexane}]{h\nu,\ 253K} (CO)_4Mn-\text{[allyl]} + CO \qquad (81)$$

(78)

irradiated under these conditions in the presence of 1,3-pentadienes the cis-alkyl–diene complexes **79** were spectroscopically identified (Scheme 13). The diene ligand in **79** was readily displaced by PMe_3 or $AsMe_3$ to give the cis-monosubstituted complexes **80** in low yield. When a solution of **79** was heated to 342 K, the η^3-allyl complex **78** was formed, presumably via insertion of the coordinated diene into the Mn—alkyl bond of **79**.

These various results appear to be most consistent with a primary photochemical step involving CO loss from the starting alkyl complex, a process which was confirmed by low-temperature matrix isolation studies (*69, 70*). In a brief communication, Ogilvie (*69*) reported that irradiation of $Mn(CO)_5CH_3$ in an argon matrix at 17 K gave IR signals assigned to free CO and to various carbonyl compounds derived from the parent materials after loss of carbon monoxide. Prolonged photolysis led to the formation of acetyl complexes of manganese, indicated by CO-stretching frequencies at 1767 and 1763 cm^{-1}, but the nature of these species was not determined.

On the other hand, Rest and co-workers (*70*) studied the photochemistry of $Mn(CO)_5CH_3$ in frozen gas matrices at 12 K and found that acetyl complexes were *not* produced under their conditions. Instead, photolysis of $Mn(CO)_5CH_3$ with 222 and 274 nm light cleanly gave the coordinatively unsaturated species **81** [Eq. (82)]. Annealing the matrix regenerated the starting complex. The 16-electron species **81** was shown to have a fluxional trigonal-bipyramidal structure with the methyl group in an equatorial position. Irradiation with higher energy light produced additional IR bands attributed to secondary photolysis products such as $Mn(CO)_nCH_3$ ($n = 1-3$). No evidence for the formation of radicals was obtained under

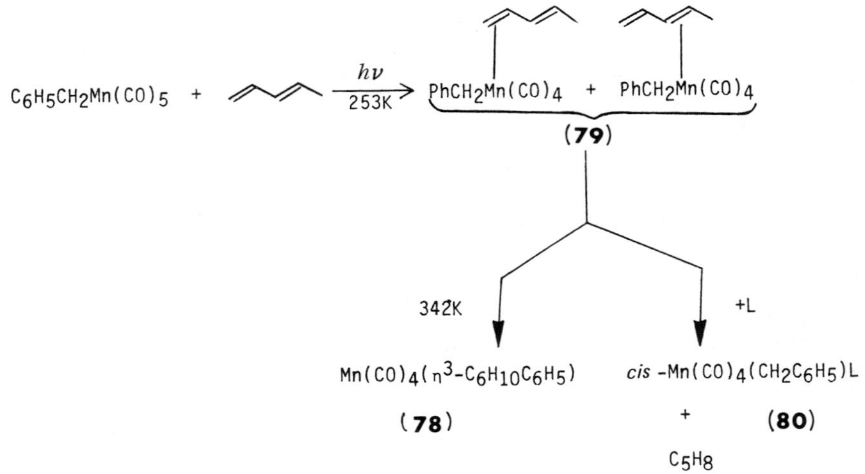

$$C_6H_5CH_2Mn(CO)_5 \; + \; \underset{}{\diagup\!\!\diagdown\!\!\diagup\!\!\diagdown} \; \xrightarrow[253K]{h\nu} \; PhCH_2Mn(CO)_4 \; + \; PhCH_2Mn(CO)_4$$

$$(79)$$

$$Mn(CO)_4(\eta^3\text{-}C_6H_{10}C_6H_5) \qquad\qquad cis\text{-}Mn(CO)_4(CH_2C_6H_5)L$$

$$342K \qquad\qquad\qquad\qquad +L$$

$$(78) \qquad\qquad\qquad\qquad\qquad + \qquad (80)$$

$$C_5H_8$$

SCHEME 13

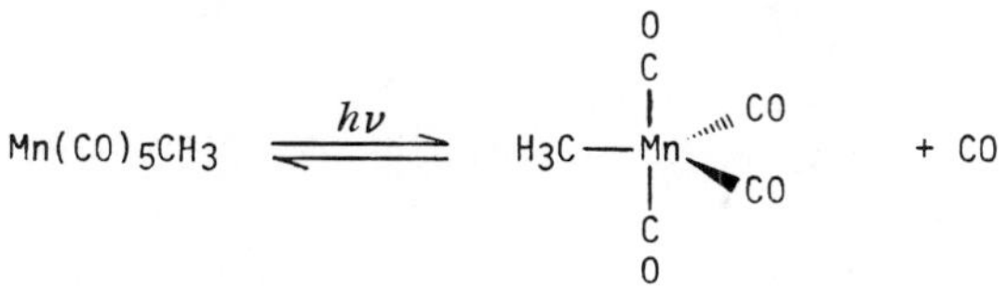

$$\text{Mn(CO)}_5\text{CH}_3 \xrightleftharpoons{h\nu} \begin{array}{c} \text{O} \\ \text{C} \\ | \\ \text{H}_3\text{C}-\text{Mn}\overset{\ldots\ldots\text{CO}}{\underset{\overset{|}{\text{C}}}{\diagdown}}\text{CO} \\ \text{O} \end{array} + \text{CO} \qquad (82)$$

(81)

these conditions (*69, 70*) and we must conclude that Mn—R bond homolysis is not a major primary photochemical decay path for these complexes.

2. $Mn(CO)_5(C_6H_5)$

Kreiter and co-workers (*68*) studied the photochemistry of this phenyl complex in the presence of fulvenes and found that labile C_6H_5Mn-$(CO)_4(\eta^2$-fulvene) complexes were formed [Eq. (83)]. These thermally

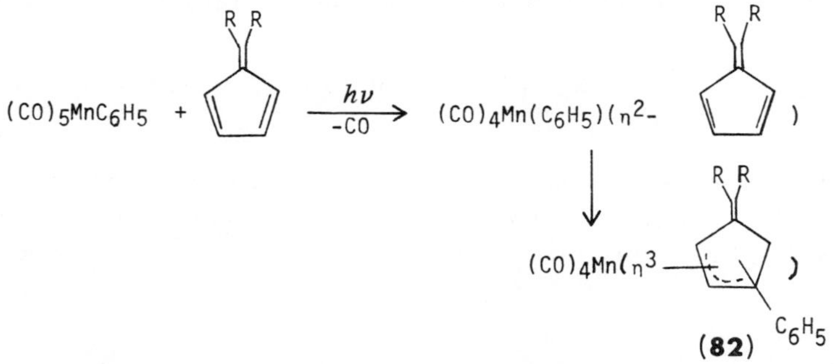

$$\text{(CO)}_5\text{MnC}_6\text{H}_5 + \underset{R\ R}{\overset{}{\bigtriangleup}} \xrightarrow[-\text{CO}]{h\nu} \text{(CO)}_4\text{Mn(C}_6\text{H}_5)(\eta^2- \underset{R\ R}{\overset{}{\bigtriangleup}})$$

$$\downarrow \qquad (83)$$

$$\text{(CO)}_4\text{Mn(}\eta^3- \underset{R\ R}{\overset{}{\bigtriangleup}}_{C_6H_5})$$

(82)

rearrange upon workup to give the stable η^3-fulvene complexes **82**, presumably via insertion of the double bond of the coordinated fulvene into the Mn—aryl bond. These results are consistent with a primary process involving CO loss from the starting Mn–aryl complex.

3. $Re(CO)_5(\eta^1\text{-}C_3H_5)$

Irradiation of the title complex in pentane solution was reported (*71*) to lead to decarbonylation and formation of the η^3-alkyl complex $Re(CO)_4$ $(\eta^3\text{-}C_3H_5)$ in 55% yield. No other details were provided.

4. $Re(NCMe_3)_2(R)_3$ ($R = CH_2SiMe_3$, CH_2Ph, or CH_3)

Schrock and co-workers (*72*) have noted that these complexes are photosensitive and that irradiation of the trimethylsilyl complex in pentane solution gave the alkylcarbene complex $Re(NCMe_3)_2(CHSiMe_3)$ (CH_2SiMe_3) in quantitative yield. Irradiation of the other two complexes gave intractable products but the species $Re(NCMe_3)(CHPh)(CH_2Ph)$ was identified spectroscopically. No mechanistic details were given but the authors suggested that the products arose by photoinduced α elimination of alkane from the starting complexes.

F. Fe, Ru, and Os Complexes

1. [Fe(i-C₃H₇)₃]

In a series of reports, Fischer and Müller (71–76) described a method for the synthesis of olefin and arene complexes through photolysis of $[Fe(i-C_3H_7)_3]$. No information was given concerning the photochemical steps, but the overall reaction appears to be replacement of alkyl radicals by olefin or aromatic molecules [Eq. (84)]. The equimolar product mixture of

$$[Fe(i\text{-}C_3H_7)_3] + 2 \;\bigcirc\; \xrightarrow{h\nu} [Fe(C_6H_6)(C_6H_8)] + \tfrac{3}{2} C_3H_8 + \tfrac{3}{2} C_3H_6 + H_2 \quad (84)$$

propane and propene suggests that the reaction may proceed through photoinduced β-hydride elimination to produce a metal hydride intermediate such as **83**. The latter could then collapse to give propane and propene.

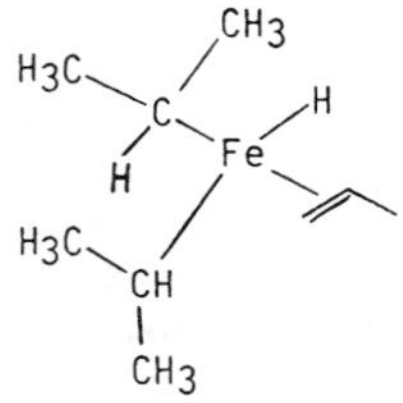

(**83**)

2. [Fe(C₁₆H₂₁N₂O₈)(CO)₄]

Irradiation of **84** in the presence of substrate leads to a variety of products, all apparently arising through initial photoelimination of CO (77). With PPh_3, $[FeR(CO)_3PPh_3]$ results, [Eq. (85)], but irradiation of **84** in the presence of diphenylacetylene and 2,3-dimethylbutadiene resulted in incorporation of the acetylene or diene into the organic portion of the molecule.

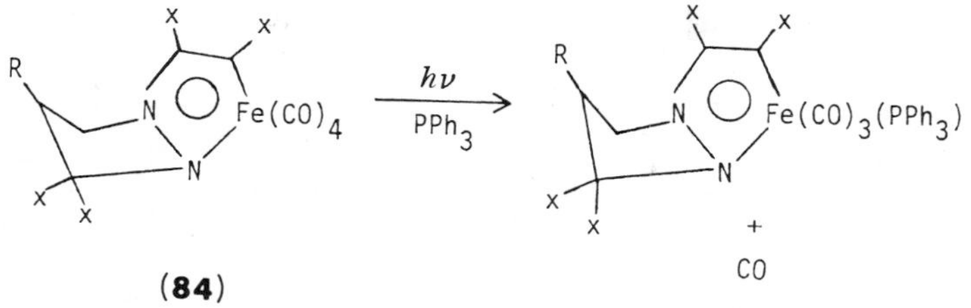

$$(85)$$

(84)

3. CpM(CO)₂R and CpM(PR₃)(CO)R (M = Fe, Ru, or Os)

The photochemistry of the iron complexes in this family has been extensively investigated and four different processes have been proposed by different authors for the primary photochemistry step. These are CO dissociation [Eq. (86)], $\eta^5 - \eta^3$ shift of the Cp ring [Eq. (87)], CO insertion into the metal–alkyl bond [Eq. (88)], and metal–alkyl bond homolysis [Eq. (89)].

$$\text{CpFe(CO)}_2\text{R} \xrightarrow{h\nu} \text{CpM(CO)R} + \text{CO} \tag{86}$$

(85)

$$\text{CpFe(CO)}_2\text{R} \xrightarrow{h\nu} \text{[cyclopentadienyl]} \text{—Fe(CO)}_2\text{R} \tag{87}$$

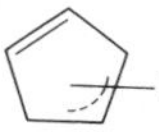

(86)

$$\text{CpFe(CO)}_2\text{R} \xrightarrow{h\nu} \text{CpFe(CO)[C(O)R]} \tag{88}$$

(87)

$$\text{CpFe(CO)}_2\text{R} \xrightarrow{h\nu} \text{CpF·e(CO)}_2 + \cdot\text{R} \tag{89}$$

(88)

However, recent results have shown that CO dissociation is the primary photoprocess under most experimental conditions. Comparatively few studies on the analogous Ru (77a) and Os (78) complexes have appeared, but their photochemistry seems to parallel that of the Fe complexes.

In fact, photoinduced elimination of CO from CpM(CO)₂R complexes has been used extensively to synthesize mono- and disubstituted derivatives with a number of different entering ligands [Eq. (90)]. The nature and

$$\text{CpM(CO)}_2\text{R} + \text{L} \xrightarrow[-\text{CO}]{h\nu} \text{CpM(CO)(L)R} \xrightarrow[-\text{CO}]{h\nu, \text{L}} \text{CpML}_2\text{R} \tag{90}$$

(89)   **(90)**   **(91)**

relative yields of the products obtained from Eq. (90) depend on the nature of the alkyl ligand in the starting complex **89** and on the reaction conditions. High yields of the monosubstituted derivative **90** have resulted when the irradiations were carried out at low temperature in the presence of 1 equivalent of entering ligand and when R = CH₃ in **89** (Table III) (*78a–88*). Irradiation at higher temperature gave significant quantities of the acetyl complexes CpFe(CO)L[C(O)CH₃], with these presumably formed via competing thermal (*81*) or photochemical (*79*) CO insertion into the metal–methyl bond. At short irradiation times, however, the major product has been the monosubstituted derivatives **90** (*77a, 80, 81, 83, 84*). Longer irradiation times, more basic entering ligands, and higher concentrations of the latter gave higher yields of the disubstituted derivatives CpFeL₂R (*82, 83, 86*). Thermal reactions of the coordinated alkyl ligand with the entering substrate following the photosubstitution of CO have been reported (*89, 90*). These involved the insertion of a coordinated acetylene into the Fe—R bond (R = Me, CH₂Ph, or C₃H₅) (*90*) and the nucleophilic displacement of the —OMe group from the alkoxy ligand of Cp(CO)(PPh₂NHCH₂R)Fe—CHPh(OMe) by the nitrogen atom of the coordinated phosphine (*89*).

Good yields of monosubstituted derivatives **90** can also be obtained from complexes which have alkyl ligands with β hydrogens, provided that the irradiation times are short (*77a, 86, 87*). However, at low concentrations of entering ligand, the products CpM(CO)LH and [CpM(CO)₂]₂ are also formed from such complexes. These arise from competitive β-hydride elimination from the photogenerated 16-electron complex **93** (Scheme 14) (*77a*). As with the isoelectronic CpW(CO)₃R complexes (*52–52c*), irradiation of the starting alkyl complex **92** in inert matrices at low temperature was shown by Kazlauskas and Wrighton (*77a*) to give the 16-electron species **93** which was spectroscopically observed only for M = Ru (*77a*).

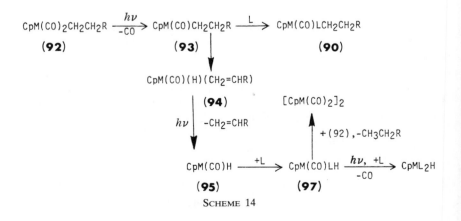

Scheme 14

TABLE III

Photolysis Products of $CpM(CO)_2R$ Complexes in the Presence of Added Ligand L

Complex		Added ligand		Irradiation Temperature (°C)	Irradiation time	Solvent	Major product yield (%)	By-products and yield (%)	References
M	R	L	Excess						
Fe	CH_3	(S)-PPh_2NRR'	—	-70		Toluene	$CpM(CO)LR_3$ (72–90%)	—	78a
Fe	CH_3	$E(Ph)_3$	4-fold	20	30 minutes	CH_3CN	$CpM(CO)L(COCH_3)$ (42%)	—	79
Fe	CH_3	$E(Ph)_3$	1.5- to 8-fold	27–30	15 minutes	Various	$CpM(CO)LR$ (NMR)	—	80
Fe	CH_3	PPh_3	1.3	90–100	6 hour	Pet. ether[a]	$CpM(CO)LR$ (67%)	$CpM(CO)L(COR)$ (16%)	81
Fe	CH_3	PPh_3	1.3	90–100	12 hour	Pet. ether	$CpM(CO)LR$ (63%)	$CpM(CO)L(COR)$ (31%)	81
Fe	CH_3	PPh_3	1.3	90–100	20 hour	Pet. ether	$CpM(CO)L(COR)$ (66%)	$CpM(CO)LR$ (33%)	81
Fe	CH_3	PMe_3	1.6	R.T.[a]	6.5 hour	Pet. ether	$CpML_2R$ (81%)	—	82
Fe	CH_3	PMe_3	3.3	R.T.	60 minutes	Pentane	$CpM(CO)LR$ (49%)	—	82
Fe	CH_3	$P(OMe)_3$	3.3	R.T.	20 minutes	Pentane	$CpM(CO)LR$ (38%)	$CpML_2R$ (33%)	83
Fe	CH_3	PPh_3	5, 50	25		Isooctane	$CpM(CO)LR$	—	77a
Fe	CH_3	P(O-o-tolyl)	5, 50	25		Isooctane	$CpM(CO)LR$	—	77a
Fe	CH_3	$P(OCH_2)_3R$	5, 50	25		Isooctane	$CpM(CO)LR$	—	77a
Fe	CH_3	^{13}CO	10	25		Isooctane	$CpM(CO)LR$	—	77a
Ru	CH_3	$P(OR')_3$	5–50	25		Isooctane	$CpM(CO)LR$	—	77a
Fe	$CH_2RC{=}CRH$	^{13}CO	100	-285	60 minutes	Ar	$CpM(CO)LR$	—	84
Fe	CH_3	$P(OR)_3$	Equimolar	—	15–30 minutes	Pet. ether	$CpM(CO)LR$ (70–77%)	$Cp_2M_2(CO)_3L$ (0–29%)	85
Fe	CH_3	$P(OPh)_3$	2	30–40	60 hour	THF	$[CpML_2]_2$ (66%)	$[CpM(CO)_2]_2$ (2%)	86
Fe	C_2H_5	PPh_3	Equimolar	30–35	3 hour	Pet. ether	$CpM(CO)LR$ (40%)	$CpM(CO)LCl$ (7%)	86
Fe	C_2H_5	PPh_3	500	25	10 µsec	Isooctane	$CpM(CO)LR$	—	77a
Fe	C_2H_5	PPh_3	5.5	-25	25 minutes	Isooctane	$CpM(CO)LR$ (40.1%)	$CpM(CO)(H)$ (59.9%)	77a
Fe	C_2H_5	PMe_3	4	-30		None	$CpM(CO)LR$ (37%)	$CpM(CO)LH$ (20%)	89
Fe	C_2H_5	PMe_3	7	-30	10 minutes	Pentane	$CpML_2R$ (87%)	$CpML_2R$	87
Fe	C_2H_5	$P(OMe)_3$	3.4	-30	20 minutes	Pentane	$CpML_2R$ (22%)	$CpM(CO)LR$	87
Fe	C_2H_5	$P(OMe)_3$	5	-30	5 hour	Pentane	$CpML_2H$ (70%)	$CpM(CO)LH$	87
Fe	$CH_2CH(CH_3)(Ph)$	PPh_3	Equimolar	25	1 hour	Benzene	$CpM(CO)LR$ (53.3%)	$[CpM(CO)_2]_2$	88

[a] R.T., room temperature; pet. ether, petroleum ether.

By warming the matrices containing high concentrations of entering ligand to 298 K, this intermediate could be efficiently trapped to give the monosubstituted derivative **90** with the less basic phosphine PPh$_3$, and the disubstituted derivative **91** in high yield with PMe$_3$ as the entering ligand (*87*). The hydride–olefin intermediate **94** was observed spectroscopically (*87*) at low temperature but could not be isolated due to its thermal and photochemical instability. Continued irradiation at 77 K resulted in the formation of CpRu(CO)$_2$H (**97**, L = CO) which decomposed to the dimer **96** upon prolonged photolysis. No evidence for M—R bond homolysis, CO insertion into the M—R bond, or Cp-ring slippage was found under these conditions for either the Fe or Ru complexes. When the starting alkyl ligand had a coordinating atom such as nitrogen in the β position as in the complex Cp(CO)$_2$Fe(η^1-CH$_2$NMe$_2$), the open coordination site generated by photoinduced CO loss could be trapped efficiently to give Cp(CO)Fe-(η^2-CH$_2$NMe$_2$) in 63% yield (*91*).

The above results are consistent with those of Gerhartz *et al.* (*78*) who showed that irradiation of complex **98** in a low-temperature argon matrix cleanly gave the hydride complex **101**, presumably via the sequence of steps shown in Eq. (91). Again, the 16-electron intermediate **99** was not

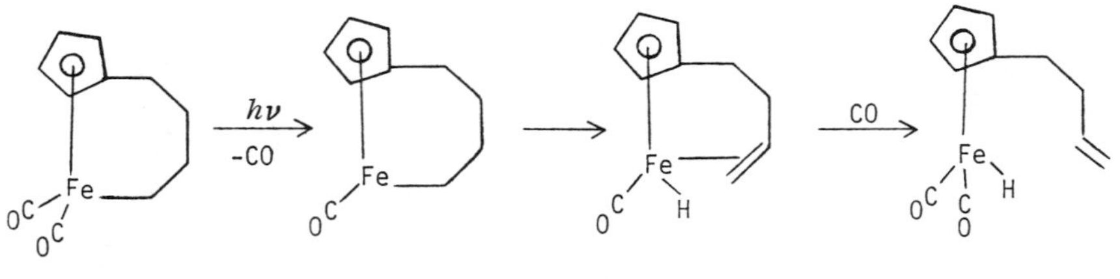

| (98) | (99) | (100) | (101) |

$$(91)$$

observed nor was the olefin–hydride complex **100**. This suggests (1) that β-hydride elimination from **99** was very rapid and (2) that the photoejected CO did not diffuse away from **94** in the matrix; it rapidly recombined with the olefin–hydride complex **100** to give the observed hydride **101**.

Similar results were obtained using the hydroxymethyl complex **102** (*92*). Irradiation of this complex in THF-d_8 or cyclohexane-d_{12} solution gave the hydride complex **103** in essentially quantitative yield [Eq. (92)].

$$\text{Cp*Os(CO)}_2\text{CH}_2\text{OH} \xrightarrow{h\nu} \text{Cp*Os(CO)}_2\text{H} + \text{H}_2 + \text{CO} \qquad (92)$$

| (102) | (103) |

Deuterium labeling studies showed that the hydride originated from the methylene carbon and the authors suggested the mechanism of Scheme 150 for its formation.

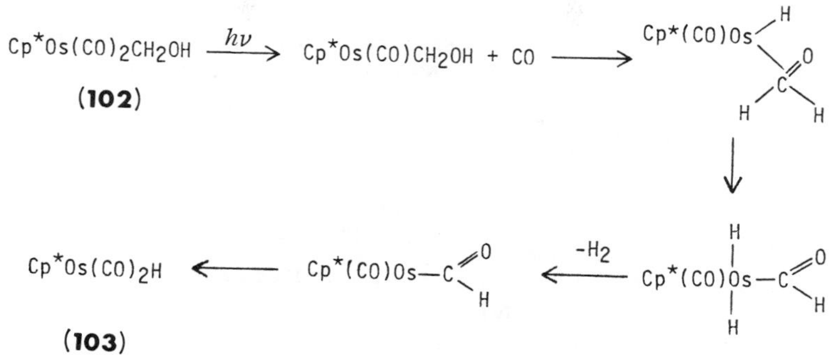

$$Cp^*Os(CO)_2CH_2OH \xrightarrow{h\nu} Cp^*Os(CO)CH_2OH + CO \longrightarrow$$

(102)

(Scheme structure: $Cp^*(CO)Os$ bonded to H and to a C with O, H, H)

$$Cp^*Os(CO)_2H \longleftarrow Cp^*(CO)Os-C\overset{\diagup O}{\diagdown H} \xleftarrow{-H_2} Cp^*(CO)Os-C\overset{\diagup O}{\diagdown H}$$

(103)

SCHEME 15

Also consistent with the transformations of Scheme 14 are the results by Knox and co-workers (*93a*) who irradiated the complexes [Cp(CO)$_2$M{μ-(CH$_2$)$_3$}M(CO)$_2$Cp] (M = Fe or Ru). The only organometallic products isolated after prolonged irradiation were the [CpM(CO)$_2$]$_2$ dimers, except in the diruthenium case where additional products resulting from the photolysis of [CpRu(CO)$_2$]$_2$ were obtained (*93a*). The major organic product was propene, with a trace of cyclopropane, when M was Fe [Eq. (93)]. The authors interpreted these results in terms of a mechanism

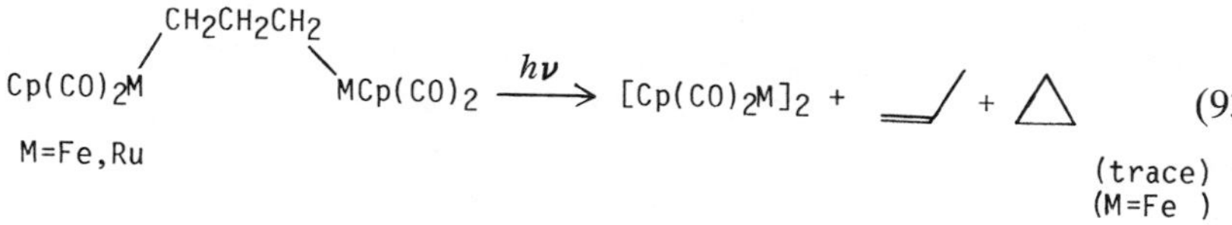

$$Cp(CO)_2M\overset{\diagup CH_2CH_2CH_2}{\diagdown}MCp(CO)_2 \xrightarrow{h\nu} [Cp(CO)_2M]_2 + \diagup\!\!\!\!\diagdown + \triangle \qquad (93)$$

M=Fe,Ru

(trace)
(M=Fe)

involving initial photoinduced CO loss, β-hydride elimination of the alkyl at one metal center, followed by reductive elimination of the olefin at the other metal center. Indirect evidence for the metallacyclic intermediate **104** was provided.

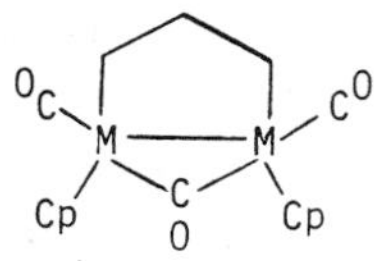

(Structure 104: metallacyclic dimer with two M centers, each bearing CO and Cp, bridged by a C=O and a (CH$_2$)$_3$ chain)

(104)

However, Rest and co-workers (*84*) showed that irradiation of CpFe(CO)$_2$CH$_3$ at high dilution in CO matrices at 12 K gave rise to infrared signals which they attributed to the new species (η^3-C$_5$H$_5$)Fe(CO)$_3$CH$_3$. No evidence for the species CpFe(CO)CH$_3$ was found. The authors concluded that irradiation of the starting alkyl complex in the presence of high concentrations of CO causes the ring-slippage reaction of Eq. (87) to be observed. Presumably, this is because the excess CO causes efficient recombination with photogenerated $CpFe(CO)CH_3$ to suppress this path and allow observation of the ring-slippage reaction, which is probably of minor importance in the overall photochemistry of this class of compounds.

It is now clear that photoinduced metal–alkyl bond homolysis does not readily occur in $CpM(CO)_2R$ complexes when $R = CH_3$, C_2H_5, or C_2H_4R. The only direct evidence for the formation of free radicals upon photolysis of this class of complexes is an ESR study of $CpFe(CO)_2CH_3$ (59) in which only very weak signals attributable to the methylnitrosodurene spin adduct were observed upon photolysis in the presence of the spin trap. On the other hand, irradiation of the related $CpFe(CO)_2CH_2Ph$ and $CpFe(CO)_2CH_2SiMe_3$ complexes gave rise to strong ESR signals assignable to the adducts derived from both $R\cdot$ and $CpFe(CO)_2\cdot$ radicals. However, Kupletskaya and co-workers (94) showed that irradiation of $CpFe(CO)_2$(1-naphthylmethyl) led to CO loss followed by a $\sigma-\pi$ rearrangement to give the η^3-naphthylmethyl complex. This suggests that metal–alkyl bond homolysis in $CpFe(CO)_2R$ complexes is significant only if R forms a relatively stable radical such as benzyl or trimethylsilyl and that CO dissociation is either the initial or the predominant photoreaction.

Consistent with this are the results of Nelson and co-workers (95) who recently showed that irradiation of $CpFe(CO)_2CH_2Ph$ (105) in the presence of 1 equivalent of PPh_3 gave the monosubstituted derivative 106 in addition to $[CpFe(CO)_2]_2$ and a trace of $Cp_2Fe_2(CO)_3PPh_3$ [Eq. (94)]. The

$$PPh_3 + CpFe(CO)_2CH_2Ph \xrightarrow[C_6H_6]{h\nu}$$
$$(105)$$
$$CpFe(CO)(PPh_3)CH_2Ph + [CpFe(CO)_2]_2 + Cp_2Fe_2(CO)_3(PPh_3) \qquad (94)$$
$$(106)$$

formation of 106 suggests CO loss upon photolysis of the starting alkyl complex. In the absence of added ligand, however, only $[CpFe(CO)_2]_2$ and bibenzyl were recovered from the reaction mixtures (94, 96). Furthermore, irradiation of $CpFe(CO)_2CH_2Ph$ in the presence of the hydrogen atom donor 9,10-dihydroanthracene gave mostly toluene [Eq. (95)], which could

$$\qquad\qquad\qquad\qquad\qquad\qquad\qquad\qquad\qquad\qquad (95)$$

CH3Ph + PhCH2CH2Ph

80%   20%

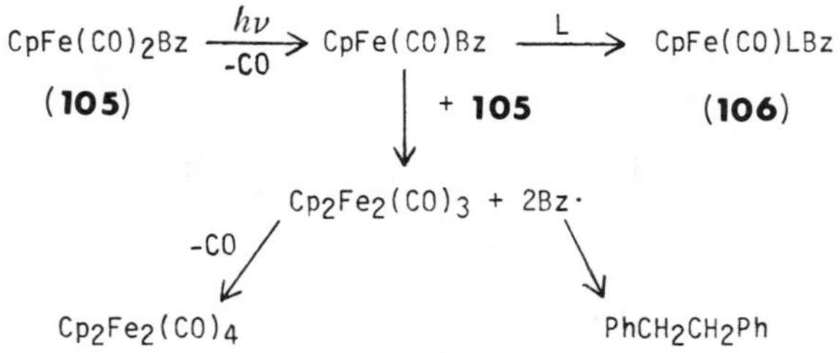

$$CpFe(CO)_2Bz \xrightarrow[-CO]{h\nu} CpFe(CO)Bz \xrightarrow{L} CpFe(CO)LBz$$

$$(\mathbf{105}) \qquad\qquad\qquad \Big\downarrow +\ \mathbf{105} \qquad\qquad (\mathbf{106})$$

$$Cp_2Fe_2(CO)_3 + 2Bz\cdot$$

$$-CO \nearrow \qquad\qquad\qquad \searrow$$

$$Cp_2Fe_2(CO)_4 \qquad\qquad\qquad PhCH_2CH_2Ph$$

SCHEME 16

be construed as evidence supporting formation of benzylradicals. Nonetheless, the authors concluded that CO loss occurred in the primary photoprocess and that benzyl radicals were formed in a subsequent thermal, bimolecular reaction as shown in Scheme 16 (Bz = benzyl). These could be efficiently trapped by 9,10-dihydroanthracene to produce or combine to give bibenzyl.

Sim and co-workers (97) reported that irradiation of $CpFe(CO)_2CH_2Ph$ in benzene solution and in the presence of PPh_3 gave the complex $Fe(CO)_2(PPh_3)(exo\text{-}\eta^4\text{-}C_5H_5CH_2Ph)$ in which the benzyl ligand had migrated via an *intramolecular* reaction to the Cp ring. However, no yield or mechanistic details were provided.

4. FeCp(CO)₂Ar

The photochemistry of these aryl complexes has been briefly investigated by Nesmeyanov and co-workers (98, 98a) and is quite similar to that of the analogous alkyl complexes. Irradiation of $CpFe(CO)_2Ar$ complexes in the presence of tertiary phosphines at room temperature led to the formation of the corresponding monosubstituted derivatives (98) [Eq. (96)]. Prolonged irradiation in the presence of triphenylphosphite

$$CpFe(CO)_2Ar + PR_3 \xrightarrow[-CO]{h\nu} CpFe(CO)(PR_3)Ar + CO \qquad (96)$$

$$(\mathbf{107}) \qquad\qquad\qquad (\mathbf{108})$$

gave a species originally identified as $[CpFe[P(OPh)_3]_2]_2$ (98a). However, recent studies (99, 99a) showed that irradiation of 107 in the presence of 1 equivalent of triphenylphosphite in benzene solution did not give the monosubstituted derivative 108 but instead the orthometallated complex 109 (99). Furthermore, the species originally thought to be $[CpFe(P(OPh)_3)_2]_2$ has now been identified as the orthometallated species 110(99). The authors also showed that the monosubstituted derivative 108

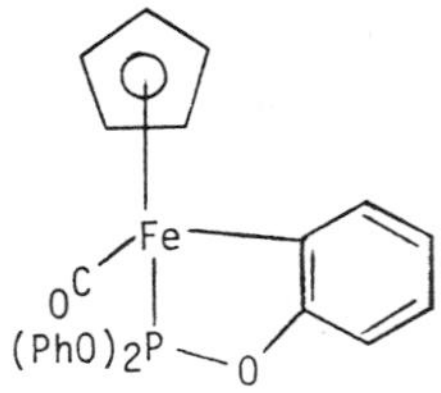

(109) 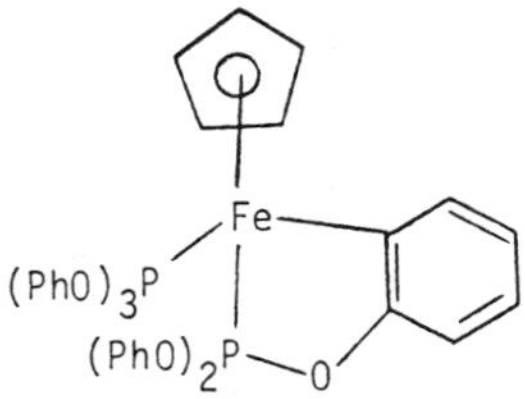   **(110)**

(R = OPh) is initially formed but then rapidly converts **109** via an unknown mechanism [Eq. (97)].

$$CpFe(CO)_2C_6H_5 + P(OPh)_3 \xrightarrow{h\nu} CpFe(CO)[P(OPh)_3]C_6H_5 + CO \xrightarrow{h\nu} \mathbf{109} + PhH \qquad (97)$$

However, Kupletskaya and co-workers (*100*) have shown that the products of photodecomposition of $CpFe(CO)_2R$ (R = Ph, 1-naphthyl, 1-azulenyl, or acenaphthyl) in air-saturated solution were those expected from the trapping of Ar radicals by oxygen. These results could be construed as evidence for photoinduced Fe—Ar bond homolysis, but it is more likely that these products arise from a free radical reaction initiated by the oxygen present in solution. A similar free radical-initiated homolysis of an Fe—$(\eta^1\text{-}C_5H_5)$ bond has been postulated for $CpFe(CO)_2(\eta^1\text{-}C_5H_5)$ (*101*).

5. $Os(CH_3)_2(CO)_4$

Norton and co-workers (*102*) briefly noted that photolysis of $Os(CH_3)_2(CO)_4$ in hexane solution led to the formation methane but produced no ethane. Presumably, methane derives from photoinduced homolysis of an osmium–methyl bond and scavenging of hydrogen by the resultant methyl radical.

G. Co and Rh Complexes

1. $[Co(CN)_5(CH_2C_6H_5)]^{3-}$

Vogler and Hirschmann (*103*) observed that $[Co(CN)_5(CH_2C_6H_5)]^{3-}$ was quite photosensitive, producing $[Co(CN)_5]^{3-}$ and bibenzyl when irradiated with 313 nm light in deaerated water solution [Eq. (98)]. The 313 nm quantum yield of disappearance of $[Co(CN)_5(CH_2C_6H_5)]^{3-}$ was

$$2[Co(CN)_5(CH_2C_6H_5)]^{3-} \xrightarrow{h\nu} 2[Co(CN)_5]^{3-} + C_6H_5CH_2CH_2C_6H_5 \qquad (98)$$

0.13. The electronic absorption spectrum of the complex shows an intense band at 295 nm ($\varepsilon = 18{,}000$), but ligand field (LF) bands were not observed at lower energy. The intense band was assigned as a benzyl-to-cobalt CT transition, and it was proposed that the photoreaction proceeds through population of such a CT state leading to homolytic cleavage of the cobalt–carbon bond.

Irradiation of $[Co(CN)_5(CH_2C_6H_5)]^{3-}$ in a strongly alkaline solution that was subsequently saturated with oxygen gave formation of $[(CN)_5CoO_2Co(CN)_5]^{3-}$ through a thermal reaction between O_2 and $[Co(CN)_5]^{3-}$. Irradiation in the presence of oxygen gave different results. The superoxo complex was not formed, and benzaldehyde was the predominant organic product. It was suggested that under these conditions the reactions shown in Eqs. (99) and (100) occur. It was suggested that

$$[Co(CN)_5(CH_2C_6H_5)]^{3-} + O_2 \xrightarrow{h\nu} [Co(CN)_5(O_2CH_2C_6H_5)]^{3-} \qquad (99)$$

$$[Co(CN)_5(O_2CH_2C_6H_5)]^{3-} + H_2O \xrightarrow{h\nu} [Co(CN)_5H_2O]^{2-} + C_6H_5CHO + OH^- \qquad (100)$$

reaction (99) proceeds by initial generation of benzyl radicals which are scavenged by O_2 and then react with photogenerated $[Co(CN)_5]^{3-}$ to give the peroxo complex; the 313 nm quantum yield for disappearance of $[Co(CN)_5(CH_2C_6H_5)]^{3-}$ under these conditions was 0.15. The similarity of the quantum yields in aerated and deaerated solutions provides strong support for the hypothesis that the same primary photoreaction, homolytic cleavage of a Co—C bond, occurs in both cases.

Sheats and McConnell (*104*) demonstrated the use of this photochemical reaction as a biophysical tool to trap nitroxides. For example, they used reactions of the type shown in Eq. (101) to measure rates of lateral

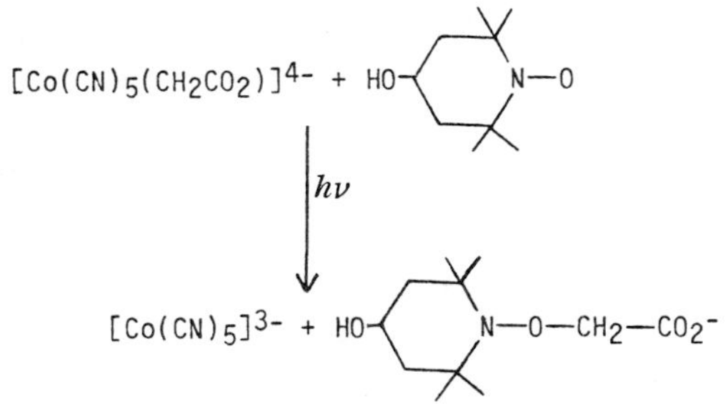

$$\text{[Co(CN)}_5\text{(CH}_2\text{CO}_2\text{)]}^{4-} + \text{HO—}\bigcirc\text{N—O} \qquad (101)$$

diffusion in phospholipid bilayers and to determine the number of nitroxide labels on the outer surface of liposomes. This type of reaction has found numerous applications in the study of phospholipid membranes (*104a–107*).

2. Methylcobalamin, Coenzyme B_{12}, Cobaloximes, and Related Derivatives

These complexes, all of which contain cobalt within a macrocyclic ligand and an axial cobalt–carbon bond, have been found to be photosensitive and have been the subject of numerous photochemical studies. Space does not permit a detailed summary of all the studies which have been conducted on these classes of compounds, and we present here only a general summary of the various observations and discuss pertinent articles from the most recent literature. The reader is referred to an excellent review of the subject by Koerner von Gustorf et al. (108) which presents a detailed discussion of reports that appeared prior to 1975.

The various photochemical studies of these compounds have been conducted primarily because of the interest of researchers in the synthesis, properties, and biological activity of vitamin B_{12} and its derivatives. The structure of vitamin B_{12}, as determined by Crowfoot-Hodgkin et al. (109), is shown in Fig. 2. It consists of cobalt in a corrin ring complexed axially by an α-5,6-dimethylbenzimidazole nucleotide and by cyanide ion. Replacement of the axial CN^- by a methyl group gives methylcobalamin, and by 5'-deoxyadenosine gives coenzyme B_{12}. The formal oxidation state of

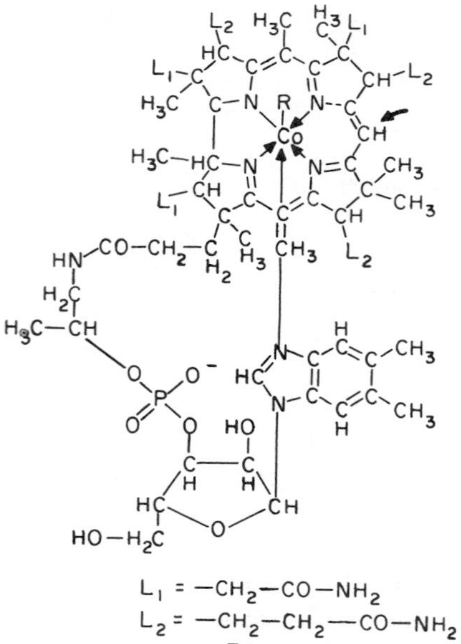

$$L_1 = -CH_2-CO-NH_2$$
$$L_2 = -CH_2-CH_2-CO-NH_2$$
$$R = CN^-$$

FIG. 2.   Structure of vitamin B_{12} (108, 109).

cobalt is $+3$. The metal under certain conditions can be reduced to the $+2$ state, and these derivatives are labeled B_{12r}. Numerous complexes have been prepared to model various aspects of B_{12} chemistry, and the majority of these have centered around bis(dimethylglyoximato) (111), N,N'-ethylenebis(salicylideneiminato) (112), and related complexes. These pos-

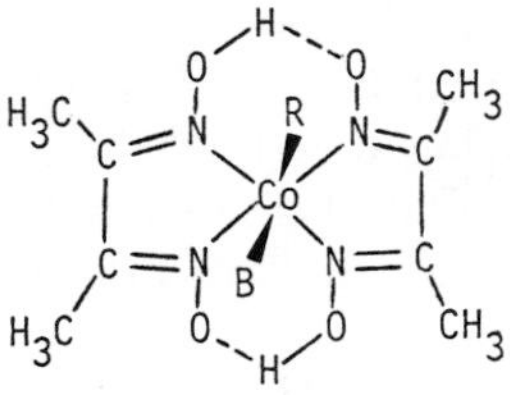

(111) 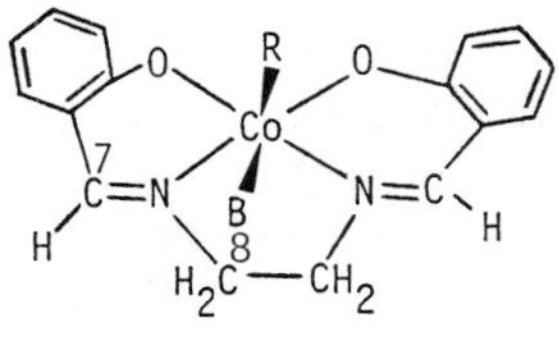　　　　　　　　　　　　(112)

sess a donor ligand B and an alkyl ligand R on the axis and have been given the common names alkylcobaloxime and alkylsalen, respectively.

Most alkylcobalamin derivatives are thermally stable but are photosensitive. When irradiated in the presence of O_2, methylcobalamin gives formation of formaldehyde and a cobalt (III) aquo complex [Eq. (102)].

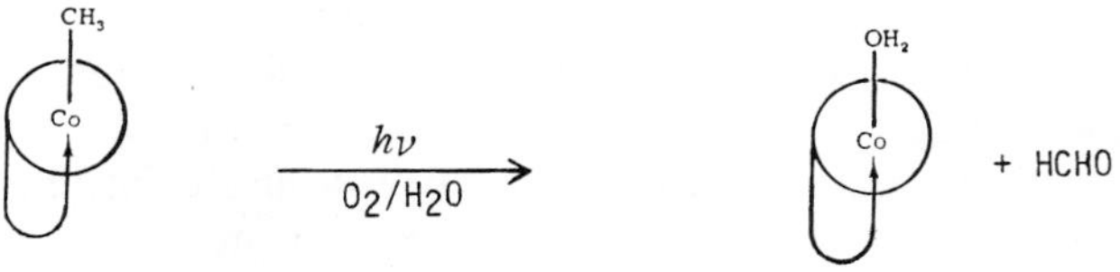

$$\xrightarrow[\text{O}_2/\text{H}_2\text{O}]{h\nu}$$

$+$ HCHO　　　(102)

The quantum yield varies with pH and with irradiation wavelength, but ranges between 0.2 and 0.5 (*108*). In the absence of oxygen, photolysis yields B_{12r}, and a mixture of methane and ethane [Eq. (103)]. Deuteration

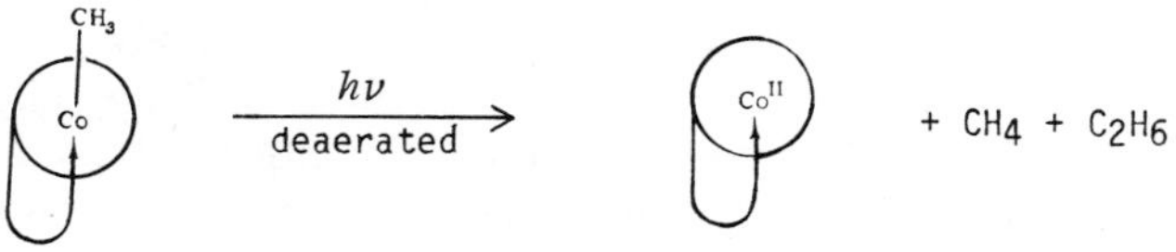

$$\xrightarrow[\text{deaerated}]{h\nu}$$

$+$ CH_4 $+$ C_2H_6　　　(103)

studies showed that the ethane arose primarily from coupling of two methyl radicals from methylcobalamin (*110*). Quantum yields under anaerobic conditions are much smaller than those obtained in the presence of O_2, presumably due to rapid recombination of methyl radicals with B_{12r}. Similar results were obtained with several other alkylcobalamin complexes with bulky alkyl groups (*111*).

Both aerobic and anaerobic processes are consistent with the primary photochemical reaction being homolytic cleavage of the cobalt–carbon bond, initially producing B_{12r} and methyl radicals. Endicott and co-workers (*112–112b*) have observed this process in flash-photolysis studies. In all flash-photolyzed samples–oxygenated or deoxygenated, water or isopropanol solution—they observed generation of B_{12r}. In deaerated solution, recombination of some of the methyl radicals with B_{12r} to regenerate a portion of the methylcobalamin was demonstrated (*112*). Most of the methyl radicals, however, dimerized to ethane, giving a relatively large net generation of B_{12r}. It was noted that in continuous photolysis experiments, in which the irradiation intensities would be several orders of magnitude lower than in in the flash experiments, the stationary-state concentrations of methyl radicals would rarely be as high as 10^{-8} M. Under those conditions in deaerated solutions the, B_{12r} concentration would rapidly become high enough to scavenge efficiently most of the methyl radicals produced, and the resultant quantum yields would be low. Similar results were obtained with the adenosylcobalamin analogue in fluid solution (*112c*) but it has been shown by ESR Spectroscopy that irradiation of frozen propane-1,2-diol solutions of adenosylcobalamin produced only low concentration of B_{12r} (*113*). However, when the irradiated solutions were allowed to warm up, substantial signals due to B_{12r} were observed, indicating that some precursor was formed upon irradiation of the frozen medium. The nature of this precursor was not determined.

ESR studies (*112, 114–118a*) are also consistent with the formation of free radicals upon photolysis of alkylcobalamin and coenzyme B_{12}. For example, Lappert and co-workers (*116*) demonstrated that homolysis of the cobalt–alkyl bond occurs upon photolysis of coenzyme B_{12} and ethylcobalamin by trapping the 5'-deoxyadenosyl and ethyl radicals produced with $(CH_3)_3CNO$. They were able to detect the spin-trapped $(CH_3)_3CN(O)R$ radicals by ESR spectroscopy. Homolysis of the cobalt–methyl bond was also shown to occur upon anaerobic photolysis of methylcobalamin (*117*). However, the presence of traces of oxygen in the methanol solvent was shown to affect significantly the photochemistry of methylcobalamin (*118*). Indeed, under those conditions, the 5,5-dimethyl-1-pyrroline N-oxide (DMPO) spin adducts of both the methyl and hydrogen radicals, **113** and **114**, respectively, were detected by ESR spectro-

(**113**)   (**114**)

scopy. Irradiation of both coenzyme B_{12} and methylcobalamin in H_2O or D_2O solutions at pH 7 in the presence of 5,5-dimethyl-1-pyrroline-N-oxide (DMPO) also gave rise to signals attributed to the spin adducts **113** and **114**. The authors (**114**) concluded that two main photochemical processes follow light absorption by alkylcobalamins and coenzyme B_{12}: cobalt–carbon bond homolysis and loss of a hydrogen atom. However, the authors could not determine if these events occur simultaneously or if one precedes the other, nor could they determine the relative importance of each process. The source of the hydrogen atom was not clearly established but it was suggested that it originated from the C-10 position of the corrin ring (see arrow in Fig. 2) or from the adenosyl ligand in coenzyme B_{12}. Presumably, light absorption causes electron transfer from the corrin ring to cobalt, which causes a weakening of the carbon–hydrogen bond at the C-10 position of the corrin ring and generation of a hydrogen radical (Scheme 17). Alternatively, the electron can populate the $Co(d_z^2)\sigma^*$ orbital and induce the homolytic cleavage of the cobalt–alkyl bond to generate an alkyl radical and B_{12r}.

The photochemistry of the cobaloximes is similar to that of the cobalamins except that the quantum yields are much lower (*108*). The nature of the primary photochemical processes is also less clear. Giannotti and co-workers (*112c, 118a*), for example, have proposed a mechanism different from that for methylcobalamin. They obtained evidence that photoaquation of the axial ligand B precedes oxygen insertion when

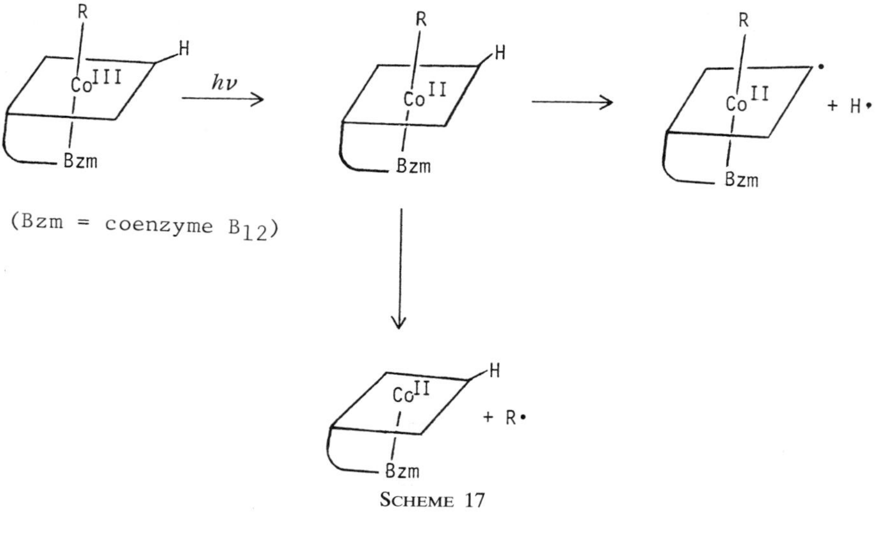

(Bzm = coenzyme B_{12})

SCHEME 17

methylcobaloxime is irradiated in aerated solution [Eq. (104)]. After the initial photoaquation, another photon is necessary to form the peroxo

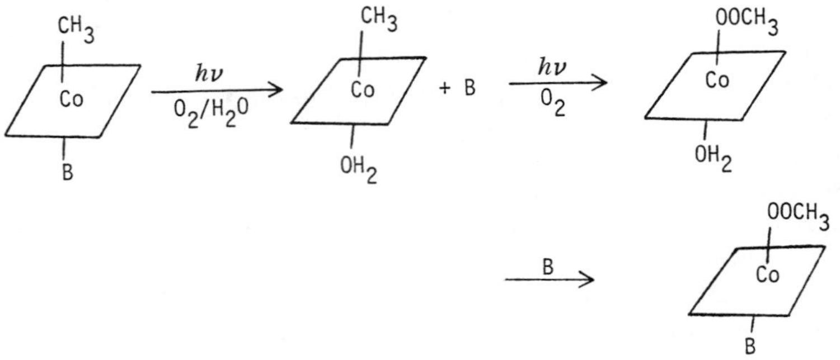

(104)

complex, and it is likely that excitation induces homolytic cleavage of the Co—CH$_3$ bond. The resultant methyl radicals could scavenge O$_2$ to give OOCH$_3$ which could then recombine with cobaloxime to give the peroxo products, Giannotti and co-workers (*112c*) actually isolated a series of alkylperoxycobaloximes by photolysis in the presence of O$_2$ at temperatures ranging from -20 to -120°C. An interesting question is why photoaquation of the axial ligand B has to precede the peroxy formation. Perhaps, as has been suggested by Koerner von Gustorf et al. (*108*), a competition for the excitation energy exists. With a base such as pyridine, photosubstitution of the base must be the preferred reaction, but with water, cleavage of the Co—CH$_3$ bond is favored.

Rétey and co-workers (*120*) obtained similar results upon irradiation of the bridged alkylcobaloxime **115** in methanol solution under strictly

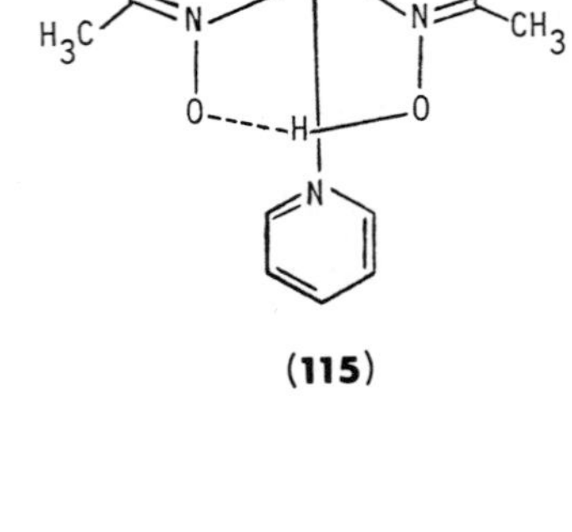

(**115**)

anaerobic conditions. The authors showed that short-term (1 hour) irradiation of **115** caused the exchange of the axial pyridine ligand by a solvent
molecule affording the methylmethoxycobalamin analogue in 90% isolated
yield. However, long-term irradiation in methanol or ethanol led to the
irreversible cleavage of the Co—C bond. The photolysis product was not
identified but alkaline hydrolysis afforded methylsuccinic acid in 82–95%
yield and led the authors to suggest a reversible cobalt–carbon bond
homolysis followed by the irreversible intramolecular abstraction of hydrogen atoms [Eq. (105)]. The authors noted that these results did not rule

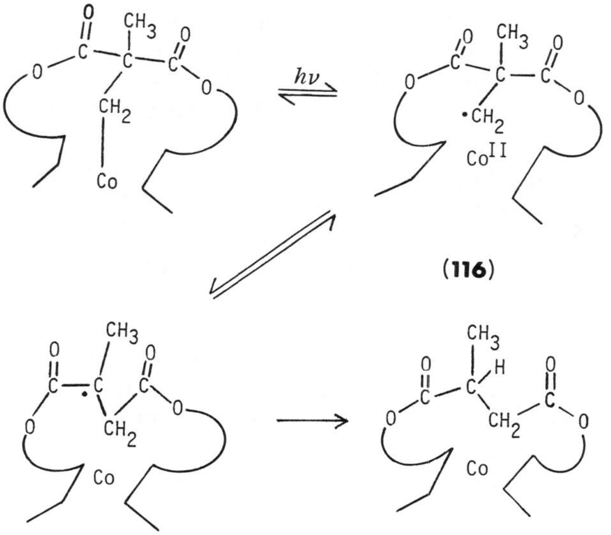

(116)   (105)

out Co—C bond homolysis in the primary photochemical step since the
rapid, intramolecular recombination of the radical pair in **116** could
compete efficiently with the rearrangement and hydrogen abstraction
reactions (*120*). In a related study Rétey and Fountoulakis (*120a*) showed
that the yield of methylsuccinic diester obtained upon irradiation of
cholestanocobaloximes and -rhodoximes in the polar solvents ethanol and
propanol increased with the hydrophilicity of the alkyl ligand. The authors
suggested that the photogenerated radical pairs recombine more readily if
they can be maintained in close proximity by hydrophobic interactions
between the alkyl radical and the surroundings as is thought to occur at the
active site of methylmalonyl—CoA-mutase. Recent photochemical studies
(119, 121) on organocobaloximes have provided evidence that the 1,2-migra-
tion of a thioester group catalyzed by methylmalonyl-CoA—mutase proceeds via radical mechanism involving the migration of the acyl group of
the 5′-deoxyadenosyl group of coenzyme B_{12} to the adjacent radical center.

Giannotti and co-workers (*114*) studied the photochemistry of a series of alkylcobaloximes, alkylsalen, and alkylnioxime derivatives using spin-trapping techniques under anaerobic conditions and found considerable evidence for the formation of alkyl and hydrogen radicals upon irradiation of these complexes in various protic and aprotic solvents. The nature of the spin adducts detected was dependent on the spin-trapping agent and solvent. Irradiation of these complexes in the presence of mixtures of the spin-trapping agents DMPO, ND, and PBN confirmed that methyl and hydrogen radicals were being generated simultaneously by two competing reactions, consistent with the overall mechanism shown in Scheme 18 (*114*). The exact source of the hydrogen atom is not clear but deuterium labeling studies (*114*) led the authors to suggest that it originates from the hydroxyl groups of the oxime ring of cobaloximes and probably from the C-7 and C-8 positions of the equatorial ligand in the salen complexes (**112**).

Flash photolysis studies (*121*) showed that the rate of homolysis of the cobalt–carbon bond in methylcobaloxime in CHCl₃ solution was fast ($k = 1.2 \times 10^5 \text{ sec}^{-1}$) and that the addition of a spin trap affected only the radical recombination step. Giannotti and co-workers (*118*) also showed that polar solvents such as water and methanol or the presence of oxygen strongly favoured the homolytic cleavage of the Co(III)—C bond over the departure of a hydrogen atom. This is presumably because these molecules interact strongly with the photogenerated alkyl radicals and hinder the otherwise efficient recombination reaction. Kemp and co-workers (*122*) studied the effects of pH and a variety of free radical scavengers upon the distribution of alkyl-derived products in the aqueous photochemistry of several alkylaquocobaloximes. In the absence of added scavengers, they found that elimination of alkene predominated at pH 7 whereas at pH 2 free radical-type chemistry was observed. The authors suggested that the distribution of organic products was consistent with the formation of *two* different excited states upon irradiation, both of which decayed to yield products.

Irradiation of the related C₆H₅Co(bae) [bae = dianion of *N,N'*-ethylenebis(acetylacetoneimine)] complex in the solid state under anerobic conditions also led to Co(III)—C bond homolysis as evidenced by ESR signals corresponding to a Co(II)(bae) molecule and a C₆H₅· radical trapped in some type of lattice cage (*123*). Giannotti and co-workers (*124*, *125*) also showed that Co(III)—C bond homolysis in alkylcobaloximes was greatly accelerated when these complexes were irradiated as mixed micelles with the detergents sodium dodecylsulfate and cetylmethylammonium bromide. The exact role of the detergents was not clearly established in these studies which, nonetheless, confirmed the generality and ease of

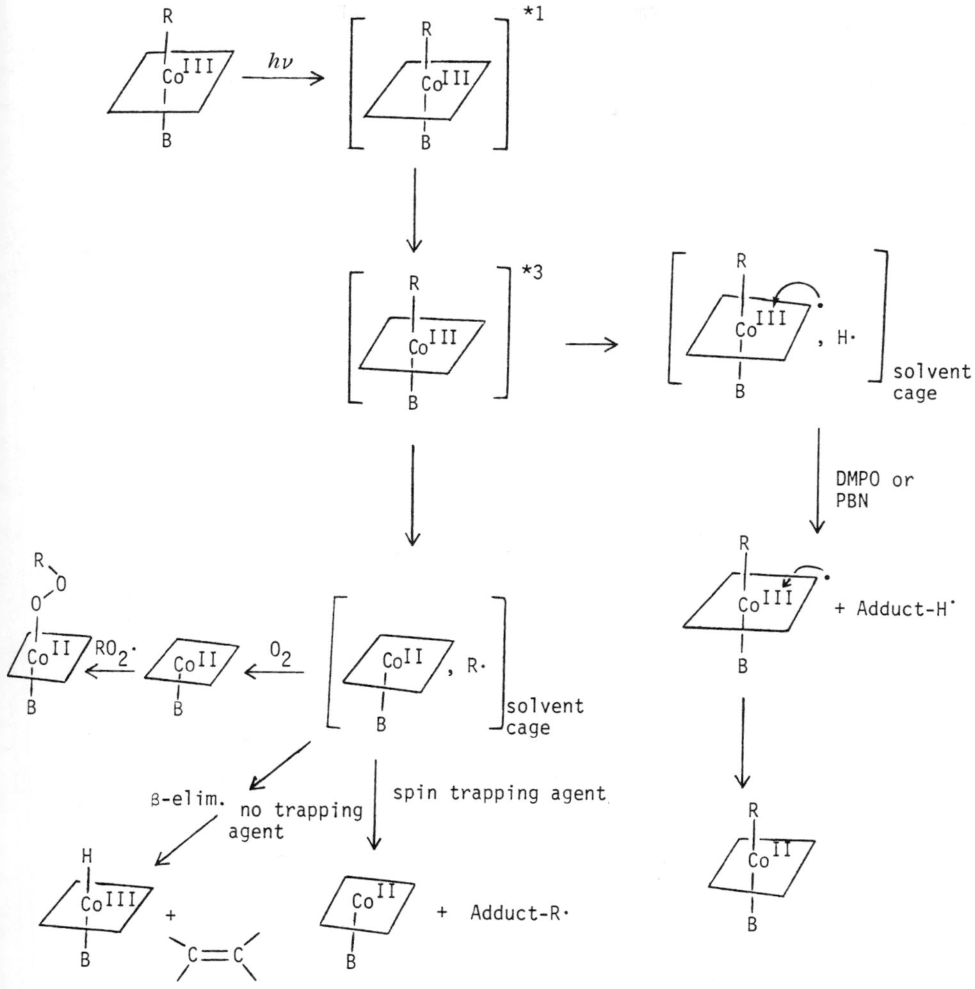

SCHEME 18

Co(III)—C bond homolysis in alkylcobaloximes in the presence of the appropriate substrates.

This property of alkylcobaloximes and alkylcobalamins has been used to model the conversion of 1,2-diols to aldehydes catalyzed by the adenosyl-cobalamin-dependent enzyme diol dehydratase. For example, both organo-cobalamins (*126, 126a*) and alkylcobaloximes (*126a, b*) were shown to convert 1,2-diols to aldehydes under photolytic conditions. Lappert *et al.*

(*126a*) showed that the pH dependence of the yield of aldehyde paralleled that obtained in ·OH induced reactions. They (*126a*) and Golding *et al.* (*126b*) suggested a mechanism involving an initial hydrogen atom abstraction from substrate by a carbon-centered radical. Other applications include their use as quenchers for triplet excited states of organic molecules and excited states of inorganic complexes (*127*), and the high-yield syntheses of a series of trichloroethylcyclopropanes, trichlorobutenes (*128*), and organosulfides and selenides (*128a*) by reaction of the photogenerated alkyl radicals with the appropriate organic substrates [Eqs. (106) and (107)].

$$Cl_3CSO_2Cl + RCo(dmgH)_2L \xrightarrow{h\nu} RCCl_3 + ClCo(dmgH)_2L + SO_2 \qquad (106)$$
(dmgH = dimethylglyoxime)

$$Ar_2S_2 + 2RCo(dmgH)_2py \xrightarrow{h\nu} 2ArSR + 2 Co(II)(dmgH)_2py \qquad (107)$$
(py = pyridyl)

Several other alkylcobalt complexes with different chelating ligands have been synthesized and have photochemistry similar to that of alkylcobaloximes and alkylsalen complexes. For example, Mok and Endicott (*112a*) showed that irradiation of the macrocyclic cobalt–alkyl complex $[CoCH_3([14]aneN_4)OH_2]^+$ ($[14]aneN_4$ = 1,3,8,11-tetraazacyclotetradecane) gave homolytic cleavage of the Co—CH_3 bond with a quantum yield of 0.30. The same photochemistry was observed (*59*) with the porphyrin complexes $[CoR'(L)(oep)]$ (oep = octaethylporphyrin; R' = Me, L = pyridine; and R' = Et, L = OH_2), which gave clear ESR signals arising from the RN(O)R' spin adducts when irradiated in the presence of ND and which is again consistent with Co—C bond homolysis. Jäger and co-workers (*129*) reported that complexes **117** and **118** were photosensitive.

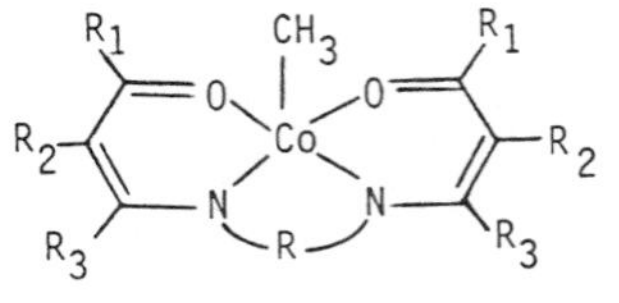

(**117**) 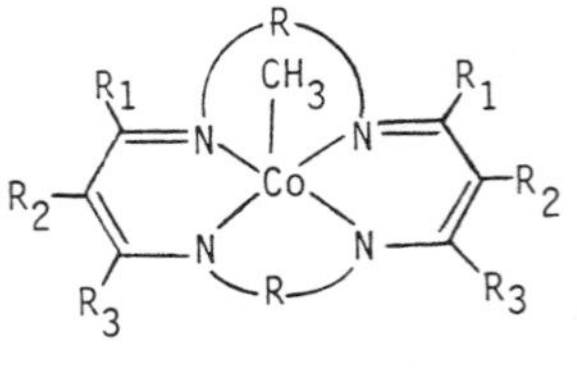 (**118**)

Irradiation of **118** in deoxygenated benzene solution gave an insoluble cobalt(II) complex which was not further characterized. Irradiation in the presence of O_2 gave a clean photoreaction to yield an organometallic complex tentatively formulated as an alkylperoxo complex. Irradiation of dimethylcobalt complexes such as $[(CH_3)_2Co(bipy)_2]^+$ (bipy = bipyridyl) (*130*) and $(CH_3)_2Co(bdm1,3pn)$ [bdm1,3pn = bis-1,3-(diacetylmonoximeimino)propane] **(119)** *(131)* also led to the generation of methyl radicals. It is not clear whether one or both Co—CH$_3$ bonds are cleaved in

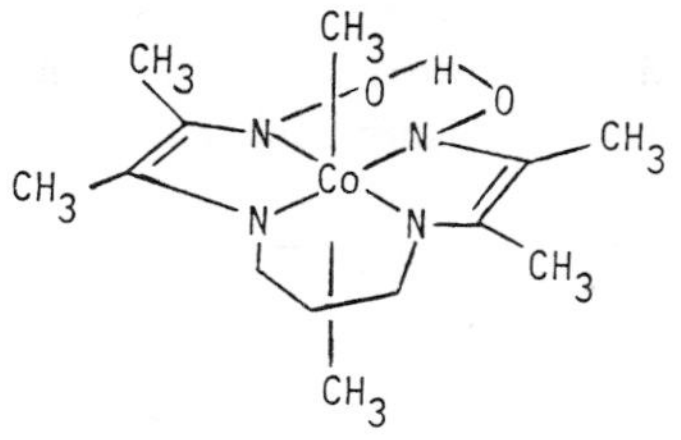

(119)

the [(CH$_3$)$_2$Co(bipy)$_2$]$^+$ complexes *(130)* but Tamblyn and Kochi *(131)* showed that irradiation of **119** in halogenated solvents caused the sequential homolysis of both Co—C bonds to give the corresponding dihalocobalt complexes and methane. The authors also showed that approximately 10% of the hydrogen incorporated into methane originated from the starting complex **119**.

Examples of heterolytic cleavage of the Co—C bond to give Co(III) and R$^-$ in cobalt macrocycles are scarce but this mechanism has been suggested for the photoassisted, base-catalyzed methane formation from methyl(a-quo)cobaloxime *(132)* and the photolysis of alkylcobalt(bdhc) complexes (bdhc = 1,19 -dimethyl-AD-bisdehydrocorrinato) *(133, 134)*. Although a homolytic mechanism could not be positively ruled out in the cobaloxime case *(132)*, Murakami and co-workers *(133, 134)* provided good evidence that alkylcobalt(bdhc) complexes with bulky alkyl groups undergo heterolytic cleavage of the cobalt–carbon bond under photolytic conditions. For example, unlike the methyl and ethyl complexes, the presence of oxygen in solutions of the isopropylcobalt(bdhc) complex did not affect the photolysis rate *(132)*. Furthermore, the product obtained was identified as a Co(III) complex and not the Co(II)(bdhc) complex obtained with the methyl and ethyl complexes [Eq. (108)]. The alkyl ligand was recovered as the alkane in most cases studied.

$$(Me_2HC)Co(II)(bdhc) \xrightarrow[H_2O]{h\nu} (H_2O)_2Co(III)(bdhc) + Me_2CH_2 \qquad (108)$$

3. *[Rh(C$_2$H$_5$)(NH$_3$)$_5$]$^{2+}$ and trans-[Rh(C$_2$H$_5$)(OH$_2$)(NH$_3$)$_4$]$^{2+}$*

Inoue and Endicott *(135)* have studied the photochemical properties of [Rh(C$_2$H$_5$)(NH$_3$)$_5$]$^{2+}$ in aqueous solution. The *trans*-[Rh(C$_2$H$_5$)(OH$_2$) (NH$_3$)$_4$]$^{2+}$ complex was examined in greater detail because of complicating thermal reactions with [Rh(C$_2$H$_5$)(NH$_3$)$_5$]$^{2+}$. The experimental evidence was consistent with photoinduced homolytic cleavage of the Rh—C$_2$H$_5$

bond [Eq. (109)]. The quantum yield varied with wavelength of irradiation and approached a limiting value of 0.4 as the excitation energy increased. In aerated solution the final product was suggested to be $[Rh(O_2C_2H_5)(OH_2)(NH_3)_4]^{2+}$ formed either through the sequence of reactions (110) and (111) or through the reactions (112) and (113).

$$trans\text{-}[Rh(C_2H_5)(OH_2)(NH_3)_4]^{2+} \xrightarrow[H_2O]{h\nu} [Rh(OH_2)_2(NH_3)_4]^{2+} + C_2H_5 \quad (109)$$

$$[Rh(OH_2)_2(NH_3)_4]^{2+} + O_2 \rightarrow [Rh(O_2)(OH_2)(NH_3)_4]^{2+} \quad (110)$$

$$[Rh(O_2)(OH_2)(NH_3)_4]^{2+} + C_2H_5 \rightarrow [Rh(O_2C_2H_5)(OH_2)(NH_3)_4]^{2-} \quad (111)$$

$$C_2H_5 + O_2 \rightarrow O_2C_2H_5 \quad (112)$$

$$[Rh(OH_2)_2(NH_3)_4]^{2+} + O_2C_2H_5 \rightarrow [Rh(O_2C_2H_5)(OH_2)(NH_3)_4]^{2+} \quad (113)$$

H. Ni, Pd, and Pt Complexes

1. CpNiR(L)

The reaction shown in Eq. (114) was observed when CpNi(CO)(CH$_2$C$_3$H$_5$) was irradiated, but no quantitative or mechanistic details were

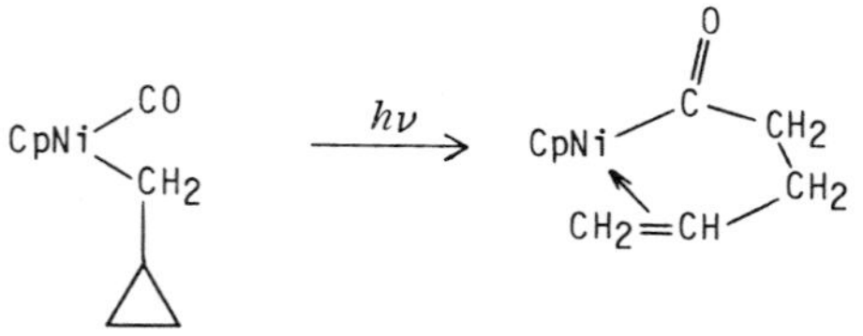

$$(114)$$

given (136). It was suggested that the reaction proceeded via CO insertion into the Ni—alkyl bond followed by Ni-assisted opening of the cyclopropane ring.

Emad and Rausch (137) examined the photoinduced decomposition of CpNiR(PPh$_3$) (R = CH$_3$, C$_2$H$_5$, C$_6$H$_5$, or C$_6$H$_4$CH$_3$), which was found to depend markedly on the R group involved. In all cases, solutions of the complexes turned brown upon irradiation, but the organometallic photoproducts were not identified. However, the organic products were characterized and various labeling studies were conducted. These showed that irradiation of NiCp(CH$_3$)(PPh$_3$) gave mostly methane (96%) with a small amount of ethane (4%) produced. Labeling studies showed that most of the methane produced formed via abstraction of hydrogen from the Cp ligand with a small amount of solvent and CH$_3$ hydrogen abstraction also indicated. However, no hydrogen abstraction from the PPh$_3$ ligand was detected. Curiously, a relatively large amount of CD$_2$H$_2$ (26–29%) was formed when $CpNi(CD_3)(PPh_3)$ was irradiated in C_6H_6 and C_6D_6 solutions. This suggests the probable formation of metal–methylene intermediates. Overall, the results imply photoinduced homolysis of the $Ni—CH_3$ bond, although they could be explained equally well by photoinduced PPh_3 loss followed by thermal degradation of the coordinatively unsaturated $CpNi(CH_3)$ complex.

The latter conclusion is supported by results obtained for the ethyl complex [Eq. (115)] (*137*). Ethane and ethylene were the only organic

$$CpNi(C_2H_5)PPh_3 \xrightarrow[\text{12 hours}]{h\nu} CH_2{=}CH_2 + CH_3CH_3 \qquad (115)$$
$$\phantom{CpNi(C_2H_5)PPh_3 \xrightarrow[\text{12 hours}]{h\nu}} (60\%) \quad (40\%)$$

products and labeling studies showed that the ethane formed did not result from hydrogen abstraction from solvent. These data suggest a β-elimination process in which the ethyl complex yields a hydride–olefin complex. Olefin loss followed by bimolecular reaction between the resultant hydride complex and $CpNi(C_2H_5)(PPh_3)$ would yield ethane. However, the initial ethyl complex is coordinatively saturated and β-hydride elimination requires an open coordination site. We suggest that the reaction probably proceeds via photoinduced PPh_3 dissociation to open up that site, followed by the sequence of reactions in Scheme 19.

When the corresponding phenyl complex was irradiated in benzene solution, biphenyl formation was observed but, surprisingly, most of this came from coupling of two solvent molecules and not from the initial phenyl ligand (*137*). The fate of the latter was not determined but presumably benzene was formed. Irradiation of $CpNi(C_6H_4CH_3)(PPh_3)$ gave biphenyl (3%) and 4,4'-dimethylbiphenyl (45%) as the major products, but no production of toluene was detected. The latter observation

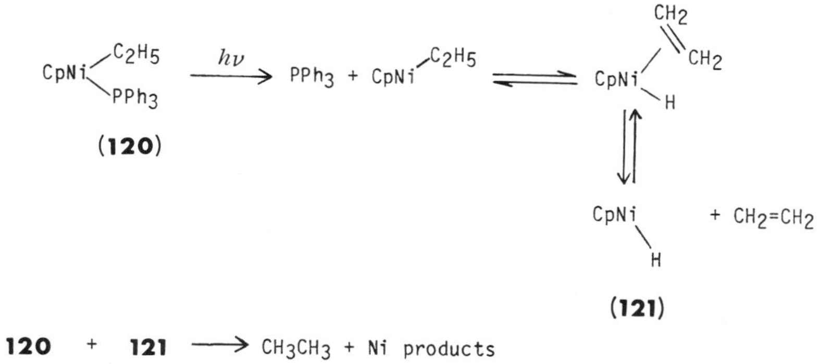

$$\mathbf{120} + \mathbf{121} \longrightarrow CH_3CH_3 + Ni \text{ products}$$

Scheme 19

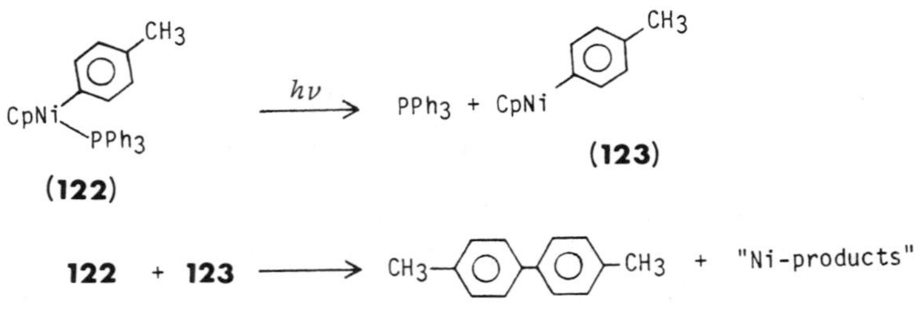

(122) → $h\nu$ → PPh_3 + CpNi ... **(123)**

122 + 123 → CH_3—◯—◯—CH_3 + "Ni-products"

SCHEME 20

appears to rule out a radical process whereas the high yield of 4,4′-dimethylbiphenyl strongly suggests a bimolecular reaction between two Ni complexes, a reaction which can also be explained by initial photoinduced dissociation of PPh_3 (Scheme 20). Thus, although no definitive mechanistic conclusions can yet be drawn, we feel that photoinduced PPh_3 dissociation best explains all the results for this class of compounds.

2. $L_2\overline{MCH_2CH_2CH_2CH_2}$ (M = Ni or Pt)

Grubbs et al. have briefly explored the photochemistry of the nickelacyclopentanes **124–126** (138, 138a). Photolysis of each gave formation of

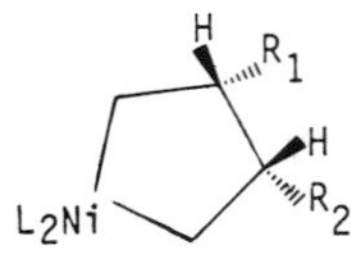

(124), L=PPh_3, **(a)** $R_1=R_2=H$, **(b)** R_1, $R_2=(CH_2)_4$

(125), L=$PMePh_2$, $R_1=R_2=H$

(126), L=$Ph_2PCH_2CH_2PPh_2$, $R_1=R_2=H$

ethylene and butane, but in significantly different ratios (Table IV). These compounds also decompose thermally but the ethylene/butene ratios are much different, with the yield of ethylene increasing upon photolysis. Particularly striking was the diphos complex **(126)**, which gave no C_2H_4 upon thermolysis, whereas C_2H_4 was the major product upon irradiation. The authors did not state whether photolysis increased the *rate* of complex decomposition, although that is a probable result. Irradiation of the related nickelahydrindane complex **124b** in toluene solution at $-40°C$, followed by acid hydrolysis of the products, gave 1,7-octadiene (37%), *trans*-1,2-dimethylcyclohexane (15%), and *cis*-1,2-dimethylcyclohexane (41%). This led the authors (138a) to suggest that 37% of **124b** had decomposed and

TABLE IV

PHOTOINDUCED DECOMPOSITION OF NICKELACYCLOPENTANE COMPLEXES

Complex	Temperature (°C)	Light	% C_2H_4	% C_4H
124	0	UV	21	79
	0	Dark	6	94
125	5	UV	19.5	81
	5	Dark	15	86
126	0	UV	59	41
	0	Dark	0	99.9

29% of the remaining metallacycle **124b** had isomerized to the thermo-dynamically favored isomer with the R groups in trans positions.

The corresponding Pt(II) metallacyclopentane complexes [L_2 = Ph_2-$PCH_2CH_2PPh_2$, $(PPh_3)_2$, or $(PMe_2Ph)_2$] were studied by Puddephatt and co-workers (*139*). Ethylene and 1-butene were the dominant volatile photochemical products except when a good hydrogen donor such as PPh_2H was present. Butane became the major product under such circumstances. The ethylene/butene ratio was dependent upon the solvent used and additives employed, but ethylene generally predominated. For example, irradiation of $PtCH_2CH_2CH_2CH_2(PMe_2Ph)_2$ in CD_2Cl_2 solution gave 95% ethylene and 5% 1-butene. Recall that enhanced ethylene yields also resulted upon photolysis of the corresponding nickelacyclopentane complexes discussed above (*138a, b*). The Pt results were explained by the competing decay paths of Scheme 21 which invoke β-hydride elimination to give a butenyl complex (path A) and concerted cleavage of the

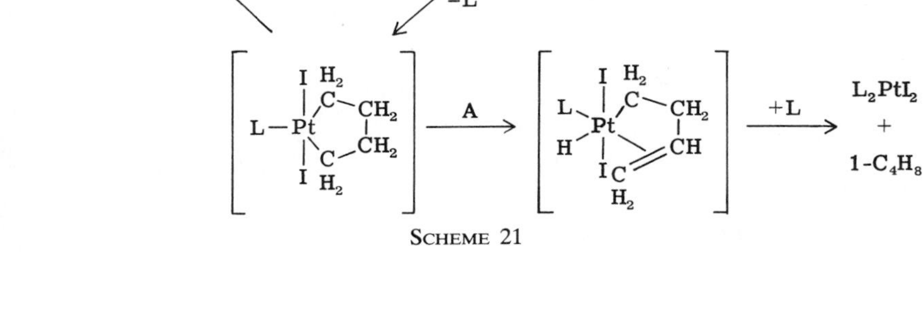

SCHEME 21

metallacyclopentane ring into two ethylene ligands (path B). The excited complex is already coordinatively unsaturated and ligand loss is not necessary for β-hydride elimination to occur. Thus, there was little effect of added excess L on the 1-butene/ethylene ratio. It was also noted that irradiation of these complexes in CH_2X_2 (X = Cl or Br) solution gave small amounts of 1-pentene. It was suggested that this forms via oxidation addition of CH_2X_2 to the excited molecule followed by eventual coupling of a CH_2 fragment to the original C_4H_8 ligand.

3. $PtCH_2CH_2CH_2(N-N)X_2$

The title Pt(IV) metallacyclobutane derivatives [X = Cl or Br; N–N = 1,10-phenanthroline, 2,2-bipyridine, $(CH_2NMe_2)_2$, $(CH_3CN)_2$ or $(py)_2$] have been well studied by Puddephatt and co-workers (140–142). They irradiated these complexes in several different solvents, with and without a variety of additives, and assayed the organic products produced. Propene and cyclopropane were the predominant volatile products under nearly all conditions, although in some cases substantial amounts of ethylene were found. In solvents of low dielectric constant (e.g., CH_2Cl_2, THF), propene formed in >85% yield, independent of the presence of a large excess of neutral ligand. This implied an intramolecular 1,2-hydrogen shift in the excited molecule and no involvement of ligand dissociation in the formation of propene (path A of Scheme 22). In contrast, irradiation in

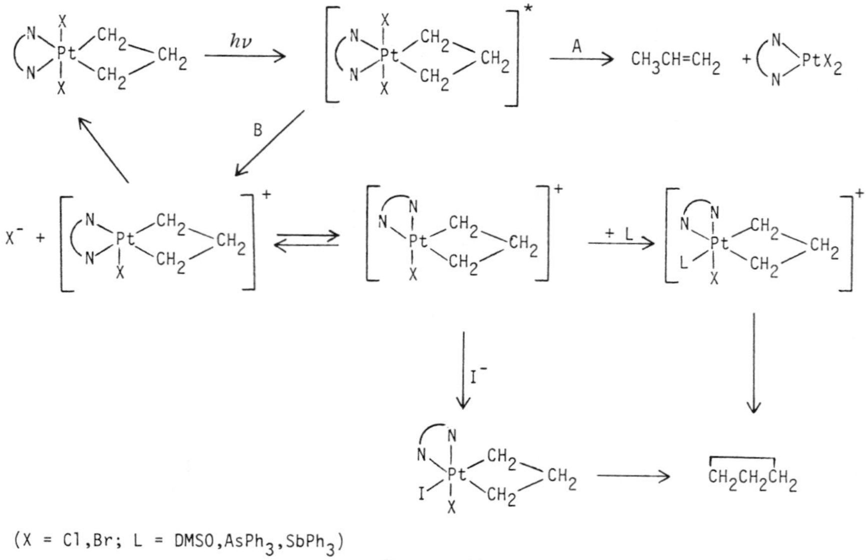

(X = Cl,Br; L = DMSO,AsPh$_3$,SbPh$_3$)

SCHEME 22

solvents of high dielectric constant [e.g., dimethyl sulfoxide (DMSO)], which can stabilize ionic products, gave cyclopropane in much higher yield. The yield of cyclopropane (1) was independent of added N–N ligand, (2) increased if excess ligand with a high trans effect such as I⁻ was present, and (3) decreased when excess [Me₄N]Cl was present. These results were explained by involving a competing decay path B (Scheme 22) for the excited complex which involves dissociation of halide. Subsequent coordination of the high trans effect ligand leads to *thermal* cyclopropane formation whereas excess Cl⁻ scavenges the coordinatively unsaturated intermediate to return to starting complex. Ethylene was also believed to derive from a competing thermal decay of the products resulting from photoinduced halide loss. An excited state rationalization of the two pathways was not attempted, except that it was noted that the UV spectrum of $PtC_3H_6(phen)Cl_2$ is almost identical to free phenanthroline (phen) itself with a 270 nm absorption maximum (*142*).

4. $PtCH_2CH_2CH_2CH_2(L)_2I_2$

Puddephatt and co-workers (*139, 140*) also studied the title Pt(IV) metallacyclopentane derivatives (L = PMe_2Ph or PPh_3) under a variety of conditions. Their results and mechanistic conclusions are summarized in Scheme 23 with two competing decay paths (A and B) for the excited molecules. No cyclobutane was formed in either case, in contrast to the production of cyclopropane upon irradiation of the corresponding metallacyclobutane complexes. However, as noted above, cyclopropane was a thermal rather than photochemical product. The gaseous photolysis products of the metallacyclopentane complexes were exclusively ethylene and

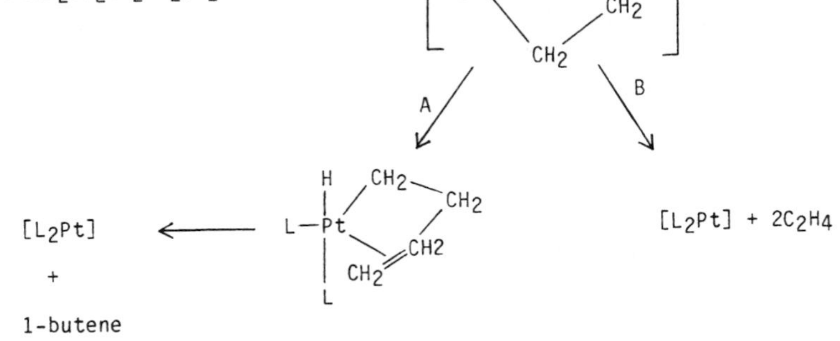

SCHEME 23

1-butene. Addition of excess L suppressed the formation of 1-butene, indicating that its formation required ligand loss as indicated in Scheme 23. Likewise, the relative yield of ethylene increased when L = PMePh$_2$ was changed to PPh$_3$, implying a steric effect on the decay of the excited complex. In the presence of a good hydrogen donor such as PPh$_2$H, n-butane became a predominant product, as this presumably formed by coupling of the active hydrogen of the PPh$_2$H ligand with the C$_4$H$_8$ ligand.

Irreproducible results were obtained upon irradiation of $\overline{PtCH_2CH_2CH_2CH_2}$(Me)I(PMe$_2$Ph)$_2$, but CH$_4$ was a predominant product (139). The authors suggested that excitation induced hydrogen abstraction from the Pt(CH$_2$)$_4$ ring by the CH$_3$ ligand to give CH$_4$, and then a complicated decay pattern for the resultant intermediate followed.

5. MR$_2$L$_2$ (M = Pd or Pt)

The photochemistry of a series of *cis*- and *trans*-PdEt$_2$L$_2$ (L = PMe$_2$Ph, PEt$_2$Ph, or PEt$_3$) complexes has been well studied by Yamamoto, Grubbs, and co-workers (143). The reactions were conducted at −10 and 2°C in the presence of PhC≡CPh to trap the photogenerated PdL$_2$ complexes. In the absence of added tertiary phosphine, irradiation of both the cis and trans isomers gave a 2:2:1 ratio of ethylene, ethane, and butane [Eq. (116)]. No

$$cis\text{- or } trans\text{-PdEt}_2\text{L}_2 \xrightarrow{h\nu} \text{CH}_3\text{CH}_3 + \text{CH}_2{=}\text{CH}_2 + \text{CH}_3\text{CH}_2\text{CH}_2\text{CH}_3 \quad (116)$$
$$ 2 \quad : \quad 1 \quad : \quad 1$$

deuterium incorporation from solvent or other ligands was observed in the organic products and no exchange of alkyl ligands between complexes was detected. These data rule out a radical process. Addition of excess L to the cis complex slowed the photoreaction significantly and decreased the yield of butane. Also, cis–trans isomerization of both PdEt$_2$(PMe$_2$Ph)$_2$ and PdMe$_2$(PMe$_2$Ph)$_2$ was found to be photoinduced at a rate substantially greater than the reaction of Eq. (116). The isomerization reaction was inhibited by the presence of added phosphine. Irradiation of the palladacyclopentane complex shown in Eq. (117) gave largely ethylene with some

$$(\text{PMe}_2\text{Ph})_2\text{Pd} \begin{array}{c} \text{CH}_2{-}\text{CH}_2 \\ | \quad\quad | \\ \text{CH}_2{-}\text{CH}_2 \end{array} \xrightarrow{h\nu} \text{CH}_2{=}\text{CH}_2 \quad + \quad \square \quad + \quad \diagup\!\!\!\diagup \quad\quad (117)$$
$$ 75\text{-}80\% \quad\quad 2\text{-}3\% \quad\quad {\sim}20\%$$

1-butene and cyclobutane. This is in marked contrast to 70°C thermolysis of the complex which gave 2-butene in >95% yield with no ethylene detected.

   To explain these disparate results the authors proposed that irradiation of either the cis or trans isomers gave an excited state with tetrahedral geometry (Scheme 24). Such a species could efficiently relax to give either the cis or trans complex which would result in the observed photoisomerization. Such an isomerization mechanism proceeding via a tetrahedral excited state had been earlier proposed to explain photoinduced cis–trans isomerization in $Pd(PR_3)_2Cl_2$ complexes (144). It was suggested that some fraction of these excited states decayed by either β-hydride elimination or loss of L (Scheme 24). The $LPdEt_2$ intermediate generated by the latter process could yield butane by coupling of the ethyl ligands, analogous to the mechanism of butane formation upon thermolysis of cis-$PdEt_2L_2$ (145, 146). As indicated in Scheme 24, added L would suppress this decay path, as observed. A β-hydride elimination reaction would generate the hydride-ethylene-ethyl complex shown in Scheme 24 which could decay to yield ethylene and ethane, as occurs upon thermolysis of trans-$PdEt_2L_2$ by a similar mechanism.

   Earlier, Van Leeuwen and co-workers (147) briefly investigated the photochemical properties of a series of cis-MR_2L_2 (M = Pd or Pt, L = PPh_3 or $L_2 = Ph_2PCH_2CH_2PPh_2$, R = CH_3 or C_2H_5) complexes in $CHCl_3$ solutions. The reactions were monitored using the chemically induced dynamic nuclear polarization (CIDNP) technique, and most attention was focused on $Pt(CH_3)_2L_2$. These authors suggested the sequence of events shown in Scheme 25. They did not speculate on the details of the first step, but the absence of a fast reaction in an inert solvent such as C_2D_6 suggested that simple homolysis of the $Pt—CH_3$ bond did not occur. Intermediate 127 was invoked to account for the eventual formation of propane derivatives. It was noted that irradiation of $Pd(CH_3)_2(PEt_3)_2$ in C_6D_6 solution led to CH_4 as the only observed product, with no deuterium incorporation. This suggested an α-hydride abstraction from the adjacent

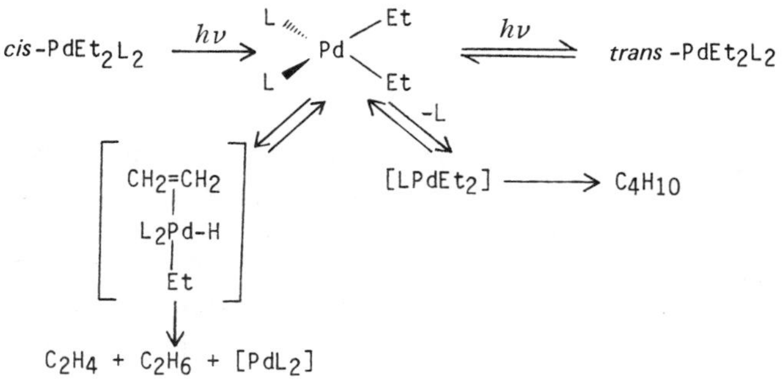

SCHEME 24

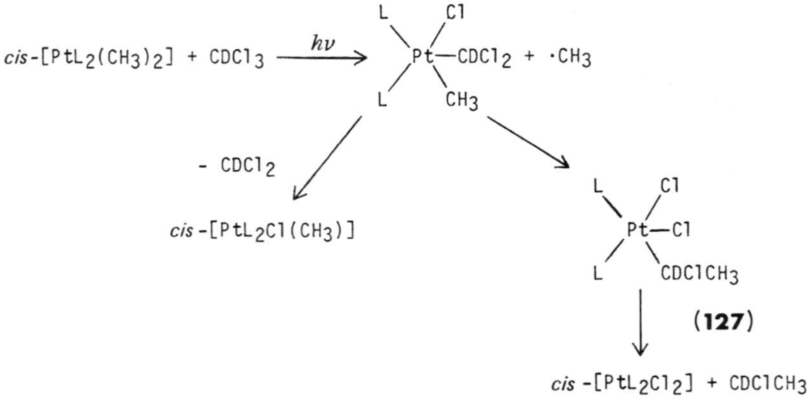

$$cis\text{-}[PtL_2(CH_3)_2] + CDCl_3 \xrightarrow{h\nu}$$

$$cis\text{-}[PtL_2Cl(CH_3)]$$

$$cis\text{-}[PtL_2Cl_2] + CDClCH_3$$

SCHEME 25

methyl ligand to give CH_4 and a transient $M{=}CH_2$ complex. Similar results were obtained for $Pd(CH_3)_2(diphos)$. Although two adjacent CH_3

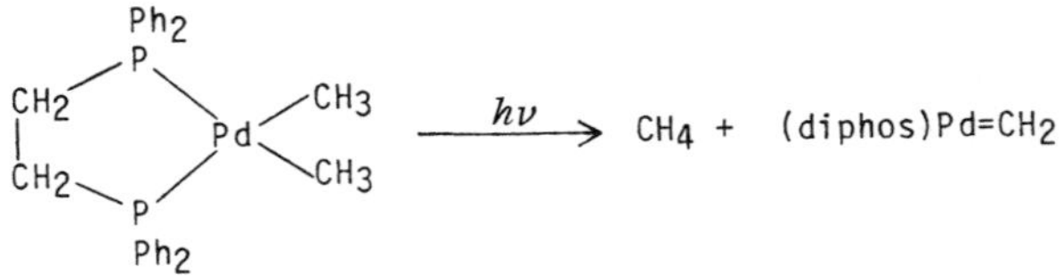

$$\xrightarrow{h\nu} CH_4 + (diphos)Pd{=}CH_2 \qquad (118)$$

ligands could couple to give ethane as in the ethyl complexes discussed above, they apparently do not. β-Hydride elimination is not a possibility, so α-hydride elimination (abstraction) occurs instead, but radicals are not required to account for the CH_4 product. It was later mentioned that irradiation of $cis\text{-}PtR_2(PMePh_2)_2$ ($R = CH_2SiMe_3$ and CH_2CMe_3) in the presence of spin traps ($R'NO$) gave rise to ESR signals due to $RN(\dot{O})R'$, implying the formation of free radicals, but no further details were given (148).

6. $MR(X)L_2$ ($M = Pd$ or Pt)

There have been several brief reports on the photochemistry of this class of compounds in the literature, but a clear picture of their properties has not emerged. Van Leeuwen and co-workers (147) reported that when $Pd(CH_3)Cl(diphos)$ was irradiated in $CDCl_3$ solution, CIDNP spectra were observed that implied the primary products shown in Eq. (119). No further details were given.

$$(diphos)Pd(CH_3)Cl \xrightarrow[CDCl_3]{h\nu} (diphos)PdCl_2 + CH_3\cdot + \cdot CDCl_2 \qquad (119)$$

Suzuki and co-workers prepared a series of $Pd(CH_2CN)X(PPh_3)_2$ complexes [X = Cl, N_3, NCO, NO_2, $N(CO)_2C_6H_4$, or I^-]. All were photochromic in the solid state, but not in solution, changing color when exposed to light but reverting to the original color upon standing in the dark (149). In an effort to understand this effect, McCrindle, Ferguson, and co-workers (150) determined the crystal structure of $Pd(CH_2CN)Cl(PPh_3)_2$ and subjected the crystals to UV irradiation. They, too, observed photochromic behavior, although repeated photolysis led to deterioration of the crystals. Cleavage of irradiated crystals after irradiation showed that the reaction was confined to the crystal surfaces. They reported $Pd(CH_2CN)Cl(PPh_3)_2$ to be photosensitive in C_6H_6 solution, giving the reaction shown in Eq. (120) when irradiated in the absence

$$Pd(CH_2CN)Cl(PPh_3)_2 \xrightarrow{h\nu} CH_3CN + PdCl_2(PPh_3)_2 + \text{reduced Pd complexes} \quad (120)$$

of O_2. They believed that the CH_3CN formed by reaction with traces of H_2O present, and they attributed the red-orange color to "reduced Pd complexes," possibly $Pd(PPh_3)_2$. Irradiation in the presence of O_2 gave a more complicated reaction yielding $OPPh_3$, $[PdCl(CH_2CN)(PPh_3)]_n$, and $[PdCl_2(PPh_3)]_2$, in addition to the products of Eq. (120).

On the other hand, the Pt(II) complexes $trans$-$PtH(CH_2CN)(PPh_3)_2$, $trans$-$PtH(C_3H_6CN)(PPh_3)_2$, and $trans$-$PtH(CF_3)(PPh_3)_2$ were reported (151) to undergo a photoinduced reductive elimination of CH_3CN, $CH_3(CH_2)_2CN$, and CHF_3, respectively, with formation of $Pt(PPh_3)_2$. Spectroscopic studies (151b) showed these reactions to be concerted.

The complex $Pt(Ph)Cl(PEt_3)_2$ has been reported to undergo the clean cis–trans photoisomerization shown in Eq. (121) (151a). The quantum

$$cis\text{-}Pt(Ph)Cl(PEt_3)_2 \underset{h\nu}{\overset{h\nu}{\rightleftharpoons}} trans\text{-}Pt(Ph)Cl(PEt_3)_2 \quad (121)$$

yield for the cis → trans isomerization measured in CH_3CN solution was not wavelength dependent and averaged 0.096 for 254, 280, and 313 nm irradiation. In contrast, the trans → cis quantum yield decreased from 0.048 at 254 nm to 0.023 at 280 nm to $<10^{-3}$ at 313 nm. According to the author's spectral interpretation, 313 nm irradiation was into LF bands of both complexes and the higher energy excitation populated LMCT states, predominantly $Cl^- \to Pt$ charge transfer. However, doubt was cast later on the validity of these assignments and thus the following conclusions must be viewed with caution (152). On the basis of their assignments, the authors concluded that the cis → trans isomerization occured through an LF state which had a tetrahedral geometry and then relaxed to the trans isomer, in accord with the photochemistry of other square-planar Pd and Pt complexes (143, 144). However, they believed that the trans → cis isomerization took place via a Cl $\rightarrow$ Pt charge transfer state which involved an intermediate solvent-trapped radical pair.

7. $PtR_3I(bipy)$ and $PtR_2I_2(bipy)$

Irradiation of PtIMePh$_2$(bipy), (**128**), PtIMe$_3$(bipy) (*129*), and PtIMe$_2$(bipy) (**130**) was found by Puddephatt and co-workers (*153*) to give products best explained by photoinduced homolysis of Pt—CH$_3$ and Pt—Ph bonds. The production of CH$_4$, C$_2$H$_6$, C$_6$H$_6$, and C$_6$H$_5$CH$_3$ were all observed, but particularly revealing was the formation of ethane upon photolysis of **128** and production of PhC$_6$H$_4$CH$_3$ upon irradiating a mixture of **128** and PtIMe(C$_6$H$_4$CH$_3$)$_2$(bipy). The latter two organic products cannot arise by an intramolecular coupling process, unless alkyl/aryl exchange were to occur between complexes. However, the incorporation of deuterium into the organic products upon irradiation in deuterated solvents evidences the production of radicals via homolysis, as does the observed polymerization of methyl methacrylate upon irradiation of **129**.

8. $Pt(cod)(i\text{-}C_3H_7)_2$

Irradiation of the title complex in the presence of excess cyclooctadiene (COD) has been reported to yield Pt(cod)$_2$ but no further details have been given (*154*). This reaction may proceed via photoinduced β-hydride elimination as in Eq. (122).

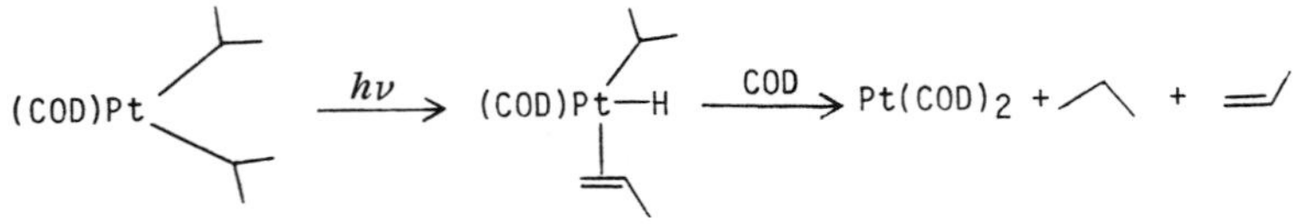

$$(COD)Pt \xrightarrow{h\nu} (COD)Pt-H \xrightarrow{COD} Pt(COD)_2 + \diagdown\!\diagup + =\!\diagup \quad (122)$$

9. $PtCp(Me)_3$

Irradiation of this complex was shown by Hackelberg and Wojcicki (*155*) to give homolysis of the Pt—CH$_3$ bonds and the products shown in Eq. (123). The CH$_3$ radicals produced upon irradiation were trapped by

$$CpPt(CH_3)_3 \xrightarrow[\psi_{350} = 0.0044]{h\nu} CH_4 + C_2H_6 + C_2H_4 \quad (123)$$
$$(90\%) \quad (5\%) \quad (5\%)$$

various spin adducts and were themselves identified by low-temperature irradiation using ESR monitoring. Labeling studies showed the CH$_3$ radicals to be very non selective in abstracting hydrogen, with >45%

abstraction from other CH_3 ligands, ~5% from the Cp ligands, and <50% from solvent.

10. $Pt_2(dppm)_2(Me)H_2$, $Pt_2(dppm)_2(Me)_2H$, and $Pt_2(dppm)_2(Me)_3$

Puddephatt and co-workers (156) examined the photochemistry of these binuclear dppm complexes (dppm = $Ph_2PCH_2PPh_2$) and found different results in each case. Photolysis of **131** gave elimination of H_2, but not methane [Eq. (124)], whereas irradiation of **132** gave formation of CH_4

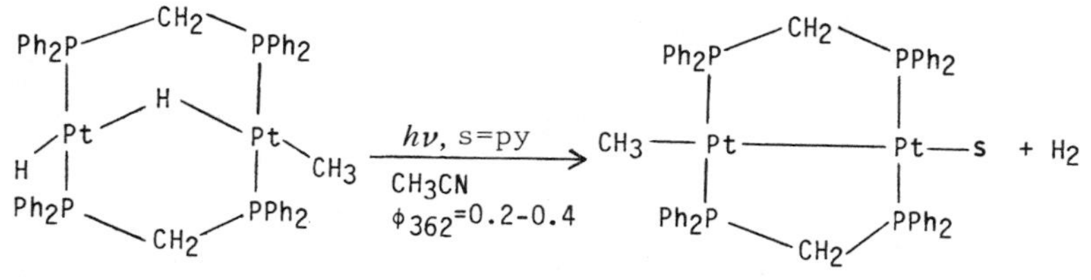

$$\text{(131)} \xrightarrow[\substack{CH_3CN \\ \phi_{362}=0.2-0.4}]{hv,\ s=py} CH_3 - Pt - Pt - s\ +\ H_2 \quad (124)$$

(131)

with $\phi_{362} = 0.30$ and 0.34 in pyridine and CH_3CN solutions, respectively [Eq. (125)].

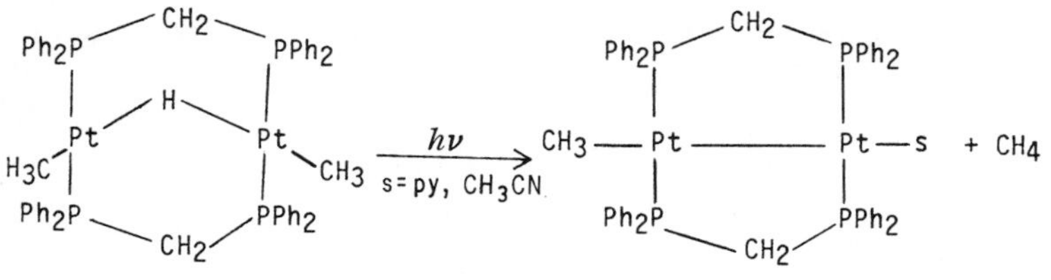

$$\text{(132)} \xrightarrow[s=py,\ CH_3CN]{hv} CH_3 - Pt - Pt - s\ +\ CH_4 \quad (125)$$

(132)

The photochemistry of the trimethyl complex **133** was more complicated (Scheme 26). Irradiation in pyridine gave the mixture of mononuclear products **134** and **135** with a 362 nm quantum yield of 0.6. It was suggested that $d \rightarrow \sigma^*(PtPt)$ excitation led to cleavage of the donor–acceptor Pt—Pt bond with subsequent coordination of the good σ donor pyridine and cleavage into mononuclear complexes. Irradiation of **133** in the presence of weak donor solvents such as acetone, benzene, and CH_2Cl_2 gave reductive elimination of ethane. In acetone, the resultant coordinatively unsaturated binuclear complex was stabilized and the overall 362 nm quantum yield of disappearance of **133** was 1.2 ± 0.3. Labeling experiments showed that at least part of the reductive-elimination reaction is intermolecular, involving

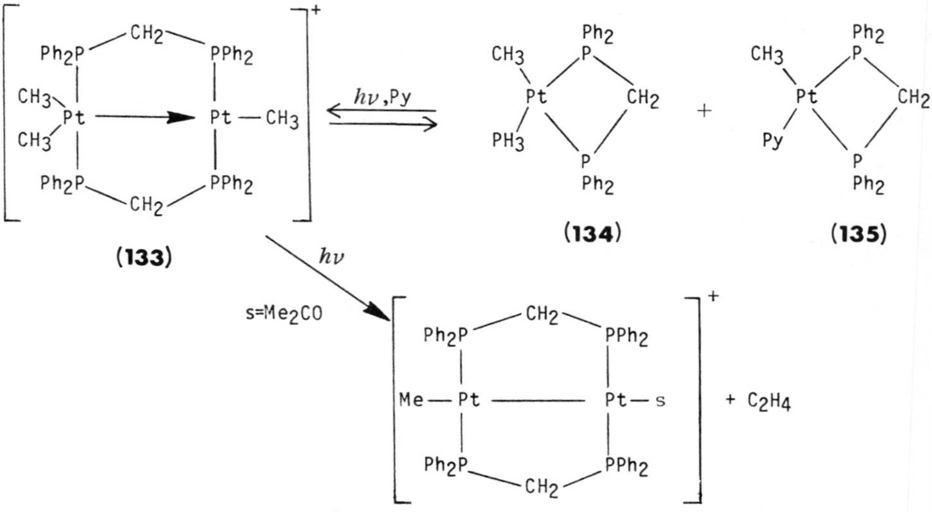

SCHEME 26

CH_3 groups from different molecules. The reaction is still more complicated in CH_2Cl_2 solution. The quantum yield for reductive elimination of C_2H_6 was 1.0 but the product was **136** [Eq. (126) with

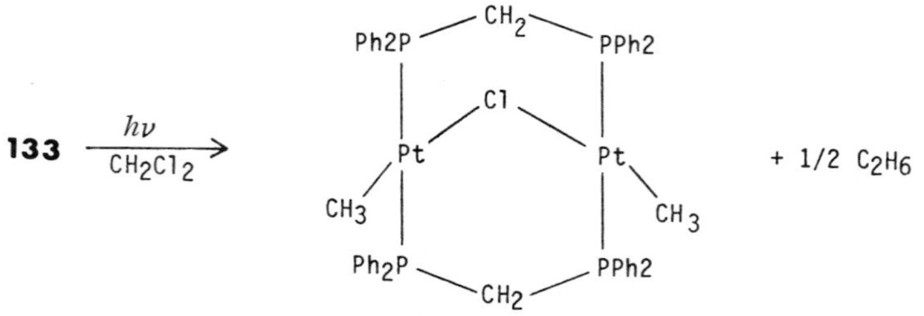

exactly $\frac{1}{2}$ mol of C_2H_6 formed per mol of **133** decomposed. The product
(**136**) is formed by reaction of the photogenerated coordinatively unsaturated intermediate with **133**, followed by Cl atom abstraction from solvent.
Thus, one photon induces loss of 2 mol of **133** and the overall quantum
yield of disappearance of **133** is 2.0. The significance of this study is the
demonstration of photoinduced coupling of two CH_3 groups to give
ethane, a reaction not commonly observed in metal–alkyl photochemistry.

I. *Cu, Ag, and Au Complexes*

1. $Cu(CH_3)H$

Ozin and co-workers (*157*) suggested that the photoinduced activation of methane by Cu atoms in 12 K matrices to produce CuH, CH_3, $CuCH_3$, and H atoms proceeds via a mechanism which involves the photolytic decomposition of the intermediate $HCuCH_3$. This complex was not isolated but was identified by its IR and ESR spectra. The authors suggested that Cu—CH_3 bond homolysis to give CuH and $\cdot CH_3$ radicals is the principal mode of photodecomposition.

2. $Ag(n\text{-}C_4H_9)[P(n\text{-}C_4H_9)]_3$

Whitesides and co-workers (*158*) have mentioned that pyrex-filtered irradiation of the title complex gave the organic products shown in Eq. (127). This product distribution strongly suggests the generation of

$$Ag(n\text{-}C_4H_9)[P(n\text{-}C_4H_9)]_3 \xrightarrow{h\nu} \text{1-butene} + C_4H_{10} + C_8H_{18} + C_4H_9CH(CH_3)C_3H_5 \quad (127)$$

$$\qquad\qquad (77\%) \qquad (39\%) \quad (35\%) \qquad\qquad (1\%)$$

n-butyl radicals by photoinduced homolysis of the Ag—butyl bonds.

3. $Au(CH_3)L$ $(L = PPh_3$ or $PPh_2Me)$

Irradiation of $Au(CH_3)L$ ($L = PPh_3$ or PPh_2Me) in the presence of the fluoroolefins C_2F_4, C_3F_6, and C_2F_3Cl has been reported to give insertion products of the type **137** [Eq. (128)] (*159*). Similar reaction with hexa-

$$AuCH_3L + F_2C{=}CF_2 \xrightarrow{h\nu} L{-}Au{-}CF_2CF_2CH_3 \quad (128)$$

$$\textbf{(137)}$$

fluoro-2-butyne gave the corresponding vinyl complex. Mechanistic details were not presented, although a free radical process was suggested.

Van Leeuwen and co-workers (*160*) later noted that irradiation of $AuCH_3(PPh_3)$ in $CDCl_3$ solution gave formation of $AuCl(PPh_3)$ and CH_3D [Eq. (129)], and their results strongly support the intermediacy of

$$Au(CH_3)PPh_3 + CDCl_3 \xrightarrow{h\nu} AuCl(PPh_3) + CH_3D \quad (129)$$

free radicals produced by homolysis of Au—CH_3 bonds. They observed strong CIDNP effects when monitoring by ^1H NMR, which supports this contention. Interestingly, one of the CIDNP polarizations observed was attributed to the methyl-exchange reaction of Eq. (130).

$$CH_3\cdot + CH_3{-}Au{-}PPh_3 \rightarrow CH_3{-}Au{-}PPh_3 + CH_3\cdot \quad (130)$$

Lappert's (161) trapping of CH_3 and CH_2SiMe_3 radicals upon irradiation of the $AuR(PPh_3)$ complexes in the presence of spin traps also implicates photoinduced homolysis of Au—R bonds.

4. $Au_2(CH_3)I\{(CH_2)_2PR_2\}_2$

The binuclear complex (138) has been reported by Fackler and Basil (162) to be extremely photosensitive, giving the products in Eq. (131). The

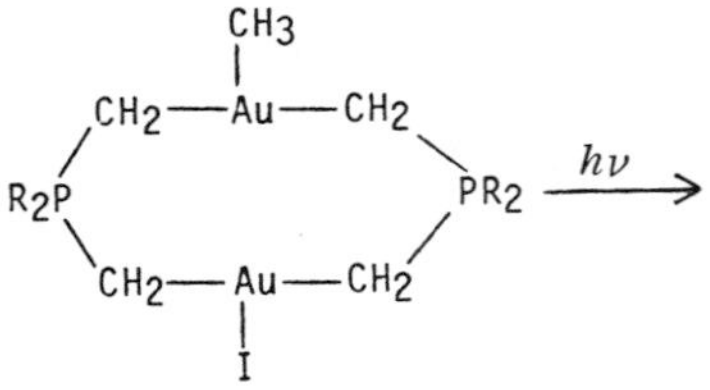

(138)

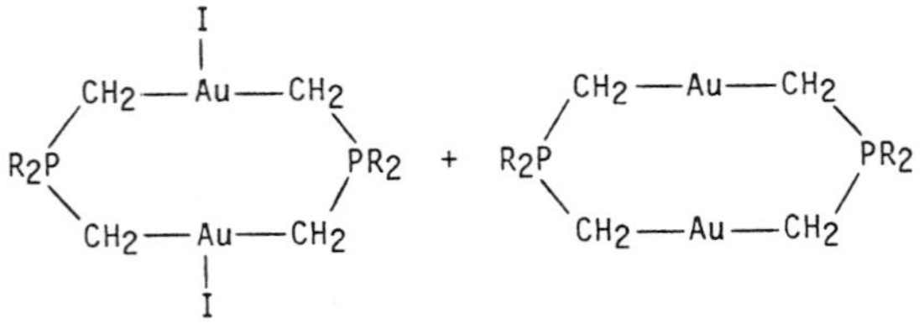

$$ \text{(139)} \qquad\qquad \text{(140)} \tag{131} $$

ratio of products 139 and 140 was reported to depend on the presence of excess CH_3I, with the yield of 139 apparently increasing with $[CH_3I]$. This reaction proceeds in toluene, THF, and halocarbon solutions. Photolysis of 138 in $CDCl_3$ solution gave formation of $CDCl_3$ which was taken as evidence for a free radical process with irradiation inducing homolysis of the Au—CH_3 bond.

III

PHOTOCHEMISTRY OF ALKYLIDENE AND VINYLIDENE COMPLEXES

These two different sets of complexes are discussed together because of their obvious structural resemblance.

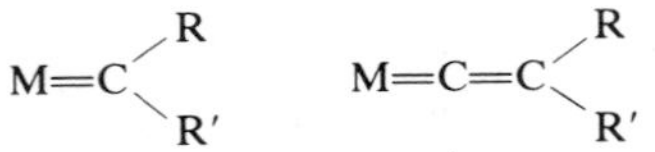

(alkylidene)   (vinylidene)

For the carbene complexes, only the $M(CO)_5$(alkylidene) (M = Cr, Mo, or W) class has been studied extensively, allowing a comprehensive photochemical picture to be assembled only for that class.

A. Cr, Mo, and W Complexes

1. $M(CO)_5$(alkylidene) (M = Cr, Mo, or W)

In this series of compounds, the tungsten complexes have been studied more extensively than the chromium analogues, and the less stable molybdenum derivatives have been scarcely examined. While making generalizations for the entire triad from one member is dangerous, all of the reported studies appear consistent with the comprehensive picture which now exists for the tungsten compounds. The electronic absorption data, where reported for members of this class of compounds, appear similar, with each complex generally showing a metal-to-ligand (carbene) charge-transfer (MLCT) transition in the visible spectral region with less intense LF transitions lying at higher energy (163–167). A representative spectrum of $W(CO)_5\{C(OMe)Ph\}$ is shown in Fig. 3 along with that of $W(CO)_6$ for comparison (165). The W → carbene charge-transfer band is centered at 402 nm with LF transitions appearing as shoulders in the range 300–370 nm.

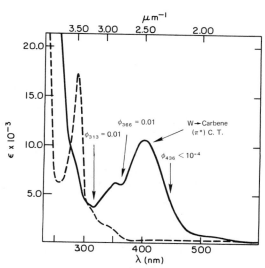

FIG. 3.   Electronic absorption spectrum of $W(CO)_5\{C(OMe)Ph\}$ (———) and $W(CO)_6$ (---) at 300 K in hexane solution.

A one-electron energy level diagram appropriate for the compounds in this series is shown in Fig. 4 and illustrates both $\sigma[a(\sigma)]$- and $\pi[1b_1(\pi)]$-bonding of the carbene to the metal (165). The MLCT transition is represented by the $b_2^2 \to b_2^1 2b_1^1$ one-electron transition whereas LF transitions can involve population of both the $2a_1$ and $3a_1$ orbitals. The position of the MLCT transitions in these complexes is a very sensitive function of the carbene substituents (163–166), implying that the position of the $2b_1(\pi^*)$ orbital can vary considerably.

A number of results now indicate that irradiation of these complexes can give CO loss but that CO loss does *not* occur from the MLCT state. The higher energy LF states are responsible for elimination of CO, but the quantum efficiency is poor with $\Phi = 0.01$–0.02. The latter contrasts with the high-efficiency CO loss from the parent hexacarbonyls ($\Phi = 0.7$–1.0) (167) and is apparently a reflection of rapid internal conversion from the active LF states to the inactive MLCT state. For example, photo-substitution of CH_3CN for CO in $W(CO)_5\{C(OMe)Ph\}$ was shown to occur cleanly and quantitatively to yield the cis-substituted product shown

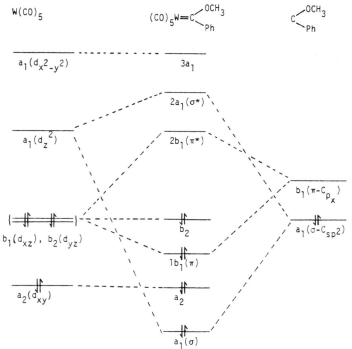

FIG. 4. Simplified one-electron energy level diagram for $W(CO)_5\{C(OMe)Ph\}$.

in Eq. (132) (*165*) but with a strong wavelength dependence. Irradiation

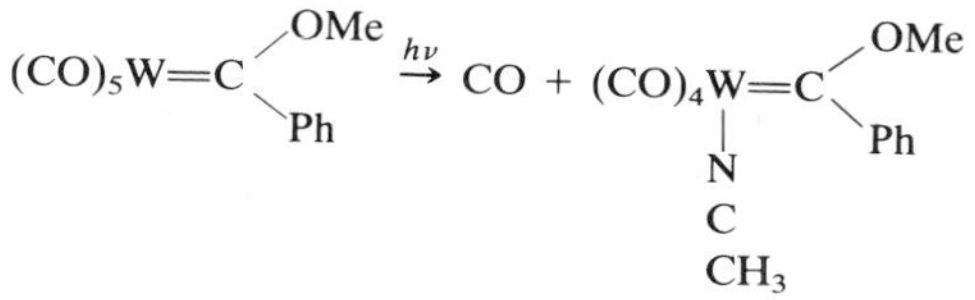

$$(CO)_5W{=}C\underset{\diagdown Ph}{\overset{\diagup OMe}{}} \xrightarrow{h\nu} CO + (CO)_4W{=}C\underset{\diagdown Ph}{\overset{\diagup OMe}{\underset{\underset{CH_3}{C}}{N}}} \qquad (132)$$

into the LF bands with 313 and 366 nm light gave photosubstitution of CO by CH_3CN with $\Phi = 0.01$. However, irradiation into the MLCT band with 436 nm light gave slow photosubstitution but with an immeasurably low quantum yield ($\Phi < 0.0001$) (*165*). Qualitatively similar results were obtained by Fong and Cooper (*166*) who did not measure quantum yields for the alkylidene metathesis reaction shown in Eq. (133) but did show that

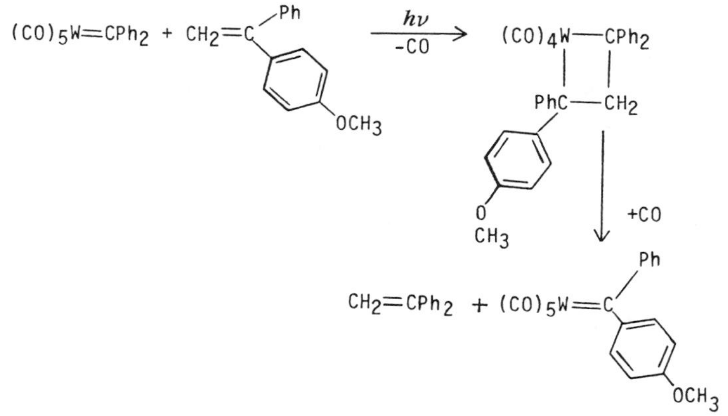

$$(CO)_5W{=}CPh_2 + CH_2{=}C\underset{\diagdown}{\overset{\diagup Ph}{}}\!\!\!\!\!\!\!\!\!\!\begin{smallmatrix}\\OCH_3\end{smallmatrix} \xrightarrow[-CO]{h\nu} \begin{smallmatrix}(CO)_4W{-\!\!-}CPh_2\\ |\quad\quad |\\ PhC{-\!\!-}CH_2\\ \\ \end{smallmatrix} \qquad (133)$$

$$CH_2{=}CPh_2 + (CO)_5W{=}C\underset{\diagdown}{\overset{\diagup Ph}{}}\!\!\!\!\!\!\!\begin{smallmatrix}\\OCH_3\end{smallmatrix}$$

it must proceed via CO loss and that the MLCT state is inactive toward this reaction. The reaction apparently proceeds via the intermediacy of the metallacyclobutane intermediate shown in Eq. (133).

Earlier results also indicated that irradiation of this class of compounds gave CO loss. Thus Fischer and Fischer (*168*) noted in 1974 that irradiation of Cr– and W–alkylidene complexes in the presence of a series of tertiary phosphines led to CO loss and formation of a mixture of *cis-* and *trans*-M(CO)$_4$(PR$_3$)(carbene) products [Eq. (134)]. Low-temperature

$$(CO)_5M{=}CR'R'' + PR_3 \xrightarrow{h\nu} CO + \textit{cis-} \text{ and } \textit{trans-}(CO)_4(PR_3)M{=}CR'R'' \quad (134)$$

irradiation experiments indicated that the cis derivatives formed first and that these subsequently isomerized to the trans products.

Ofele, Herberhold, and co-workers (*169–172*) have examined the photochemistry of a series of cyclic carbene complexes of Cr, Mo, and W, and their results appear consistent with CO dissociation. It was first noted that irradiation of complex **141** in boiling tetrahydrofuran led to the disproportionation reaction of Eq. (135). While the mechanism of this

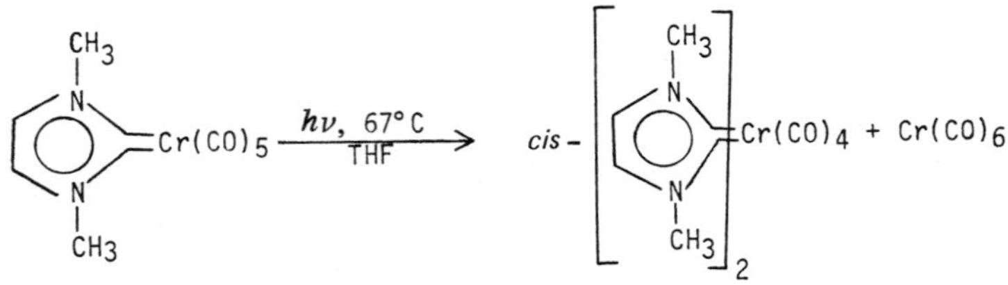

(141)

reaction is unknown, that outlined in Scheme 27 appears most reasonable. Stone (*173*) and co-workers elegantly demonstrated that a coordinatively unsaturated metal fragment (such as **142** in Scheme 27) will add across a metal–carbene double bond to give isolable bridging-carbene complexes (such as **143** in Scheme 27). Earlier, Fischer and Beck also (*174*) had observed the transfer of a carbene ligand from $Cp(CO)(NO)Mo=CR'R''$ to "$Fe(CO)_4$" to yield $(CO)_4Fe=CR'R''$ and "$Cp(CO)(NO)Mo$."

These workers also noted that irradiation of the cis-substituted biscarbene complexes **144a–144c** produced isomerization to the less stable trans derivatives [Eq. (136)] (*169, 170*). The latter thermally reconvert to the more stable cis forms. In the presence of pyridine, some photosubstitution of pyridine for CO was detected in the case of the Mo derivative and the

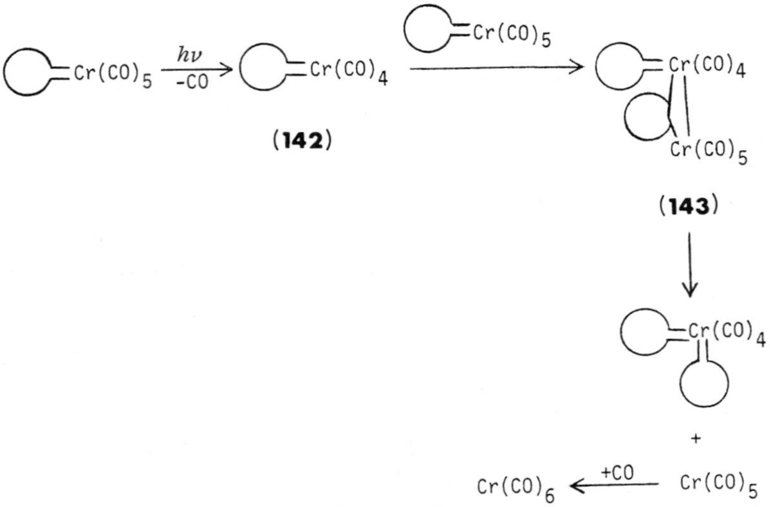

SCHEME 27

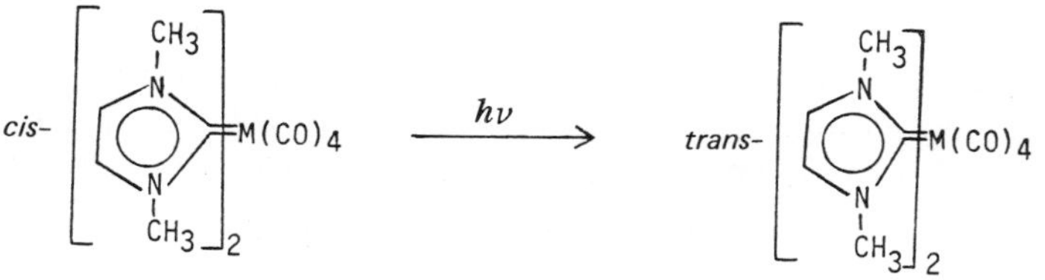

$$cis-\left[\underset{\underset{CH_3}{N}}{\overset{\underset{CH_3}{N}}{\bigcirc}}=M(CO)_4\right]_2 \xrightarrow{h\nu} trans-\left[\underset{\underset{CH_3}{N}}{\overset{\underset{CH_3}{N}}{\bigcirc}}=M(CO)_4\right]_2 \quad (136)$$

M=Cr (**144 a**)

M=Mo (**144 b**)

M=W (**144 c**)

authors proposed a photoinduced CO dissociation process to explain all of their results (*169*). Similar photoisomerization results were also noted by Lappert *et al.* (*175*) on the same Mo compound and these latter workers also observed photosubstitution of Co by phosphine and phosphite ligands to give $Mo(CO)_3L_2$(carbene) derivatives.

Dahlgren and Zink (*176*) also examined the photochemistry of $(CO)_5W\{C(OMe)Ph\}$ and $(CO)_5W\{C(OMe)Me\}$ using a 405 nm irradiation source. Consistent with the above discussions they detected little or no photosubstitution chemistry, since irradiation was directly into the MLCT band. They also observed no detectable replacement of the carbene ligand upon photolysis under a CO atmosphere.

Neumann (*177*) showed that irradiation of the carbene complexes $(CO)_5W\{CPh_2\}$, $(CO)_5W\{C(OCH_3)Ph\}$, and $(CO)_5Cr\{CO(CH_2)_2CH_2\}$ under ^{13}CO atmosphere led to ^{13}CO incorporation into the complexes, a result which implies photoinduced CO dissociation. They also irradiated these complexes under H_2 atmosphere and observed slow, if any, hydrogenation of the carbene ligands in the tungsten complexes but more rapid hydrogenation for the Cr complex [Eq. (137)] (*177*).

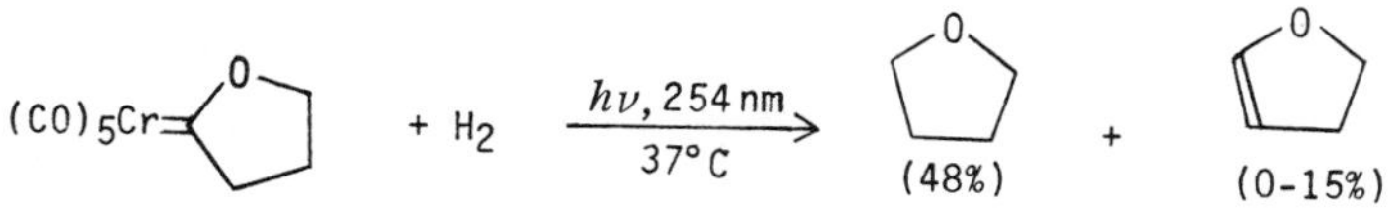

$$(CO)_5Cr=\overset{O}{\underset{}{\bigvee}} + H_2 \xrightarrow[37°C]{h\nu, 254\,nm} \underset{(48\%)}{\overset{O}{\bigcirc}} + \underset{(0-15\%)}{\overset{O}{\bigcirc}} \quad (137)$$

Casey and co-workers (*178, 179*) also studied the photochemistry of olefin-substituted carbene complexes and observed CO dissociation. For example, irradiation of a 1.38:1 mixture of the *Z* and *E* isomers of **145** led to CO substitution by the allyl-carbene group to give the isolable carbene–olefin complex **146** [Eq. (138)], a reaction which demonstrates CO dissociation. (*Z*)-**145** has the proper stereochemistry to yield **146** upon photoinduced CO loss. However, the *E* isomer does not. Yet, the authors observed simultaneous disappearance of the *E* isomer upon irradiation of the E/Z mixture, but at a rate half as fast as that of the Z isomer. The authors proposed that photoisomerization of the carbene ligand occurred and that this was a consequence of population of a MLCT excited state. Since there is restricted thermal rotation about the carbene–nitrogen bond in **145**, the resonance form given in Eqs. (138) and (139) must be

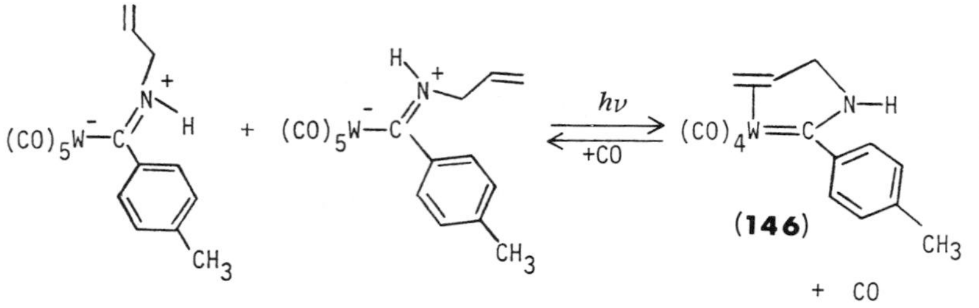

$$(138)$$

(Z)-(**145**) (E)-(**145**)

important. Tungsten-to-nitrogen charge transfer would yield an excited state with carbene–nitrogen single bond character in which free rotation about this bond could occur readily [Eq. (139)]. Since irradiation was into

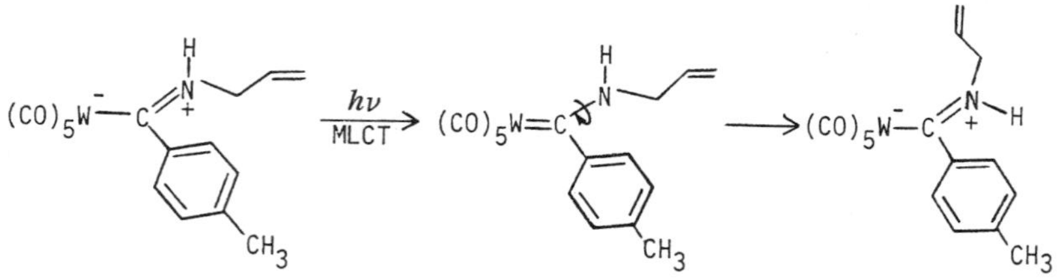

$$(139)$$

the spectral region dominated by LF states (254 and 350 nm), CO loss from the LF states competed with internal conversion to the MLCT state that gave ligand isomerization, and hence both reactions were observed.

It has also been noted by Edwards and Rausch (*180*) that the diphenyl-carbene complex $(CO)_5W{=}CPh_2$ is photosensitive, decomposing upon irradiation to give $W(CO)_6$ and tetraphenylethylene as primary products, along with products formed by hydrogen abstraction, namely, Ph_2CH_2 and $Ph_2CHCHPh_2$ [Eq. (140)]. Similar products result when the carbene

$$(CO)_5W{=}CPh_2 \xrightarrow{h\nu} W(CO)_6 + Ph_2C{=}CPh_2 + Ph_2CH_2 + Ph_2CHCHPh_2 \quad (140)$$

complex thermally decomposes. In view of Neumann's (*177*) photoinduced $^{13}CO/^{12}CO$ exchange in this complex, it is likely that CO dissociation is the primary photoreaction and that the products of Eq. (140)

probably arise via subsequent transformations of the photogenerated $(CO)_4W{=}CPh_2$ intermediate.

The observed photochemistry of this class of compounds has been exploited for both synthetic and mechanistic applications. In our study of $(CO)_5W{=}C(OMe)Ph$, we observed that $-40°C$ irradiation of the complex in the presence of alkynes led to the formation of alkyne–carbene complexes [Eq. (141)] (165). The $PhC{\equiv}CPh$ adduct complex was isolated

$$(CO)_5W{=}C\begin{smallmatrix}OMe\\\\Ph\end{smallmatrix} \quad \xrightarrow[\substack{-40°C\\-CO}]{h\nu} \quad (CO)_4W{=}C\begin{smallmatrix}OMe\\\\Ph\end{smallmatrix} \quad \rightleftharpoons \quad (CO)_4W{-}\overset{Ph}{\underset{}{C}}{-}OMe$$

$$+ \ R{-}C{\equiv}C{-}R'$$

$$\underset{R}{C}{\equiv}\underset{R'}{C} \qquad\qquad \underset{R}{C}{=}\underset{R'}{C} \tag{141}$$

$$\textbf{(147a)} \qquad\qquad\qquad \textbf{(147b)}$$

and characterized, but the $PhC{\equiv}CH$, $BuC{\equiv}CH$, $CH_3C{\equiv}CCH_3$, and $PhC{\equiv}CH_3$ adducts were only characterized by low-temperature IR spectroscopy. The spectral data were not sufficient to determine if these complexes existed as discrete alkyne–carbene complexes **(147a)** or as metallacyclobutene complexes **(147b)**, although the subsequent thermal behavior of the compounds suggests the intermediacy of structure **147b**. These complexes are not stable above 0°C, and warm-up to room temperature led to their decomposition to give either indene derivatives or polymer, depending on the alkyne substituents. Similar products resulted upon irradiation at 25°C. The reaction sequences shown in Scheme 28 were proposed to account for the observed products.

These results illustrate the utility of photochemical methods for preparing and characterizing thermally unstable organometallic intermediates via low-temperature irradiation. In many cases, photoinduced ligand loss can occur at temperatures low enough that intermediates thus formed do not have enough internal energy to further react so that they can be characterized. The subsequent thermal decomposition of such compounds can then be monitored upon warm-up of the solutions.

McGuire and Hegedus (181) demonstrated the synthesis of a series of β-lactams by the photolytic reactions of chromium–carbene complexes with imines [Eq. (142)]. The yields of the organic products were good

$$(CO)_5Cr{=}C\begin{smallmatrix}OMe\\\\R'\end{smallmatrix} \quad + \quad \underset{R^2}{\overset{H}{C}}{=}N\begin{smallmatrix}R^3\\\\\end{smallmatrix} \quad \xrightarrow{h\nu} \quad \begin{array}{c} Me \quad R^2\\ \text{(β-lactam ring)} \end{array} \tag{142}$$

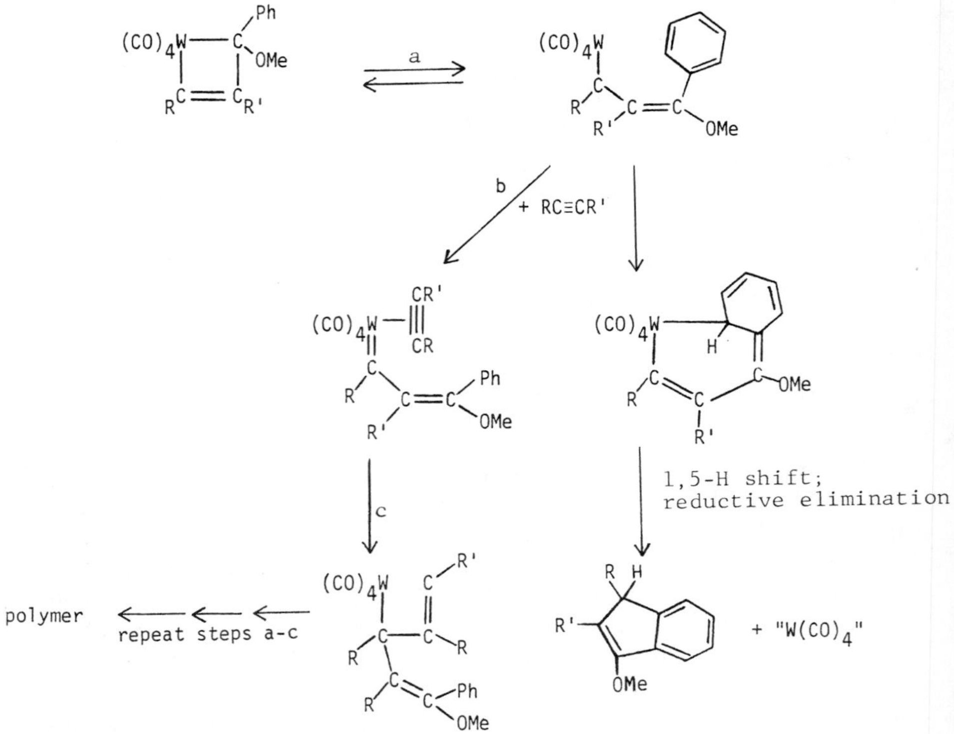

SCHEME 28

(38–81%). The mechanism of the reaction is not known; photoinduced CO loss and photoinduced CO–carbene coupling to yield a ketene intermediate were considered.

2. $Cp(NO)(CO)Mo{=}C(OCH_3)Ph$

Fischer and co-workers (174, 181a) reported that irradiation of a mixture of the title complex with $Fe(CO)_5$ resulted in the carbene-transfer reaction shown in Eq. (143). While the details of this reaction are not known, we

$$Cp(NO)(CO)Mo{=}C(R)Ph + Fe(CO)_5 \xrightarrow{h\nu} (CO)_4Fe{=}C(R)Ph + Cp(NO)(CO)_2Mo \quad (143)$$
$$(R = OCH_3, OC_2H_5, \text{ or } N(CH_3)_2)$$

suspect that it proceeds via photoinduced CO loss from $Fe(CO)_5$ with subsequent addition of photogenerated $Fe(CO)_4$ across the Mo=carbene double bond to give an intermediate such as (148). This would collapse to

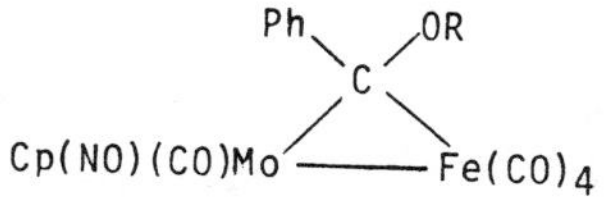

$$
\begin{array}{c}
Ph\diagdown \quad \diagup OR \\
C \\
\diagup \quad \diagdown \\
Cp(NO)(CO)Mo \diagdown \quad \diagup Fe(CO)_4
\end{array}
$$

(148)

yield the Fe–carbene product and an Mo fragment which could capture CO to yield the observed $CpNO(CO)_2Mo$.

3. $Cp(CO)_2W[\eta^1\text{-}C(Ph)N(Ph)]$

Irradiation of the title complex was reported (182) to lead to CO loss followed by an η^1–η^2 rearrangement of the carbene-like iminoacyl ligand to give the complex $Cp(CO)W[\eta^2\text{-}C(Ph)N(Ph)]$, in which the iminoacyl ligand is bound to the metal through both the carbon and the lone pair of the nitrogen atom. No further details were given.

B. *Mn and Re Complexes*

1. $Cp(CO)_2Mn{=}CRR'$

In a synthetic study, Fischer and Besl (183) noted that they could photoinduce substitution of a diphosphine ligand for the two carbonyls in complex **149** [Eq. (144)]. The product **150** was isolated in 82% yield and

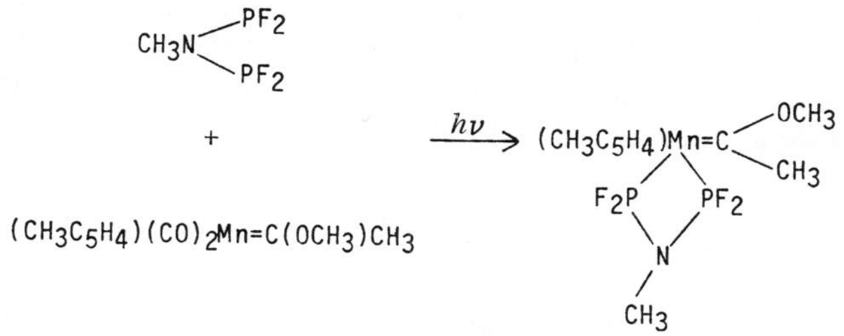

$$
\begin{array}{c}
CH_3N\diagdown PF_2 \\
\diagdown PF_2 \\
+ \\
(CH_3C_5H_4)(CO)_2Mn{=}C(OCH_3)CH_3
\end{array}
\xrightarrow{h\nu}
(CH_3C_5H_4)Mn{=}C\diagdown ^{OCH_3}_{CH_3}
\begin{array}{c}
F_2P \quad PF_2 \\
\diagdown N \diagup \\
CH_3
\end{array}
\qquad (144)
$$

(149)   **(150)**

fully characterized, but no further details of the photochemistry of **149** were given. This reaction likely proceeds via photoinduced CO loss from **149**.

The acyl-substituted carbene complex **151** has also been examined by Herrmann (184) and found to rearrange to give the diphenylketene complex **152** in very low yield [Eq. (145)], but no mechanistic details were reported.

$$(C_5H_4R)(CO)_2Mn{=}C\underset{\text{Ph}}{\overset{\text{O}}{\underset{\parallel}{\overset{\parallel}{C-Ph}}}} \xrightarrow{h\nu} (C_5H_4R)(CO)_2Mn-\underset{\underset{\text{O}}{\parallel}}{\underset{\parallel}{\overset{\text{Ph}}{\overset{\diagdown}{\underset{C}{\parallel}}}}}\overset{\text{Ph}}{\diagup} \qquad (145)$$

(151)   R=H, CH$_3$ **(152)** (3–5%)

The photochemistry of $Cp(CO)_2Mn{=}CPh_2$ has been studied by Wright and Vogler (*185*). This complex shows an intense absorption band at 380 nm ($\varepsilon \simeq 7000$) which was logically attributed to an Mn-to-carbene charge-transfer transition. Ligand-field transitions were thought to lie at high energy as they do for the isoelectronic complex $CpMn(CO)_3$. Irradiation of solutions of $Cp(CO)_2Mn{=}CPh_2$ led to disappearance of the UV-visible and IR bands of the complex and to general decomposition with deposition of a brown undefined solid. The quantum yield for this process was 0.10 ± 0.01 and independent of solvent (THF, EtOH, CH_3CN, $CHCl_3$, or cyclohexane). The fate of the carbene ligand was not determined. Irradiation in either a THF or EtOH glass at 77 K gave a color change from yellow to red and the appearance of a weak emission characteristic of the diphenylmethyl radical $Ph_2CH\cdot$. The ESR spectra of such irradiated 2-MeTHF glasses showed the presence of Mn(II), solvent radicals, and an incompletely characterized material which showed a doublet signal, implying a free electron on carbon coupled to an α hydrogen.

This study did not lead to an unambiguous definition of the photochemistry of $Cp(CO)_2Mn{=}CPh_2$, but the authors suggested that their results were consistent with initial formation of a high-spin Mn(II) complex and a diphenylcarbene radical anion [Eq. (146)]. Such reaction is reasonable

$$Cp(CO)_2Mn{=}C\underset{\text{Ph}}{\overset{\text{Ph}}{\diagup}} \xrightarrow{h\nu} \cdot CPh_2^- + Mn(II) \qquad (146)$$

from an active Mn → carbene charge-transfer excited state but is quite unprecedented in known carbene complex chemistry. The authors state that the high-yield formation of $CpMn(CO)_2(t\text{-BuNC})$ upon irradiation in the presence of excess t-BuNC is not consistent with CO loss. However, the actual yield of the t-BuNC complex was not given and thus an evaluation of the validity of this conclusion cannot be made. Also, it is not obvious how the formation of $CpMn(CO)_2(t\text{-BuNC})$ can be reconciled with reaction (146). The ESR conclusions also have to be viewed with great caution since it is well known that the very high sensitivity of this technique allows detection of radical species in low concentration that can arise via competing photoreactions of minor importance. The sequence of reactions shown in Scheme 29 could explain the general decomposition observed and the *t*-BuNC influence. The carbene-coupling reaction is consistent with thermal decomposition pathways of other carbene complexes.

2. $Cp_2Mn_2(CO)_4(\mu\text{-}C{=}CHR)$

Several vinylidene complexes containing terminal and bridging vinylidene ligands are known, but the title complex and simple derivatives are the only ones whose photochemistry has been studied. Caulton and co-workers (*186*) observed that irradiation of **153** in the presence of cyclohexene gave disruption of the dimer and formation of the mononuclear vinylidene and cyclohexane complexes **154** and **155** [Eq. (147)], along

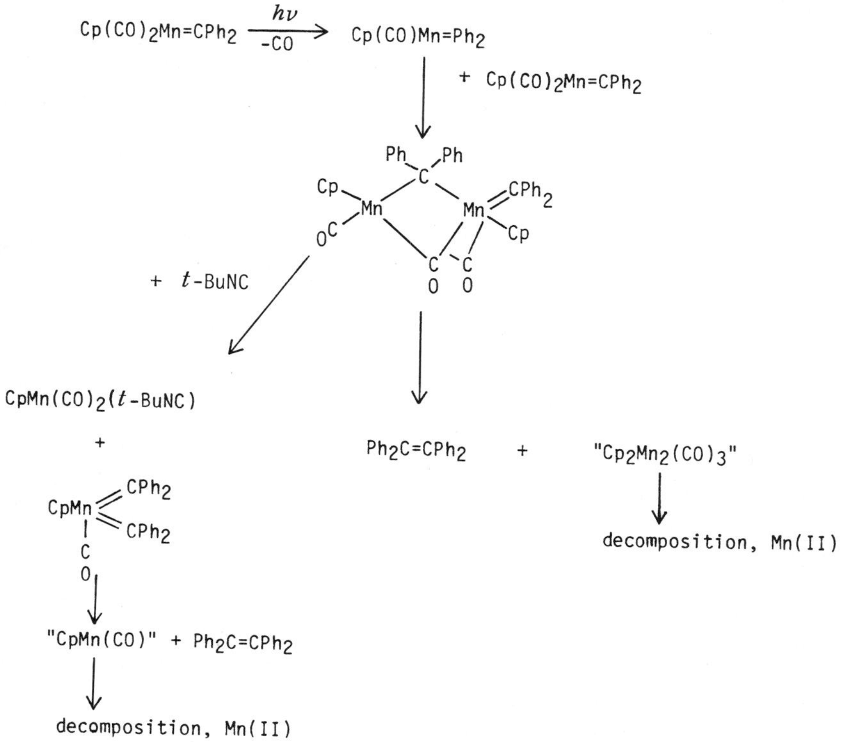

Cp(CO)$_2$Mn=CPh$_2$ $\xrightarrow[-CO]{h\nu}$ Cp(CO)Mn=Ph$_2$

+ Cp(CO)$_2$Mn=CPh$_2$

+ *t*-BuNC

CpMn(CO)$_2$(*t*-BuNC)

+

CpMn(CPh$_2$)(CPh$_2$)(C)(O)

"CpMn(CO)" + Ph$_2$C=CPh$_2$

decomposition, Mn(II)

Ph$_2$C=CPh$_2$ +   "Cp$_2$Mn$_2$(CO)$_3$"

decomposition, Mn(II)

SCHEME 29

with some $CpMn(CO)_3$ and an unidentified compound. No further details were given. A reaction similar to that in Eq. (147) had been observed

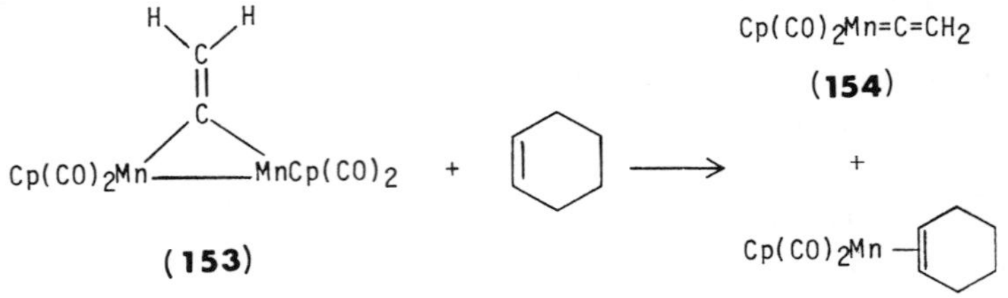

$$Cp(CO)_2Mn=C=CH_2 \tag{154}$$

$$(147)$$

(153)

$Cp(CO)_2Mn$ —⬡

(155)

earlier by Antonova *et al.* (*187*) who irradiated $Cp_2(CO)_4Mn_2(\mu\text{-}C=CHPh)$ in the presence of PPh_3 and observed formation of $Cp(CO)_2Mn=C=CHPh$ and $Cp(CO)_2MnPPh_3$.

3. *[Cp(NO)(PPh₃)Re=CHPh]*

This complex, which exists as a $> 99:1$ mixture of anticlinal (**156a**) and synclinal (**156b**) geometric isomers at 25°C, was shown by Gladysz and co-workers (*188*) to undergo the novel photoisomerization reaction shown in Eq. (148).

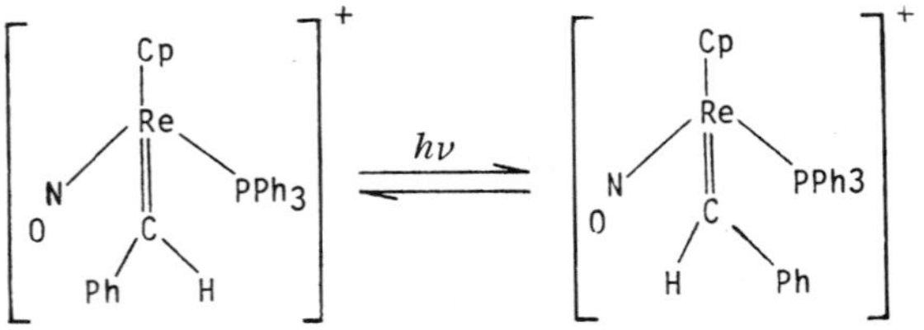

$$(148)$$

(156a)   (156b)

Irradiation between -20 and $-78°C$ gave a photostationary 55:45 mixture of **156a** and **156b** after 3 hours. Continual photolysis did not change this ratio, but, in the dark, slow thermal isomerization occurred to give the original 99:1 mixture. Similar results were obtained on the corresponding propylidene complex $[Cp(NO)(PPh_3)Re=CHCH_2CH_3]^+$. The absorption spectrum of **156a** showed an intense band at 365 nm ($\varepsilon = 13,000$) which is probably an Re-to-carbene change-transfer transition. In accord with such an assignment, the corresponding propylidene complex does not show this band but instead only a featureless tail into the visible region. Charge transfer to benzylidene should be facilitated by the Ph substituent as compared to propylidene. Photoisomerization from an MLCT state is entirely reasonable, as such an excited state would formally have only an Re—carbon single bond about which free rotation could occur [Eq. (149)].

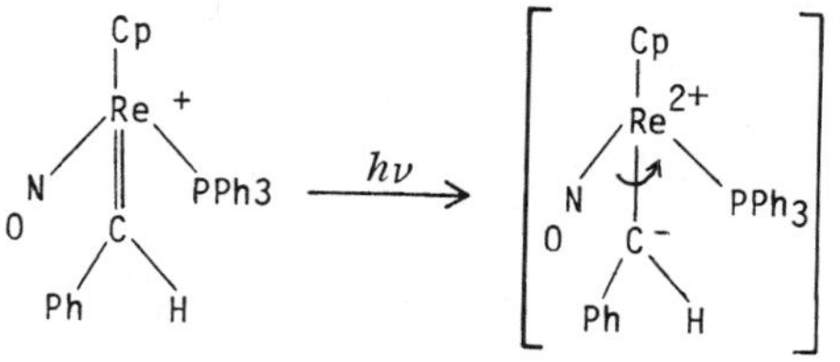

(149)

C. Fe and Ru Complexes

1. $FeCp(CO)_2[C(Ph)N(R)]$

Adams and co-workers (*182, 189*) showed that irradiation of the title complexes led to CO loss and formation of various products depending on the nature of R. When the substituent R on the nitrogen atom was an alkyl group, either migration with formation of an isocyanide ligand or formation of an η^2-iminoacyl occurred. When the substituent was an aryl group, orthometallation of the *N*-aryl ring occurred to give the compound $CpFe(CO)[C(Ph)N(H)(p\text{-}CH_3C_6H_3)]$, in which the *p*-tolyl ring is bonded to the iron atom in an ortho position.

2. $[Cp(CO)(PPh_3)Fe{=}CHPh]^+$

Gladysz and co-workers (*188*) attempted to study the photochemistry of this complex but they concluded that it was not stable under photochemical conditions. It rapidly decomposed upon $-78°C$ irradiation with complete disappearance of the benzylidene NMR resonances.

3. $Cp_2(CO)_3M_2(\mu\text{-}CRR')$ $(M = Fe$ or $Ru)$

Knox and co-workers (*190*) reported that irradiation of the Fe–and Ru–carbene complexes **157** in the presence of alkynes results in loss of CO and insertion of the alkyne into a metal–carbene bond to give **158** [Eq. (150)]. A variety of alkynes ($R''{=}H$, CH_3, or CO_2Me) and

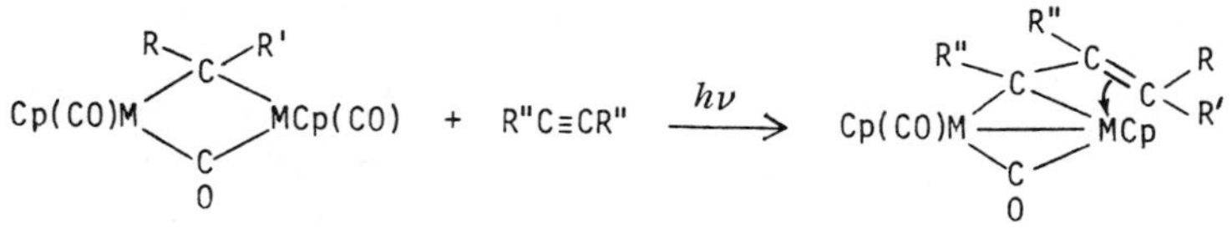

(150)

(157)   **(158)**

carbene substituents can be tolerated in this chemistry. The overall course of the reaction is unknown but it likely proceeds via photoinduced CO loss followed by coordination of the alkyne and then alkyne–carbene coupling. Photoinduced CO loss was also indicated by the reaction shown in Eq. (151). Irradiation of $Cp_2(CO)_3Ru_2(\mu\text{-}CMe_2)$ in the presence of

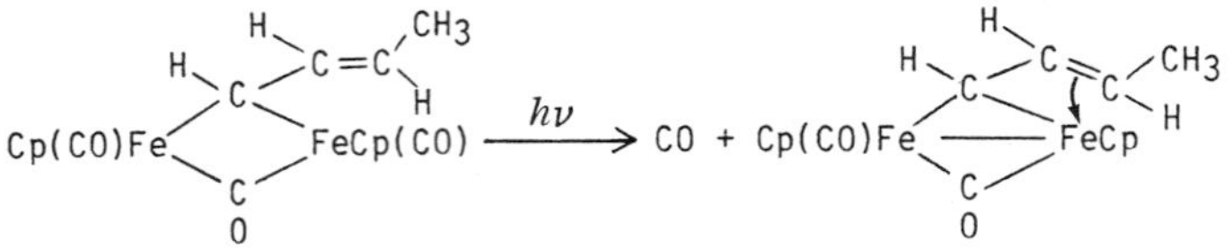

$$(151)$$

PMePh$_2$ was found to yield the simple substituted product in Eq. (152), a result which further supports photoinduced CO dissociation from the intact dimer (190).

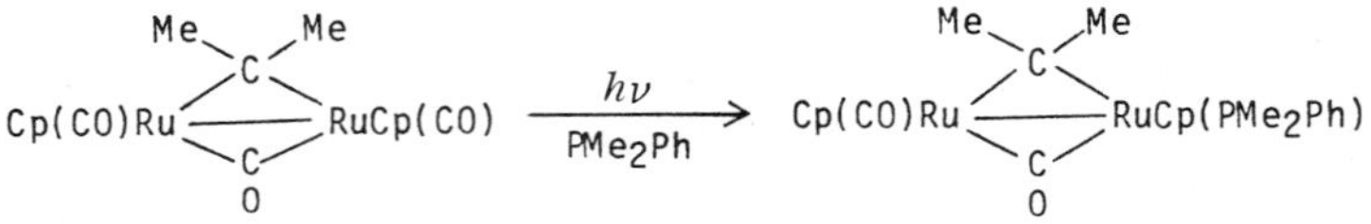

$$(152)$$

4. $Cp_2(CO)_2Ru_2(\mu\text{-}CHMe)_2$

Knox et al. reported that photolysis of the bis(carbene) complex 159 in the presence of acetylene gives ejection of the two carbene ligands and formation of an acetylene complex [Eq. (153)] (191). However, no further

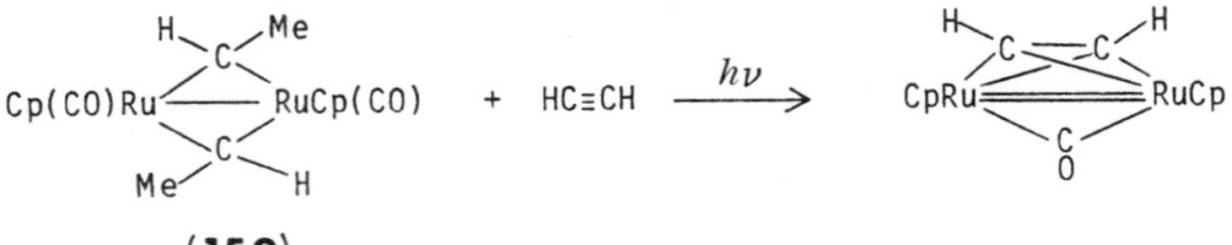

$$(153)$$

details of the photochemistry were given nor was the fate of the carbene ligands determined.

IV

PHOTOCHEMISTRY OF ALKYLIDYNE COMPLEXES

Only a few alkylidyne complexes have had their photochemical properties examined and this is clearly an area in need of further development. The reported studies involve both mononuclear and polynuclear alkylidyne complexes.

A. W Complexes

1. trans-$Cl(CO)_4W\equiv CC_6H_4CH_3$

The title complex has been shown to be photosensitive in the presence of acetylacetone (Hacac) [Eq. (154)] (192). The hydroxyarylacetylene ligand

$$\textit{trans-}Cl(CO)_4W\equiv CC_6H_4CH_3 + \text{Hacac} \xrightarrow[-60°C]{h\nu} \textit{trans-}Cl(CO_2(acac)W-\underset{\underset{C_6H_4CH_3}{|}}{\overset{\overset{OH}{|}}{\overset{C}{\|}}} \tag{154}$$

in the crystallographically characterized product was the first of its type reported, and it resulted from coupling of a coordinated CO with the alkyne ligand, followed by protonation of the carbonyl oxygen in a subsequent step. The role of light in this process was not elucidated but photoinduced CO loss is a likely reaction.

B. Os Complexes

1. $(CO)_2(PPh_3)_2Os\equiv CPh$ and $(CO)(PPh_3)_2(Cl)Os\equiv CPh$

Vogler *et al.* (193) reported that irradiation of either $(CO)_2$-$(PPh_3)_2Os\equiv CPh$ (**160**) or $(CO)(PPh_3)_2(Cl)Os\equiv CPh$ (**161**) in the presence of HCl gives protonation of the alkylidyne ligand and formation of the alkylidene complex **162** [Eq. (155)]. Both starting complexes show

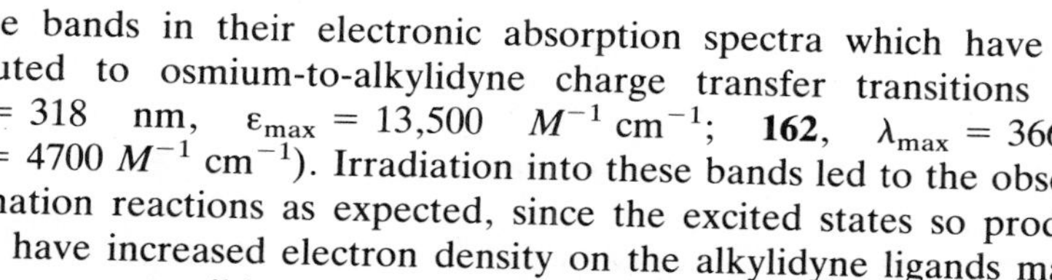

$$
\begin{array}{c}
\text{HCl} + (CO)_2(PPh_3)_2Os\equiv CPh \xrightarrow[\phi_{365}=0.1]{h\nu} \\
\text{(160)} \\[2ex]
\text{HCl} + (CO)(PPh_3)_2(Cl)Os\equiv CPh \xrightarrow[\phi_{313}=0.2]{h\nu} \\
\text{(161)}
\end{array}
\longrightarrow
\begin{array}{c}
\underset{\text{(162)}}{
\begin{array}{c}
Ph\diagdown_{C}\diagup^{H} \\
\underset{Ph_3P}{\overset{OC}{\diagdown}}\underset{|}{\overset{\|}{Os}}\underset{Cl}{\overset{\diagup PPh_3}{\diagdown Cl}}
\end{array}
}
\end{array}
\tag{155}
$$

intense bands in their electronic absorption spectra which have been attributed to osmium-to-alkylidyne charge transfer transitions (**160**, $\lambda_{max} = 318$ nm, $\varepsilon_{max} = 13{,}500$ M^{-1} cm^{-1}; **162**, $\lambda_{max} = 366$ nm, $\varepsilon_{max} = 4700$ M^{-1} cm^{-1}). Irradiation into these bands led to the observed protonation reactions as expected, since the excited states so produced would have increased electron density on the alkylidyne ligands making them more susceptible to proton addition [Eq. (156)]. The authors further suggested that the MLCT excited states of these complexes possess "bent"

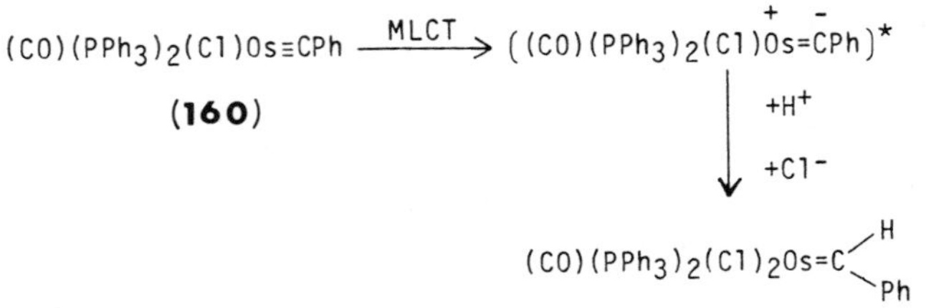

$$(CO)(PPh_3)_2(Cl)Os\equiv CPh \xrightarrow{MLCT} ((CO)(PPh_3)_2(Cl)\overset{+}{Os}\overset{-}{=}CPh)^*$$

(160)

$$\downarrow \begin{array}{l} +H^+ \\ +Cl^- \end{array}$$

(156)

$$(CO)(PPh_3)_2(Cl)_2Os=C\overset{\displaystyle H}{\underset{\displaystyle Ph}{\big\langle}}$$

(162)

carbyne ligands as illustrated below and that the photochemistry of carbyne and nitrosyl complexes may be similar.

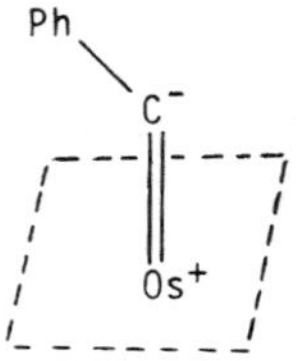

2. $HCCo_3(CO)_9$ and $CH_3CCo_3(CO)_9$

Of the large number of μ_3-alkylidyne clusters known, only the title complexes have had their photochemical properties reported (*194*). The structure of $CH_3CCo_3(CO)_9$ (**163**) consists of a triangle of metal–metal

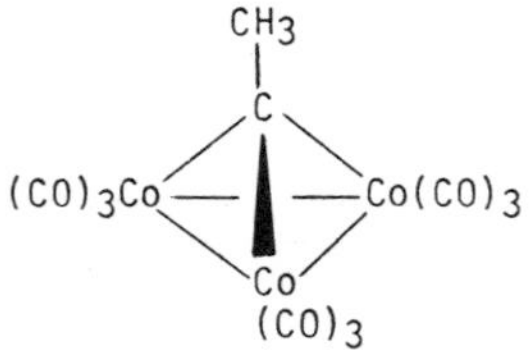

(163)

bonded Co atoms with a symmetrically bridging μ_3-CCH_3 ligand. All CO ligands are terminal. The electronic absorption spectrum of **163** is reproduced in Fig. 5 (*194*). The low-lying bands in this and related compounds have been attributed to transitions localized within the Co_3 metal framework (*194*), a conclusion also supported by a detailed theoretical treatment (*195*), as well as photoelectron (*195, 196*) and ESR (*197*) data.

The observed photochemistry of the two title compounds is summarized in Scheme 30. In the absence of added ligand, **163** and **164** are photostable. However, photolysis in the presence of H_2 (1 atm) gives rapid cluster fragmentation. If excess CO is present, $Co_2(CO)_8$ froms in quantitative

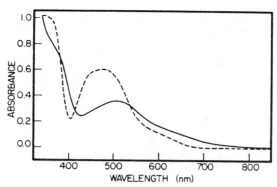

FIG. 5. Electronic absorption spectra of $CH_3CCo_3(CO)_9$ at 300 K (——) and 77 K (– – –) in EPA solution.

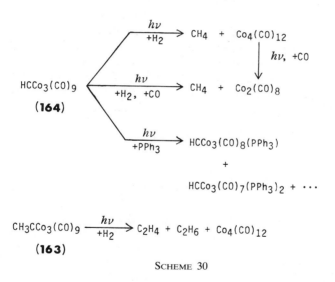

SCHEME 30

yield, but if excess CO is not available, $Co_4(CO)_{12}$ is produced. Photolysis of **164** in the presence of D_2 gave only CHD_3 as the organic product. The conversion of **163** and **164** into $Co_4(CO)_{12}$ is obviously not direct and must occur through several intermediates. The transient formation of $HCo(CO)_3$ is implicated by the observation that irradiated solutions of **164** are effective at catalytically isomerizing 1-hexene to a mixture of *cis*- and *trans*-2-hexene. The isomerization is *photocatalytic* since it continued long after the light was turned off.

The detailed mechanism of these reactions is not known, but the authors considered both photoinduced cleavage of a Co—Co bond to give a diradical intermediate such as 165 and photoinduced CO loss as possible

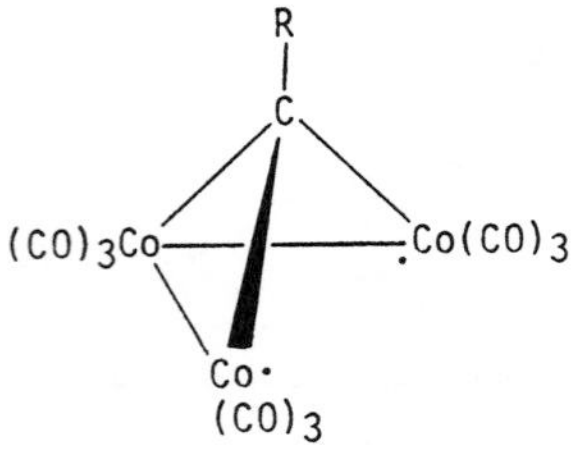

(165)

primary photoreactions. The former is consistent with the nature of the low-lying electronic transitions, whereas the latter is implicated by the photoinduced PPh_3 substitution results (Scheme 30). Carbon monoxide loss could also occur subsequent to the formation of 165 since 17-electron organometallics have precedent for being substitution labile. In any case, a coordination site must be opened for addition of H_2 to the cluster. These results demonstrate the necessity of having H_2 present in order to induce cluster fragmentation since the latter can proceed only by removal of the μ_3-CR group, and that is best accomplished through hydrogenation.

V

SUMMARY

The photochemistry of metal alkyl, alkylidene, and alkylidyne complexes has advanced considerably since the earlier review on this subject (9), in which the reported studies of Cp_2MR_2 (M = Ti, Zr, Hf, U, Nb, or Ta), $CpM(CO)_3R$ (M = Cr, Mo, or W), and $CpM(CO)_2R$ (M = Fe or Ru) complexes presented a confusing picture of the photochemical properties of these compounds. More recent studies have clarified the situation significantly. On the basis of all the results reported herein, it can now be generally assumed that if an alkylmetal complex possesses carbonyl ligands, then photolysis will likely induce CO dissociation and not metal–alkyl bond homolysis. Of course, as with any generalization, there are apparent exceptions and there will certainly be more as further studies are conducted. Metal–alkyl bond homolysis appears to dominate the photochemistry of complexes such as Cp_2TiMe_2 and $Cr(norbornyl)_4$ that do not possess carbonyl ligands. In these complexes, however, the photoinduced radical pairs so generated can recombine so efficiently that they may appear to give nonradical products. Furthermore, it is now clear that great caution must be used in interpreting the photochemistry of metal alkyl complexes. No single experimental approach is sufficient to define unambiguously the photochemistry of a given compound. Electron spin resonance studies, in particular, must be interpreted with great caution because of the extreme sensitivity of this technique: it can detect trace quantities of radicals formed by minor photoreactions. Deuterium labeling studies can be very informative, but they alone are not sufficient to define primary photochemical paths, nor are matrix isolation studies sufficient to determine the initial effect of light since the matrix used can suppress certain types of photochemical transformations. In this day and age, a proper photochemical study requires the correct combination of several of these and related approaches.

Since only a few alkylidene and alkylidyne complexes have been investigated, generalizations cannot presently be made. However, it does appear that these complexes will often show low-lying metal-to-ligand charge-transfer states which do not appear to be active toward CO or other ligand dissociation. These charge transfer states do appear to show interesting photochemistry of their own, giving isomerization of carbene ligands as well as bimolecular reactions of the excited molecules. These are processes which take advantage of the greatly different electron density on the organic ligand as compared to the ground state complex. The alkylidene and alkylidyne complexes represent a fertile, relatively unexplored area of investigation which should lead to significant advances in the years ahead.

REFERENCES

1. B. R. James, *Adv. Organomet. Chem.* **17**, 319 (1979).
2. R. L. Pruett, *Adv. Organomet. Chem.* **17**, 1 (1979).
3. B. R. James, in "Comprehensive Organometallic Chemistry" (G. Wilkinson and F. G. A. Stone, eds.), Vol. 8, pp 353–360. Pergamon, Oxford, 1982.
4. D. Forster, *Adv. Organomet. Chem.* **17**, 255 (1979).
5. K. H. Grubbs, in "Comprehensive Organometallic Chemistry" (G. Wilkinson and F. G. A. Stone, ed.), Vol. 8, p. 499. Pergamon, Oxford, 1982.
6. M. Brookhart, J. R. Tucker, and G. R. Husk, *J. Am. Chem. Soc.* **105**, 258 (1983), and references therein.
7. M. Bottrill, P. D. Davens, J. W. Kelland, and J. McMeeking, in "Comprehensive Organometallic Chemistry" (G. Wilkinson and F. G.A. Stone, eds.), Vol. 3, p. 475.
8. R. H. Grubbs, in "Comprehensive Organometallic Chemistry" (G. Wilkinson and F. G. A. Stone, eds.), Vol. 8, p. 548.
9. G. L. Geoffroy and M. S. Wrighton, "Organometallic Photochemistry." Academic Press, New York, 1979.
10. J. W. Bruno, D. G. Kalina, E. A. Mintz, and T. J. Marks, *J. Am. Chem. Soc.* **104**, 1860 (1982).

10a. D. G. Kalina, T. J. Marks, and W. A. Wachter, *J. Am. Chem. Soc.* **99**, 3877 (1977).

11. T. J. Marks, *Acc. Chem. Res.* **98**, 703 (1976), and references therein.

12. E. Klähne, C. Giannotti, H. Marquet-Ellis, G. Folcher, and R. D. Fisher, *J. Organomet. Chem.* **201**, 399 (1980).

12a. M. Burton, H. Marquet-Ellis, G. Folcher, and C. Giannotti, *J. Organomet. Chem.* **229**, 21 (1982).

13. D. Rehorek and H. Hennig, *Can. J. Chem.* **60**, 1565 (1982); P. K. Wong, K. S. Y. Lau, and J. K. Stille, *J. Am. Chem. Soc.* **96**, 5956 (1974); G. M. Whitesides, D. E. Bergbreiter, and P. E. Kendall, *ibid.* **96**, 2806 (1974).

14. H. A. Zinnen, J. J. Pluth, and W. J. Evans, *J. Chem. Soc., Chem. Commun.* p. 810 (1980).

15. U. Zucchini, E. Albizatti, and U. Giannini, *J. Organomet. Chem.* **26**, 357 (1971).

16. D. G. H. Ballard and P. W. Van Lienden, *Makromol. Chem.* **154**, 177 (1972).

17. J. C. W. Chien, J. -C. Wu, and M. D. Rausch, *J. Am. Chem. Soc.* **103**, 1180 (1981).

18. H. B. Abrahamson and M. E. Martin, *J. Organomet. Chem.* **238**, C58 (1982).

19. H. De Vries, *Recl. Trav. Chim. Pay-Bas.* **80**, 866 (1961).

20. J. H. Wengrovius and R. R. Schrock, *J. Organomet. Chem.* **205**, 319 (1981).

21. R. W. Harrigan, G. S. Hammond, and H. B. Gray, *J. Organomet. Chem.* **81**, 79 (1974).

22. M. D. Rausch, W. H. Boon, and H. G. Alt, *J. Organomet. Chem.* **141**, 299 (1979).

23. E. Samuel, H. G. Alt, D. C. Hrncir, and M. D. Rausch, *J. Organomet. Chem.* **113**, 331 (1976).

24. P. W. N. M. Van Leeuwen, H. Van Der Heijden, C. F. Roobeek, and J. H. G. Frijns, *J. Organomet. Chem.* **209**, 169 (1981).

25. E. Samuel, P. Maillard, and C. Giannotti, *J. Organomet. Chem.* **142**, 289 (1977).

26. C. H. Bamford, R. J. Puddephatt, and D. M. Slater, *J. Organomet. Chem.* **159**, C31 (1978).

27. H. G. Alt and M. D. Rausch, *J. Am. Chem. Soc.* **96**, 5936 (1974).

28. W. H. Boon and M. D. Rausch, *J. Chem. Soc., Chem. Commun.* p. 397 (1977).

29. A. Hudson, M. F. Lappert, and R. Pichon, *J. Chem. Soc., Chem. Commun.* p. 374 (1983).

30. E. Samuel, *J. Organomet. Chem.* **198**, C65 (1980).

31. H. G. Alt and M. D. Rausch, *Z. Naturforsch., B: Anorg. Chem., Org. Chem.* **30B**, 813 (1975).

32. E. Samuel and C. Giannotti, *J. Organomet. Chem.* **113**, C17 (1976).

32a. M. Pankowski and E. Samuel, *J. Organomet. Chem.* **221**, C21 (1981).

33. P. E. Matkovskii, L. L. Chernaya, and F. S. D'yachkovskii, *Dokl. Akad. Nauk SSSR* **244**, 1351 (1971).

34. A. R. Lepley and G. L. Closs, eds., "Chemically Induced Magnetic Polarization." Wiley, New York, 1973.

35. M. R. M. Bruce, A. Kenter, and D. R. Tyler, *J. Am. Chem. Soc.* **106**, 639 (1984).

36. M. Peng and C. H. Brubaker, Jr., *Inorg. Chim. Acta.* **26**, 231 (1978).

37. M. D. Rausch, W. H. Boon, and E. A. Mintz, *J. Organomet. Chem.* **160**, 81 (1978).

38. G. Erker, *J. Organomet. Chem.* **134**, 189 (1977).

39. H. -S. Tung and C. H. Brubaker, JR., *Inorg. Chim. Acta.* **52**, 197 (1981).

40. V. Skibbe and G. Erker, *J. Organomet. Chem.* **241**, 15 (1983).

40a. G. Erker, J. Wicher, K. Engel, and C. Krüger, *Chem. Ber.* **115**, 3300 (1982).

40b. G. Erker, J. Wicher, K. Engle, F. Rosenfeldt, W. Dietrich, and C. Krüger, *J. Am. Chem. Soc.* **102**, 6344 (1980).

40c. S. Pouly, G. Tainturier, and B. Gautheron, *J. Organomet. Chem.* **232**, C65 (1982).

40d. G. Erker, K. Kropp, J. L. Atwood, and W. E. Hunter, *Organometallics* **2**, 1555 (1983).

41. P. Czisch and G. Erker, *J. Organomet. Chem.* **253**, C9 (1983).

41a. G. Erker, K. Kropp, J. L. Atwood, and W. E. Hunter, *Organometallics* **2**, 1555 (1983).

42. D. F. Foust and M. D. Rausch, *J. Organomet. Chem.* **226**, 47 (1982).

42a. D. F. Foust, M. D. Rausch, and E. Samuel, *J. Organomet. Chem.* **193**, 209 (1980).

43. J. L. Migot, J. Sala-Pala, and J. -E. Guerchais, *J. Organomet. Chem.* **243**, 427 (1983).

44. P. R. Sharp and R. R. Schrock, *J. Organomet. Chem.* **171**, 43 (1979).

44a. L. W. Messerle, P. Jennische, R. R. Schrock, and G. Stucky, *J. Am. Chem. Soc.* **102**, 6744 (1980).

45. H. B. Abrahamson and E. Dennis, *J. Organomet. Chem.* **201**, C19 (1980).

46. E. A. Mintz and M. D. Rausch, *J. Organomet. Chem.* **171**, 345 (1979).

47a. M. H. Chisholm, M. W. Extine, R. L. Kelly, W. C. Mills, C. A. Murillo, L. A. Rankel, and W. W. Reichert, *J. Am. Chem. Soc.* **17**, 1673 (1978).

47a. M. H. Chisholm, M. W. Extine, R. L. Kelly, W. C. Mills, C. A. Murillo, L. A. Rankel, and W. W. Reichert, *J. Am. Chem. Soc.* **17**, 1673 (1978).

48. S. M. B. Costa, A. R. Dias, and F. J. S. Pina, *J. Organomet. Chem.* **175**, 193 (1979).

48a. S. M. B. Costa, A. R. Dias, and F. J. S. Pina, *J. Chem. Soc., Dalton Trans.* p. 314 (1981).

49. M. Ephritikine and M. L. H. Green, *J. Chem. Soc., Chem. Commun.* p. 926 (1976).

49a. G. J. A. Adam, S. G. Davies, K. A. Ford, M. Ephritikine, P. F. Todd, and M. L. H. Green, *J. Mol. Catal.* **8**, 15 (1980).

50. K. W. Barnett and P. M. Treichel, *Inorg. Chem.* **6**, 294 (1976).

51. H. G. Alt, *J. Organomet. Chem.* **124**, 167 (1977).

51a. H. G. Alt, J. A. Schwärzle, and C. G. Kreiter, *J. Organomet. Chem.* **153**, C7 (1978).

51b. H. G. Alt and J. A. Schwärzle, *J. Organomet. Chem.* **162**, 45 (1978).

51c. H. G. Alt and M. E. Eichner, *J. Organomet. Chem.* **212**, 397 (1981).

51d. R. Korswagen and M. L. Ziegler, *Z. Naturforsch., B: Anorg. Chem., Org. Chem.* **35B**, 1196 (1980).

52. R. J. Kazlauskas and M. S. Wrighton, *J. Am. Chem. Soc.* **102**, 1727 (1980).

52a. R. J. Kazlauskas and M. S. Wrighton, *J. Am. Chem. Soc.* **104**, 6005 (1982).

52b. B. Klein, R. J. Kazlauskas, and M. S. Wrighton, *Organometallics* **1**, 1338 (1982).

52c. C. Lewis and M. S. Wrighton, *J. Am. Chem. Soc.* **105**, 7768 (1983).

53. R. G. Severson and A. Wojcicki, *J. Organomet. Chem.* **157**, 173 (1978).

54. H. G. Alt, *Angew Chem., Int. Ed. Engl.* **15**, 759 (1976); *Z. Naturforsch., B: Anorg. Chem., Org. Chem.* **32B**, 1139 (1977); H. G. Alt, M. E. Eichner, and B. M. Jansen, *Angew. Chem., Int. Ed. Engl.* **21**, 861 (1982).

55. K. A. Mahmoud, R. Narayanaswamy, and A. J. Rest, *J. Chem. Soc., Dalton Trans.* p. 2199 (1981).

56. D. R. Tyler, *Inorg. Chem.* **20**, 2257 (1981).

57. M. D. Rausch, T. E. Gismondi, H. G. Alt, and J. A. Schwärzle, *Z. Naturforsch.*, *B: Anorg. Chem., Org. Chem.* **32B**, 998 (1977).

58. E. Samuel, M. D. Rausch, T. E. Gismondi, E. A. Mintz, and C. Giannotti, *J. Organomet. Chem.* **172**, 309 (1979).

59. A. Hudson, M. F. Lappert, P. W. Lednor, J. J. Macquitty, and B. K. Nicholson, *J. Chem. Soc., Dalton Trans.* p. 2159 (1981).

60. K. A. Mahmoud, A. J. Rest, and H. G. Alt, *J. Chem. Soc., Chem. Commun.* p. 1011 (1983).

60a. R. B. Hitam, R. H. Hooker, K. A. Mahmoud, R. Narayanaswamy, and A. J. Rest, *J. Organomet. Chem.* **222**, C9 (1981).

61. H. G. Alt and M. E. Eichner, *Angew. Chem., Int. Ed. Engl.* **21**, 78 (1982).

61a. H. G. Alt and M. E. Eichner, *Angew. Chem., Int. Ed. Engl.* **21**, 205 (1982).

62. S. -C. H. Su and A. Wojcicki, *Organometallics* **2**, 1296 (1983).

63. J. Chetwynd-Talbot, P. Grebenik, and R. N. Perutz, *Inorg. Chem.* **21**, 3647 (1982).

63a. P. Grebenik, A. J. Downs, M. L. H. Green, andd R. N. Perutz, *J. Chem. Soc., Chem. Commun.* p. 742 (1979).

64. K. W. Chiu, R. A. Jones, G. Wilkinson, A. M. R. Galas, M. B. Hursthouse, and K. M. Abdul Malik, *J. Chem. Soc., Dalton Trans.* p. 1204 (1981).

65. A. Hudson, M. F. Lappert, P. W. Lednor, and B. K. Nicholson, *J. Chem. Soc., Chem. Commun.* p. 966 (1974).

66. C. H. Bamford and S. U. Mullik, *J. Chem. Soc., Faraday Trans. 1.* p. 2562 (1979).

67. W. Lipps and C. G. Kreiter, *J. Organomet. Chem.* **241**, 185 (1983).

68. C. G. Kreiter and W. Lipps, *J. Organomet. Chem.* **253**, 339 (1983).

69. J. F. Ogilvie, *Chem. Commun.* p. 323 (1970).

70. T. M. McHugh and A. J. Rest, *J. Chem. Soc., Dalton Trans.* p. 2323 (1980).

71. B. J. Brisdon, D. A. Edwards, and J. W. White, *J. Organomet. Chem.* **175**, 113 (1979).

72. D. S. Edwards, L. V. Biondi, J. W. Ziller, M. R. Churchill, and R. R. Schrock, *Organometallics* **2**, 1505 (1983).

73. E. O. Fischer and J. Müller, *Z. Naturforsch., B: Anorg. Chem., Org. Chem., Biochem., Biophys., Biol.* **17B**, 776 (1962); 413, 1137 (1963); *Chem. Ber.* **96**, 3217 (1963).

74. E. O. Fischer and J. Müller, *J. Organomet. Chem.* **1**, 89 (1963).

75. E. O. Fischer and J. Müller, *J. Organomet. Chem.* **1**, 464 (1964).

76. E. O. Fischer and J. Müller, *J. Organomet. Chem.* **5**, 275 (1966).

77a. R. J. Kazlauskas and M. S. Wrighton, *Organometallics* **1**, 602 (1982).

78. W. Gerhartz, G. Ellerhorst, P. Dahler, and P. Eilbracht, *Liebigs Ann. Chem.* p. 1296 (1980).

78a. H. Brünner and H. Vogt, *J. Organomet. Chem.* **210**, 223 (1981).

79. A. C. Gingell and A. J. Rest, *J. Organomet. Chem.* **99**, C27 (1975).

80. C. R. Folkes and A. J. Rest, *J. Organomet. Chem.* **136**, 355 (1977).

81. P. M. Treichel, R. L. Shubkin, K. W. Barnett, and D. Reichard, *Inorg. Chem.* **5**, 1177 (1966).

82. P. M. Treichel and D. A. Komar, *J. Organomet. Chem.* **206**, 77 (1981).

83. H. G. Alt, M. Herberhold, M. D. Rausch, and B. H. Edwards, *Z. Naturforsch., B: Anorg. Chem., Org. Chem.* **34B**, 1070 (1979).

84. D. J. Fettes, R. Narayanaswamy, and A. J. Rest, *J. Chem. Soc., Dalton Trans.* p. 2311 (1981).

85. M. Rosenblum and P. S. Waterman, *J. Organomet. Chem.* **187**, 267 (1980).

86. S. R. Su and A. Wojcicki, *J. Organomet. Chem.* **27**, 231 (1971).

87. H. G. Alt, M. E. Eichner, B. M. Jansen, and U. Thewalt, *Z. Naturforsch., B: Anorg. Chem., Org. Chem.* **37B**, 1109 (1982).

88. P. Reich-Rohrwig and A. Wojcicki, *Inorg Chem.* **13**, 2457 (1974).

89. H. Brünner and G. O. Nelson, *J. Organomet. Chem.* **173**, 389 (1979).

90. M. Bottrill, M. Green, E. O'Brien, L. E. Smart, and P. Woodward, *J. Chem. Soc., Dalton Trans.* p. 292 (1980).

91. E. Kent Barefield and D. J. Sepelak, *J. Am. Chem. Soc.* **101**, 6542 (1979).

92. C. J. May and W. A. G. Graham, *J. Organomet. Chem.* **234**, C49 (1982).
93. M. Cooke, N. J. Forrow, and S. A. R. Knox, *J. Chem. Soc., Dalton Trans.* p. 2435 (1983).
93a. M. Cooke, N. J. Forrow, and S. A. R. Knox, *J. Organomet. Chem.* **222**, C21 (1981).
94. N. B. Kupletskaya, P. G. Buyanovskaya, and S. S. Churanov, *Zh. Obshch. Khim.* **51**, 1110 (1981).
95. G. O. Nelson and M. E. Wright, *J. Organomet. Chem.* **239**, 353 (1983).
96. A. N. Nesmeyanov, T. B. Chenskaya, G. M. Babakhina, and I. I. Kritskaya, *Bull. Acad. Sci. USSR, Div. Chem. Sci. (Engl. Transl.)* p. 1129 (1970).
97. G. A. Sim, D. I. Woodhouse, and G. R. Knox, *J. Chem. Soc., Dalton Trans.* p. 629 (1979).
98. A. N. Nesmeyanov, Yu. A. Chapovsky, and Yu. A. Ustiynyuk, *J. Organomet. Chem.* **9**, 345 (1967).
98a. A. N. Nesmeyanov, L. G. Makarova, and I. V. Polovyanyuk, *J. Organomet. Chem.* **22**, 707 (1970).
99. R. P. Stewart, Jr., J. J. Benedict, L. Isbrandt, and R. S. Ampulski, *Inorg. Chem.* **14**, 2933 (1975).
99a. R. P. Stewart, Jr., L. R. Isbrandt, J. J. Benedict, and J. G. Palmer, *J. Am. Chem. Soc.* **98**, 3215 (1976).
100. N. B. Kupletskaya, P. G. Buyanovskaya, and S. S. Churanov, *Dokl. Akad. Nauk SSSR* **248**, 111 (1979); *Zh. Obshch. Khim.* **49**, 1556 (1979).
101. B. D. Fabian and J. A. Labinger, *J. Am. Chem. Soc.* **101**, 2239 (1979).
102. J. Evans, S. J. Okrasinski, A. J. Pribula, and J. R. Norton, *J. Am. Chem. Soc.* **99**, 5835 (1977).
103. A. Vogler and R. Hirschmann, *Z. Naturforsch., B: Anorg. Chem., Org. Chem.* **31B**, 1082 (1976).
104. J. R. Sheats and H. M. McConnell, *J. Am. Chem. Soc.* **99**, 7091 (1977).
104a. J. R. Sheats and H. M. McConnell, *Biophys. J.* **21**, 126a (1978); **25**, 64a (1979).
105. J. R. Sheats and H. M. McConnell, *J. Am. Chem. Soc.* **101**, 3272 (1979).
106. M. A. Schwartz and H. M. McConnell, *Biochemistry* **17**, 837 (1978).
107. J. R. Sheats and H. M. McConnell, *Proc. Natl. Acad. Sci. U.S.A.* **75**, 4661 (1978).
108. E. A. Koerner von Gustorf, L. H. G. Leenders, I. Fischler, and R. N. Perutz, *Adv. Inorg. Chem. Radiochem.* **19**, 65 (1976).
109. D. Crowfoot-Hodgkin, J. Pickworth, J. H. Robertson, K. N. Trueblood, R. J. Prosen, and J. G. White, *Nature (London)* **175**, 325 (1955).
110. G. N. Schrauzer, J. W. Sibert, and R. J. Windgassen, *J. Am. Chem. Soc.* **90**, 6681 (1968).
111. G. N. Schrauzer and J. H. Grate, *J. Am. Chem. Soc.* **103**, 541 (1981), and references therein.
112. J. F. Endicott and G. J. Ferraudi, *J. Am. Chem. Soc.* **99**, 243 (1977).
112a. C. Y. Mok and J. F. Endicott, *J. Am. Chem. Soc.* **99**, 1276 (1977).
112b. J. F. Endicott and T. L. Netzel, *J. Am. Chem. Soc.* **101**, 4000 (1979).
112c. C. Giannotti, C. Fontaine, A. Chiaroni, and C. Riche, *J. Organomet. Chem.* **113**, 57 (1976).
113. D. J. Lowe, K. N. Joblin, and D. J. Cardin, *Biochim. Biophys. Acta* **539**, 398 (1979).
114. P. Maillard and C. Giannotti, *Can. J. Chem.* **60**, 1402 (1982), and references therein.
115. D. N. Ramakrishna Rao and M. C. R. Symons, *J. Chem. Soc., Perkin Trans. 2* p. 187 (1983), and references therein.
116. K. N. Joblin, A. W. Johnson, M. F. Lappert, and B. K. Nicholson, *J. Chem. Soc., Chem. Commun.* p. 441 (1975).

117. J. M. Pratt, "The Inorganic Chemistry of Vitamin B12." Academic Press, New York, 1972.
118. P. Maillard and C. Giannotti, *J. Organomet. Chem.* **182**, 225 (1979), and references therein.
118a. C. Giannotti, C. Fontaine, and B. Septe, *J. Organomet. Chem.* **71**, 107 (1974).
119. M. Okabe, T. Osawa, and M. Tada, *Tetrahedron Lett.* **22**, 1899 (1981), and references therein.
120. H. Flohr, W. Pannhorst, and J. Rétey, *Helv. Chim. Acta* **61**, 1565 (1978), and references therein.
120a. M. Fountoulakis and J. Rétey, *Chem. Ber.* **113**, 650 (1980).
121. D. A. Lerner, R. Bonneau, and C. Giannotti, *J. Photochem.* **11**, 73 (1979).
122. B. T. Golding, T. J. Kemp, and H. H. Sheena, *J. Chem. Res. Synop.* p. 34 (1981).
123. V. D. Ghanekar and R. E. Coffman, *J. Organomet. Chem.* **198**, C15 (1980).
124. D. A. Lerner, F. Ricchiero, and C. Giannotti, *J. Colloid Interface Sci.* **68**, 596 (1979).
125. D. A. Lerner, F. Ricchiero, and C. Giannotti, *J. Phys. Chem.* **84**, 3007 (1980).
126. I. P. Rudakova, T. E. Ershova, A. B. Belikov, and A. M. Yurkevich, *J. Chem. Soc., Chem. Commun.* p. 592 (1978).
126a. A. J. Hartshorn, A. W. Johnson, S. M. Kennedy, M. F. Lappert, and J. J. McQuitty, *J. Chem. Soc., Chem. Commun.* p. 643 (1978).
126b. B. T. Golding, T. J. Kemp, C. S. Sell, P. J. Sellars, and M. P. Watson, *J. Chem. Soc., Perkin Trans. 2* p. 839 (1978).
127. H. Y. Al-Saigh and T. J. Kemp, *J. Chem. Soc., Perkin Trans. 2* p. 615 (1983).
128. P. Bougeard, M. D. Johnson, and G. M. Lampman, *J. Chem. Soc., Perkin Trans. 1* p. 849 (1982).
128a. J. Deniau, K. N. V. Duong, A. Gaudemer, P. Bougeard, and M. D. Johnson, *J. Chem. Soc., Perkin Trans. 2* p. 393 (1981).
129. E. G. Jäger, P. Renner, and R. Schmidt, *Z. Chem.* p. **17**, 307 (1977).
130. G. Roewer and D. Rehorek, *J. Prakt. Chem.* **320**, 566 (1978); S. Komiya, M. Bundo, T. Yamamoto, and A. Yamamoto, *J. Organomet. Chem.* **174**, 343 (1179).
131. W. H. Tamblyn and J. K. Kochi, *J. Inorg. Nucl. Chem.* **43**, 1385 (1981).
132. K. L. Brown, *J. Am. Chem. Soc.* **101**, 6600 (1979).
133. Y. Murukami, Y. Aoyama, and K. Tokunaga, *Inorg. Nucl. Chem. Lett.* **15**, 7 (1979).
134. Y. Murakami, Y. Aoyama, and K. Tokunaga, *J. Am. Chem. Soc.* **102**, 6736 (1980).
135. I. Inoue and J. F. Endicott, private communication.
136. J. M. Brown, J. A. Connelly, and K. Mertes, *J. Chem. Soc., Dalton Trans.* p. 905 (1974).
137. A. Emad and M. D. Rausch, *J. Organomet. Chem.* **191**, 313 (1980).
138. R. H. Grubbs, A. Miyashita, M. Liu, and P. Burk, *J. Am. Chem. Soc.* **100**, 2418 (1978).
138a. R. H. Grubbs and A. Miyashita, *J. Organomet. Chem.* **161**, 371 (1978).
139. D. C. L. Perkins, R. J. Puddephatt, and C. F. H. Tipper, *J. Organomet. Chem.* **191**, 481 (1980).
140. D. C. L. Perkins, R. J. Puddephatt, and C. F. H. Tipper, *J. Organomet. Chem.* **154**, C16 (1978).
141. D. C. L. Perkins, R. J. Puddephatt, and C. F. H. Tipper, *J. Organomet. Chem.* **186**, 419 (1980).
142. G. Phillips, R. J. Puddephatt, and C. F. H. Tipper, *J. Organomet. Chem.* **131**, 467 (1977); R. J. Al-Essa, R. J. Puddephatt, D. C. L. Perkins, M. C. Rendle, and C. F. H. Tipper, *J. Chem. Soc., Dalton Trans.* p. 1738 (1981); D. C. L. Perkins and

R. J. Puddephatt, *Organometallics* **2**, 1472 (1983); D. C. L. Perkins, R. J. Puddephatt, M. C. Rendle, and C. F. H. Tipper, *J. Organomet. Chem.* **195**, 105 (1980).

143. F. Ozawa, A. Yamamoto, T. Ikariya, and R. H. Grubbs, *Organometallics* **1**, 1481 (1982).

144. N. W. Alcock, T. J. Kemp, and F. J. Winner, *J. Chem. Soc., Dalton Trans.* p. 635 (1981); P. Hoake and T. A. Hylton, *J. Am. Chem. Soc.* **84**, 3374 (1962); S. H. Mastin and P. Haake, *J. Chem. Soc., Chem. Commun.* p. 202 (1980); M. Cusumano, G. Gugliem, V. Ricevuto, O. Traveso, and T. J. Kemp, *J. Chem. Soc., Dalton Trans.* p. 302 (1981).

145. F. Ozawa, T. Ito, Y. Nakmura, and A. Yamamoto, *Bull. Chem. Soc. Jpn.* **54**, 1968 (1981).

146. A. Gillie and J. K. Stille, *J. Am. Chem. Soc.* **102**, 4933 (1980).

147. P. W. N. M. Van Leeuwen, C. F. Roobeek, and R. Huis, *J. Organomet. Chem.* **142**, 243 (1977).

148. A. Hudson, M. F. Lappert, P. W. Lednor, J. J. MacQuitty, and B. K. Nicholson, *J. Chem. Soc., Dalton Trans.* p. C159 (1981).

149. K. Suzuki and H. Yamamoto, *J. Organomet. Chem.* **54**, 385 (1973); K. Suzuki, J. Ooyame, and M. Sakurai, *Bull. Chem. Soc. Jpn.* **49**, 464 (1976).

150. R. McCrindle, G. Ferguson, A. J. McAlees, M. Parvez, and P. J. Roberts, *J. Chem. Soc., Dalton Trans.* p. 1699 (1982).

151. L. L. Constanzo, S. Giuffrida, G. Condorelli, and R. Romeo, *Atti Congr. Naz. Chim. Inorg., 12th*, p. 404 (1979).

151a. L. L. Costanzo, S. Giuffrida, and R. Romeo, *Inorg. Chim. Acta* **38**, 31 (1980).

152. D. A. Roberts, W. R. Mason, and G. L. Geoffroy, *Inorg. Chem.* **20**, 789 (1981).

153. D. C. L. Perkins, R. J. Puddephatt, and C. F. H. Tipper, *J. Organomet. Chem.* **166**, 261 (1979).

154. J. Müller and P. Göser, *Angew. Chem., Int. Ed. Engl.* **6**, 364 (1967).

155. O. Hackleberg and A. Wojcicki, *Inorg. Chim. Acta* **44**, L63 (1980).

156. R. J. Hill and R. J. Puddephatt, *Organometallics* **2**, 1472 (1983); M. P. Brown, S. J. Cooper, A. A. Frew, L. Manojlović-Muir,, K. W. Muir, R. J. Puddephatt, and M. A. Thomson, *J. Chem. Soc., Dalton Trans.* p. 299 (1982).

157. G. A. Ozin, D. F. McIntosh, S. A. Mitchell, and J. Garcia-Prieto, *J. Am. Chem. Soc.* **103**, 1574 (1981).

158. G. M. Whitesides, D. E. Bergbreiter, and P. E. Kendall, *J. Am. Chem. Soc.* **96**, 2806 (1974).

159. C. M. Mitchell and F. G. A. Stone, *J. Chem. Soc., Dalton Trans.* p. 102 (1976).

160. P. W. N. M. Van Leeuwen, R. Kaptein, R. Huis, and C. F. Roobeek, *J. Organomet. Chem.* **104**, C44 (1976).

161. A. Hudson, M. F. Lappert, P. W. Lednor, J. J. MacQuitty, and B. K. Nicholson, *J. Chem. Soc., Dalton Trans.* p. 2159 (1981).

162. T. P. Fackler, Jr. and J. D. Basil, *Organometallics* **1**, 871 (1982).

163. M. Y. Darensbourg and O. J. Darensbourg, *Inorg. Chem.* **9**, 32 (1970).

164. C. P. Casey and T. J. Burkhardt, *J. Am. Chem. Soc.* **95**, 5833 (1973).

165. H. C. Foley, L. M. Strubinger, T. S. Targos, and G. L. Geoffroy, *J. Am. Chem. Soc.* **105**, 3064 (1983).

166. L. K. Fong and N. J. Cooper, *J. Am. Chem. Soc.* **106**, 2595 (1984).

167. G. L. Geoffroy and M. S. Wrighton, "Organometallic Photochemistry." Academic Press, New York, 1979.

168. E. O. Fischer and H. Fischer, *Chem. Ber.* **107**, 657 (1974).

169. K. Ofele and M. Herberhold, *Angew. Chem. Int. Ed. Engl.* **9**, 739 (1970).
170. K. Ofele and M. Herberhold, *Z. Naturforsch., B: Anorg. Chem., Org. Chem.* **28B**, 306 (1973).
171. K. Ofele, E. Roos, and M. Herberhold, *Z. Naturforsch., B: Anorg. Chem., Org. Chem.* **31B**, 1070 (1976).
172. R. D. Rieke, H. Kojima, and K. Ofele, *J. Am. Chem. Soc.* **98**, 6735 (1976).
173. F. G. A. Stone, *Angew. Chem., Int. Ed. Engl.* **23**, 89 (1984).
174. E. O. Fischer and H. J. Beck, *Chem. Ber.* **104**, 3101 (1971).
175. M .F. Lappert, P. L. Pye, and G. M. McLaughlin, *J. Chem. Soc., Dalton Trans.* p. 1273 (1977).
176. R. M. Dahlgren and J. I. Zink, *Inorg. Chem.* **16**, 3154 (1977).
177. S. M. Neumann, Ph. D. Dissertation, University of Wisconsin, Madison (1978).
178. C. P. Casey, A. J. Shusterman, N. W. Vollendorf, and K. J. Haller, *J. Am. Chem. Soc.* **104**, 2417 (1982).
179. C. P. Casey and A. J. Shusterman, *J. Mol. Catal.* **8**, 1 (1980).
180. B. H. Edwards and M. D. Rausch, *J. Organomet. Chem.* **210**, 9 (1981).
181. M. A. McGuire and L. S. Hegedus, *J. Am. Chem. Soc.* **104**, 5538 (1982).
181a. E. O. Fischer and H. -J. Beck, *Angew. Chem., Int. Ed. Engl.* **9**, 72 (1970).
182. R. D. Adams, D. F. Chodosh, and N. M. Golembeski, *J. Organomet. Chem.* **139**, C39 (1977).
183. E. O. Fischer and G. Besl. *J. Organomet. Chem.* **157**, C33 (1978).
184. W. A. Herrmann, *Chem. Ber.* **108**, 486 (1975).
185. R. E. Wright and A. Vogler, *J. Organomet. Chem.* **160**, 197 (1978).
186. K. Folting, J. C. Huffman, L. N. Lewis, and K. G. Caulton, *Inorg. Chem.* **18**, 3483 (1979).
187. A. B. Antonova, N. E. Kolobova, P. V. Petrovsky, B. V. Lokshin, and N. S. Obezyuk, *J. Organomet. Chem.* **137**, 55 (1977).
188. F. B. McCormick, W. A. Kiel, and J. A. Gladysz, *Organometallics* **1**, 405 (1982).
189. R. D. Adams, D. F. Chodosh, N. M. Golembeski, and E. C. Weissman, *J. Organomet. Chem.* **172**, 251 (1979).
190. A. F. Dyke, S. A. R. Knox, P. J. Naish, and G. E. Taylor, *J. Chem. Soc., Chem. Commun.* p. 803 (1980); R. E. Colborn, A. F. Dyke, S. A. R. Knox, K. A. Mead, and P. Woodward, *J. Chem. Soc., Dalton Trans.* p. 2099 (1983).
191. M. Cooke, D. L. Davies, J. E. Guerchais, S. A. R. Knox, K. A. Mead, J. Roue, and P. Woodward, *J. Chem. Soc., Chem. Commun.* p. 862 (1981).
192. E. O. Fischer and P. Friedrich, *Angew. Chem., Int. Ed. Engl.* **18**, 327 (1979).
193. A. Vogler, J. Kisslinger, and W. R. Roper, *Z. Naturforsch., B: Anorg. Chem., Org. Chem.* **38B**, 1506 (1983).
194. G. L. Geoffroy and R. A. Epstein, *Inorg. Chem.* **16**, 2795 (1975).
195. P. T. Chesky and M. B. Hall, *Inorg. Chem.* **20**, 4419 (1981).
196. G. Granozzi, E. Tondello, D. Ajo, M. Casarin, S. Aime, and D. Osella, *Inorg. Chem.* **21**, 1081 (1982).
197. T. W. Matheson, B. M. Peake, B. H. Robinson, J. Simpson, and D. J. Watson, *J. Chem. Soc., Chem. Commun.* p. 894 (1973); B. M. Peake, B. H. Robinson, J. Simpson, and D. J. Watson, *Inorg. Chem.* **16**, 405 (1977).

ADVANCES IN ORGANOMETALLIC CHEMISTRY, VOL. 24

X-Ray Structural Analyses of Organolithium Compounds

WILLIAM N. SETZER[1]
PAUL VON RAGUÉ SCHLEYER

Institut für Organische Chemie
Friedrich-Alexander-Universität Erlangen-Nürnberg
Erlangen, Federal Republic of Germany

[1] Present address: Department of Chemistry, University of Utah, Salt Lake City, Utah 84112.

I

INTRODUCTION

Lithium compounds, perhaps the most widely used organometallic synthetic intermediates, have fascinating structures. Nevertheless, organic chemistry textbooks often treat organolithium compounds merely as "carbanions" or as monomers. The X-ray structures presented in this article demonstrate both of these descriptions to be grossly inaccurate oversimplifications for these associated, polar organometallic species. The chemical behavior of lithium compounds cannot be understood without a knowledge of the structures, the degree of aggregation, and the extent and nature of solvation.

The typical "inorganic" descriptions of lithium chemistry do not reflect its importance. Lithium, as the first metal in the periodic table, is expected to exhibit properties and chemical behavior characteristic of this group of over 80 elements. A few typical structures are presented, but these do not allow comparisons or overall conclusions to be drawn.

Although the structures of lithium compounds seldom follow classical bonding considerations, a few generalizations and trends may be noted. Lithium bonds have a high degree of ionic character; this leads to large energies of association. Dimers, tetramers, and other oligomeric structures are common. Similarly, two or more lithium atoms often bridge the same set of atoms simultaneously; this is the intramolecular equivalent of aggregation. In addition, lithium crystal structures often show interactions with electron-rich molecules. They are often complexed by ligands, such as N,N,N',N'-tetramethylethylenediamine (TMEDA),[2] or by solvents, e.g., ether or tetrahydrofuran (THF). Lithium often interacts with hydrocarbon π systems (olefins, arenes, or acetylenes) at many different sites simultaneously. Indeed, lithium tends to engage in multicenter covalent bonding, but the ionic character still dominates. The propensity for aggregation makes it particularly difficult to predict the structures of organolithium compounds. The steric size of the "carbanion" and the nature and degree

[2] Abbreviations: acac, Acetylacetonate; cdt, cyclododecatrienyl; cod, cyclooctadienyl; dme, dimethoxyethane; dmp, 1,4-dimethylpiperazine; edta, ethylenediaminetetraacetate; edds, ethylenediamine-N,N'-disuccinate; eddda, ethylenediamine-N,N'-diacetate-N,N'-di-3-proppionate; en, $NH_2CH_2CH_2NH_2$; hmpt, hexamethylphosphoric triamide [$(Me_2N)_3P(O)$]; HOMO, highest occupied molecular orbital; MNDO, modified neglect of differential overlap; MO, molecular orbital; NAD$^+$, nicotinamide adenine dinucleotide (oxidized); PMDETA, pmdeta, $(CH_3)_2N(CH_2)_2N(CH_3)(CH_2)_2N(CH_3)_2$; pyr, pyridine; salen, N,N'-ethylenebis-(salicylidenimine); THF, thf, $O(CH_2)_4CH_2$; TMEDA, tmeda, $(CH_3)_2NCH_2CH_2N(CH_3)_2$; tmhda, $(CH_3)_2N(CH_2)_6N(CH_3)_2$(N,N,N',N'-tetramethyl-1,6-diaminohexane); tmpda, $(CH_3)_2N(CH_2)_3N(CH_3)_2$.

of "solvation" or complexation are factors determining the extent of association and geometries around the lithium atom.

As NMR studies show, the degree of association often changes with the temperature or the medium. This is reflected in some X-ray structures, where both dimers and tetramers of the same compound are represented. The X-ray structures often may be reasonably good approximations of the nature of many complexes in solution. There are exceptions, e.g., dynamic processes observable by NMR, frozen out in the solid state.

While lithium is often tetracoordinate and adopts tetrahedral geometries, exceptions are common. The coordination sphere around the lithium seems to be governed largely by steric effects. No octet rule considerations apply, although this is a common interpretation. Lone pair ligands do not donate two electrons to lithium; the bonding is largely ion-dipole in character. It is hard enough for lithium to bind two or three electrons; eight is quite out of the question! There are many examples where lithium coordinates with five or six donor atoms, but very few electrons actually are shared by the metal.

Lithium structural chemistry is a relatively recent development. The first X-ray investigation of an organolithium compound (the tetrahedral ethyl-lithium tetramer) was reported by H. Dietrich in 1963. This was followed a year later by E. Weiss's similar methyllithium tetramer structure. Both investigators continue to be principal contributors to this area. G. D. Stucky made many notable advances during the 1970s. The singular locations of lithium counterions associated with aromatic carbanions and dianions were interpreted by means of molecular orbital (MO) theory. The structures of many transition metal compounds, which also contain lithium atoms in revealing locations, have been determined by C. Krüger, K. Jonas, and their co-workers at the Max-Planck-Institute, Mülheim/Ruhr. While both Stucky and Jonas have reviewed their contributions, and general accounts of lithium chemistry (see Bibliography) include discussion of crystal structures, no comprehensive summary has been published before. The recent burst of activity—the number of available X-ray investigations of lithium compounds has doubled in the last 5 years—provides a further impetus to focus attention on a rapidly emerging area.

The restriction of this article to the X-ray structures of organolithium compounds is less artificial than it may appear. Practically no other source of detailed geometrical information is available. Experimentally, only a few lithium structures in the gas phase are known, and these molecules seldom contain carbon. An increasing number of investigations of lithium compounds in solution, notably by NMR spectroscopy, give invaluable but more general information, e.g., the degree of aggregation and the coupling of one atom to another, and this complements the X-ray data very nicely. The organometallic compounds of the other alkali metals have structures just as interesting as those of lithium, but comparatively few X-ray studies have been reported. We hope this article will stimulate research in this area as well.

Theoretical calculations have now become a primary source of detailed information for chemical compounds. Efficient computer programs and standard methods make this investigational tool readily available to all chemists. Most extensive calculations on organolithium compounds have been concerned with small, isolated monomers. While these have revealed remarkable geometries, energies, and bonding arrangements (1), few comparisons with the experimental structures of solvated aggregates were possible. However, this situation is changing rapidly. Theoretically predicted structures are now being verified experimentally and calculations are an indispensible adjunct for the interpretation of experimental results. For example, calculations help us understand why alternative geometries that seem equally probable qualitatively are not favored. Relatively large systems including solvation can be examined by MNDO (semi-empirical) theory; even data for isolated monomers, which can be calculated more precisely at *ab initio* levels, are proving to be relevant to solution and solid state chemistry (1). Nevertheless, the objective of this article is not to compare theory with experiment—this is still somewhat premature—but to organize and to present the existing X-ray data.

We survey the crystal structures of lithium compounds which were available through 1983. The Cambridge Crystallographic Database provided the major source of references, but many colleagues kindly provided information prior to publication. Organolithium compounds (structures having lithium–carbon contacts) are our major concern, but other molecules of interest to lithium chemists are summarized for completeness. The article is organized in terms of the nature of the lithium–carbon contacts, the degree of association, and whether other metals are involved.

II

SIMPLE ORGANOLITHIUM COMPOUNDS

A. Terminally Bonded Organolithium Compounds

"Terminally bonded" organolithium compounds afford, in principle, the simplest examples of lithium–carbon σ-bonds. This class comprises lithium compounds with a single contact between lithium and a carbon atom.

However, the available examples, **1–4**, have stabilized carbanion moieties and the lithiums also are stabilized by coordination with electron pair donor atoms. Two compounds in the literature involve an sp^3-hybridized carbon atom σ-bonded to a lithium atom [**1** (2) and **3** (3)]. Compound **2** (4) also involves a terminally bonded lithium atom but the more or less planar benzylic carbon can perhaps be described as being sp^2 hybridized.

Compounds **1** and **2** are monomeric species, probably owing to the large amount of steric bulk in the neighborhood of the lithium atom: coordinating solvents [N,N,N',N',N''-pentamethyldiethylenetriamine (PMDE-TA) in **1** and TMEDA and THF in **2**] complex with lithium to complete its "coordination sphere." The geometry about the lithium atom is nearly tetrahedral in both of these compounds.

Compound **3** is a dimeric structure, but the interactions of the lithium atom are analogous to those in **1** and **2**. A sulfur atom from a second

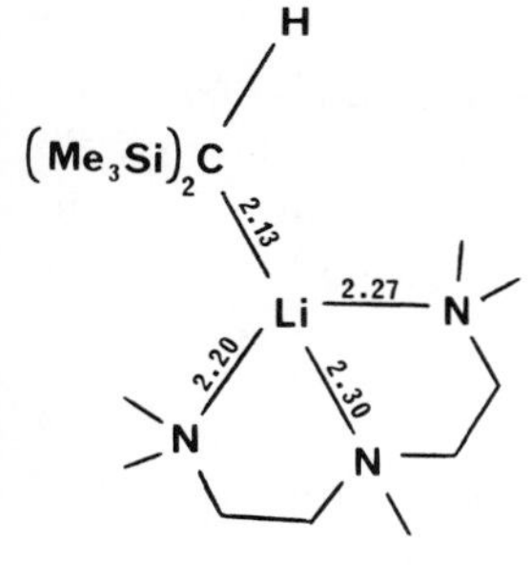

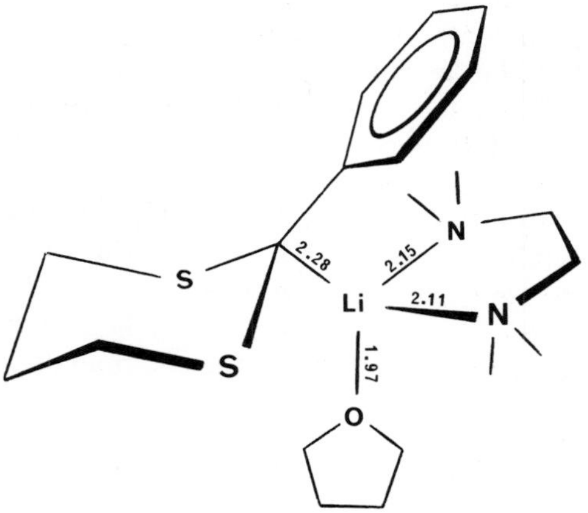

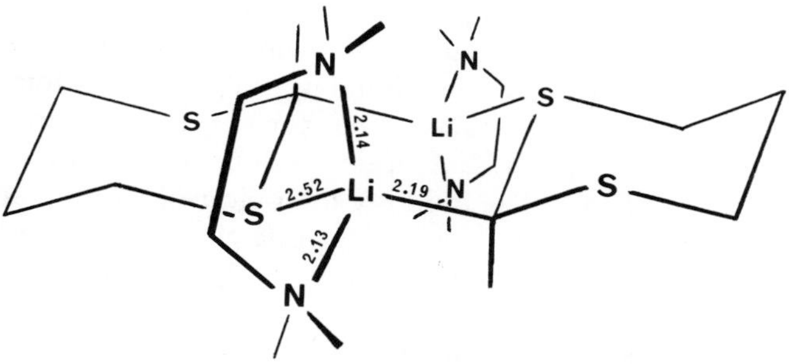

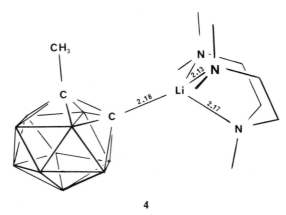

4

dithiane molecule and a molecule of TMEDA complete the tetrahedral coordination about the lithium atom. The lithium–carbon bond lengths for such compounds average about 2.16 Å for **1** and **3**; a 2.20 Å distance is found in **2**. This is substantially longer than the 2.00 Å C—Li bond length calculated consistently at various basis set levels for monomeric CH_3Li (5). The difference is attributable to the lithium coordination and greater anionic stabilization in **1–3**.

Complex **4** (6), $(MeC_2B_{10}H_{10})Li(pmdeta)$ [pmdeta = $(CH_3)_2N(CH_2)_2$-$N(CH_3)(CH_2)_2N(CH_3)_2$], a lithiated carborane, also is "terminally bonded" and represents a simple example of a carbon–lithium σ-bond. The C—Li distance is 2.18 Å. The structural features at the lithium atom are analogous to those for compounds **1–3**.

B. Organolithium Compounds with Bridging Lithium Atoms

In situations not precluded by steric interactions, organolithium compounds crystallize adopting dimeric, tetrameric, or oligomeric structures involving bridging lithium atoms. Such geometries maximize favorable electrostatic and orbital interactions (5). These structures illustrate what may be termed "electron-deficient" bonding of the lithium atoms. Dimeric structures involve bridging of two carbon atoms by two lithium atoms. The coordination sphere of the lithium atom is often completed by a chelating solvent such as TMEDA. The dimer of lithiobicyclo[1.1.0]butane (**5**) (7), and of lithiodimethylsulfide (**6**) (8), illustrate the bridging between sp^3-hybridized carbon atoms, while phenyllithium (**7**) (9), exemplifies bridging between sp^2 carbon atoms.

More recent examples of lithium atoms bridging sp^2 carbon atoms are the dimers 8-(dimethylamino)-1-lithionaphthalene (**8**) (10), and 2-lithio-

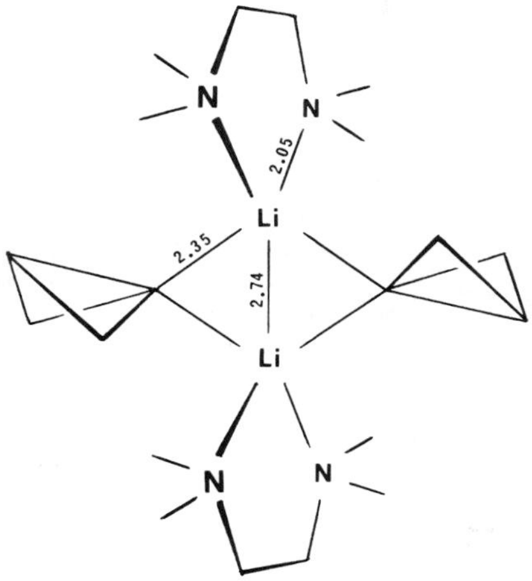

5

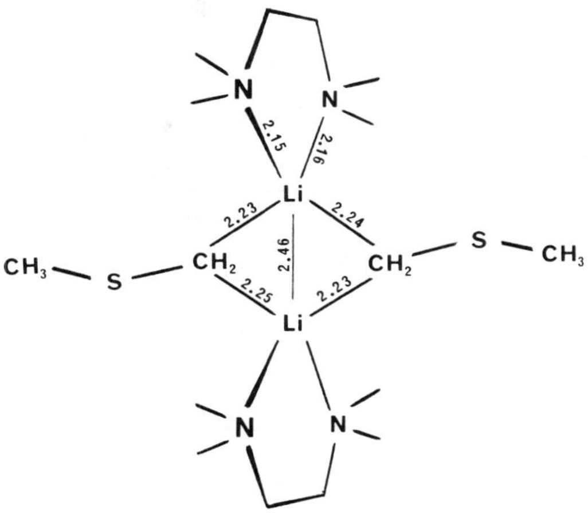

6

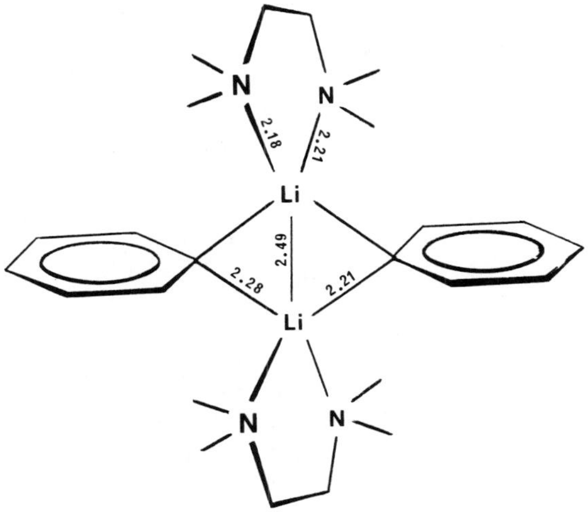

7

biphenyl (**9**) (*11*). The lithium atom in **8** lies in a distorted tetrahedral environment provided by the two carbon atoms, the amino nitrogen, and the ether oxygen. An interesting structural feature of **8** is that C(2), C(9), Li(1), and Li(2) are nearly coplanar, that is, C(1) is a nearly planar tetracoordinate carbon (*12*).

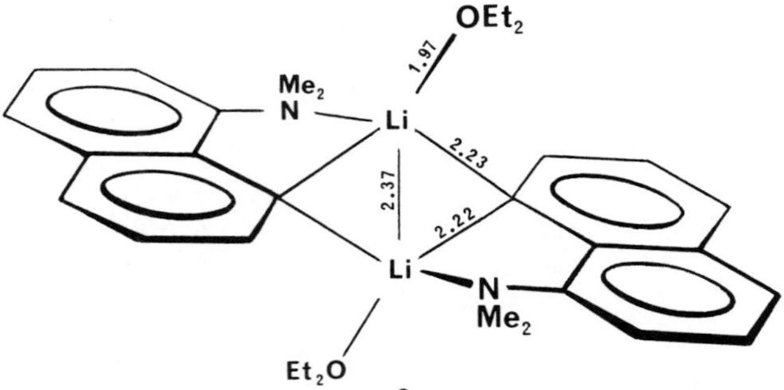

8

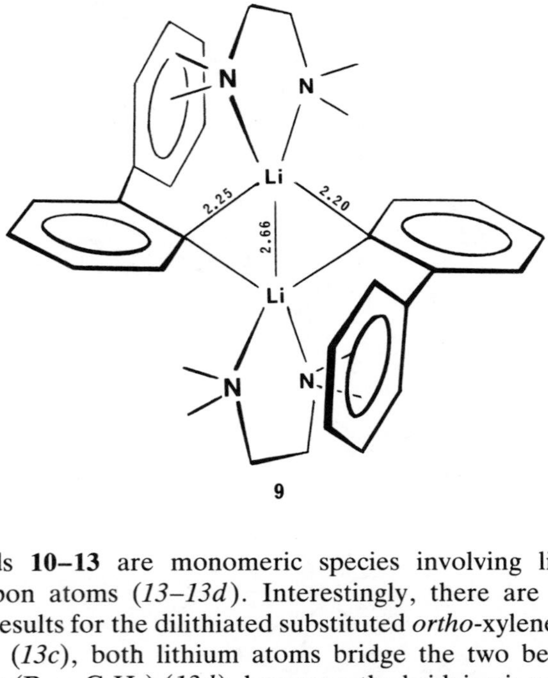

9

Compounds **10–13** are monomeric species involving lithium atoms bridging carbon atoms *(13–13d)*. Interestingly, there are two different geometrical results for the dilithiated substituted *ortho*-xylenes (**10**). In **10a** (R = SiMe₃) *(13c)*, both lithium atoms bridge the two benzylic carbon atoms. In **10b** (R = C₆H₅) *(13d)*, however, the bridging is unsymmetrical; one lithium atom bridges the benzylic carbons and one bridges a benzylic carbon and the C(3) ring carbon. The different modes of bridging in these two compounds can be interpreted by electrostatic arguments. Trimethylsilyl groups are electron donating and tend to concentrate the negative charge of **10a** on the benzylic carbons. The positively charged lithium atoms, then, associate with this concentration of negative charge by bridging these two carbon atoms. On the other hand, phenyl substituents tend to remove electron density from the benzylic carbons and the lithium atom associates with another center of electron density, namely the C(3) ring carbon. Model MNDO calculations are in agreement with these geometrical results (*13a*).

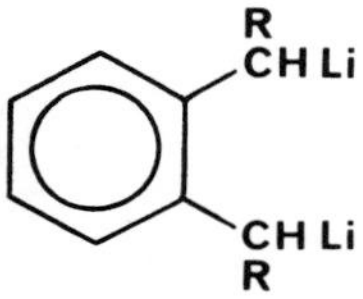

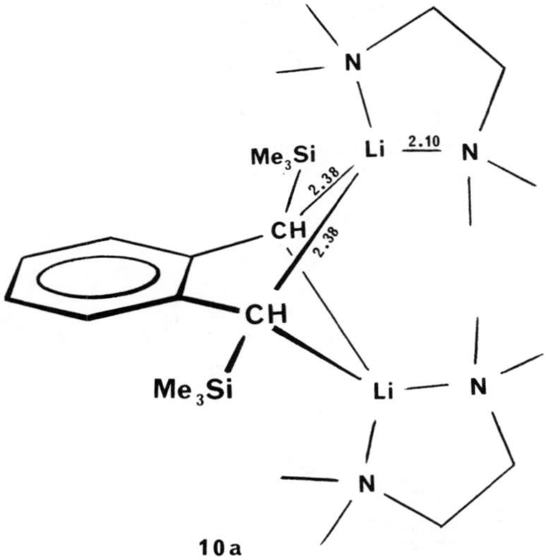

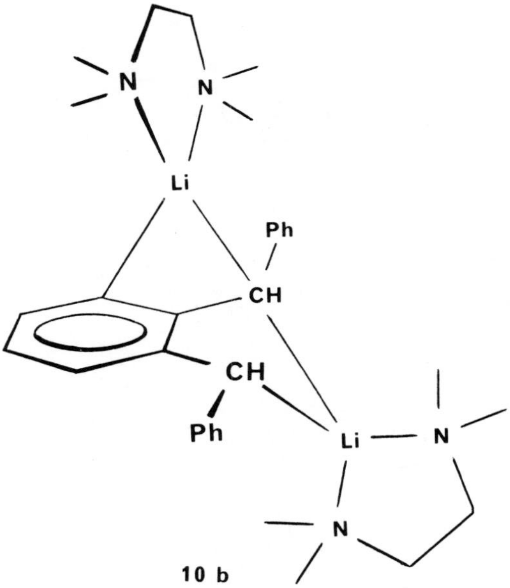

Compounds **10a** (*13c*), **11** (*13e*), **12** (13g), and **13** (*13*), while monomeric, have the same general geometrical features as those found in **5–9**, and can be regarded as examples of "intramolecular association." That is, two lithium atoms are involved in a symmetrical double-bridging arrangement between benzylic (**10a** and **11**) and aryl (**12** and **13**) sp^2 carbon atoms, respectively.

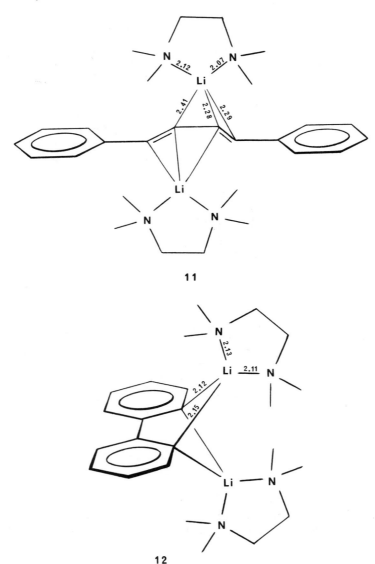

11

12

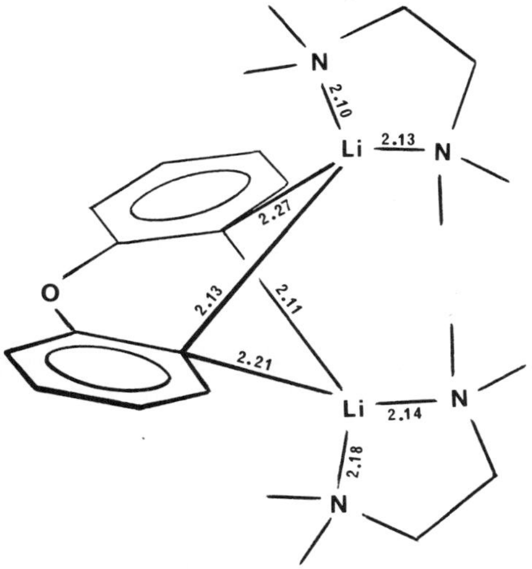

13

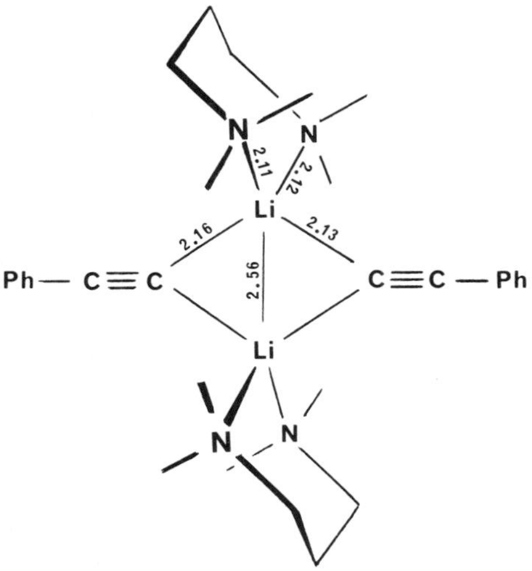

14

However, there are small differences due to carbanion orbital orientations and the degree of charge delocalization (*13a*). Both **10** and **12** have essentially the largest symmetry possible (C_{2v}); although all the carbons in **11** lie in the same plane, the lithiums are displaced to opposite sides. Symmetry C_2 is also found in **13**, but both rings are twisted.

Wise (*14*) reported the characterization of a dimeric acetylenic compound (PhC≡CLi·tmpda)₂ (**14**) [tmpda = $(CH_3)_2N(CH_2)_3N(CH_3)_2$], with bridging *sp*-hybridized carbon atoms.

The lithium–carbon bond lengths for sp^3-hybridized carbon atoms bonded to bridging lithium atoms average 2.29 Å, somewhat longer than the average value for nonbridging lithium atoms bonded to sp^3 carbons. The lithium–carbon bond lengths for aryl carbon atoms bonded to bridging lithium atoms range from 2.12 to 2.28 Å with an average value of 2.21 Å. Since the covalent radius for lithium is 1.34 Å (*15*), the lithium–lithium distances in these bridged structures (which range from 2.37 to 2.74 Å) suggest the possibility of some lithium–lithium interactions. However, *ab initio* MO calculations on model dimers indicate that this is usually not significant. These include methyllithium, vinyllithium, and ethynyllithium dimers, which are indicated to have C—Li bond lengths of 2.15, 2.13, and 2.08 Å, respectively, with the 3-21G basis set (*13a*); all are about 0.16 Å longer than the lengths in the corresponding monomers (Table I).

Similar trends are found in the available experimental structures. Bond lengths involving terminally bonded lithium are somewhat shorter than those for the bridging lithium cases, a result attributable to the electron-deficient bonding (and therefore lower bond order) of the bridging lithium

TABLE I

VARIATIONS OF LITHIUM–CARBON BOND LENGTHS (Å) DUE TO ASSOCIATION, SOLVATION, AND CARBON HYBRIDIZATION. COMPARISON OF X-RAY AND THEORETICAL RESULTS[a]

Compound	Monomer				Dimer			Tetramer	
	Unsolvated	Solvated			Unsolvated	Solvated		Unsolvated	Monosolvated
		Mono	Di	Tri		Mono	Di		
Alkyllithium (sp^3) Average X Ray[b]	2.03	—	—	2.16	—	—	2.29	2.28	2.27
Methyllithium (sp^3)									
Ab initio	2.00[c]	2.03	—	—	2.15	—	—	2.19	—
MNDO	1.82	1.84–1.86	1.87–1.92	1.88–1.94	2.04	2.07–2.09	2.11–2.13	2.20	2.24–2.25
Vinyllithium (sp^2)									
Ab initio	1.98[c]	—	—	—	2.13	—	—	—	—
MNDO	1.78[d]	—	—	—	—	—	—	—	—
Aryllithium (sp^2) Average X ray	—	—	—	—	—	—	2.20	—	2.31
Phenyllithium (sp^2) MNDO	1.80[e]	—	1.86[e]	—	2.03[e]	2.05[e]	—	—	—

364

Allyllithium (sp^2)								
Ab initio	2.13[f]	—	2.13[g]	2.20[g]	—	—	—	—
MNDO	2.07[g]	—	—	—	—	—	—	—
Benzyllithium (sp^2)								
Average X ray	1.84[h]	2.21	2.28	—	—	—	2.38	—
MNDO	—	—	—	—	—	—	—	—
Ethynllithium (sp)								
Average X ray	—	—	—	—	2.08	—	2.15	2.20
Ab initio	1.92	—	—	—	2.08	—	2.08	—
MNDO	1.74[d]	—	—	—	—	—	—	—

[a] *Ab initio* (3-21G//3-21G) and MNDO results from E. Kaufmann, Diplomarbeit, Erlangen (1982), and unpublished results; cf. T. Fjeldberg, P. B. Hitchcock, M. S. Lappert, and A. J. Thorne, *J. Chem. Soc., Chem. Commun.*, pp. 821–824 (1984).

[b] Average taken for collections of compounds of each type presented in this article; for the unsolvated monomer see J. L. Atwood, P. Fjeldberg, M. S. Lappert, N. T. Luong-Thi, R. Shakir, and A. J. Thorne, *J. Chem. Soc., Chem. Commun.*, pp. 1163–1165, 1984.

[c] Carnegie–Mellon Quantum Chemistry Archive, R. A. Whiteside, R. J. Frisch, and J. A. Pople, eds., 3rd Ed., 1983, Carnegie–Mellon University.

[d] J. Andrade and P. von R. Schleyer, unpublished results.

[e] J. Chandrasekhar and P. von R. Schleyer, *J. Chem. Soc., Chem. Commun.* pp. 260–261 (1981).

[f] T. Clark, C. Rhode, and P. von. R. Schleyer, *Organometallics* **2**, 1344–1351 (1983).

[g] G. Decker and G. Boche, *J. Organomet. Chem.* **259**, 31–36 (1983).

[h] A. J. Kos and P. von R. Schleyer, unpublished results.

atoms as compared to terminally bonded. The lithium–carbon bond lengths involving sp^3-hybridized alkyl carbons are slightly longer than corresponding bond lengths involving aryl carbon atoms, as expected from the higher s character. Similarly, Li—C bond lengths for the acetylenic dimer are shorter than those for aryl dimers. Carbon–lithium bond lengths for benzylic organolithium compounds are unusually long, however.

C. Organolithium Compounds with Lithium Tetrahedra

When steric hindrance is minimal, organolithium compounds often form aggregates (oligomers); tetramers are frequently observed. Their structures may be described as tetrahedral arrays of lithium atoms with the substituents occupying each face of the tetrahedron or, alternatively, as cubic arrays with alternating lithium and carbon atoms (**15**). The Li—Li distances are shorter than the C—C separations, however.

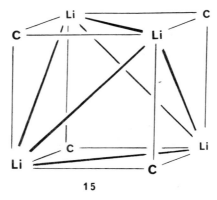

15

The simple alkyllithium compounds (MeLi)$_4$ (*16*, *17*), (MeLi)$_4$·2tmeda [tmeda = $(CH_3)_2NCH_2CH_2N(CH_3)_2$] (*18*), and (EtLi)$_4$ (*19*, *20*) adopt such tetrahedral structures.[2] The lithium–carbon bond lengths range from 2.19 to 2.53 Å, with an average value of 2.29 Å, while the lithium–lithium distance lies in the range 2.42–2.63 Å (average 2.56 Å). These lithium–carbon bond lengths in these tetrameric alkyllithium compounds are slightly longer than those for the terminally bonded organolithium compounds (Table I); this is due to the multicenter bonding in the oligomeric structures.

Similarly, 3-lithio-1-methoxybutane (**16**) (*25*) adopts a tetrameric structure, with each triangular lithium face occupied by a C(3) carbon atom. Each oxygen atom coordinates to lithium. The complex (LiBr)$_2$·

---

[2] Methylsodium(*21*) also adopts a tetrahedral structure, but methylpotassium (*22*), methylrubidium (*23*), and methylcesium (*24*) do not.

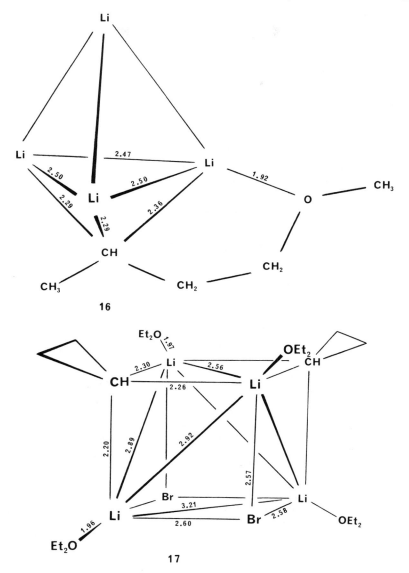

16

17

($\triangleright$—Li)$_2 \cdot 4$Et$_2$O (**17**) (*26*), adopts an analogous (albeit distorted) cubic structure. Compound **17** may be envisioned as a complex between two dimers, ($\triangleright$—Li)$_2 \cdot 2$Et$_2$O and (LiBr)$_2 \cdot 2$Et$_2$O. The lithium–carbon bond lengths (average value of 2.25 Å) and lithium–lithium bond length (2.56 Å) for the "organic" part of the complex are comparable to those of other dimeric or tetrameric organolithium compounds.

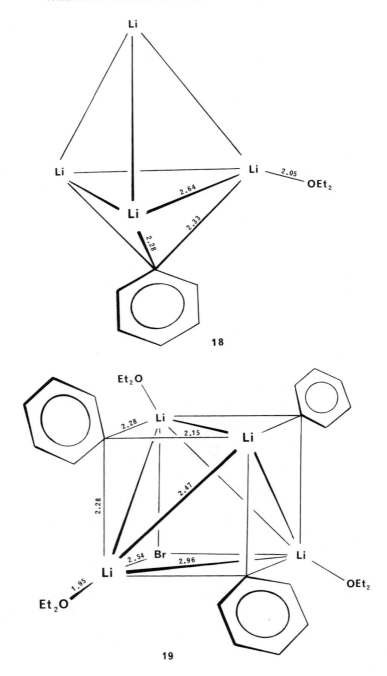

18

19

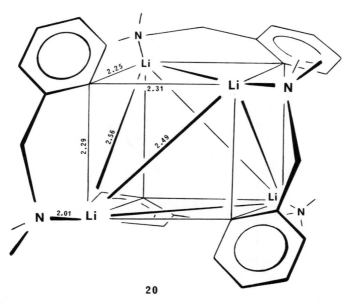

20

Aryllithium compounds have also been found to form tetrameric complexes. The structures of phenyllithium monoetherate $[(PhLi \cdot Et_2O)_4]$ **(18)** and the lithium bromide complex of phenyllithium monoetherate $[(PhLi \cdot Et_2O)_3 \cdot LiBr]$ **(19)** have been determined (27). In each of these compounds the phenyl moiety occupies a triangular face of the tetrahedron of lithium atoms and the diethyl ether molecules are complexed to the lithium atoms at the apices. The tetrameric structure of 1-lithio-2-dimethyl-aminomethylbenzene **(20)** (28) is very similar to that of **18**. The lithium atoms are arranged in a distorted tetrahedron with the aryl groups occupying the faces of the tetrahedron. Each amino group coordinates to a lithium atom. This tetramer may also be visualized as consisting of two **6**- or **7**-like dimers which have associated further.

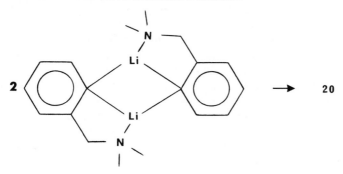

Two substitued lithium acetylides, $(PhC\equiv CLi)_4(tmhda)_2$ (**21a**) (*29*) and $(t-BuC\equiv CLi)_4(thf)_4$ (**21b**) (thf = tetrahydrofuran) (*30*), also adopt

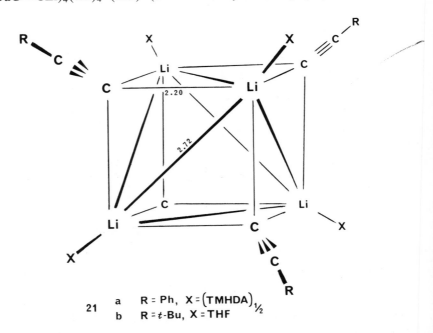

21	a	R = Ph, X = (TMHDA)₁/₂
	b	R = *t*-Bu, X = THF

similar tetrameric structures. The acetylide moieties occupy the faces of the tetrahedron; each lithium atom is further complexed by a donor atom. The lithium–carbon bond distances in **21** are comparable to those in the simple tetrameric alkyllithium compounds. The lithium–lithium interatomic distances are somewhat longer, i.e., greater than the sum of the covalent radii for lithium (2.68 Å) (*14*).

Structural features in the tetrameric organolithium compounds are comparable to those found in the analogous dimeric organolithium compounds. That is, the alkyllithium compounds (sp^3 carbon atom bonded to lithium) have similar carbon–lithium and lithium–lithium bond lengths in the dimers (**5** and **6**) and in the tetramers (**15–17**). Similarly, the dimeric aryllithium compounds (**7–9**) have C—Li and Li—Li bond lengths comparable to those for the tetrameric aryllithium compounds (**18–20**); the same is true for the acetylenic compounds **14** and **21**. Due to multicenter bonding, oligomeric organolithiums, either dimeric or tetrameric, have longer Li—C bond lengths than monomeric lithium compounds. *Ab initio* MO calculations (*13b*) (Table I) also indicate Li—C bond lengthening upon oligomerization.

D. *Higher Oligomeric Organolithium Compounds*

A few organolithium compounds crystallize in aggregates with more than four subunits. The structure of hexameric cyclohexyllithium $[(C_6H_{11}Li)_6 \cdot 2C_6H_6]$ (*31*) involves an octahedral array of lithium atoms with the cyclohexyl moieties occupying six of the eight triangular faces (**22**).[3] The two benzene solvent molecules seem to lie at the other two transoid faces of the octahedron. An "average" triangular face for this complex also is shown below.

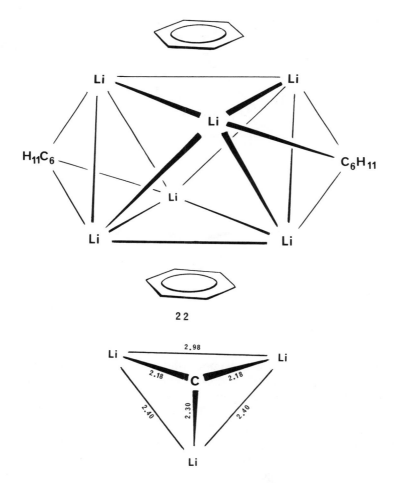

2 2

[3] (LiSiMe₃) (*32*) and [LiN=C(*t*-Bu)₂]₆ (*33*) also adopt structures involving octahedral arrangements of lithium atoms (see Section IV).

1-Lithiomethyl-2,2,3,3-tetramethylcyclopropane crystallizes in a similar hexameric octahedral structure (**23**) (*34*). The hydrocarbon groups occupy six of the eight triangular faces; the two remaining transoid faces are unoccupied. The bonding in hexameric **22** and **23**, as with the tetrameric organolithium compounds, is electron deficient, and involves four-center two-electron orbitals. Each carbon atom interacts with a triangle of lithium atoms.

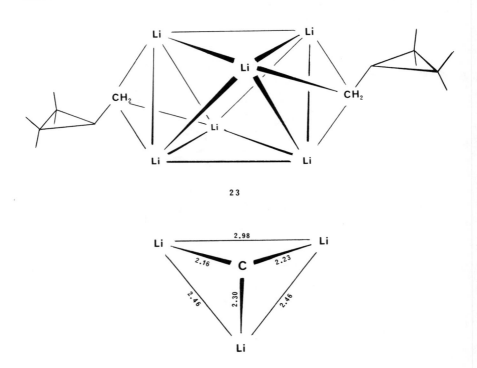

2 3

The structure (2,6-dimethoxyphenyllithium)$_6$·Li$_2$O (*35*) may be envisioned as two tetrahedral arrays of lithium atoms each having a face complexed to the unique oxygen atom. This forms an octahedral array of lithium atoms around the oxygen. The aryl groups occupy the other three exposed tetrahedral faces (**24**). The geometrical arrangement of carbon atoms with respect to lithium triangular faces in this complex is similar to those for other oligomeric organolithium compounds (Li—C = 2.28– 2.48 Å, Li—C (average) = 2.39 Å; Li—Li (average) = 2.55 Å). The bonding in **24** has been postulated to involve asymmetric three-center

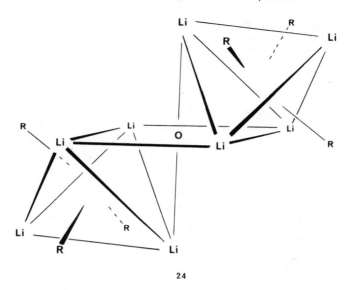

24

two-electron, rather than four-center two-electron bonding (28). The important structural features of this complex are illustrated below:

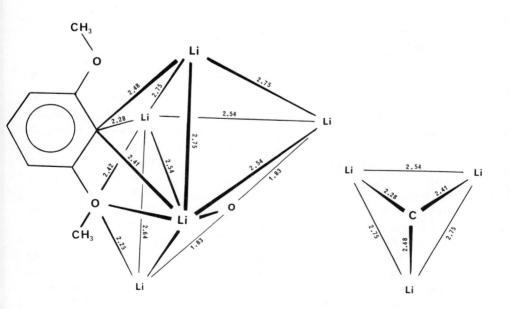

E. *Organolithium Compounds with Lithium–π Interactions*

Many organolithium compounds involve π-delocalized hydrocarbon "anions" or "polyanions" interacting with lithium cations. The lithiums are within "bonding distance" of two or more carbon atoms of the π system. Stucky and co-workers (*7, 31, 36, 37*) were pioneers in this area, and determined the X-ray structures of many organolithium π complexes. These will be presented in detail later in this section. The geometries (location and orientation of the lithium atom with respect to the hydrocarbon) of these organolithium π complexes were interpreted in terms of multicenter covalent interactions between the unoccupied atomic orbitals of lithium and the appropriate occupied molecular orbitals of the hydrocarbon anion. This conceptually simple model rationalized the often unexpected locations of the lithium atoms quite convincingly. The geometries of such organolithium π complexes were not considered by Stucky's group to be consistent with simple electrostatic interpretations (*36, 37*).

However, more recent analyses have emphasized the ionic character of lithium bonding to a greater extent (*38–38b*). Streitwieser (*38*), in particular, has insisted that lithium bonds have very little, if any, covalent character. Before presenting details of the crystal structures, the two interpretations will be compared with simple examples: allyllithium and cyclopentadienyllithium.

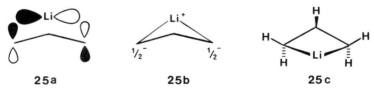

25a   25b   25c

Allyllithium. The highest occupied allyllithium molecular orbital, shown in **25a**, is composed of the allyl highest occupied molecular orbital, (HOMO) and a vacant *p* orbital on lithium. This interaction is favored when lithium adopts the bridging position indicated by NMR studies (*39*). (The available allyllithium X-ray structure is that of a polymer (*40*), but it may be possible with suitable lithium ligands to obtain the monomer.) The alternative ion pair formation (**25b**) would also place the lithium cation equidistant between the two negatively charged carbon centers at C(1) and C(3).

Both the interpretations thus lead to the same qualitative rationalization in this case. The geometrical details of allyllithium are more revealing: these are currently available only through theoretical computations, although analogous deformations are found for allylnickel derivatives (*41*).

While the isolated allyl anion is calculated to have a planar structure (*42*), interaction with lithium results in movement of the three of the hydrogens markedly out-of-plane (*43–45*). The *ab initio* (3-21G basis set) allyllithium geometry is shown in **25c**; the central hydrogen is bent *toward* and the two inner hydrogens at C(1) and C(3) are bent *away from* the lithium.

While these out-of-plane hydrogen distortions were originally ascribed to better *p* overlap (as in **25a**) (*43*), electrostatic effects were later found to be responsible. When the valence orbitals on lithium were omitted (forcing Li^+ to be an ion) and the calculations repeated, the same out-of-plane hydrogen distortions were found. A point charge in place of Li produced the same effect on the geometry when all bond lengths and angles were allowed to attain their optimum positions. In contrast, when "counterpoise" calculations were carried out, that is, the lithium *p* orbitals were included *without the charge* (Li^+ nucleus), the resulting geometry was quite different. Ion pair interactions thus clearly dominate. However, multicenter interactions involving lithium valence orbitals do contribute to the total bonding energy, but to a more limited extent (*45*).

Cyclopentadienyllithium. The π charge in the aromatic cyclopentadienyl anion is distributed equally to all five carbon atoms. A lithium counterion should thus electrostatically favor a central location (C_{5v}, **26a**) over the π face. The same conclusion is reached on the basis of MO considerations (*46*). The six interstitial electron interactions involving the three cyclopentadienyl π orbitals and those of corresponding symmetry on lithium (one of these is shown in **26b**) also favor structure **26a**.

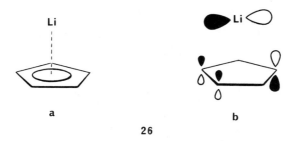

a   b

26

A calculational search of the potential energy surface carried out by moving the lithium about shows that $^5\eta$ bonding is favored over $^4\eta$, $^3\eta$, $^2\eta$, and $^1\eta$. Calculations suggest that the hydrogens of the ring are bent out of the carbon plane, away from the metal (*46*). While this is consistent with the diffuse character of the lithium *p* orbitals (the ring orbitals will give better overlap when bent outward), point charge on Li^+ cation models gives a similar result (*47*). Multicenter covalent bonding involving lithium may make a minor contribution, but ionic interactions dominate.

Two cyclopentadienyllithium compounds, **27** (*48*) and **28** (*49*), have been prepared and characterized. The structural features of these compounds are in agreement with theoretical predictions (*46*).

These X-ray structures illustrate another important point. Note that each lithium is formally associated with *ten* electrons (six from the

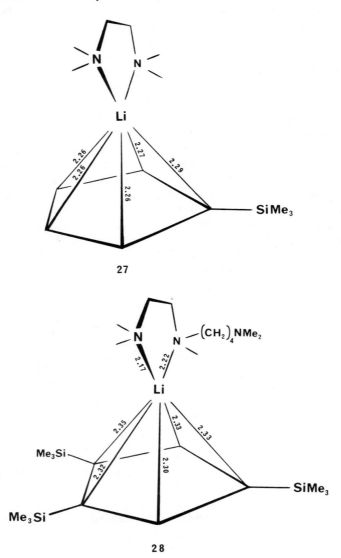

27

28

cyclopentadienyl anion and four from the two nitrogens). However, octet rule considerations are not significant in lithium chemistry. Lithium is largely ionic and gains, at best, only a few tenths of an electron from the total environment. The coordination number of lithium, e.g., the number of solvent molecules which can be attached, is governed primarily by steric factors.

In benzyllithium (**29**) (*50*), the lithium atom is within bonding distance of the benzylic carbon atom and both C(1) and C(2) of the phenyl ring. The interaction of the unoccupied 2*p* orbital of the lithium cation and the HOMO of the benzyl anion (Stucky's rationalization of this geometry) is depicted below.

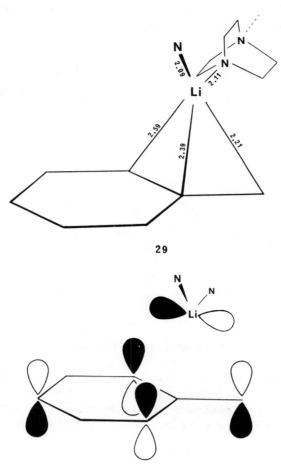

29

The nitrogen atoms occupy the "tetrahedral" positions expected on this basis. An alternative location of lithium over the center (approximately) of the phenyl ring would also have favorable orbital interactions, and actually is found to be more stable by MNDO calculations when no solvent is present (51). Stucky's rationalization has been challenged by Bushby (38b).

In the triphenylmethyllithium–tetramethylethylenediamine complex **30** (52), the lithium atom is not located directly over the central carbon, as might have been expected, but rather has four close contacts to the trityl anion. Besides the sp^2-hybridized central carbon, these involve two carbon atoms of one nearly coplanar phenyl group, and one carbon atom of a somewhat twisted second ring. The third phenyl is nearly perpendicular, so that the negative charge is largely delocalized over only two of the phenyls.

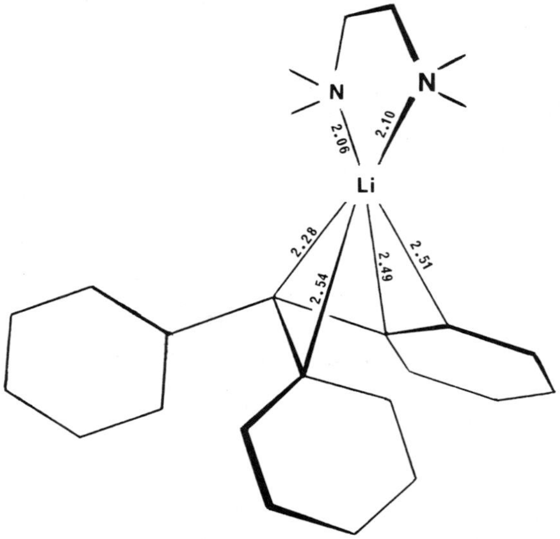

30

The geometry of fluorenyllithium bisquinuclidiene (**31**) (36), also can be visualized by means of a simple orbital interaction picture, shown along with **31**. In the δ HOMO, the two largest Hückel coefficients are at C(1) and C(9). Alternative locations over the five- and the six-membered rings, found for some transition metal derivatives, are not favored here. Albright et al. (53) provide a lucid analysis of the interactions in polycyclic complexes.

In dilithionaphthalene (**32**) (54), indenyllithium (**33**) (55), and dilithio-anthracene (**34**) (56), each lithium atom is associated with the ring faces.

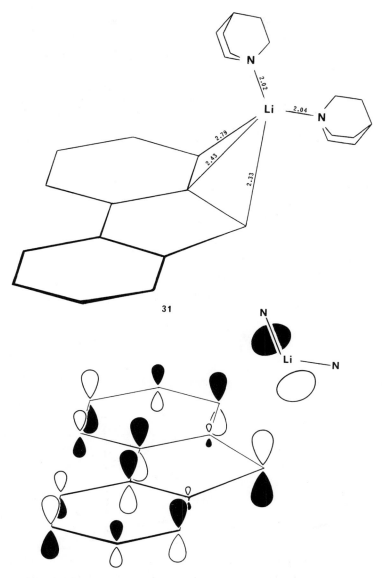

31

The positioning of the lithiums and the orientations of the nitrogen ligands have also been interpreted by Stucky (37) in terms of interactions of the carbanion molecular orbitals with the orbitals on lithium. However, unlike the benzyllithium and fluorenyllithium cases, the simplified interpretation involving interaction of the HOMO of the carbanion with the

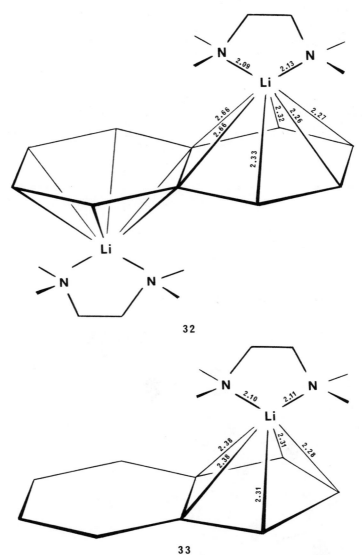

32

33

empty $2p$ orbital of lithium does not adequately explain the geometries. In our view, factors other than those involving lithium p orbitals are responsible. Bushby has shown that many of these structures can be rationalized electrostatically. (*38b*).

Dilithioacenaphthalenebis(tetramethylethylenediamine) (**35**) (*57*) has the lithium atoms located above and below the five-membered ring of the carbanion.

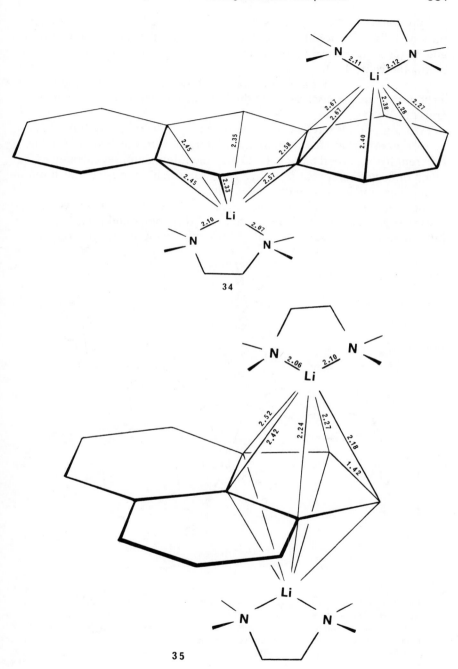

34

35

Additional examples of π-complexed organolithium compounds are provided by dilithiobifluorene (**36**) (*58*)[4] and the dilithiostilbenes **37** and **38** (*60*). In both of these systems the two lithium atoms lie above and below the central ethylene.

Arora *et al.* (*61*) have prepared and determined the structure of a nonaromatic π complex, dilithiohexadienebis(tetramethylethylenediamine) (**39**).

Two compounds have been prepared and characterized by Rewicki and co-workers in which the lithium atom interacts with the π systems of two different hydrocarbon anions. In **40** (*62*) the lithium atom is solvated by an ether molecule, whereas in **41** (*63*), a "sandwich" complex, the lithium atom does not interact with solvent at all.

Dietrich *et al.* (*64*) have deduced the structure of dilithiodibenzylketonebis(tmeda), a dienolate.[5] The geometry of this compound is such that the lithium atoms are resting in a "nest" provided by a phenyl group and the

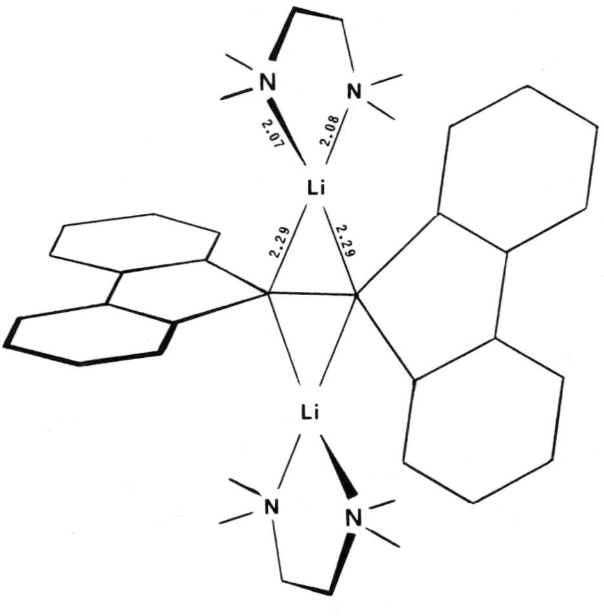

36

---

[4] It has been demonstrated that removal of the second proton from 9,9′-bifluorene is nearly as easy as removal of the first (*59*). It has been suggested that this phenomenon is due to electrostatic stabilization provided by the geometrical arrangement of point charges as provided by a structure such as that observed for **36** (*38a*).

[5] See Section IV for structures of lithium enolates without lithium–carbon contacts.

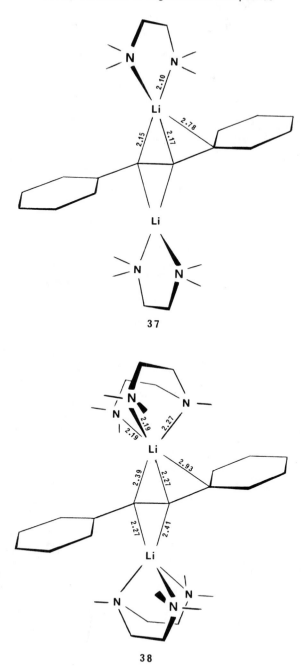

37

38

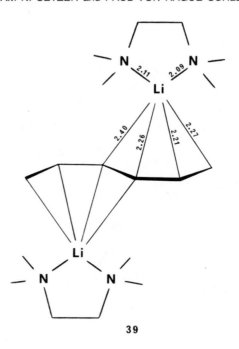

39

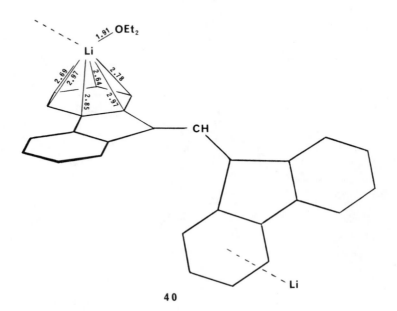

40

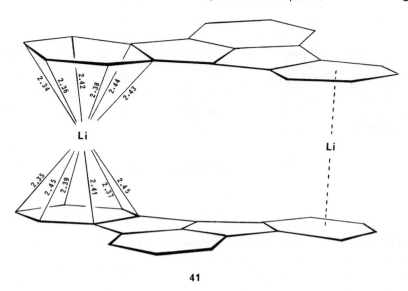

41

carbonyl group (**42**). The structure of the compound is analogous to that for benzyllithium (**29**), in that each lithium atom is associated with C(1) and C(2) of a phenyl ring as well as the benzylic carbon. The lithium atoms further interact with the carbon and oxygen atoms of the carbonyl group. Note, however, that the carbonyl group and the phenyl rings are not coplanar.

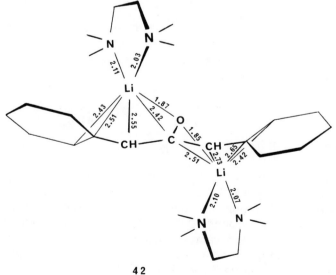

4 2

F. *Organolithium Compounds Involving Cumulenic Systems*

Such compounds, derived from substituted acetylenic substrates (excluding simple acetylides), form oligomeric structures involving bridging lithium atoms and possible interactions of the lithium atoms with the acetylenic π system. The Erlangen group has prepared and characterized a number of organolithium compounds derived from 5-substituted-2,2,8,8-tetramethyl-3-6-nonadiyne (**43**).

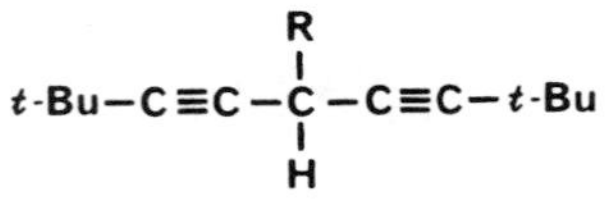

$$t\text{-Bu}-C\equiv C-\underset{\underset{H}{|}}{\overset{\overset{R}{|}}{C}}-C\equiv C-t\text{-Bu}$$

43

Monolithiation of the parent hydrocarbon (R= H) gave an unstable product which underwent a 1,3-hydrogen rearrangement. The X-ray structure of this product (**44**) (*1, 13*) shows it to be an allenyllithium

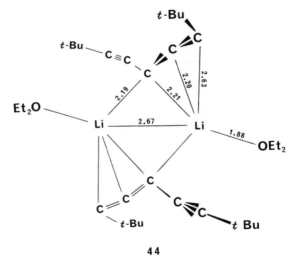

44

derivative. The compound is dimeric and involves bridging lithium atoms analogous to structures **5–9**. Each lithium atom, rather than being coordinated to two donor atoms, coordinates with one solvent molecule and a carbon–carbon double bond of the allenic π system. The important structural features of this compound, i.e., the bent allenic system with bridging lithium atom, bond lengths, and bond angles, were predicted on the basis of prior *ab initio* calculations (*1, 65*).

In order to preclude the rearrangement observed in **44**, substituents were introduced at the central carbon. Monolithiation of **43** (R = CH₃) gave a product which was indicated by NMR measurements to have a rapidly fluctuating unsymmetrical structure. The X-ray structure (**45** (*1, 13*) confirmed this conclusion, and also showed the compound to be a dimeric allenyllithium derivative with bridging lithiums.

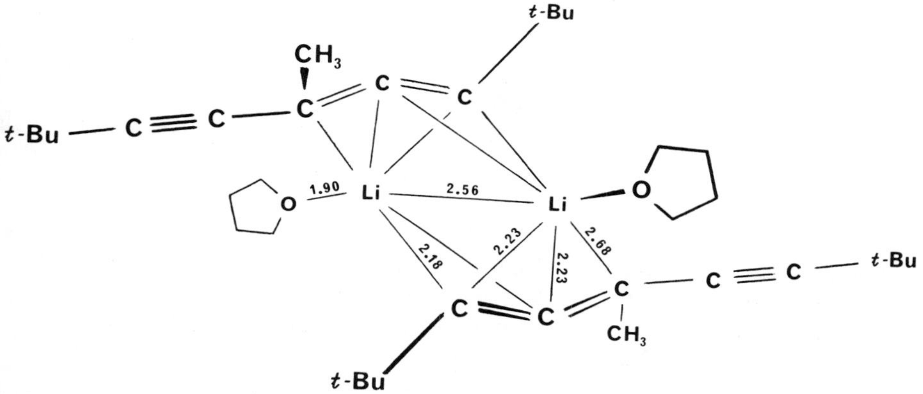

45

The monolithiated derivative of **43** (R = SiMe₃), on the other hand, was indicated by NMR studies to have, in solution, the symmetrical structure **46**, found in the solid (*1, 13*). This structure is also a dimer with bridging

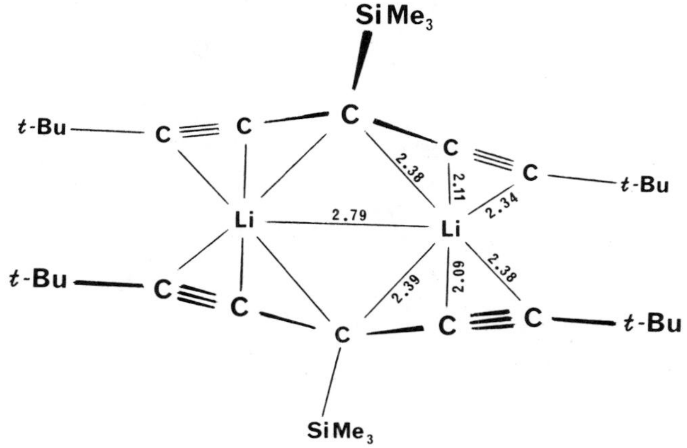

46

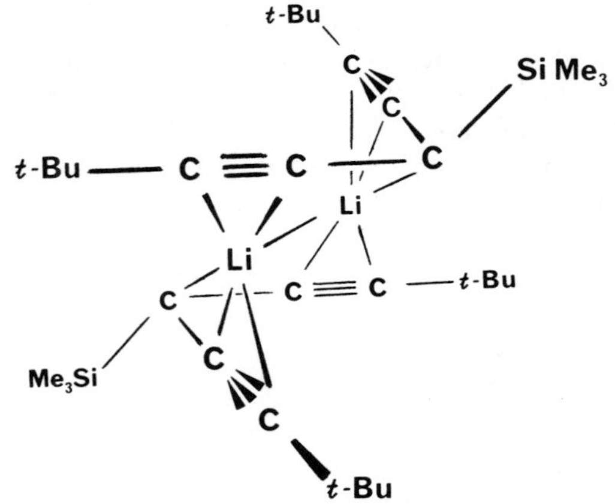

46b

lithiums (see structure **46b**). The lithium atoms, however, are not solvated, but rather interact with two acetylenic π systems.

Dilithiation of **43** (R = H) gave the tetrameric complex **47** (66). This structure, involving two crystallographically different diyne molecules in

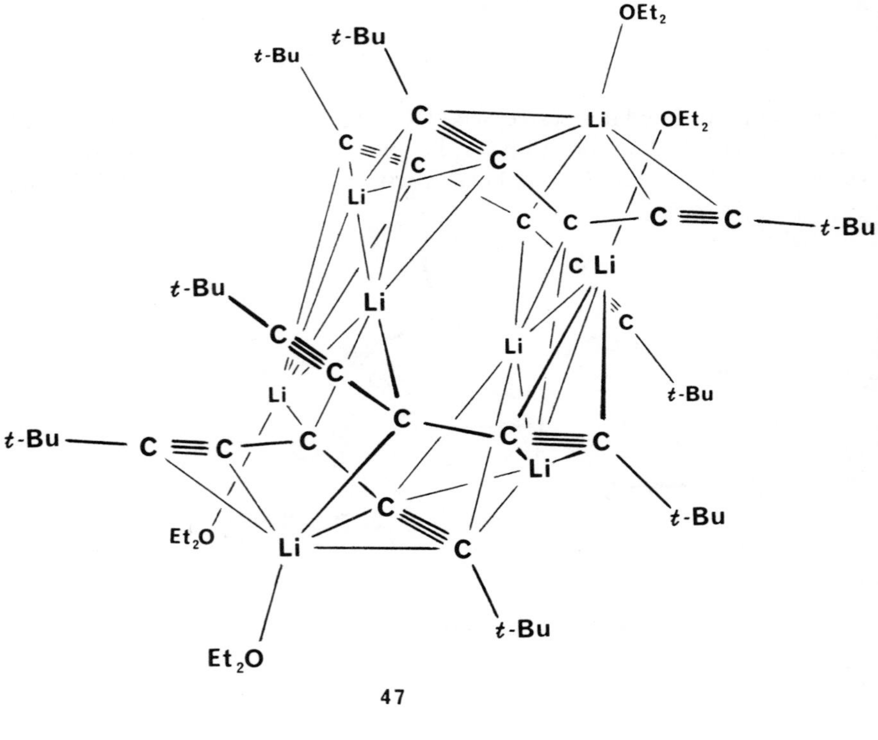

47

different environments, seems to have an octahedral array of six lithium atoms, and two lithium atoms without lithium–lithium contacts. The complex appears to be held together by lithium–acetylenic π interactions as well as lithium–lithium interactions. In one of the diyne molecules a lithium atom occupies a position between the two halves of the molecule, complexed to the two acetylenic π systems. This geometrical arrangement was anticipated by MNDO geometry optimization of C_5Li_4:

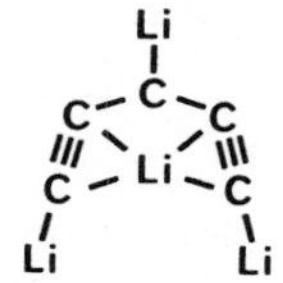

The structure of the dilithiobutatriene (**48**) (*67*), prepared by lithium reduction of 1,4-di-*t*-butyldiacetylene, is a dimer with nearly parallel, C_4 cumulene units. The four lithium atoms serve as bridges between the units.

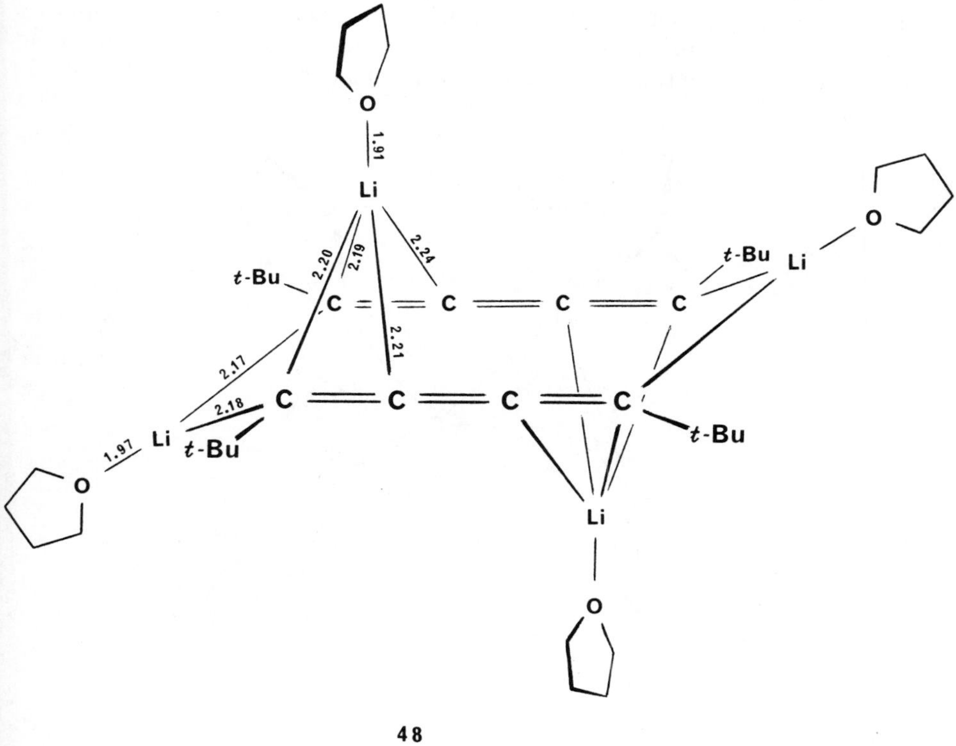

48

Two lithiums are coordinated to only two carbons while the other two bind simultaneously on opposite sides to four carbons. There are no lithium–lithium contacts in this system. This structure, predicted by MNDO calculations (67), may be envisioned as a dilithiated derivative of:

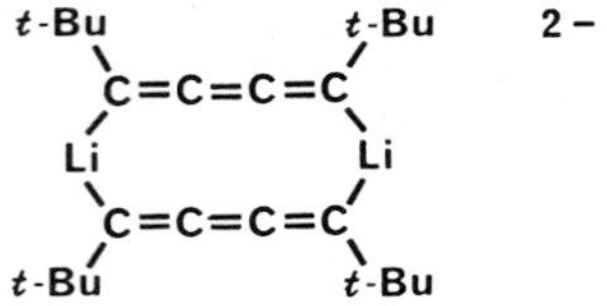

By analogy with dilithionaphthalene (**32**), the additional lithium atoms occupy opposite faces of this system, interacting with as many carbons as possible and avoiding competition (**48**).

Dilithiodibenzylacetylenebis(tmeda) (**49**) (*68*) is monomeric. Each lithium atom is bonded to a benzylic carbon and has further interactions with a phenyl C(1) and the acetylenic π system. The structure can be understood as being derived from a perpendicular 2-butyne-1,4-diyl dianion, $^-CH_2$—C≡C—CH_2^-, in which each negative charge is stabilized in its own orthogonal π system. The lithiums bridge each of these propargyl (or allenyl) anion systems in a 1,3 manner.

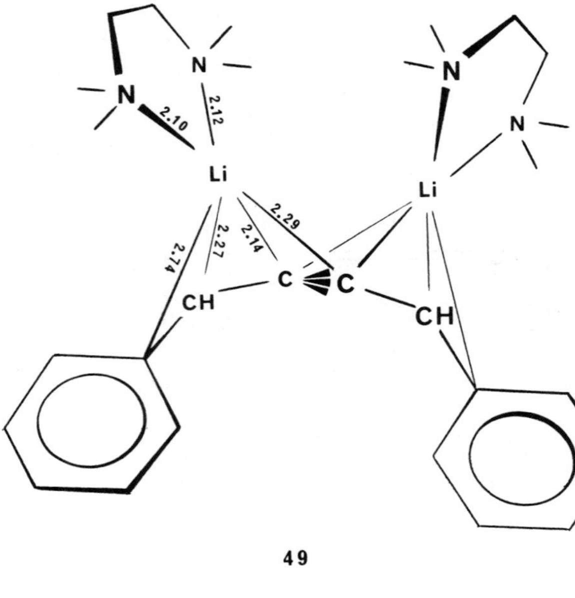

49

G. *Miscellaneous Organolithium Compounds*

The benzophenonedilithium compound **50**, formed by reduction of benzophenone with lithium metal, crystallizes as a dimer (*69*). The four lithium atoms in the structure are divided into two different sets. The two benzophenone moieties are bridged, through the carbonyl oxygen atoms, by two symmetry-equivalent lithium atoms. Each of the two other lithiums is bonded to one phenyl ring and the ketone functionality reminiscent of that observed in benzyllithium (**29**), dilithiodibenzyl ketone (**42**), and dilithiodibenzylacetylene (**49**). The two different types of lithium atoms are complexed further to THF and TMEDA.

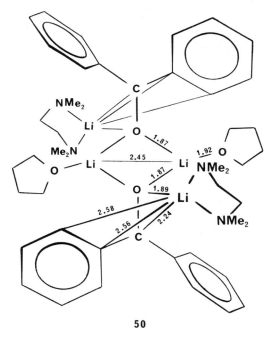

50

The lithium "ate" complex $[Li(thf)_4]^+[Li\{C(SiMe_3)_3\}_2]^-$ (**51**) (*70*) is unique in that the anion has a lithium atom which is linearly double coordinated. Due to steric and electronic factors, this lithium is un-complexed by solvent. The second Li^+, coordinated to four THF molecules, is present as a solvent-separated ion pair. Although "ate" complexes are written with formal negative charges on the central atoms, MO calculations indicate the Li in **51** to be positively charged. The negative charge is delocalized in the $(Me_3Si)_3C^-$ carbanion moieties, which are stabilized by the $SiMe_3$ groups.

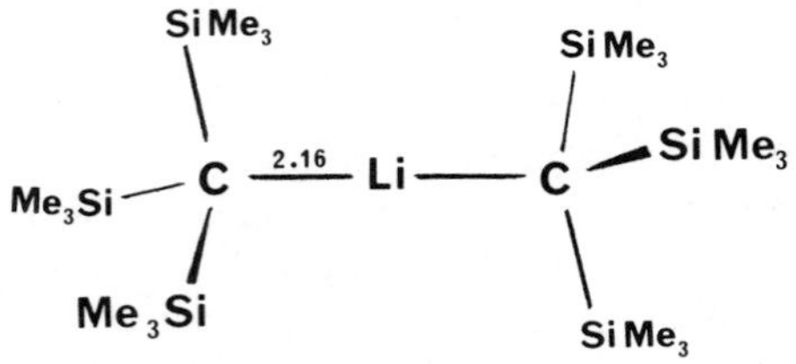

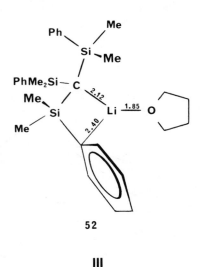

51

The related compound $LiC(SiMe_2Ph)_3(thf)$ (**52**) (*71*) also is monomeric but the lithium atom adopts a distorted trigonal-planar geometry. The lithium atom interacts with one of the phenyl rings and is solvated by a single THF molecule. Steric factors may be responsible for structural differences between **51** and **52**.

52

III

MIXED-METAL ORGANOLITHIUM COMPOUNDS

A. *Mixed-Metal Organolithium Compounds with Lithium–Carbon–Metal Bridges*

A large number of mixed-metal organometallic compounds have been prepared in which the hydrocarbon ligands bridge the lithium and the other metal atoms. These mixed-metal analogues of type **II**, *B* organolithium compounds involve multicenter electron-deficient bonding which may include lithium–metal as well as Li—H—C interactions.

The magnesium complexes **53** (*72*) and **54** (*73*) are structurally very similar to **6** and **7**. Methyl and phenyl groups bridge the two metal atoms. The lithium–magnesium interatomic distances in **53** and **54** (2.62 and 2.94 Å, respectively) are less than the sum of the atomic radii for these metals (3.15 Å), (*74*), indicating that some degree of metal–metal interaction may be present.

The structure of LiAl(C_2H_5)$_4$ (**55**) (*75*) is similar to **53** and **54**, but consists of linear chains of alternating lithium and aluminum metal atoms.

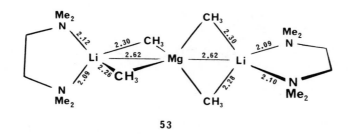

53

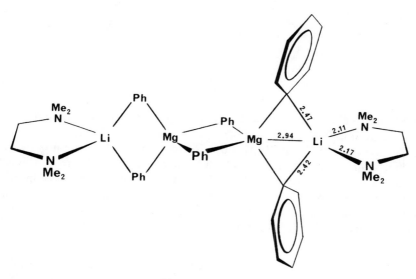

54

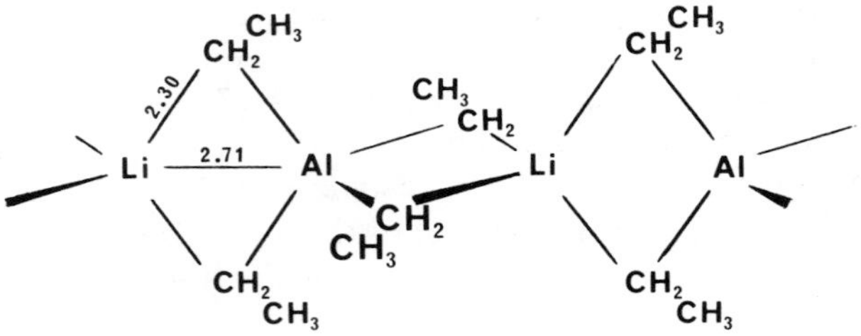

55

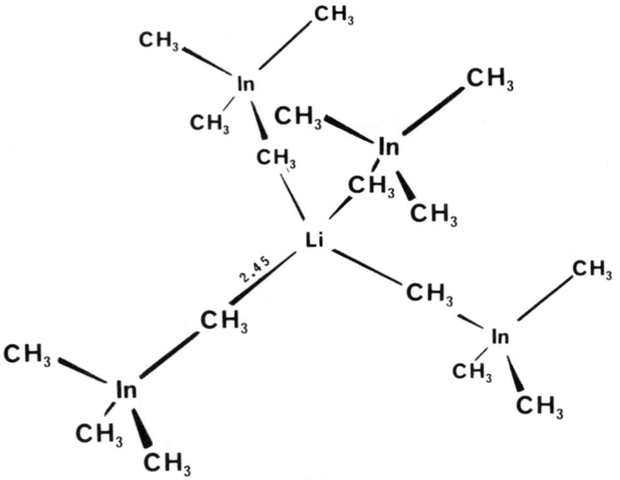

56

The lithium–aluminum distance in this compound also is less than the sum of their atomic radii (2.98 Å) (74).

The structure of LiIn(CH$_3$)$_4$ (56) (76) is a three-dimensional network. Each lithium atom is surrounded by a tetrahedral array of carbon atoms.[6]

The structure of LiB(CH$_3$)$_4$ (57) (79) consists of planar sheets of lithium atoms bridged by tetramethylboron groups. A unique feature is the presence of both linear and bent Li—C—B units. A neutron diffraction study of LiB(CH$_3$)$_4$ (79) illustrates the Li—H—C interactions present in

[6] Li$_2$Zn(CH$_3$)$_4$ (77) and Li$_2$Be(CH$_3$)$_4$ (78) are isostructural and involve three-dimensional network arrangements analogous to 56. These compounds were determined by X-ray powder techniques; there is some ambiguity as to the space group, and hence, in the geometries as well. The Li—C distances are about 2.5 Å in each of these compounds.

this complex. Both dibridged,

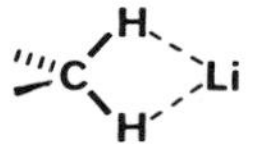

and tribridged,

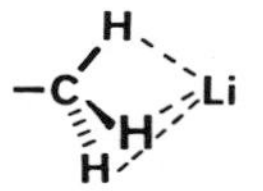

configurations are present.

The lithium–metal distances in **55–57** are all greater than the sum of their respective atomic radii suggesting lithium–metal interactions to be relatively unimportant.

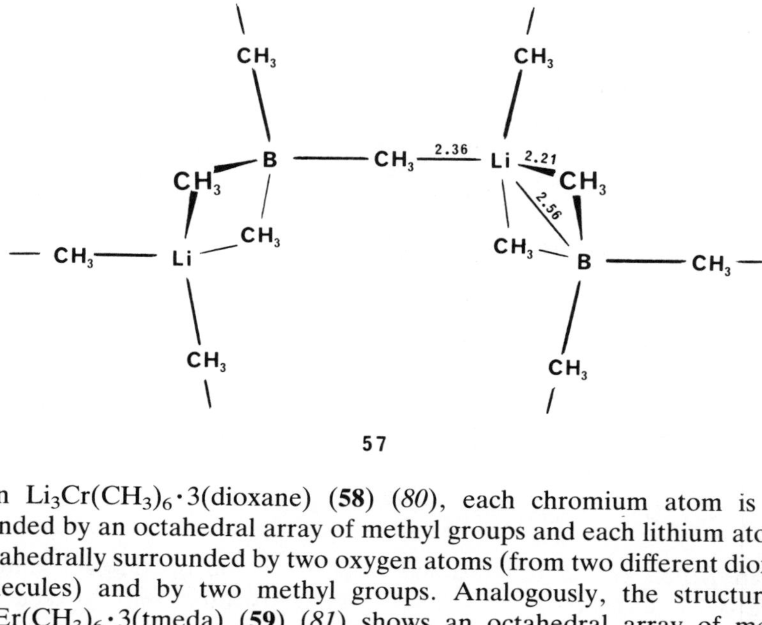

57

In $Li_3Cr(CH_3)_6 \cdot 3$(dioxane) (**58**) (*80*), each chromium atom is surrounded by an octahedral array of methyl groups and each lithium atom is tetrahedrally surrounded by two oxygen atoms (from two different dioxane molecules) and by two methyl groups. Analogously, the structure of $Li_3Er(CH_3)_6 \cdot 3$(tmeda) (**59**) (*81*) shows an octahedral array of methyl groups around the erbium atom and each lithium atom bonded to two methyl groups and two nitrogen atoms in a tetrahedral arrangement. Lithium–carbon interactions are likely to be involved in **58** and **59**. The lithium–metal distances for these two complexes again are less than the sum of their respective atomic radii.

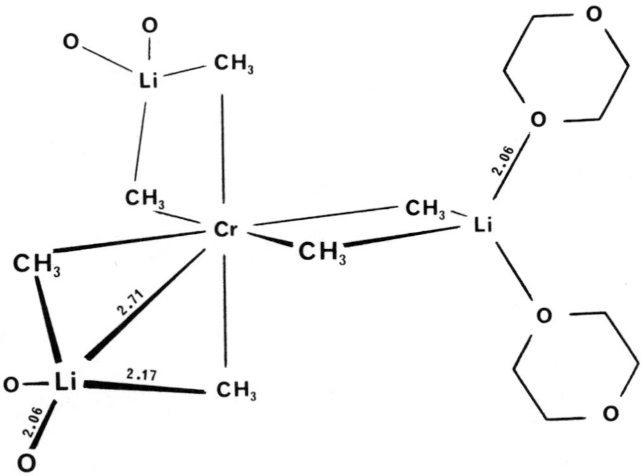

58

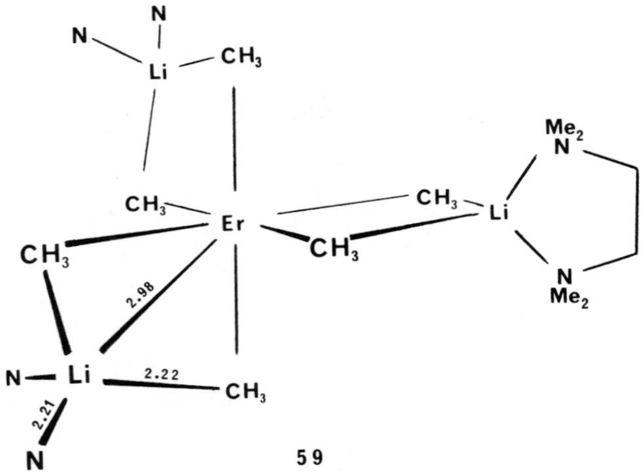

59

Van Koten *et al.* (*82*) characterized the cyclic organogold complex $Au_2Li_2[C_6H_4CH_2N(CH_3)_2]_4$. The X-ray structure (**60**) reveals the lithium and gold atoms lying in a planar four-membered ring. The environment of the lithium atoms is similar to that found in $(C_6H_5Li \cdot tmeda)_2$ (**7**) or in $[(Li \cdot tmeda)(C_6H_5)_3Mg)]_2$ (**54**); the lithium atoms bridge the carbons of two aryl groups and are further chelated by the nitrogens of the substituents. The lithium–carbon (2.42–2.52 Å) and lithium–nitrogen (2.12–2.35 Å) bond distances are similar to those of **7** and **54**.

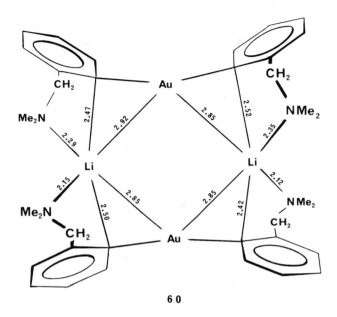

60

B. Mixed-Metal Organolithium Compounds Having Metal–Metal Multiple Bonds

A number of lithium/transition metal compounds involving metal–metal multiple bonds as well as lithium–alkyl group interactions have been structurally characterized. These compounds, having the general formula $(Li \cdot OR_2)_n M_2 R'_8$, involve a square-pyramidal geometry around each lithium atom, made up of four carbon atoms and one ether oxygen atom (**61**). The structural parameters involving lithium for these complexes are summarized in Table II (83–87). As with type **III**, A organolithium compounds, lithium–transition metal as well as Li—H—C distances are relatively short.

TABLE II

AVERAGE BOND LENGTHS (Å) FOR COMPOUNDS **61**

Compound	Li—C	Li—M	Li—O	Reference
$(Li \cdot thf)_4 Cr_2 (CH_3)_8$	2.38	2.56	2.00	83
$(Li \cdot Et_2O)_4 Cr_2 (C_4 H_8)_4$	2.47	2.63	1.98	84
$(Li \cdot thf)_4 Mo_2 (CH_3)_8$	2.43–2.56	a	1.93	85
$(Li \cdot Et_2O)_2 Re_2 (CH_3)_8$	2.45–2.53	2.75	1.94	86
$(Li \cdot Et_2O)_4 W_2 (CH_3)_8$	2.54	2.74	1.93	87

[a] Value not reported.

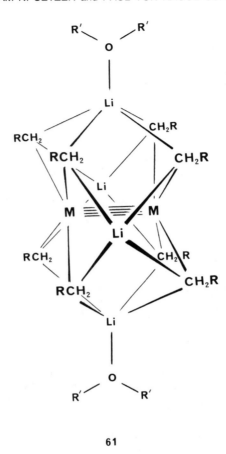

61

C. Mixed-Metal Organolithium Compounds with Unsaturated (Alkene, Arene) Ligands

Chiefly through the work of Jonas and co-workers (*88*) mixed-metal organometallic complexes also are known that involve interactions of lithium atoms with unsaturated π-bonded hydrocarbon ligands (olefin, cyclopentadienyl, arene, etc.). While reviews already are available (*88, 89*), we include examples for comparison purposes. The lithium complexes in this section show increasing complexity and diversity both in the geometries around the lithium atoms and in the degree and type of interactions involved. The common feature in these compounds is the interaction of a lithium atom with a hydrocarbon ligand which is π complexed to a transition metal. Most of these complexes also involve lithium–metal interactions.

The olefin complexes $Li(thf)_2Co(cod)_2$ (**62**) (cod=cyclooctadienyl) (*90*) and $[Li(tmeda)]_2Fe(C_2H_4)_4$ (**63**) (*91*) are relatively simple structures in which the central transition metal is tetrahedrally coordinated to four olefinic ligands. The lithium atoms in each of these structures serve to

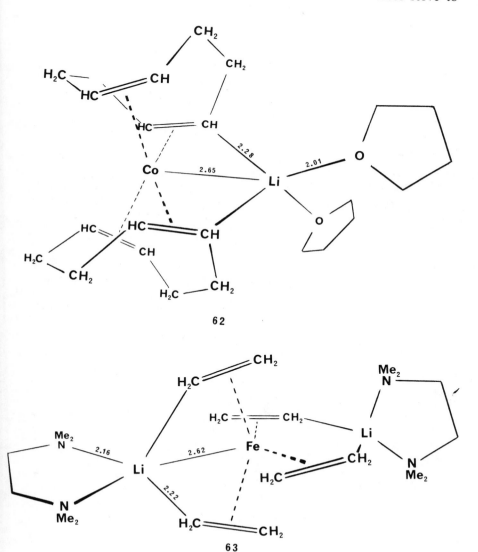

62

63

bridge two of the olefin ligands. Tetrahedral coordination of each lithium atom is achieved by further coordination with solvent molecules.

The norbornene complex [Li(tmeda)]$_2$Ni(norbornene)$_2$ (**64**) (*89*) is analogous to **62** and **63** in having a tetrahedral arrangement of carbon atoms and donor atoms around the lithium, but the geometry of the complex necessarily brings the lithium atom into close contact with the transition metal atom.

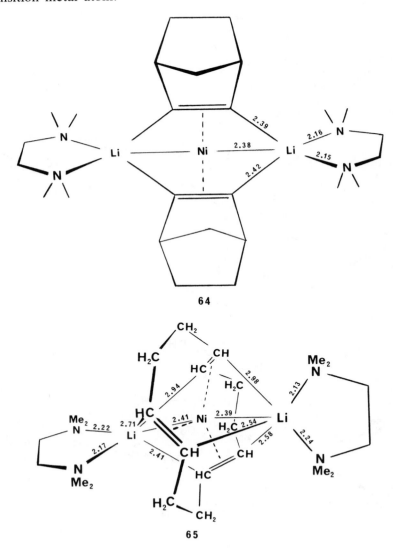

64

65

The cyclododecatrienylnickel complex [Li(tmeda)]$_2$Ni(cdt) (**65**) (*92*) is structurally analogous to **64**; each lithium atom is apparently tetrahedrally surrounded by two olefinic carbon atoms and two nitrogen atoms with a possible additional interaction with the nickel atom. The nickel atom in **65** adopts a distorted trigonal-bipyramidal geometry with the lithiums occupying the apices, whereas **64** involves a square-planar nickel atom.

The cyclopentadienyliron–olefin complexes Li(tmeda)(η^5-C$_5$H$_5$)Fe(cod) (**66**) and Li(tmeda)(η^5-C$_5$H$_5$)Fe(C$_2$H$_4$)$_2$ (**67**) (*89*) are more complex. The

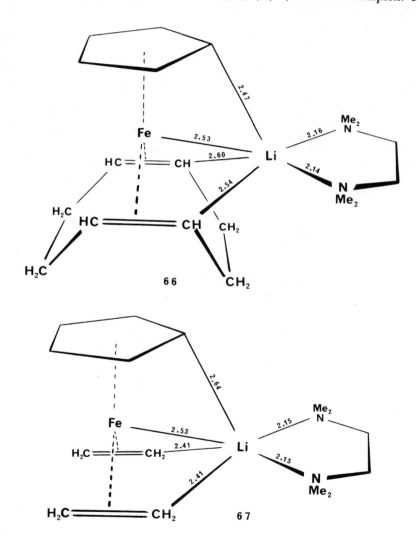

atom interacts with the cyclopentadienyl group and the iron atom as well as with the olefin carbon atoms. The lithium atom is at least pentacoordinate (three carbons and two nitrogens), and there may be iron–lithium interactions as well.

The olefin complex $Ni(C_{12}H_{17}NiLi)_2 \cdot 4thf$ (**68**) (*93*) is interesting because the tetrahedral coordination of the lithium atom is achieved by interaction with two solvent molecules, an olefinic carbon, and the central nickel atom.

The trimeric lithium–cobalt complex $[LiCo(C_2H_4)(PMe_3)_3]_3$ (**69**) (*94*), the first lithium carrier in hydrocarbon solvents, clearly involves interactions of the lithium atom with the hydrocarbon ligands as well as the cobalt atoms, but further ligation with solvent donor atoms is not required.

A dimeric dilithioferrocene compound **70** (*95*) involves terminally bonded lithium atoms analogous to structures **1–4** (type **II**, *A*) and bridging lithium atoms analogous to compounds **7–9** (type **II**, *B* structures; the bridging lithium atoms are not solvent coordinated, however). Lithium–lithium as well as lithium–iron interactions may be important in the structure of this complex.

The complicated structure of the tetrameric dilithioferrocene complex

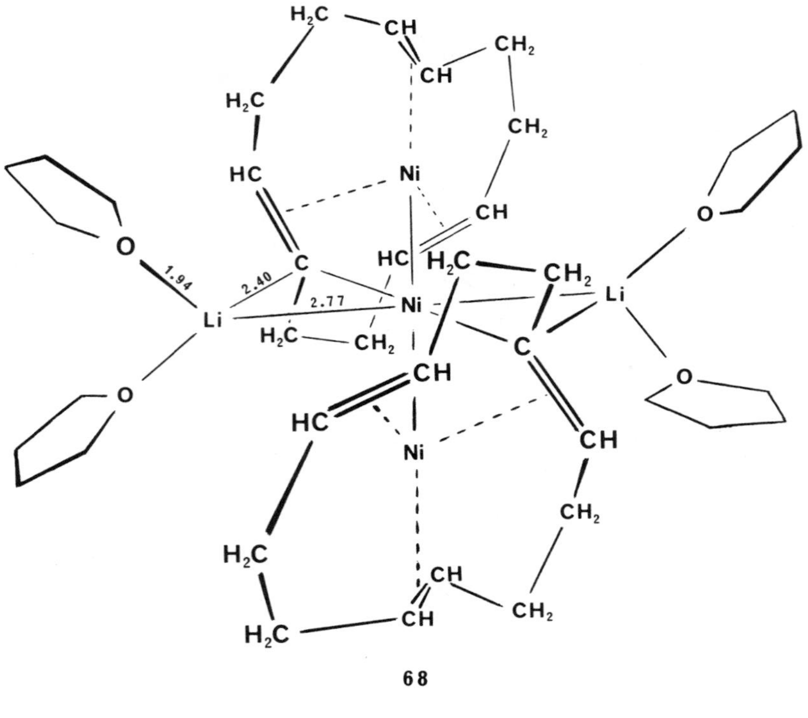

68

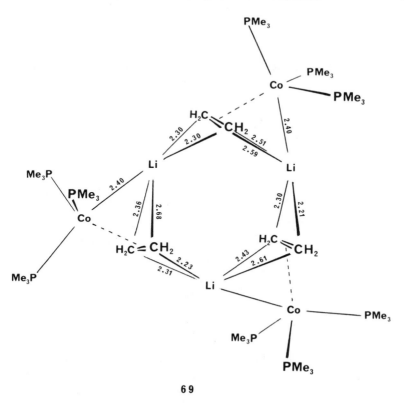

69

71, $[(\eta^5\text{-}C_5H_4Li)Fe(\eta^5\text{-}C_5H_3LiCH(Me)NMe_2]_4(LiOEt)_2(tmeda)_2$ (*96*), includes two symmetry-related tetrahedra of lithium atoms. Each lithium serves as a bridge between cyclopentadienyl units. The complex is bridged further by two additional lithium atoms. The structure may best be visualized as involving a central dimeric ferrocenyl system, analogous to **70**. An additional ferrocenyl unit is complexed at each end of the dimer. Each lithium atom appears to have a distorted tetrahedral environment. The lithium–lithium distances in this tetrahedral array range from 2.50 to 2.78 Å, with an average value of 2.64 Å.

The X-ray crystal structures of $[(\eta^5\text{-}C_5H_5)_2MHLi]_4$ (M = Mo or W) (*97*) indicate that these isomorphous complexes consist of eight-membered rings of alternate lithium and transition metal atoms (**72**). The shortest lithium–carbon interatomic distance for these complexes are in the range 2.6–2.8 Å, but the lithium environments do not correspond to ordinary coordination orientations.[7]

---

[7] The novel structure of a 3:2 compound $[(\eta^5\text{-}C_5H_4Li)_2Fe]_3[tmeda]_2$ has a central cluster composed of six lithium atoms (*96a*).

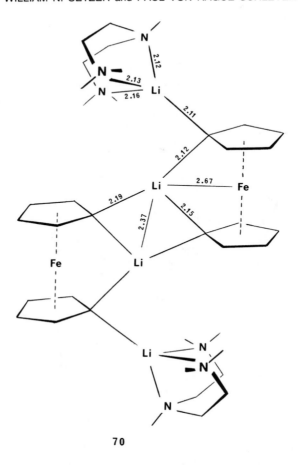

70

The nickel complex $(LiC_6H_5)_2(thf)_4Ni(C_{14}H_{10})$ (**73**) (*89*) has two tetrahedrally coordinated lithium atoms in different environments. One lithium atom is bonded to two phenyl carbons and two THF molecules, reminiscent of that seen for $[Li(tmeda)(C_6H_5)_3Mg]_2$ (**54**) (i.e., the phenyl rings are bridged by a lithium and a metal atom). The other lithium atom is somewhat analogous; it bridges a phenyl group and the phenanthrene moiety, and is solvated further by two THF molecules.

A lithium π-allenyl–cobalt complex $Li(C_6H_5Li)(Et_2O)_2Co(1,2\text{-}cod)(1,5\text{-}cod)$ (**74**) (*98*) has two lithium atoms, although in different environments, both serving as bridges between the phenyl group and the π system of the allyl ligand. One lithium atom, Li(1), interacts with an olefinic carbon and C(1) of the phenyl group, analogous to that in nickel complex **73**. The

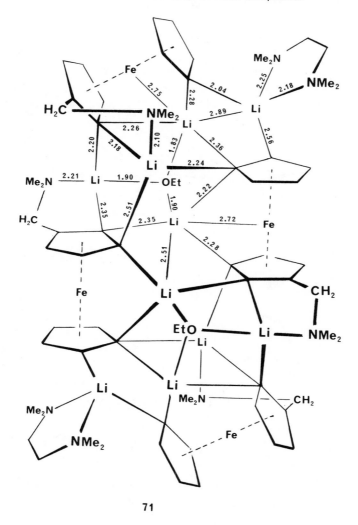

71

other lithium, Li(2), is in contact with two carbons of the allyl system and with two carbons of the phenyl group. Both lithiums are monosolvated and are within bonding distance of the cobalt.

Two π-dinitrogen–nickel complexes containing lithium, serving as models for biological nitrogen fixation, have been characterized by X-ray diffraction. Both structures are very complicated in terms of the various lithium contacts. The first compound, $[(LiC_6H_5)_6Ni_2N_2(Et_2O)_2]_2$ (*99*), has six crystallographically different lithium atoms as shown by its partial

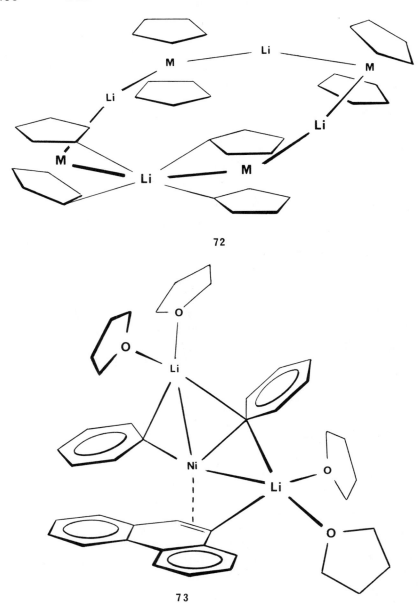

72

73

structure (**75**). Lithium(1) seems to have a tetrahedral environment, being surrounded by two nickel atoms, one nitrogen of the N_2 moiety, and an ether oxygen. Lithium(2) is in the center of a trigonal-planar grouping composed of a nickel atom, a phenyl group [bridging Li(5) of the other

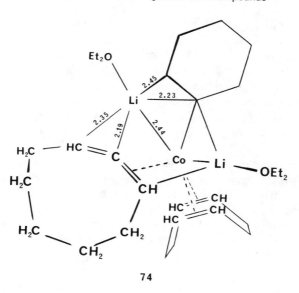

74

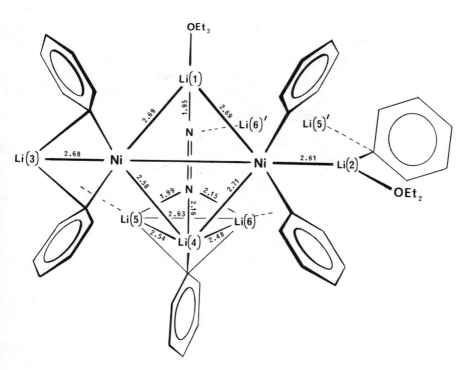

75

monomeric unit], and an ether oxygen. Lithium(3), along with the nickel atom, bridges two phenyl groups [similar to that observed in $(LiC_6H_5)_2(thf)_4Ni(C_{14}H_{10})$ (**73**)], but is unsolvated. Finally, Li(4), Li(5), and Li(6) form a trigonal bipyramid with the N_2 molecule occupying one apex and a phenyl group occupying the other. The two halves of this dimeric complex are held together by means of the Li(2)—Ph—Li(5)' bridge, mentioned above, and an Li(6)—N_2' interaction.

The second dinitrogen complex, $[Na_2Ph(Et_2O)_2(Ph_2Ni)_2N_2NaLi_6$-$(OEt)_4(Et_2O)]_2$ (*100*), has no crystallographic symmetry. However, the structure does have pseudo C_{2h} symmetry, and is built up of two $(Ph_2Ni)_2$-N_2 entities linked together by two sodium atoms and two $Li_6(OEt)_4 \cdot OEt_2$ nests. The six lithium atoms of the nest and the two sodium atoms form a distorted cube. Ethoxy groups occupy four faces of this cube; the other two faces are occupied by the $(Ph_2Ni)_2N_2$ systems. A partial structure of the inner core of this complex (**76**) shows the environments around the lithium atoms. Note that Li(1), Li(2), Li(5), and Li(6) are in chemically similar

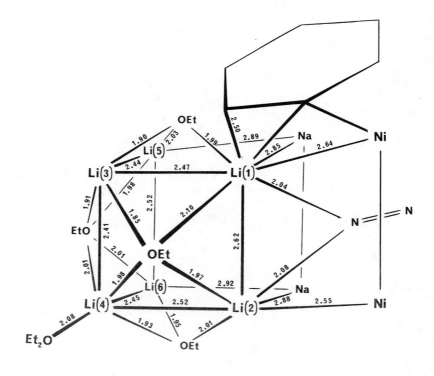

76

locations, each interacting with a nickel, a sodium, a nitrogen atom of N_2, a phenyl group, two ethoxy groups, and two lithium atoms; Li(4) occupies an "exposed" vertex of the cube, interacting with three ethoxy groups, three lithium atoms, and one molecule of ether solvent. Interestingly, the other "exposed" vertex, Li(3), is in the same chemical environment as Li(4) except that it is not solvated.

D. *Miscellaneous Mixed-Metal Organolithium Complexes*

In the tantalum–carbyne complex $Li \cdot (dmp)[(CH_3)_3CCH_2]_3Ta \equiv CC$-$(CH_3)_3$ (**77**) (dmp = 1,4-dimethylpiperazine) (*101*), tetrahedral coordination of the lithium atom is achieved by interaction with a saturated CH_2 group, the unsaturated carbyne carbon, and the two donor atoms from dimethylpiperazine. Evidently, Li—H—C interactions are involved.

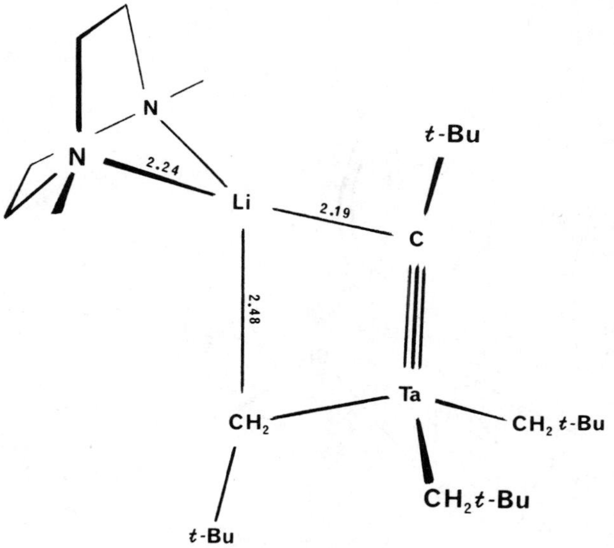

77

The monomeric LiAl $[N=C(t\text{-Bu})_2]_4$ (**78**) (*102*) is unique in that the lithium–carbon [or Li—H—C (*79*)] interactions involve hydrocarbon moieties which are removed from the other metal. That is, the carbon atom is not serving as a bridge between metal atoms. In addition, the lithium atom in this complex has a tetragonal-planar (rather than tetrahedral) environment.

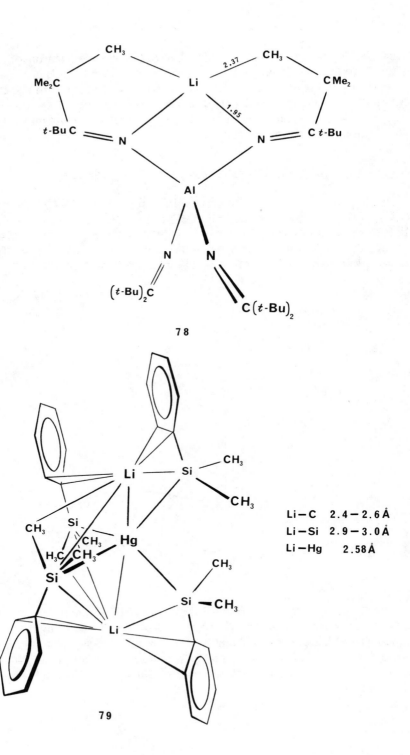

78

Li—C 2.4 — 2.6 Å
Li—Si 2.9 — 3.0 Å
Li—Hg 2.58 Å

79

80

The two mercury complexes $Li_2Hg(SiMe_2Ph)_4$ (**79**) (*103, 104*) and $Li_2Hg(SiMe_3)_4$ (**80**) (*104*) have lithium atoms which are enclosed in cages. These cages are composed of carbon, silicon, and mercury atoms, but have no electron-pair-donating atoms. Both monobridged as well as dibridged Li—H—C interactions apparently are present in **80**.

IV

STRUCTURES OF LITHIUM COMPOUNDS WITHOUT LITHIUM–CARBON BONDS

This section summarizes the structures of some lithium compounds which do not involve Li—C contacts. While not "organometallic" from this viewpoint, the structural variability and the geometrical parameters are of interest to lithium chemists. The list is not complete, but is intended to be representative.

A. Solvated Lithium Ions

1. Nonbridging Lithium Cation with Four Tetrahedrally Coordinated Donor Atoms

In these compounds, the lithium cations are surrounded by solvent or ligand molecules and can be considered to be involved in solvent-separated ion-pair structures. While these are frequently invoked to explain spectroscopic observations in solution, their appearance in crystal structures is less common.

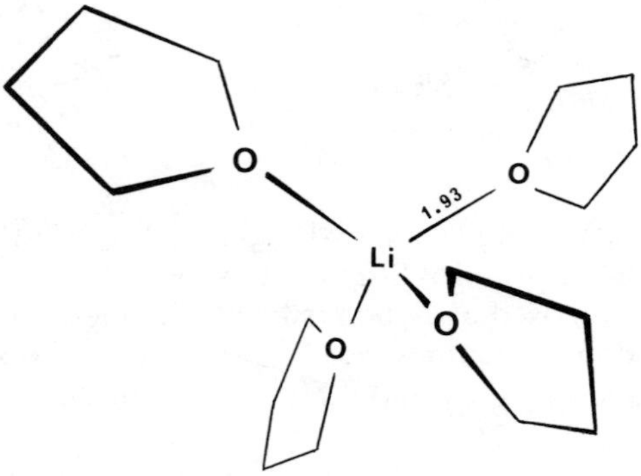

Compound	Li—X	Average bond length (Å)	Reference
[Li(thf)$_4$]$_2$[U(C$_2$B$_9$H$_{11}$)$_2$Cl$_2$]	Li—O	1.92	*105*
[Li(thf)$_4$][(C$_2$H$_4$)$_2$Ni{P(C$_6$H$_{11}$)$_2$}Ni(C$_2$H$_4$)$_2$]	Li—O	1.92	*89, 106*
[Li(thf)$_4$][Cu$_5$Ph$_6$]	Li—O	1.93	*107*
[Li(thf)$_4$][Li{C(SiMe$_3$)$_2$}$_2$]	Li—O	1.96	*70*
[Li(thf)$_4$][Lu(C$_8$H$_9$)$_4$]	Li—O	not reported (NR)	*108*
[Li(thf)$_4$][Yb{CH(SiMe$_3$)$_2$}$_3$Cl]	Li—O	NR	*109*

[Li(Ph$_3$P=O)$_4$]I·Ph$_3$P=O	Li—O	1.97	*110*
[Li(hmpt)$_4$][Li$_5$(hmpt)N=CPh$_2$)$_6$][hmpt= hexamethylphosphoric triamide, (Me$_2$N)$_3$P(O)]	Li—O	1.84	*111*
[Li(C$_6$N$_2$O$_2$H$_{10}$)$_2$]ClO$_4$(2:complex of cyclodisarcosyl with lithium perchlorate)	Li—O	1.92	*112*
LiBr·(Ala·Gly)·2H$_2$O (lithium atom is bonded to two carboxyl oxygens of one dipeptide unit, one carbonyl oxygen of another, and a water molecule)	Li—O	1.91	*113*
LiBr·(Gly·Gly·Gly)	Li—O	1.96	*114*
(LiCl)$_2$·[1,4,7,10-Tetra(2-hydroxyethyl)-1,4,7,10-tetraazacyclododecane]·2H$_2$O (lithium is coordinated to three different hydroxyl groups from three different macrocycles and one water molecule).	Li—O	1.93	*115*
LiClO$_4$·(18-crown-6)·2H$_2$O (lithium is coordinated to two ether oxygens and two water molecules)	Li—O	2.01	*116*

Compound	Li—X	Average bond length (Å)	Reference
(LiSCN)$_2$·(18-crown-6)·2H$_2$O (one lithium is coordinated to two ether oxygens and two water molecules)	Li—O	1.98	*116*
LiBr·(NH$_2$CH$_2$CH$_2$NH$_2$)$_2$	Li—N	2.07	*117*
LiCl·(NH$_2$CH$_2$CH$_2$NH$_2$)$_2$	Li—N	2.07	*117*
[Li(tmeda)$_2$][CH$_3$Ni(C$_2$H$_4$)$_2$]	Li—N	2.13	*118*
[Li(tmeda)$_2$][H$_3$BMe$_2$PCHPMe$_2$BH$_3$]	Li—N	NR	*119*

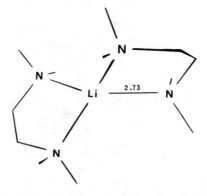

Compound	Li—X	Average bond length (Å)	Reference
LiCl·2(C$_5$H$_5$N)·2H$_2$O (lithium is surrounded by one chlorine, one oxygen (water), and two nitrogen (pyridine) atoms)	Li—Cl	2.33	*120*
	Li—O	1.94	
	Li—N	2.06	
Li(NAD$^+$)·2H$_2$O,Li(C$_{21}$H$_{26}$N$_7$O$_{14}$P$_2$)·2H$_2$O (NAD$^+$ = oxidized form of nicotinamide adenine dinucleotide) (lithium is surrounded by a nitrogen and three oxygens provided by two different NAD$^+$ molecules)	Li—O	1.89	*121*
	Li—N	2.13	
(LiSCN)$_2$·(18-crown-6)·2H$_2$O (one lithium is coordinated to one ether oxygen, one water molecule, and two NCS$^-$ ions)	Li—O	1.99	*116*
	Li—N	2.00	

2. Bridging Lithium Cation with Four Tetrahedrally Coordinated Donor Atoms

Compound	Li—X	Average bond length (Å)	Reference
LiI·2[(NH$_2$)$_2$C=O] (a 2:1 complex of urea with lithium iodide; the oxygen atoms serve as bridging ligands between lithium atoms)	Li—O Li—Li	1.92 NR	122

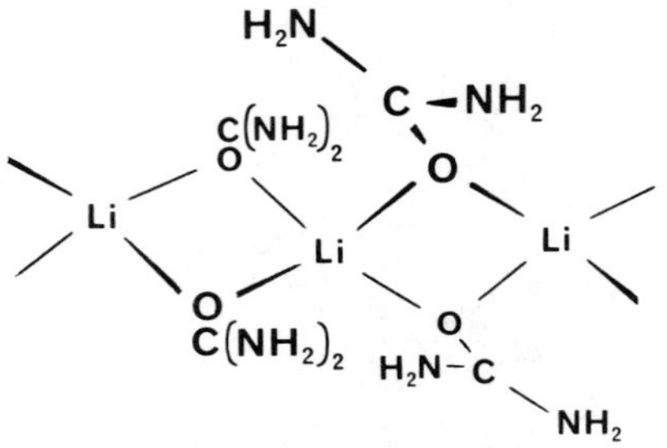

Compound	Li—X	Average bond length (Å)	Reference
Li$_2$SO$_4$·3[(NH$_2$)$_2$C=O] (the structure is built up of double polymeric chains of Li$_2$SO$_4$ joined by bridging urea molecules and monodentate urea molecules)	Li—O	Not available (NA)	123
[(H$_2$O)$_2$Li(hmpt)]$_2$(Cl)$_2$ (dimeric four-membered ring analogous to type II, B compounds)	Li—O	NA	111

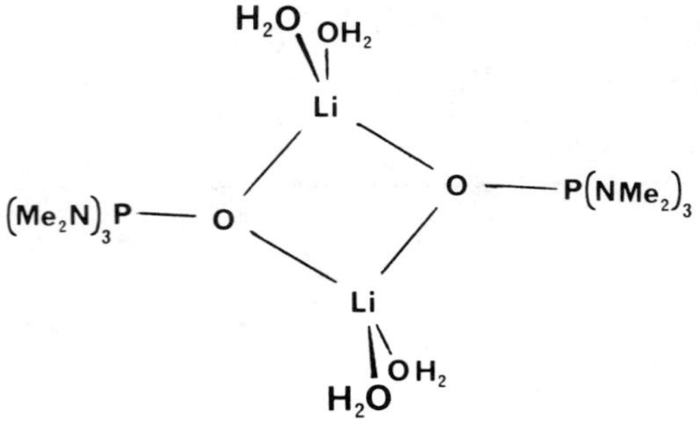

Compound	Li—X	Average bond length (Å)	Reference
(LiCl·hmpt)$_4$ (tetrameric, tetrahedral, structure analogous to type II, C compounds)	Li—O Li—Cl Li—Li	1.87 2.40 3.10	123a

Compound	Li—X	Average bond length (Å)	Reference

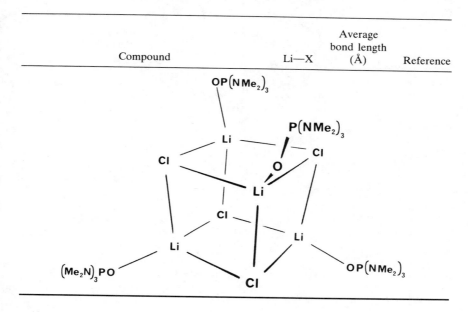

3. Pentacoordinate Lithium Cation with Tetragonal Pyramidal Geometry

Compound	Li—X	Average bond length (Å)	Reference
LiBr·antaminide			124

LiBr·antaminide
  (antaminide≡Pro—Phe—Phe—Val—Pro
                    | |
           Pro—Phe—Phe—Ala—Pro
  a cyclic decapeptide)

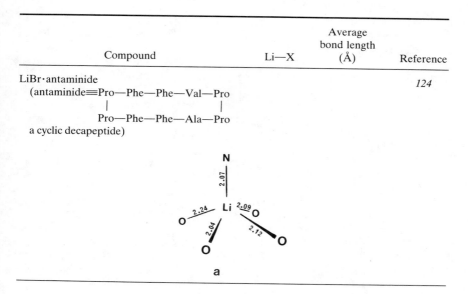

a

Compound	Li—X	Average bond length (Å)	Reference
LiNCS·(16-crown-4)			*125*

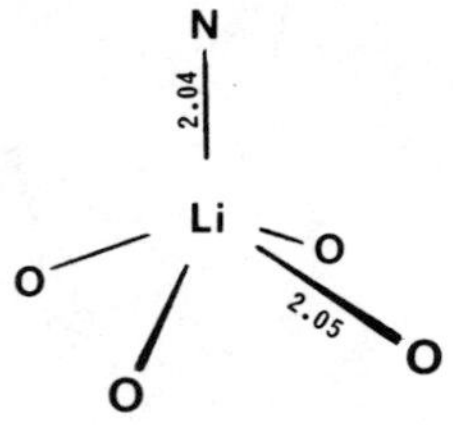

b

Compound	Li—X	Average bond length (Å)	Reference
LiNCS·(12-crown-4)			*126*

c

Compound	Li—X	Average bond length (Å)	Reference
LiNCS(benzo-13-crown-4)			*127*

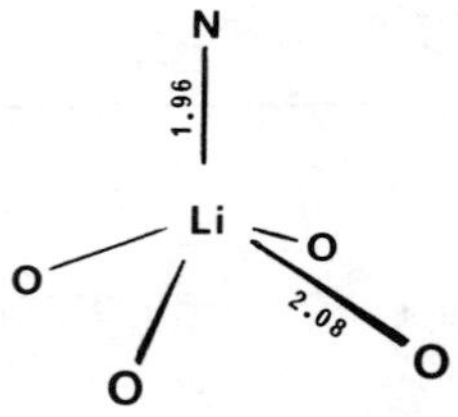

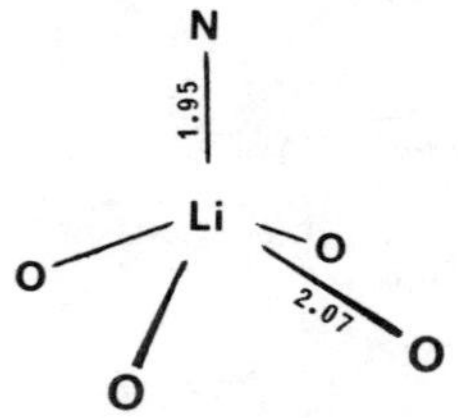

d

Compound	Li—X	Average bond length (Å)	Reference
LiCl$_2$·[1,4,7,10-tetra(2-hydroxyethyl)-1,4,7,10-tetraazacyclododecane]·2H$_2$O			*128*

Compound	Li—X	Average bond length (Å)	Reference

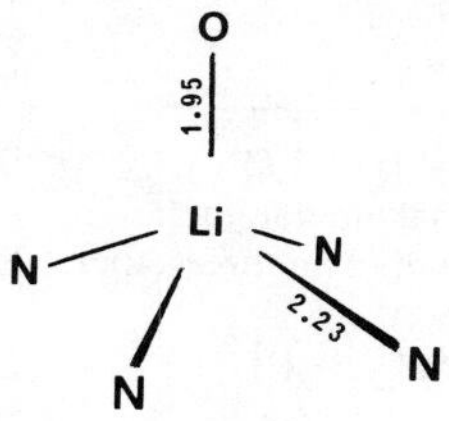

e

LiNCS·(dibenzo-14-crown-4)   129

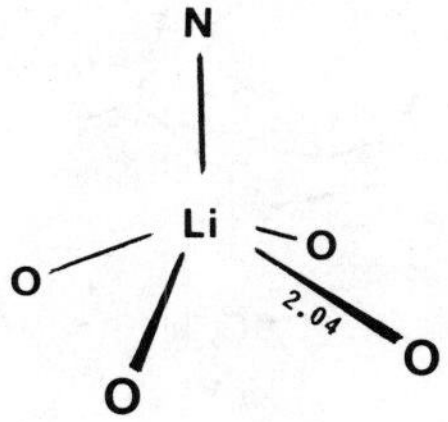

f

| LiI$_3$·I$_2$·8H$_2$O·(α-cyclodextrin)$_2$ (tetragonal pyramidal geometry of oxygen atoms) | Li—O | NR | 130 |

4. *Pentacoordinate Lithium Cation with Miscellaneous Geometries*

Compound	Li—X	Average bond length (Å)	Reference
$(LiI)_2 \cdot [18,18'\text{-spirobis-(19-crown-6)}] \cdot 4H_2O$ (lithium lies in a distorted trigonal bipyramidal geometry of oxygen atoms provided by three ether oxygens and two water molecules)	Li—O axial	1.96	*131*
	Li—O apical	2.16	
$LiFeCl_4 \cdot [(CH_3C_6H_2O)_6(CH_3)_2(CH_2CH_2CH_2)_2]$	Li—O	2.04	*132*

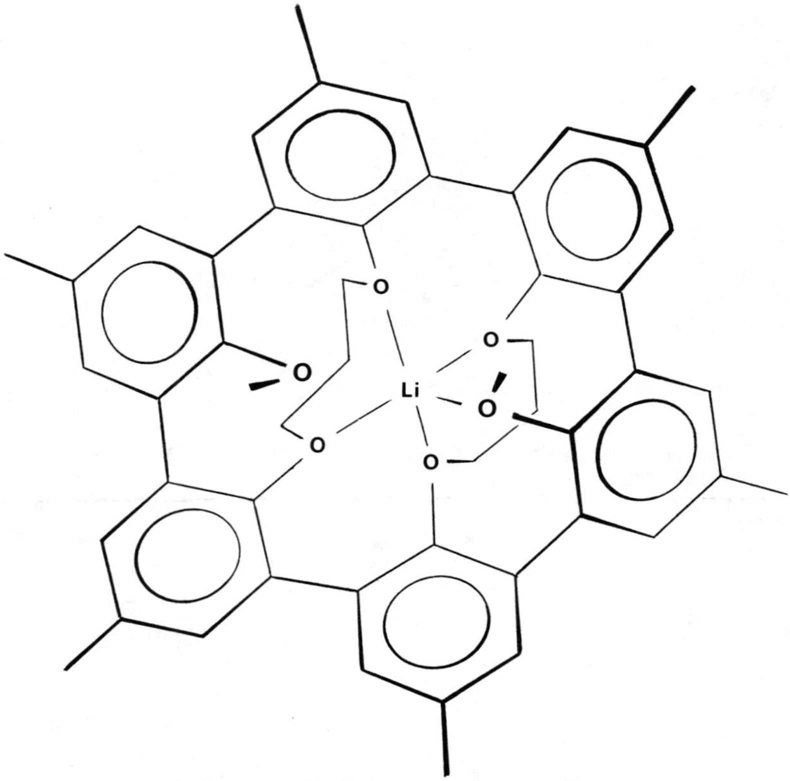

5. *Hexacoordinate Lithium Cation*

Compound	Li—X	Average bond length (Å)	Reference
LiI · (NH₂CH₂CH₂NH₂)₃ (The lithium atom lies in a distorted octahedral geometry of nitrogen atoms)	LI—N	2.26	*133*
LiI · (C₁₄H₂₈N₂O₄) (lithium lies in a distorted octahedral geometry of two nitrogen atoms and four oxygen atoms)	Li—O Li—N	2.13 2.29	*134*

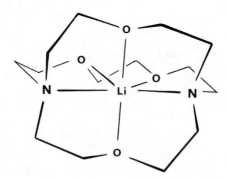

| (LiClO₄)₂·[CH₃C(O)NH₂]₃[CH₃C(O)NHC(O)CH₃] ("dimeric" structure with the lithium atom adopting a distorted octahedral geometry of oxygen atoms with bridging acetamide molecules) | Li—O
Li—Li | See structure
2.44 | *135* |

6. *Heptacoordinate Lithium Cation*

Compound	Li—X	Average bond length (Å)	Reference
$LiCl \cdot [(CH_3C_6H_2O)_6(CH_3)_2(CH_2CH_2OCH_2CH_2)_2]$	Li—O	2.26	*132*

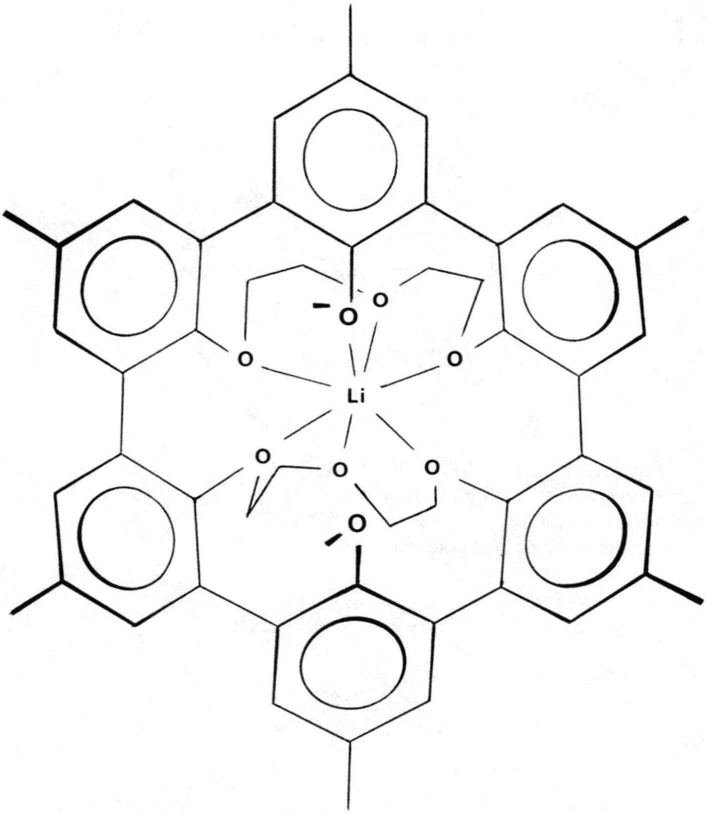

B. *Lithium Carboxylates*

1. *Tetracoordinate Lithium Carboxylates with Tetrahedral Geometries (Each Lithium Is Tetrahedrally Surrounded by Four Oxygen Atoms)*

Compound	Li—X	Average bond length (Å)	Reference
$Li_2(C_2O_4)$ (lithium oxalate)	Li—O	2.01	*136*
$Li(NH_4)H(C_6O_7H_5) \cdot H_2O$ (lithium ammonium hydrogen citrate monohydrate	Li—O	1.95	*137*
$Li(CH_3COO) \cdot 2H_2O$ (lithium acetate dihydrate)	Li—O	1.97	*138*
$Li_2(CO_2CH_2CH_2CO_2)$ (lithium succinate)	Li—O	1.95	*139*
$Li(HCO_2) \cdot H_2O$ (lithium formate monohydrate)	Li—O	1.95	*140*

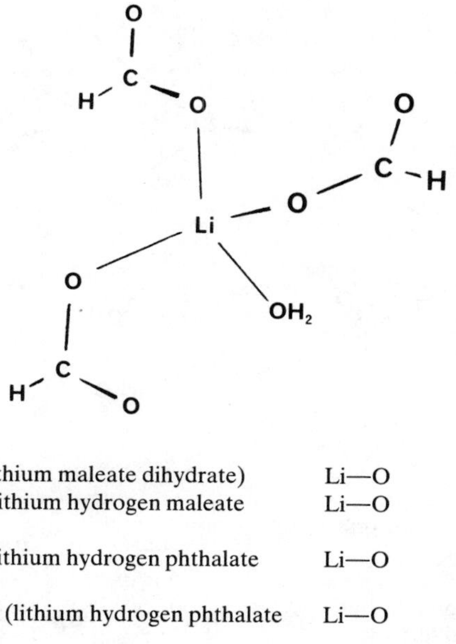

Compound	Li—X	Average bond length (Å)	Reference
$Li_2(C_4H_2O_4) \cdot 2H_2O$ (lithium maleate dihydrate)	Li—O	1.96	*141*
$LiH(C_4H_2O_4) \cdot 2H_2O$ (lithium hydrogen maleate dihydrate)	Li—O	1.99	*142*
$LiH(C_8H_4O_4) \cdot 2H_2O$ (lithium hydrogen phthalate dihydrate)	Li—O	1.94	*143*
$LiH(C_8H_4O_4) \cdot CH_3OH$ (lithium hydrogen phthalate methanol)	Li—O	1.96	*144*
$LiH(C_8H_4O_4) \cdot H_2O$ (lithium hydrogen phthalate monohydrate)	Li—O	1.96	*145*
$LiH_3CH_2(CO_2)_2)_2$ (lithium trihydrogen dimalonate)	Li—O	2.00	*146*
$Li_2(CH_2(CO_2)_2)$ (dilithium malonate)	Li—O	NA	*147*
$LiH(C_4H_4O_5)$ (lithium hydrogen (+)-1-malate)	Li—O	1.95	*148*
$LiH(C_4O_4) \cdot H_2O$ (lithium hydrogen acetylendicarboxylate monohydrate	Li—O	1.94	*149*

2. *Petacoordinate Lithium Carboxylates with Trigonal Bipyramidal Geometries (Each Lithium is Surrounded by a Trigonal Bipyramidal Arrangement of Five Oxygen Atoms)*

Compound	Li—X	Average bond length (Å)	Reference
Li(H$_2$C(OH)CO$_2$]·H$_2$O (lithium glycolate monohydrate)	Li—O$_{ap}$	2.08	*150*
	Li—O$_{eq}$	1.98	
Li[HDC(OH)CO$_2$] (lithium (S)-glycolate-2-*d*)	Li—O$_{ap}$	2.19	*151*
	Li—O$_{eq}$	1.98	

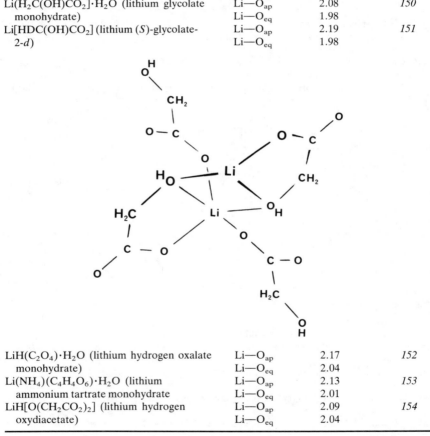

Compound	Li—X	Average bond length (Å)	Reference
LiH(C$_2$O$_4$)·H$_2$O (lithium hydrogen oxalate monohydrate)	Li—O$_{ap}$	2.17	*152*
	Li—O$_{eq}$	2.04	
Li(NH$_4$)(C$_4$H$_4$O$_6$)·H$_2$O (lithium ammonium tartrate monohydrate	Li—O$_{ap}$	2.13	*153*
	Li—O$_{eq}$	2.01	
LiH[O(CH$_2$CO$_2$)$_2$] (lithium hydrogen oxydiacetate)	Li—O$_{ap}$	2.09	*154*
	Li—O$_{eq}$	2.04	

3. *Pentacoordinate Lithium Carboxylate with Tetragonal Pyramidal Geometry*

Compound	Bond	Average bond length (Å)	Reference
$Li_2(C_2O_4) \cdot H_2O_2$ (dilithium oxalate monoperhydrate)			*155*

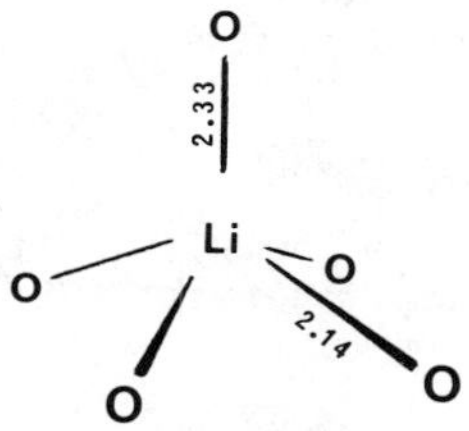

4. *Hexacoordinate Lithium Carboxylate*

Compound	Li—X	Average bond length (Å)	Reference
$LiH_2(C_6O_7H_5)$ (lithium dihydrogen citrate octahedral geometry)	Li—O	2.18	*156*

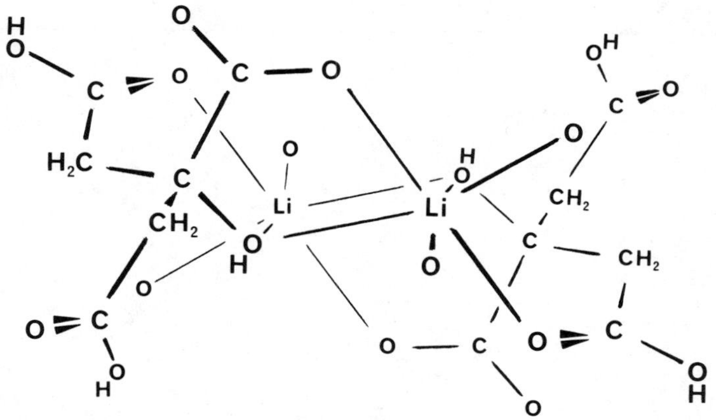

C. Lithium Alkoxides and Thiolates

Compound	Li—X	Average bond length (Å)	Reference
$Li(OC_6H_2(NO_2)_3) \cdot (benzo\text{-}15\text{-}crown\text{-}5) \cdot 2H_2O$ (monomeric; lithium atom is tetrahedrally surrounded by four oxygen atoms provided by the picrate aryl oxygen, nitro oxygen and two water molecules; lithium is *not* bonded to the crown ether)	Li—O	1.93	157
$Li[OC_6H_2Me\text{-}4\text{-}(Bu\text{-}t)_2\text{-}2,6] \cdot (Et_2O)_2$ (dimeric–analogous to type **II**, *B* compounds; lithium is tricoordinate, presumably due to steric factors)	Li—O_{aryl} Li—O_{ether} Li—Li	1.86 1.96 NR	158

| $LiOCH_3$ (polymeric; lithium is surrounded by a tetrahedral array of oxygen atoms) | Li—O
Li—Li | 1.95
2.51 | 159 |
| $LiSCH_3$ (polymeric; lithium is surrounded by a tetrahedral array of sulfur atoms) | Li—S
Li—Li | 2.4
4.37 | 160 |

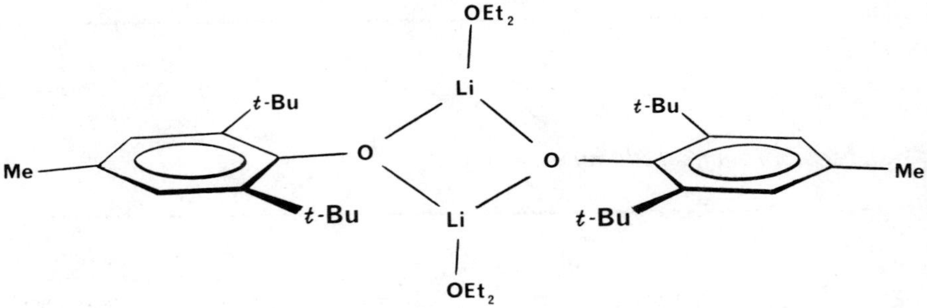

D. *Lithium Amides and Imides*

Compound	Li—X	Average bond length (Å)	Reference
Li(NPhPyr)·(PhPyrNH)·hmpt (monomeric, tetrahedrally coordinated, terminally bonded lithium atom analogous to type **II**, *A* compounds)	Li—O Li—N	1.88 2.09	*160a*

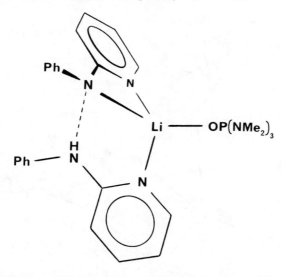

Li[N(SiMe₃)₂]·Et₂O (dimeric, analogous to type **II**, *B* compounds)	Li—O Li—N Li—Li	1.95 2.06 NR	*161*

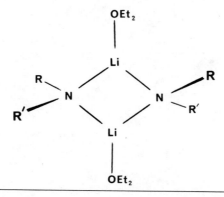

Compound	Li—X	Average bond length (Å)	Reference
Li[NHC$_6$H$_2$(t-Bu)$_3$]·Et$_2$O (dimeric, as above)	Li—O	1.91	162
	Li—N	2.01	
	Li—Li	NR	
Li[N(CH$_2$Ph)$_2$·Et$_2$O (dimeric, as above)	Li—O	2.01	$162a$
	Li—N	1.99	
Li[N(CH$_2$Ph)$_2$]·hmpt (dimeric, as above)	Li—O	1.85	$162a$
	Li—N	2.01	
Li(NPhPyr)·hmpt (dimeric, two structures)	Li—O	1.87	$162b$
	Li—N	2.10	

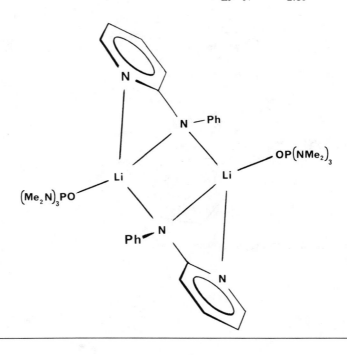

Compound	Li—X	Average bond length (Å)	Reference

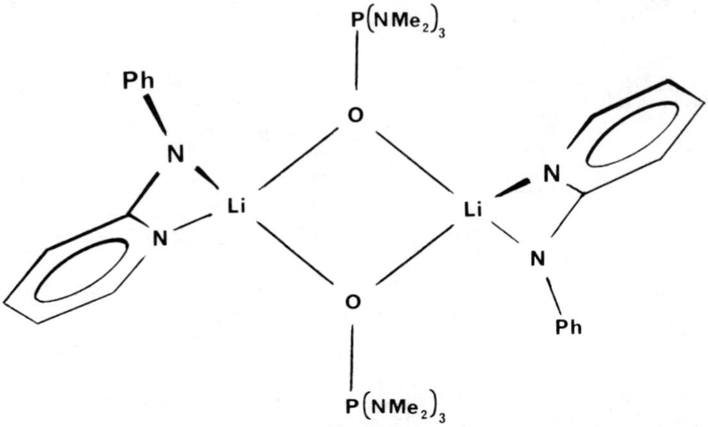

P(NMe$_2$)$_3$

Ph
N

N — Li

Li — N

O

O

P(NMe$_2$)$_3$

Ph

| | Li—O | 1.92 | |
| | Li—N | 2.08 | |

Li[N=C(Bu-t)$_2$]·hmpt (dimeric)   Li—O   1.86   *162b*
                                      Li—N   1.94

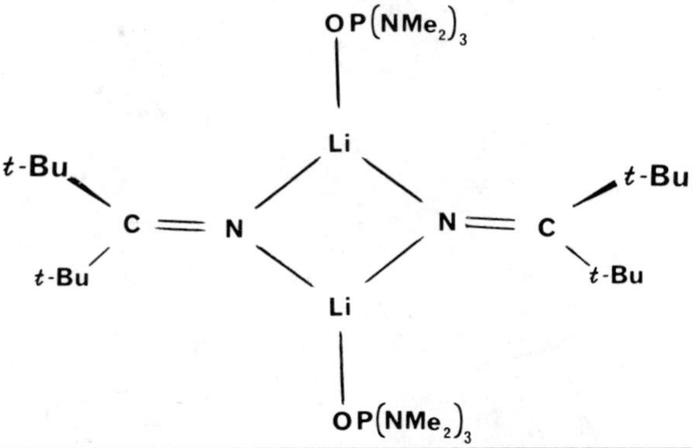

OP(NMe$_2$)$_3$

Li

t-Bu
C = N   N = C
t-Bu

t-Bu
t-Bu

Li

OP(NMe$_2$)$_3$

Compound	Li—X	Average bond length (Å)	Reference
Li[N(SiMe$_3$)$_2$](trimer, planar)	Li—N	2.00	*163*
	Li—Li	2.89	

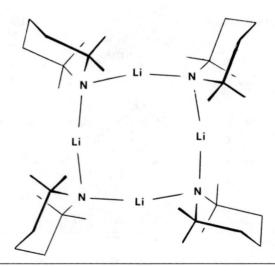

Li[N(CH$_2$Ph)$_2$] (trimer, as above)	Li—N	1.97	*162a*
Li[NC(CH$_3$)$_2$CH$_2$CH$_2$CH$_2$C(CH$_3$)$_2$] (tetrameric, planar eight-membered ring, presumably due to steric effects)	Li—N	2.00	*161*
	Li—Li	NR	

Compound	Li—X	Average bond length (Å)	Reference
Li(N=CPh₂)·pyr (pyr=pyridine) (tetrameric, tetrahedral array of lithium atoms)	Li—N$_{imido}$	2.10	*123a*
	Li—N$_{pyr}$	2.09	
	Li—Li	2.66	

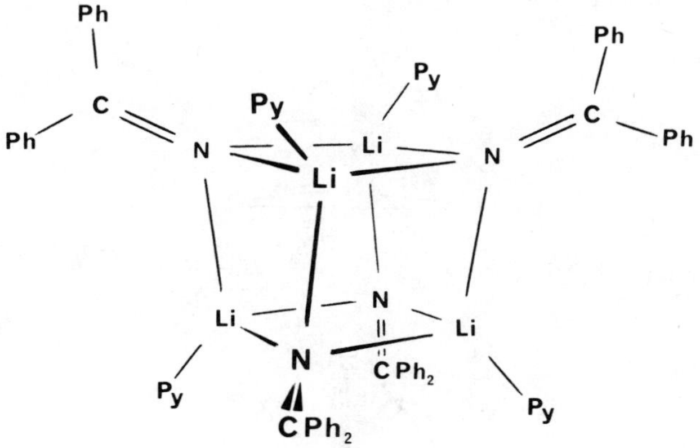

Compound	Li—X	Average bond length (Å)	Reference
[Li(hmpt)₄][Li₅(hmpt)(N=CPh₂)₆] (cluster of five lithium atoms)	Li—O	1.83	*111*
	Li—N	2.04	

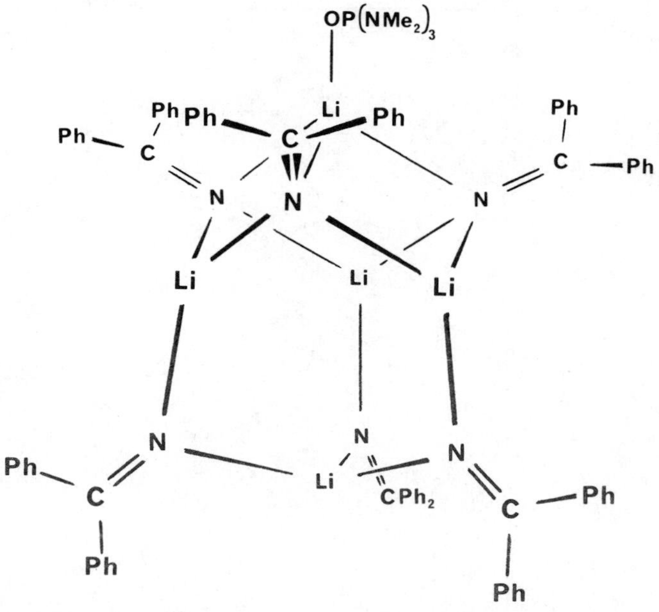

Compound	Li—X	Average bond length (Å)	Reference
LiN=C(t-Bu)₂ (hexameric, distorted octahedral array of lithiums; each nitrogen atom occupies a face of the octahedron; analogous to type **II**, *D* hexameric compounds)	Li—N Li—Li	2.06 2.56	*33, 164*

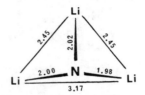

| LiN=C(NMe₂)₂ (hexameric, as above) | Li—N
Li—Li | 2.00
2.69 | *164* |

E. Silyllithium Compounds and Lithium Phosphides

Compound	Li—X	Average bond length (Å)	Reference
(LiSiMe₃)₂(tmeda)₃ (terminally bonded lithium atom analogous to type **II**, A compounds)	Li—Si Li—N	2.70 2.17	*165*

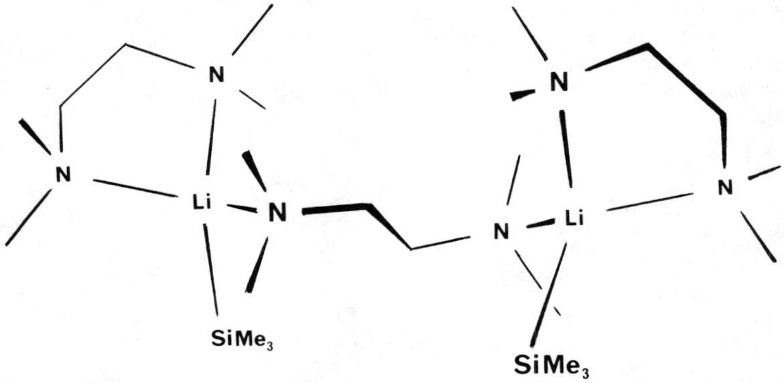

SiMe₃

LiSiMe₃ (hexameric, distorted octahedral array of lithium; each silicon atom occupies a face of the octahedron; analogous to type **II**, D hexameric compounds)	Li—Si Li—Li	2.69 2.90	*32, 166*

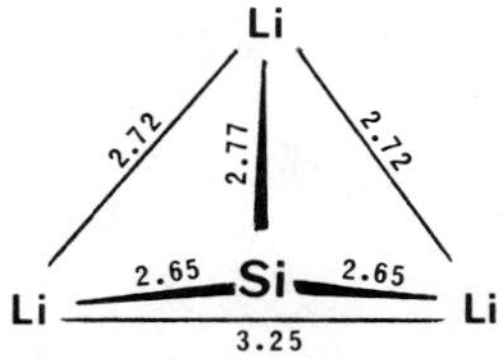			
[LiP(t-Bu)₂]₄·(thf)₂ (planar staggered Li₄ unit) 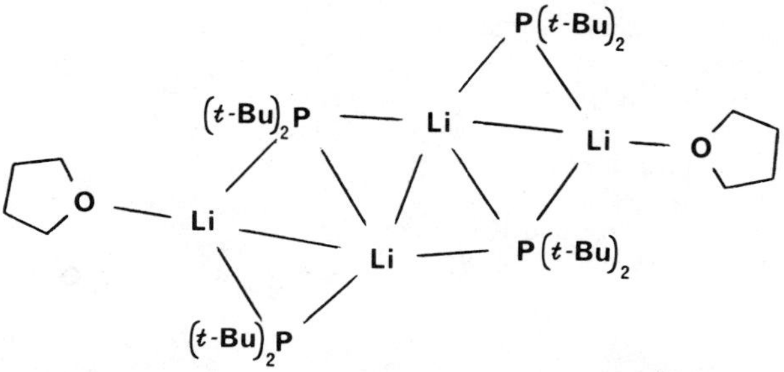	Li—P Li—O Li—Li	2.56 1.92 3.05	*167*

F. *Lithium Enolates*

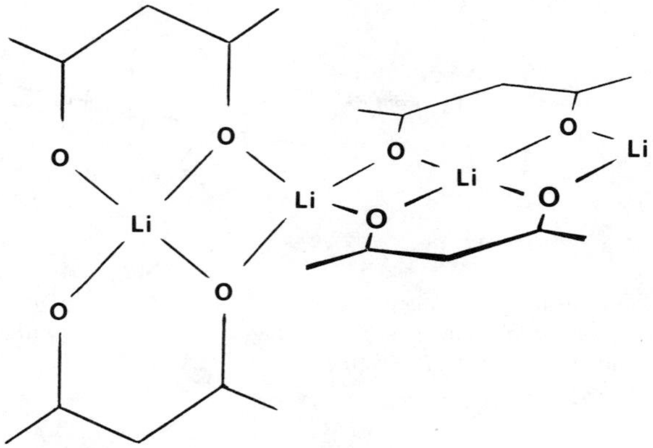

Compound	Li—X	Average bond length (Å)	Reference
LiCH[C(O)CH$_3$]$_2$ (lithium 2,4-pentanedionate) [structure consists of endless strings of pseudo Li(acac)$_2$ groups (acac = acetylacetonate), connected by bridging lithium atoms; there are two different types of lithium geometry: tetrahedral and tetragonal.]	Li—O$_{tetrahedral}$ Li—O$_{tetragonal}$ Li—Li	1.96 1.93 2.80	*168*

LiHC$_4$O$_4$ · H$_2$O (lithium squarate monohydrate) (lithium atom is tetrahedrally surrounded by four oxygen atoms)	Li—O	2.05	*169*
LiC$_5$(CO$_2$Me)$_5$ · H$_2$O [lithium pentakis(methoxycarbonylcyclopentadiene)] (lithium atom is tetrahedrally surrounded by four oxygen atoms)	Li—O	NR	*170*
[Li[CH$_3$C(O)CHCO$_2$C$_2$H$_5$]$_2$][K(C$_{12}$H$_{26}$N$_2$O$_4$)] [lithium potassium bis(ethyl acetoacetate)] (lithium atom is tetrahedrally surrounded by four oxygen atoms)	Li—O	1.94	*171*
LiC$_8$H$_4$N$_5$O$_6$ · 2H$_2$O (lithium purpurate dihydrate) (lithium atom lies in a tetragonal pyramidal geometry provided by two oxygen atoms and one nitrogen atom of one purpurate anion, an oxygen atom of a second purpurate anion, and a water molecule)			*172*

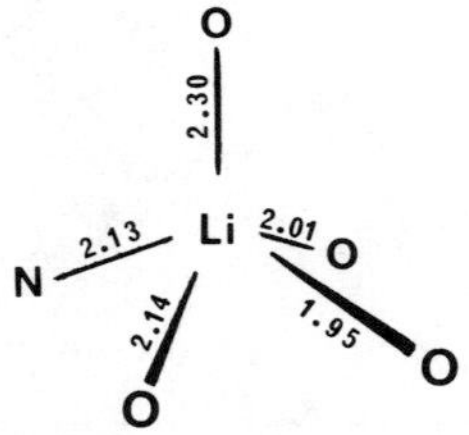

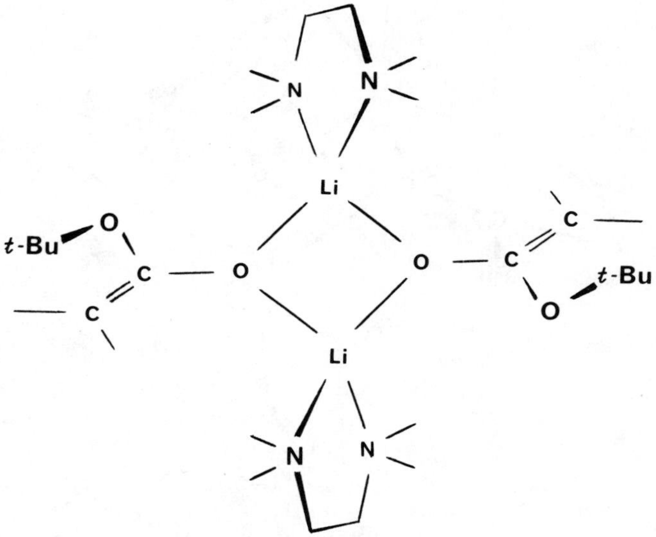

Compound	Li—X	Average bond length (Å)	Reference
Li(CH$_3$)$_2$CCO$_2$C(CH$_3$)$_3$ · tmeda (dimeric, analogous to type **II**, *B* compounds; lithium is tetrahedrally tetracoordinate)	Li—O Li—N	1.91 NR	173

LiCH$_3$CHCO$_2$C(CH$_3$)$_3$ · tmeda (dimeric, as above)	Li—O Li—N	1.92 NR	173
LiC$_7$H$_6$CON(CH$_3$)$_2$ · 2thf (dimeric, as above)	Li—O$_{thf}$ Li—O$_{bridging}$ Li—Li	1.98 1.90 1.88	174
LiP[C(O)C$_6$H$_5$)]$_2$ · dme (dme = dimethoxyethane) (dimeric; the lithium atom is pentacoordinate and lies in a distorted tetragonal-pyramidal geometry)	Li—Li	2.83	175

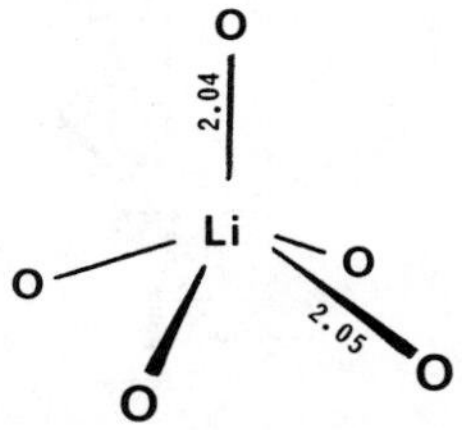

Compound	Li—X	Average bond length (Å)	Reference

Compound	Li—X	Average bond length (Å)	Reference
LiCH$_2$C(O)C(CH$_3$)$_3$ · thf (tetrameric, analogous to type **II**, *C* compounds)[8]	Li—O Li—Li	1.97 2.79	*176*

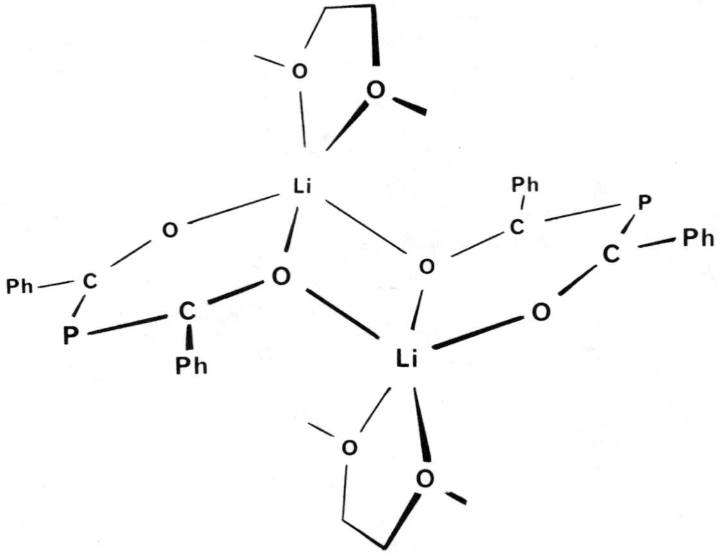

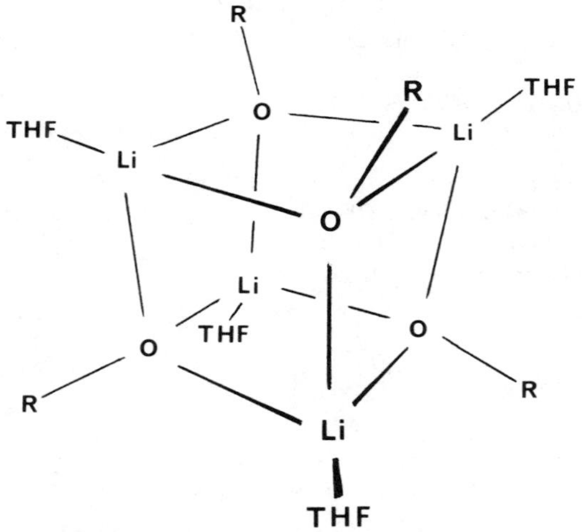

Compound	Li—X	Average bond length (Å)	Reference
LiC$_5$H$_7$O · thf (lithium cyclopentenoate) (tetrameric, as above)	Li—O Li—Li	2.00 2.31	*176*
LiCH(*t*-Bu)CO$_2$CH$_3$ · thf (tetrameric, as above)	NA		*173*

[8] The solvent-free X-ray structure is hexameric (*176a*).

G. *Miscellaneous Inorganic Lithium Compounds*

1. *Structures Having Tetracoordinate Lithium Atoms Tetrahedrally Surrounded by Four Oxygen Atoms*

Compound	Li—X	Average bond length (Å)	Reference
Li[Fe(OH$_2$)(edta)]·2H$_2$O(edta = ethylenediaminetetraacetate)	Li—O	1.95	177
Li$_2$Cu(succinimide)$_4$·H$_2$O	Li—O	1.93	178
Li[Cr(en)$_3$](d-tart)$_2$·3H$_2$O (en = ethylenediaminyl tart = tartrate)	Li—O	1.99	179
β-Li$_2$SO$_4$	Li—O	1.96	180
Li$_2$[Cu(C$_8$H$_4$O$_4$)$_2$]·4H$_2$O (C$_8$H$_4$O$_4$ = phthalate)	Li—O	1.95	181
Li$_2$Ni[NH(CH$_2$CO$_2$)$_2$]$_2$·4H$_2$O	Li—O	1.92	182
Li Co[(R,S)-edds]·3H$_2$O (edds = ethylenediamine-N,N'-disuccinate)	Li—O	1.93	183
D-(−)-Li[Cr(eddda)]·5H$_2$O (eddda = ethylenediamine-N,N'-diacetate-N,N'-di-3-propionate)	Li—O	1.96	184
Li Co[(S,S)-edds]·3H$_2$O	Li—O	1.92	185
Li[Fe$_2$(CO)$_5$[C(O)Ph](μ2-PPh$_2$)$_2$]·3thf	Li—O	1.97	186
{Li[Co(salen)]}$_2$·3thf salen = N,N'-ethylenebis(salicylidenimine)] (there are two different molecular complexes in the crystal structure)	Li—O Li—Co	1.93 NR	187

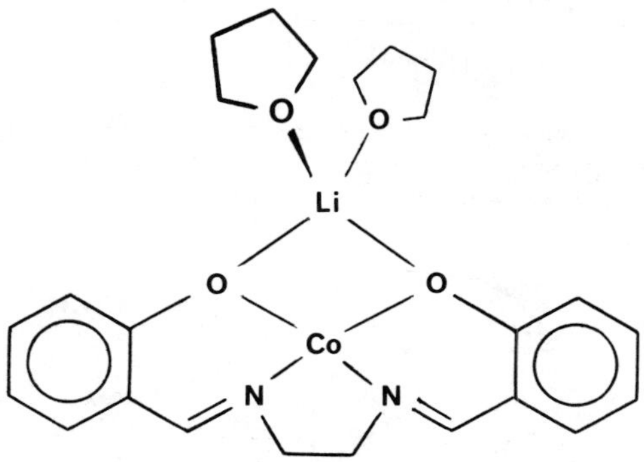

Compound	Li—X	Average bond length (Å)	Reference

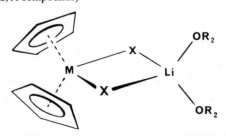

Li($C_5H_7O_4$)UO_2($C_5H_6O_4$) ($C_5H_6O_4$ = glutarate, $C_5H_7O_4$ = hydrogen glutarate)	Li—O	1.93	*188*
Li $Co_3(CO)_{10} \cdot i$-Pr_2O	Li—O	1.95	*189*
Li[Ni(edta)]·$3H_2O$	Li—O	NA	*190*

2. Structures Having Tetracoordinate Lithium Atoms Tetrahedrally Surrounded by Four Donor Atoms

Compound	Li—X	Average bond length (Å)	Reference
Li_2NCN (the lithium is surrounded by four nitrogens)	Li—N	2.07	*191*
Li[$YbCl_2${$C_5(CH_3)_5$}$_2$]·$2Et_2O$ (lithium is surrounded by two oxygens and two bridging chlorines, analogous to type **III**, A compounds)	Li—O Li—Cl	1.96 2.39	*192*

Compound	Li—X	Average bond length (Å)	Reference
Li[YbI$_2${C$_5$(CH$_3$)$_5$}$_2$]·2Et$_2$O (lithium is surrounded by two oxygens and two bridging iodines, as above)	Li—O Li—I	1.90 2.81	*192*
Li[NdCl$_2${C$_5$H$_3$(SiMe$_3$)$_2$}$_2$]·2thf (lithium is surrounded by two oxygens and two bridging chlorines, as above)	Li—O Li—Cl	NR 2.41	*193*
{Li[ReO(OPr-*i*)$_5$]LiCl·2thf}$_2$ (dimeric structure with two different lithium atoms: Li(1) is surrounded by three oxygens and one bridging chlorine; Li(2) is surrounded by two oxygens and two bridging chlorines)	Li—O Li—Cl	1.94 2.40	*194*

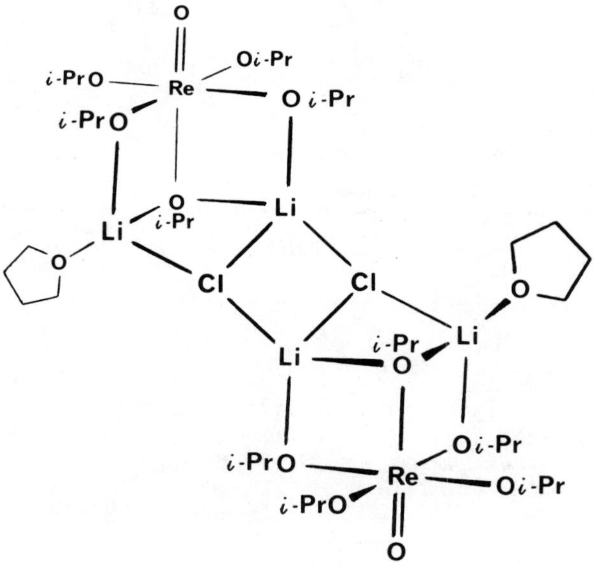

Compound	Li—X	Average bond length (Å)	Reference
trans-[PtCl$_2$(PEt$_3$)[CH(PPh$_2$O)$_2$Li]] (lithium is surrounded by an oxygen and three nitrogen atoms)	Li—O Li—Cl	1.97 2.19	*195*
Li[(C$_2$H$_4$)2NiC(O)N(CH$_3$)$_2$]·pmdeta (lithium is surrounded by an oxygen and three nitrogen atoms)	Li—O Li—N	1.84 2.13	*196*
Li[[OCH$_2$CH$_2$N(CH$_3$)CH$_2$CH$_2$N(CH$_3$)CH$_2$CH$_2$OP] (OC$_2$H$_5$)Mo(CO)$_4$(C$_6$H$_5$CO)] (lithium is surrounded by two oxygen and two nitrogen atoms)	Li—O Li—N	1.91 2.11	*197*

Compound	Li—X	Average bond length (Å)	Reference
LiH·[H(HAlNPr-*i*)₅AlH₂]·Et₂O (lithium is surrounded by one oxygen and three hydrogen atoms)	Li—O Li—H	1.92 1.92	*198*

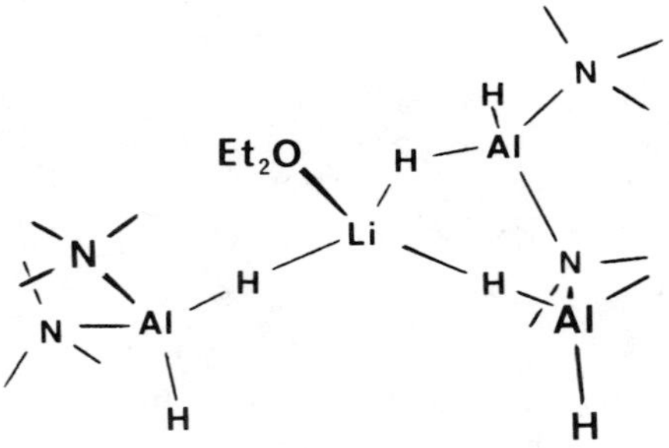

Compound	Li—X	Average bond length (Å)	Reference
(LiH)₂·[HAlN(CH₂)₃N(CH₃)₂]₆ (lithium is surrounded by one hydrogen and three nitrogen atom)	Li—N Li—H	2.10 1.92	*199*

3. *Structures Having Tricoordinate Lithium Atoms*

Compound	Li—X	Average bond length (Å)	Reference
{LiUO[C₆H₅)₂N]₃·Et₂O}₂ (lithium is surrounded by a nitrogen and two oxygen atoms, analogous to type **III**, *A* compounds)	Li—O$_{terminal}$ Li—O$_{bridging}$ Li—N Li—U	1.89 1.93 2.20 NR	*200*

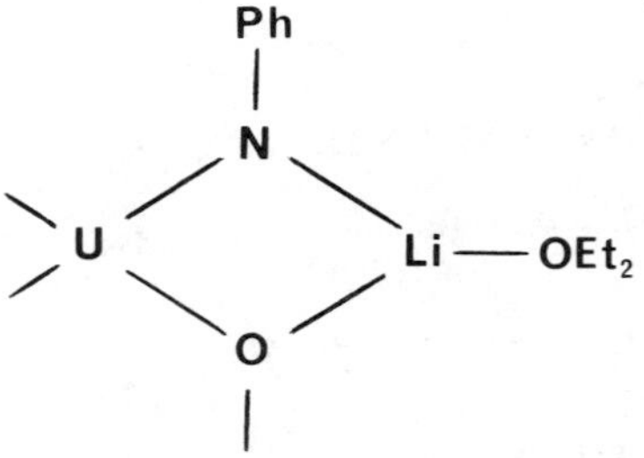

Compound	Li—X	Average bond length (Å)	Reference
$Li_6Cr_2(o\text{-}C_6H_4O)_4Br_2 \cdot 6Et_2O$ (each lithium is surrounded by a bromine and two oxygen atoms)	Li—O$_{terminal}$ Li—O$_{bridging}$ Li—Br Li—Cr	1.96 1.86 2.68 2.75	201

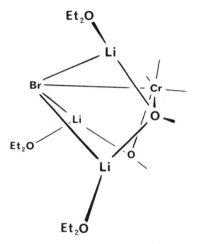

| Li Cr(OCHBu-t_2)$_3$ · thf (lithium is surrounded by three oxygen atoms, analogous to type **III**, *A* compounds | Li—O$_{thf}$ Li—O$_{bridging}$ Li—Cr | 1.94 1.85 2.73 | 202 |

| Li Fe(OCHBu-t_2)$_4$ · (*t*-Bu)$_2$CHOH (as above) | Li—O$_{terminal}$ Li—O$_{bridging}$ Li—Fe | 1.95 1.87 2.71 | |

4. *Structures Having Pentacoordinate Lithium Atoms*

Compound	Li—X	Average bond length (Å)	Reference
$Li_2Ni(S_2C_2O_2)_2 \cdot 2H_2O$ (lithium is surrounded by five oxygen atoms, the geometry of which is nearly halfway between trigonal bipyramidal and tetragonal pyramidal)	Li—O	2.07	*203*
$Mo(CO)_3(PhCOLi)[Ph_2P(OCH_2CH_2)_3OPPh_2]$ (lithium is surrounded by five oxygen atoms; three oxygens occupy "tetrahedral" positions, while two oxygens are symmetrically placed about the fourth "tetrahedral" position)	Li—O	2.04	*204*

5. *Structures Having Hexacoordinate Lithium Atoms*

Compound	Li—X	Average bond length (Å)	Reference
$Li(C_5H_5CrCl_3) \cdot 2thf \cdot dioxane$ (lithium is octahedrally surrounded by three oxygen and three chlorine atoms)	Li—O	2.01	*205*
	Li—Cl	2.90	
	Li—Cr	3.38	

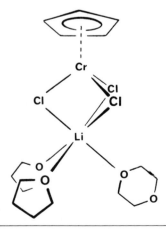

Compound	Li—X	Average bond length (Å)	Reference
LiU$_2$Cl$_5$[CH$_2$(C$_5$H$_4$)$_2$]$_2$·2thf (lithium is octahedrally surrounded by two oxygen and four chlorine atoms)	Li—O Li—Cl Li—U	2.06 2.74 NR	206

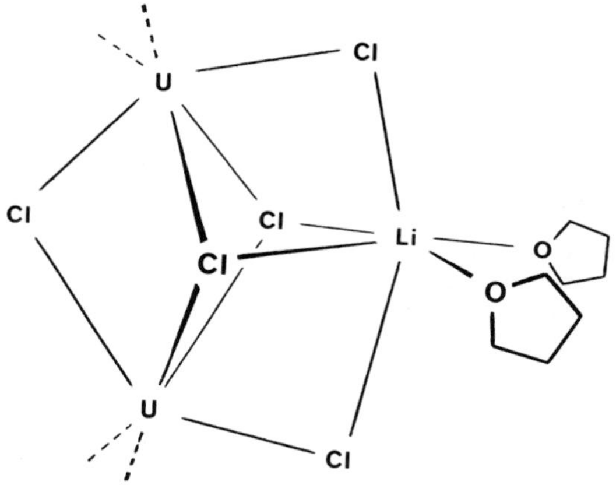

LiV$_2$F$_6$ (lithium is octahedrally surrounded by six fluorine atoms)	Li—F Li—V	2.08 3.09	207
α-LiIO$_3$ (lithium is octahedrally surrounded by six oxygen atoms)	Li—O	2.11	208
Li$_3$[Pt(C$_4$N$_2$S$_2$)$_2$]$_4$·8H$_2$O (lithium atom is hexacoordinate; no details reported)	NR		209
Li[C$_6$H$_2$CH$_3$)$_3$]$_2$BH$_2$·2dme [lithium dimesitylborohydride bis(dimethoxyethane)] (lithium is surrounded by four oxygen and two hydrogen atoms)	Li—O Li—H Li—B	2.18 2.06 2.50	210

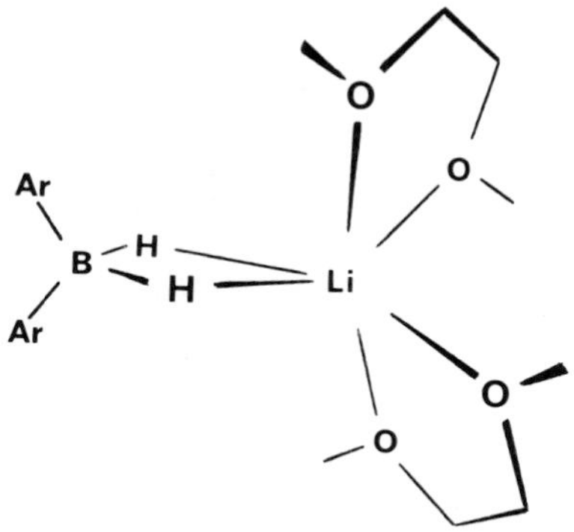

V
APPENDIX

Included in this section are some important structures of lithium compounds which have recently appeared in the literature.

Compound	Li—X	Average bond length ($\text{\AA}$)	Reference
$[(CH_3)_2PCH_2Li \cdot (tmeda)]_2$	Li—C	2.15	211
a (six-membered ring, dimeric structure with	Li—N	2.20	
terminally bonded, tetrahedrally tetracoordinate	Li—P	2.60	
lithium atoms, analogous to compound 3)			
$[Li(12\text{-crown-}4)_2][PPh_2]$, $(CHPh_2)$, and $[CPh_3]$	Li—O	2.35	212
("solvent-separated" ion-pair structures; the			
lithium cations are surrounded by eight oxygen			
atoms)			
$LiN(SiMe_3)_2 \cdot (12\text{-crown-}4)$	Li—N	1.97	213
a terminally bonded lithium amide with a	Li—O	2.23	
pentacoordinate lithium atom			
$[Li(C_{20}H_{21}O_7)] \cdot 7.5H_2O$	Li—O	2.06 (crown)	214
a crown ether complexed ion pair structure with	Li—O	1.91(H_2O)	
five oxygens around lithium in approximate C_{4v}			
symmetry			

ACKNOWLEDGMENTS

The Cambridge Crystallographic Database was an invaluable source of references for this article, but many colleagues also provided information prior to publication. We thank, in particular, Professor Hans Dietrich, Berlin; Professors Jack Dunitz and Dieter Seebach, Zürich; Professor Erwin Weiss, Hamburg; Professors Michael Lappert and Colin Eaborn, Sussex; Dr. William Clegg, Göttingen; Professor Gerard van Koten, Amsterdam; Dr. John Stezowski, Stuttgart; Professor Jürgen Daub, Regensburg; Dr. Ronald Smith, Glasgow; Professor Robert Bau, University of Southern California; Professor Philip Power, University of California, Davis; Dr. Collen Raston, Western Australia; Prof. Paul Williard, Rhode Island; Prof. William Collen, British Columbia; Prof. David Collum, Cornell; and Professor Adalbert Maercker, Siegen. We gratefully acknowledge the efforts of our many co-workers in Erlangen, especially Dr. Wolfgang Neugebauer and Dieter Wilhelm, who prepared numerous crystalline samples, and Dr. Timothy Clark, Dr. Alex Kos, Dr. Juan Andrade, and Elmar Kaufmann, who carried out extensive calculational work on lithium compounds pertinent to this article. WNS would like to thank the Alexander von Humboldt Foundation for a Fellowship during the preparation of this article.

REFERENCES

1. P. von R. Schleyer, *Pure Appl. Chem.* **56**, 151–162 (1984); **55**, 355–362 (1983).
2. M. F. Lappert, L. M. Englehardt, C. L. Raston, and A. H. White, *J. Chem. Soc., Chem. Commun.* pp. 1323–1324 (1982).

3. R. Amstutz, D. Seebach, P. Seiler, B. Schweizer, and J. D. Dunitz, *Angew. Chem.* **92**, 59–60 (1980); *Angew. Chem., Int. Ed. Engl.* **19**, 53–54 (1980).
4. R. Amstutz, J. D. Dunitz, D. Seebach, *Angew. Chem.* **93**, 487–488 (1981); *Angew. Chem., Int. Ed. Engl.* **20**, 465–466 (1981).
5. A. J. Kos and P. von R. Schleyer, *J. Am. Chem. Soc.* **102**, 7928–7929 (1980).
6. W. Clegg, D. A. Brown, S. J. Bryan, and K. Wade, *Polyhedron* **3**, 307–311 (1984).
7. R. P. Zerger and G. D. Stucky *J. Chem. Soc., Chem. Commun.* pp. 44–45 (1973).
8. R. Amstutz, T. Laube, W. B. Schweizer, D. Seebach, and J. D. Dunitz, *Helv. Chim. Acta* **67**, 224–236 (1984).
9. D. Thoennes and E. Weiss, *Chem. Ber.* **111**, 3157–3161 (1978).
10. J. T. B. H. Jastrzebski, G. van Koten, K. Goubitz, C. Arlen, and M. Pfeffer, *J. Organometl. Chem.* **246**, C75–C79 (1983).
11. D. Wilhelm, T. Clark, P. von R. Schleyer, H. Dietrich, and W. Mahdi, unpublished results.
12. J. Chandrasekhar and P. von R. Schleyer, *J. Chem. Soc., Chem. Commun.*, pp. 260–261 (1981).
13. W. Neugebauer, H. Dietrich, and P. von R. Schleyer, unpublished results.
13a. P. von R. Schleyer, A. J. Kos, D. Wilhelm, T. Clark, G. Boche, G. Decker, H. Etzodt, H. Dietrich, W. Mahdi, *J. Chem. Soc., Chem. Commun.*, 1494–1495 (1984).
13b. U. Schubert, W. Neugebauer, and P. von R. Schleyer, *J. Chem. Soc., Chem. Commun.* pp. 1184–1185 (1982); E. Kaufmann, A. J. Kos, and P. von R. Schleyer, *J. Am. Chem. Soc.* **105**, 7615–7623 (1983); E. Kaufmann and P. von R. Schleyer, unpublished results.
13c. M. F. Lappert, C. L. Raston, B. W. Skelton, and A. J. White, *J. Chem. Soc., Chem. Commun.* pp. 14–15 (1982).
13d. G. Boche, G. Decher, H. Etzrodt, H. Dietrich, W. Mahdi, A. J. Kos, and P. von R. Schleyer, *J. Chem. Soc., Chem. Commun.*, pp. 1493–1494 (1984).
13e. D. Wilhelm, T. Clark, and P. von R. Schleyer, *J. Organometal. Chem.* **280**, C6–C10 (1985); A. J. Kos, P. Stein, and P. von R. Schleyer, *ibid* **280**, C1–C5 (1985).
14. B. Schubert and E. Weiss, *Chem. Ber.* **116**, 3212–3215 (1983).
15. L. Sutton, ed., "Tables of Interatomic Distances and Configurations in Molecules and Ions;" Spec. Publ. No. 11. Chemical Society; London, 1958; No. 18 (1965); J. L. Wardell, *in* "Comprehensive Organometallic Chemistry" (G. Wilkinson, F. G. A. Stone, and F. W. Abel eds. Vol. 1, Chapter 2, p. 45. Pergamon, Oxford, 1982.
16. E. Weiss and E. A. C. Lucken *J. Organometl. Chem.* **2**, 197–205 (1964).
17. E. Weiss and G. Hencken, *J. Organometl. Chem.* **21**, 265–268 (1970).
18. H. Köster, D. Thoennes and E. Weiss, *J. Organometl. Chem.* **160**, 1–5 (1978).
19. H. Dietrich, *Acta Crystallogr.* **16**, 681–689 (1963).
20. H. Dietrich, *J. Organometl. Chem.* **205**, 291–299 (1981).
21. E. Weiss, G. Sauermann and G. Thirase, *Chem. Ber.* **116**, 74–85 (1983).
22. E. Weiss and G. Sauermann, *Chem. Ber.* **80**, 123–124 (1968); **103**, 265–271 (1970).
23. E. Weiss and H. Köster, *Chem. Ber.* **110**, 717–720 (1977).
24. E. Weiss and G. Sauermann, *J. Organometl. Chem.* **21**, 1–7 (1970).
25. G. W. Klumpp, P. J. A. Geurink, A. L. Spek and J. M. Duisenberg, *J. Chem. Soc., Chem. Commun.* pp. 814–816 (1983); further structures of similar chelated tetramers have now been determined; G. Klumpp, private communication.
26. H. Schmidbauer, A. Schier and U. Schubert, *Chem. Ber.* **116**, 1938–1946 (1983).
27. H. Hope and P. P. Power, *J. Am. Chem. Soc.* **105**, 5320–5324 (1983).
28. J. T. B. H. Jastrzebski, G. van Koten, M. Konijn, and C. H. Stam, *J. Am. Chem. Soc.* **104**, 5490–5492 (1982).

29. B. Schubert and E. Weiss, *Angew. Chem.* **95**, 499 (1983); *Angew, Chem., Int. Ed. Engl.* **22**, 496–497 (1983).
30. W. Neugebauer, E. Weiss, and P. von R. Schleyer, unpublished results.
31. R. Zerger, W. Rhine, and G. Stucky, *J. Am. Chem. Soc.* **96**, 6048–6055 (1974).
32. T. F. Schaaf, W. Butler, M. D. Glick, J. P. Oliver, *J. Am. Chem. Soc.* **96**, 7593–7594 (1974).
33. H. M. M. Shearer, K. Wade, and G. Whitehead, *J. Chem. Soc., Chem Commun.* pp. 943–945 (1979).
34. A. Maercker, M. Bsata, W. Buchmeier, and B. Engelen, *Chem. Ber.* **117**, 2547–2554 (1984).
35. H. Dietrich and D. Rewicki, *J. Organometl. Chem.* **205**, 281–289 (1981).
36. J. J. Brooks, W. Rhine, and G. D. Stucky, *J. Am. Chem. Soc.* **94**, 7339–7346 (1972).
37. G. Stucky, in "Polyamine-Chelated Alkali Metal Compounds;" A. W. Langer, ed.). Am. Chem. Soc. New York, 1974. Adv. Chem. Ser. No. 130, Chapter 3.
38. A. Streitwieser, Jr., J. E. Williams, Jr., S. Alexandratos, J. M. McKelvey, *J. Am. Chem. Soc.* **98**, 4778–4784 (1976); A. Streitwieser, Jr., *J. Organometl. Chem.* **156**, 1–3 (1978).
38a. A. Streitwieser, Jr., *Accts. Chem. Res.* **17**, 353–357 (1984).
38b. R. J. Bushby and N. P. Tyko, *J. Organomet. Chem.* **270**, 265–270 (1984).
39. M. Schlosser and M. Stähle, *Angew. Chem.* **94**, 142–143 (1982); *Angew. Chem., Int. Ed. Engl.* **21**, 145–146 (1982); M. Stähle and M. Schlosser, *J. Organometl. Chem.* **220**, 227–283 (1981); M. Schlosser and M. Stähle, *Angew. Chem.* **92**, 497–499 (1980); *Angew. Chem., Int. Ed. Engl.* **19**, 487–489 (1980); W. Neugebauer and P. von R. Schleyer, *J. Organometl. Chem.* **198**, C1-C3 (1980).
40. H. Köster and E. Weiss, *Chem. Ber.* **115**, 3422–3426 (1982).
41. R. Goddard, C. Krüger, F. Mark, R. Stansfield, and X. Zajng, *Organometallics* **4**, 285–290 (1985); B. Henc, P. W. Jolly, R. Salz, S. Stobbe, G. Wilke, R. Benn, R. Mynott, K. Seevogel, R. Goddard, and C. Krüger, *J. Organometl. Chem.* **191**, 449–475 (1980).
42. J. Chandrasekhar, J. G. Andrade, P. von R. Schleyer, *J. Am. Chem. Soc.* **103**, 5609–5612 (1981).
43. T. Clark., E. D. Jemmis, P. von R. Schleyer, J. S. Binkley, and J. A. Pople, *J. Organometl. Chem.* **150**, 1–6 (1978).
44. G. Decher and G. Boche, *J. Organometl. Chem.* **259**, 31–36 (1983).
45. T. Clark, C. Rohde, and P. von and R. Schleyer, *Organometallics* **2**, 1344–1351 (1983).
46. S. Alexandratos, A. Streitwieser, Jr., and H. F. Schaefer, III, *J. Am. Chem Soc.* **97**, 6271–6272 (1975); E. D. Jemmis and P. von R. Schleyer, *J. Am. Chem. Soc.* **104**, 4781–4788 (1982).
47. K. C. Waterman, and A. Streitwieser, Jr., *J. Am. Chem. Soc.* **106**, 3138–3140 (1984).
48. M. F. Lappert, A. Singh, L. M. Engelhardt, and A. H. White, *J. Organomet. Chem.* **262**, 271–278 (1984).
49. P. Jutzi, E. Schlüter, C. Krüger, and S. Pohl, *Angew. Chem.* **95**, 1015 (1983); *Angew. Chem. Int. Ed. Engl.* **22**, 994 (1983).
50. S. D. Patterman, I. L. Karle, and G. D. Stucky, *J. Am. Chem. Soc.* **92**, 1150–1157 (1970); for a polymeric benzyllithium structure, see M. A. Beno, H. Hope, M. M. Holmstead, and P. P. Power, *Organometallics* (in press).
51. A. J. Kos and P. von R. Schleyer, unpublished results; see also ref. 38a.
52. J. J. Brooks and G. D. Stucky *J. Am. Chem. Soc.* **94**, 7333–7338 (1972).
53. T. A. Albright, P. Hofmann, R. Hoffmannn, C. P. Lillya, and P. A. Dobosh, *J. Am. Chem. Soc.* **105**, 3396–3411 (1983).

54. J. J. Brooks, W. Rhine, and G. D. Stucky, *J. Am. Chem. Soc.* **94**, 7346–7351 (1972).
55. W. E. Rhine and G. D. Stucky, *J. Am. Chem. Soc.* **97**, 737–743 (1975).
56. W. E. Rhine, J. Davis and G. Stucky, *J. Am. Chem Soc.* **97**, 2079–2085 (1975).
57. W. E. Rhine, J. H. Davis, and G. Stucky, *J. Organomet. Chem.* **134**, 139–149 (1977).
58. M. Walczak and G. D. Stucky, *J. Organomet. Chem.* **97**, 313–323 (1975).
59. A. Streitwieser, Jr., and J. T. Swanson, *J. Am. Chem. Soc.* **105**, 2502–2503 (1983).
60. M. Walczak and G. Stucky *J. Am. Chem. Soc.* **98**, 5531–5539 (1976).
61. S. K. Arora, R. B. Bates, W. A. Beavers, and R. S. Cutler, *J. Am Chem. Soc.* **97**, 6271–6272 (1975).
62. D. Bladauski, and D. Rewicki, *Chem. Ber.* **110**, 3920–3929 (1977).
63. D. Bladauski, H. Dietrich, H. J. Hecht, and D. Rewicki, *Angew. Chem.* **89**, 490–491 (1977); *Angew. Chem., Int. Ed. Engl.* **16**, 474–475 (1977); D. Bladauski, W. Broser, H.-J. Hecht, D. Rewicki, and H. Dietrich, *Chem. Ber.* **112**, 1380–1391 (1979).
64. H. Dietrich, H. Mahdi, D. Wilhelm, P. Clark, and P. von R. Schleyer, *Angew. Chem.* **96**, 623–625 (1984).
65. E. D. Jemmis, J. Chandrasekhar, and P. von R. Schleyer, *J. Am. Chem Soc.* **101**, 2848–2856 (1979).
66. W. Neugebauer, B. Schubert, E. Weiss, and P. von R. Schleyer, unpublished results.
67. W. Neugebauer, A. J. Kos, P. von R. Schleyer, J. J. Stezowski, G. A. P. Geiger, *Chem Ber.* **118**, 1504–1516 (1985).
68. D. Wilhelm, H. Dietrich, and P. von R. Schleyer unpublished results.
69. B. Bogdanović, C. Krüger, and B. Wermeckes, *Angew. Chem.* **92**, 844–845 (1980); *Angew. Chem., Int. Ed. Engl.* **19**, 817–818 (1980).
70. C. Eaborn P. B. Hichcock, J. D. Smith, and A. C. Sullivan, *J. Chem Soc., Chem Commun.* pp. 827–828 (1983).
71. C. Eaborn, P. B. Hitchcock, J. D. Smith, and A. C. Sullivan, *J. Chem. Soc., Chem Commun.* pp. 1390–1391 (1983).
72. T. Greiser, J. Kopf, D. Thoennes, and E. Weiss, *Chem. Ber.* **114**, 209–213 (1981).
73. D. Thoennes and E. Weiss, *Chem. Ber.* **111**, 3726–3731 (1978).
74. D. R. Stranks, M. L. Hefferman, K. C. L. Dow, P. T. McTigue, and G. R. A. Withers, "Chemistry: A Structural View," p. 391, Cambridge Univ. Press, London and New York 1970.
75. R. L. Gerteis, R. E. Dicerson, T. L. Brown, *Inorg. Chem.* **3**, 872–875 (1964).
76. K. Hoffmann and E. Weiss, *J. Organomet. Chem.* **37**, 1–7 (1972).
77. E. Weiss and R. Wolfrum, *Chem. Ber.* **101**, 35–40 (1968).
78. E. Weiss and R. Wolfrum, *J. Organomet. Chem.* **12**, 257–262 (1968).
79. W. E. Rhine, G. Stucky, and S. W. Peterson, *J. Am. Chem. Soc.* **97**, 6401–6406 (1975).
80. J. Krausse and G. Marx, *J. Organomet. Chem* **65**, 215–222 (1974).
81. H. Schumann, J. Pickardt, and N. Bruncks, *Angew. Chem.* **93**, 127–128 (1981); *Angew. Chem., Int. Ed. Engl.* **20**, 120–121 (1981).
82. G. van Koten, J. T. B. H. Jastrzebski, C. H. Stam, and N. C. Niemann, *J. Am. Chem. Soc.* **106**, 1880–1881 (1984).
83. J. Krausse, G. Marx, and G. Schödl, *J. Organomet. Chem.* **21**, 159–168 (1970).
84. J. Krausse and G. Schödl, *J. Organomet. Chem.* **27**, 59–67 (1971).
85. F. A. Cotton, J. M. Troup, T. R. Webb, D. H. Williamson, and G. Wilkinson, *J. Am. Chem. Soc.* **96**, 3824–3828 (1974).
86. F. A. Cotton, L. D. Gage, K. Mertis, L. W. Shive, and G. Wilkinson, *J. Am. Chem. Soc.* **98**, 6922–6925 (1976).
87. D. M. Collins, F. A. Cotton, S. A. Koch, M. Millar, and C. A. Murillo, *Inorg. Chem.* **17**, 2017–2020 (1978).
88. K. Jonas, *Adv. Organomet. Chem.* **19**, 97–122 (1981).

89. K. Jonas and C. Krüger, *Angew Chem.* **92**, 513–531 (1980); *Angew, Chem., Int. Ed. Engl.* **19**, 520–537 (1980).
90. K. Jonas, R. Mynott, C. Krüger, J. C. Sekutowski, and Y. -H. Tsay, *Angew Chem.* **88**, 808–809 (1976) *Angew Chem., Int. Ed. Engl.* **15**, 767–768 (1976).
91. K. Jonas, L. Schieferstein, C. Krüger, and Y. -H. Tsay, *Angew Chem.* **91**, 590–591 (1979); *Angew Chem., Int. Ed. Engl.* **18**, 550–551 (1979).
92. D. J. Brauer, C. Krüger, J. C. Sekutowski, *J. Organomet. Chem.* **178**, 249–260 (1979).
93. K. Jonas, C. Krüger, and J. C. Sekutowski, *Angew. Chem.* **91**, 520–521 (1979); *Angew. Chem. Int. Ed. Engl.* **18**, 487–488 (1979).
94. H. Klein, H. Witty, and U. Schubert, *J. Chem. Soc., Chem. Commun.* pp. 231–232 (1983).
95. M. Walczak, K. Walczak, R. Mink, M. D. Rausch, and G. Stucky, *J. Am. Chem. Soc.* **100**, 6382–6388 (1978).
96. I. R. Butler, W. R. Cullen, J. Reglinski, and S. J. Rettig, *J. Organomet. Chem.* **249**, 183–194 (1983).
96a. I. R. Butler, W. R. Cullen, J. Ni, and J. J. Rettig, *Organometallics* (in press).
97. R. A. Forder and K. Prout, *Acta Crystallogr., Sect. B* **B30**, 2318–2322 (1974).
98. H. Bönnemann, C. Krüger, and Y.-H. Tsay, *Angew. Chem.* **88**, 50–51 (1976); *Angew Chem., Int. Ed. Engl.* **15**, 46–47 (1976).
99. C. Krüger and Y.-H. Tsay, *Angew Chem.* **85**, 1051–1052 (1973); *Angew. Chem., Int. Ed. Engl.* **12**, 998–999 (1973).
100. K. Jonas, D. J. Brauer, C. Krüger, D. J. Roberts, and Y.-H. Tsay, *J. Am. Chem. Soc.* **98**, 74–81 (1976).
101. L. J. Guggenberger and R. R. Schrock *J. Am. Chem. Soc.* **97**, 2935 (1975).
102. H. M. M. Shearer, R. Snaith, J. D. Sowerby, and K. Wade, *J. Chem. Soc., Chem. Commun.* pp. 1275–1276 (1971).
103. M. J. Albright, T. F. Schaaf, W. M. Butler, A. K. Hovland, M. D. Glick, and J. P. Oliver, *J. Am. Chem. Soc.* **97**, 6261–6262 (1975).
104. W. H. Ilsley, M. J. Albright, T. J. Anderson, M. D. Glick, and J. P. Oliver, *Inorg. Chem.* **19**, 3577–3585 (1980).
105. F. R. Fronczek, G. W. Halstead, and K. N. Raymond, *J. Am. Chem. Soc.* **99**, 1769–1775 (1977).
106. C. Krüger, J. C. Sekutowski, and Y.-H Tsay, *Z. Kristallogr.* **149**, 109 (1979).
107. P. G. Edwards, R. W. Gellert, M. W. Marks, and R. Bau, *J. Am. Chem. Soc.* **104**, 2072–2073 (1982).
108. S. A. Cotton, F. A. Hart, M. B. Hursthouse, and A. J. Welch, *J. Chem. Soc., Chem. Commun.* pp. 1225–1226 (1972).
109. J. L. Atwood, W. E. Hunter, R. D. Rogers, J. Holton, J. McMeeking, R. Pearce, and M. F. Lappert, *J. Chem Soc., Chem Commun.* pp. 140–142 (1978).
110. Y. M. G. Yasin, O. J. R. Hodder, and H. M. Powell, *J. Chem. Soc., Chem. Commun.* pp. 705–706 (1966).
111. D. Barr, W. Clegg, R. E. Mulvey, and R. Snaith, *J Chem. Soc., Chem Commun.* pp. 226–227 (1984).
112. N. Takahashi, I. Tanaka, T. Yamane, T. Ashida, T. Sugihara, Y. Imanishi, and T. Higashimura, *Acta Crystallogr., Sect. B* **B33**, 2132–2136 (1977).
113. J. P. Declercq, R. Meulemans, P. Piret, and M. Van Meerssche, *Acta Crytallogr., Sect. B* **B27**, 539–544 (1971).
114. R. Meulemans, P. Piret, and M. Van Meerssche, *Acta Crytallogr., Sect. B* **B27**, 1187–1190 (1971).
115. P. Groth, *Acta Chem Scand., Ser. A* **A37**, 71–77 (1983).
116. P. Groth, *Acta Chem. Scand., Ser. A* **A36**, 109–115 (1982).

117. F. Durant, P. Piret, and M. Van Meerssche, *Acta Crytallogr.* **23**, 780–788 (1967).
118. K. Jonas, K. R. Pörschke, C. Krüger, and Y. -H. Tsay *Angew. Chem.* **88**, 682 (1976); *Angew, Chem., Int. Ed. Engl.* **15**, 621–622 (1976).
119. H. Schmidbauer, E. Weiss, and B. Zimmer-Gasser, *Angew. Chem.* **91**, 848–850 (1979); *Angew. Chem., Int. Ed. Engl.* **18**, 782–783 (1979).
120. F. Durant, P. Piret, and M. Van Meerssche, *Acta Crystallogr.* **22**, 52–57 (1967).
121. W. Saenger, B. S. Reddy, K. Mühlegger, and G. Weimann, *Nature (London)* **267**, 225–229 (1977); B. S. Reddy, W. Saenger, K. Mühlegger, and G. Weimann, *J. Am. Chem. Soc.* **103**, 907–914 (1981).
122. J. Verbist, R. Meulemans, P. Piret, and M. Van Meerssche, *Bull. Soc. Chim. Belg.* **79**, 391–396 (1970).
123. Kh. Suleimanov, V. S. Sergienko, N. Kipkalova, and K. Sulaimankolov, *Koord. Khim.* **5**, 1732–1736 (1979); *Chem. Abstr.* **92**, 50423k (1980).
123a. D. Barr, W. Clegg, R. E. Mulvey, and R. Snaith, *J. Chem. Soc., Chem. Commun.* pp. 79–80 (1984).
124. I. L. Karle, *J. Am. Chem. Soc.* **96**, 4000–4006 (1974).
125. P. Groth, *Acta Chem. Scand., Ser. A* **A35**, 460–462 (1981).
126. P. Groth, *Acta Chem. Scand. Ser. A* **A35**, 463–465 (1981).
127. G. Shoham, W. N. Lipscomb, and U. Olsher, *J. Am. Chem. Soc.* **105**, 1247–1252 (1983).
128. P. Groth, *Acta Chem. Scand., Ser. A* **A37**, 71–77 (1983).
129. G. Shoham, W. N. Lipscomb, and U. Olsher, *J. Chem. Soc., Chem. Commun.* pp. 208–209 (1983).
130. M. Noltemeyer and W. Saenger, *J. Am. Chem. Soc.* **102**, 2710–2722 (1980).
131. M. Czugler and E. Weber, *J. Chem. Soc., Chem. Commun.* pp. 472–473 (1981).
132. D. J. Cram, G. M. Lein, T. Kaneda, R. C. Helgeson, C. B. Knobler, E. Maverick, and K. N. Trueblood, *J. Am. Chem. Soc.* **103**, 6228–6232 (1981).
133. H. Gillier-Pandraud and S. Jamet-Delcroix, *Acta Crystallogr., Sect. B* **B3277**, 2476–2479 (1971).
134. D. Moras and R. Weiss, *Acta Crystallogr., Sect. B* **B29**, 400–403 (1973).
135. P. S. Gentile, J. G. White, and D. D. Cavalluzzo, *Inorg. Chim. Acta* **20**, 37–42 (1976).
136. B. Beagley and R. W. H. Small, *Acta Crystallogr.* **17**, 783–788 (1964).
137. E. J. Gabe, J. P. Gluster, J. A. Minkin, A. L. Pattersen, *Acta Crystallogr.* **22**, 366–375 (1967).
138. J. L. Galigné, M. Mouvet, and J. Jalgueirettes, *Acta Crystallogr., Sect. B* **B26**, 368–372 (1970).
139. H. Klapper and H. Küppers, *Acta Crystallogr., Sect. B* **B29**, 21–26 (1973).
140. J. K. M. Rao and M. A. Viswamitra, *Ferroelectrics* **2**, 209–216 (1971); A. Enders-Beumer and S. Harkema, *Acta Crystallogr., Sect. B* **B29**, 682–685 (1973); *ibid.* J. O. Thomas, R. Tellgren, and J. Amlöf, **B31**, 1946–1955 (1975).
141. W. G. Town and R. W. H. Small *Acta Crystallogr., Sect. B* **B29**, 1950–1955 (1973).
142. M. P. Gupta, S. M. Prasad, and T. N. P. Gupta, *Acta Crystallogr., Sect. B* **B31**, 37–40 (1975).
143. W. Gonschorek and H. Küppers, *Acta Crystallogr., Sect. B* **B31**, 1068–1072 (1975).
144. G. Adiwidjaja and H. Küppers, *Acta Crystallogr., Sect. B* **B34**, 2003–2005 (1978); *ibid.* H. Küppers, A. Kvick, and I. Olovsson, **B37**, 1203–1207 (1981).
145. H. Küppers, *Acta Crystallogr., Sect. B* **B34**, 3763–3765 (1978).
146. M. Soriano-Garcia, R. Parthasarathy, *J. Chem. Soc., Perkin Trans. 2* pp. 668–670 (1978).
147. M. Soriano-Garcia, *Rev. Latinoam. Quim.* **11**, 104–106 (1980); *Chem. Abstr* **94**, 75037s (1981).

148. W. Van Havere and A. T. H. Lenstra, *Acta Crystallogr., Sect. B* **B36,** 1483–1486 (1980).
149. R. Mattes and G. Plescher, *Acta Crystallogr., Sect. B* **B37,** 697–699 (1981).
150. R. H. Colton and D. E. Henn, *Acta Crystallogr.* **18,** 820–822 (1965).
151. C. K. Johnson, E. J. Gabe, M. R. Taylor, and I. A. Rose, *J. Am. Chem. Soc.* **87,** 1802–1804 (1965); E. J. Gabe and M. R. Taylor, *Acta Crystallogr.* **21,** 418–421 (1966).
152. H. Fallner, *Z. Anorg. Allg. Chemie.* **373,** 198–203 (1970); J. O. Thomas *Acta Crystallogr., Sect. B* **B28,** 2037–2045 (1972).
153. H. Hinazumi and T. Mitsui, *Acta Crystallogr., Sect. B* **B28,** 3299–3305 (1972).
154. H. Herbertsson, *Acta Crystallogr., Sect. B* **B32,** 2381–2384 (1976).
155. B. F. Pedersen, *Acta Chem. Scand.* **23,** 1871–1877 (1969).
156. J. P. Glusker, D. van der Helm, W. E. Love, M. L. Dornberg, J. A. Minkin, C. K. Johnson, and A. L. Patterson, *Acta Crystallogr.* **19,** 561–572 (1965).
157. V. W. Bhagwat, H. Manohar, and N. S. Poonia, *Inorg. Nucl. Chem. Lett.* **16,** 373–375 (1980).
158. B. Cetinkaya, I. Gümrükcü, M. F. Lappert, J. L. Atwood, and R. Shakir, *J. Am. Chem. Soc.* **102,** 2086–2088 (1980).
159. P. J. Wheatley, *J. Chem. Soc.* pp. 4270–4274 (1960).
160. E. Weiss and U. Joergens, *Chem. Ber.* **105,** 481–486 (1972).
161. M. F. Lappert, M. J. Slade, A. Singh, J. L. Atwood, R. D. Rogers, and R. Shakir, *J. Am. Chem. Soc.* **105,** 302–304 (1983).
162. M. F. Lappert, B. Cetiniakya, P. B. Hitchcock, M. F. Lappert, M. C. Misra, and A. J. Thorne, *J. Chem. Soc., Chem. Commun.* pp. 148–149 (1984).
162a. D. Barr, W. Clegg, R. E. Mulvey, and R. Snaith, *J. Chem. Soc., Chem. Commun.* pp. 285–286 (1984).
162b. D. Barr, W. Clegg, R. E. Mulvey, and R. Snaith, *J. Chem. Soc., Chem. Commun.* pp. 700–701 (1984).
163. D. Mootz, A. Zinnius, and B. Böttcher, *Angew. Chem.* **81,** 398–399 (1969); *Angew. Chem., Int. Ed. Engl.* **8,** 378–379 (1969); R. D. Rogers, J. L. Atwood, and R. Grüning, *J. Organomet. Chem.* **157,** 229–237 (1978).
164. W. Clegg, R. Snaith, H. M. M. Shearer, K. Wade, and G. Whitehead, *J. Chem. Soc., Dalton Trans.* pp. 1309–1317 (1983).
165. B. Teclé, W. H. Ilsley, and J. P. Oliver, *Organometallics* **1,** 875–877 (1982).
166. W. H. Ilsley, T. F. Schaaf, M. D. Glick, and J. P. Oliver, *J. Am. Chem. Soc.* **102,** 3769–3774 (1980).
167. R. A. Jones, A. L. Stuart, T. C. Wright, *J. Am. Chem. Soc.* **105,** 7459–7460 (1983).
168. F. A. Schröder and H. P. Weber, *Acta Crystallogr., Sect B* **B31,** 1745–1750 (1975).
169. D. Semmingsen, *Acta Chem. Scand., Ser A* **A30,** 808–812 (1976).
170. M. I. Bruce, J. K. Walton, M. L. Williams, B. W. Skelton, and A. H. White, *J. Organometl. Chem.* **212,** C35–C38 (1981).
171. C. Cambillau, G. Bram, J. Corset, and C. Riche, *Nouv. J. Chim.* **3,** 9–11 (1979).
172. H. B. Bürghi, S. Djurić, M. Dobler, and J. D. Dunitz, *Helv. Chim. Acta* **55,** 169–170 (1972).
173. J. D. Dunitz and D. Seebach, private communication; D. Seebach, *Proc. Robert A. Welch Found. Conf. Chem. Res.* **27,** 93–145 (1984).
174. W. Bauer, T. Laube, and D. Seebach, *Chem. Ber.* **118,** 764–773 (1985).
175. G. Becker, M. Birkhahn, W. Massa, and W. Uhl, *Angew. Chem.* **92,** 756–757 (1980); *Angew. Chem., Int. Ed. Engl.* **19,** 741–742 (1980).
176. R. Amstutz, W. B. Schweizer, D. Seebach, and J. D. Dunitz, *Helv. Chim. Acta* **64,** 2617–2621 (1981).
176a. P. J. Williard and G. B. Carpenter, *J. Am. Chem. Soc.* (in press).

177. M. D. Lind, M. J. Hamor, T. A. Hamor, and J. L. Hoard, *Inorg. Chem.* **3**, 34–43 (1964).

178. T. Tsukihara, Y. Katsube, K. Kawashima, and Y. Kan-nan, *Bull. Chem. Soc. Jpn.* **47**, 1582–1585 (1974).

179. Y. Kushi, M. Kuramoto, and H. Yoneda, *Chem. Lett.* pp. 339–342 (1976).

180. A. G. Nord, *Acta Crystallogr., Sect. B.* **B32**, 982–983 (1976).

181. M. B. Cingi, A. M. M. Lanfredi, A. Tiripicchio, and M. T. Camellini, *Acta Crystallogr., Sect. B.* **B33**, 659–664 (1977).

182. N. J. Mammano, D. H. Templeton, and A. Zalkin, *Acta Crystallogr., Sect. B* **B33**, 1251–1254 (1977).

183. F. Pavelčik and J. Majer, *Z. Naturforsch., B, Anorg. Chem., Org. Chem.* **32B**, 1089–1090 F. Pavelčik, J. Garaj, and J. Majer, *Acta Crystallogr., Sect. B* **B36**, 2152–2154 (1980).

184. F. T. Helm, W. H. Watson, D. J. Radanović, B. E. Douglas, *Inorg. Chem.* **16**, 2351–2354 (1977).

185. F. Pavelčik and J. Majer, *Acta Crytallogr., Sect. B* **B34**, 3582–3585 (1978).

186. R. E. Ginsburg, J. M. Berg. R. K. Rothrock, J. P. Collman, K. O. Hodgson, and L. F. Dahl, *J. Am. Chem. Soc.* **101**, 7218–7231 (1979).

187. G. Fachinetti, C. Floriani, P. F. Zanazzi, and A. R. Zanzari, *Inorg. Chem.* **18**, 3469–3475 (1979).

188. F. Benetollo, G. Bombieri, J. A. Herrero, and R. M. Rojas, *J. Inorg. Nucl. Chem.* **41**, 195–199 (1979).

189. H.-N. Adams, G. Fachinetti, and J. Strähle, *Angew. Chem.* **92**, 411–412 (1980); *Angew. Chem., Int. Ed. Engl.* **19**, 404–405 (1980).

190. T. V. Filippova, T. N. Polynova, A. L. Il'inskii, M. A. Porai-Koshits, and L. I. Martynenko, *Zh. Neorg. Khim.* **26**, 1140–1141 (1981); and *Chem. Abstr.* **94**, 201161t. (1981).

191. M. G. Down, M. J. Haley, P. Hubbertstey, R. J. Pulham, and A. E. Thunder, *J. Chem. Soc., Chem. Commun.* pp. 52–53 (1978); *J. Chem. Soc., Dalton Trans.* pp. 1407–1411 (1978).

192. P. L. Watson, J. F. Whitney, and R. L. Harlow, *Inorg. Chem.* **20**, 3271–3278 (1981).

193. M. F. Lappert, A. Singh, J. L. Atwood, and W. E. Hunter, *J. Chem. Soc., Commun.* pp. 1191–1193 (1981).

194. P. G. Edwards, G. Wilkinson, M. B. Hursthouse, and K. M. Abdul Malik, *J. Chem. Soc., Dalton Trans.* pp. 2467–2475 (1980).

195. J. Browning, G. W. Bushnell, and K. R. Dixon, *Inorg. Chem.* **20**, 3912–3918 (1981).

196. K. Pörschke, G. Wilke, and C. Krüger, *Angew. Chem.* **95**, 564–565 (1983); *Angew. Chem., Int. Ed. Engl.* **22**, 547–548 (1983).

197. J. Powell, K. S. Ng, W. W. Ng, and S. C. Nyburg, *J. Organomet. Chem.* **243**, C1–C4 (1983).

198. M. Cesari, G. Perego, G. Del Piero, M. Corbellini, and A. Immirzi, *J. Organometl. Chem.* **87**, 43–52 (1975).

199. G. Perego and G. Dozzi, *J. Organomet. Chem.* **205**, 21–30 (1981).

200. J. G. Reynolds, A. Zalkin, D. H. Templeton, and N. M. Edelstein, *Inorg. Chem.* **16**, 1090–1096 (1977).

201. F. A. Cotton and S. Koch, *Inorg. Chem.* **17**, 2021–2024 (1978).

202. M. Bochmann, G. Wilkinson, G. B. Young, M. B. Hursthouse, and K. M. Abdul Malik, *J. Chem. Soc. Dalton Trans.* pp. 1863–1871 (1980).

203. F. Maury and A. Gleizes, *Inorg. Chim. Acta* **41**, 185–194 (1980).

204. J. Powell, A. Kuksis, C. J. May, S. C. Nyburg. and S. J. Smith, *J. Am. Chem. Soc.* **103**, 5941–5943 (1981).

205. B. Müller and J. Krausse, *J. Organometl. Chem.* **44**, 141–159 (1972).
206. C. A. Secaur, V. W. Day, R. D. Ernst, W. J. Kennelly, and T. J. Marks, *J. Am. Chem. Soc.* **98**, 3713–3715 (1976).
207. R. M. Metzger, N. E. Heimer, C. S. Kuo, R. F. Williamson, and W. O. J. Boo, *Inorg. Chem.* **22**, 1060–1064 (1983).
208. C. Svensson, J. Albertsson, R. Liminga, Å. Kvick and S. C. Abrahams, *J. Chem. Phys.* **78**, 7343 (1983).
209. A. Kobayashi, Y. Sasaki, H. Kobayashi, A. E. Underhill, and M. M. Ahmad, *J. Chem. Soc., Chem. Commun.* pp. 390–391 (1982).
210. J. Hooz, S. Akiyama, F. J. Cedar, M. J. Bennett, and R. M. Tuggle, *J. Am. Chem. Soc.* **96**, 274–276 (1974).
211. L. M. Engelhardt, G. E. Jacobsen, C. L. Raston, and A. H. White, *J. Chem. Soc., Chem. Commun.* pp. 220–222 (1984).
212. H. Hope, M. M. Olmstead, P. P. Power, and X. Xu, *J. Am. Chem. Soc.* **106**, 819–821 (1984); M. M. Olmstead and P. P. Power, *ibid.* **107**, 2174–2175 (1985).
213. P. P. Power and X. Xiaojie, *J. Chem. Soc. Chem. Commun.* pp. 358–359 (1984).
214. G. Shoham, D. W. Christianson, R. A. Bartsch, G. S. Heo, U. Olsher, and W. N. Lipscomb, *J. Am. Chem. Soc.* **106**, 1280–1285 (1984).

SELECTED BIBLIOGRAPHY

D. R. Armstrong and P. G. Perkins, Calculations on the electronic structures of organometallic compounds and homogeneous catalytic processes. Part I. Main group organometallic compounds *Coord. Chem. Rev.* **38**, 139–175 (1981).

R. B. Bates and C. A. Ogle, "Carbanion Chemistry," Springer-Verlag; and New York, Berlin 1983.

P. Beak and D. B. Reitz, Dipole-stabilized carbanions: Novel and useful intermediates; *Chem. Rev.* **78**, 275–316 (1978).

T. L. Brown, The structures of organolithium compounds. *Adv. Organomet. Chem.* **3**, 365–395 (1965).

T. L. Brown, Structures and reactivities of organolithium compounds. *Pure Appl. Chem.* **23**, 447–462 (1970).

E. Buncel and T. Durst, eds., "Comprehensive Carbanion Chemistry." Elsevier, Amsterdam: Part A, Structure and Reactivity, 1980; Part B, Selectivity in Carbon–Carbon Bond-forming Reactions, 1984.

G. E. Coates, M. L. H. Greene, and K. Wade, "Organometallic Compounds," Vol. 1, Chapter 1 Methuen; London, 1967.

D. J. Cram, "Fundamentals of Carbanion Chemistry." Academic Press; New York, 1965.

H. Gilman and J. W. Morton, Jr., The metallation reaction with organolithium compounds. *Org. React.* **8**, 258–304 (1954).

H. W. Gschwend and H. R. Rodriguez, Heteroatom-facilitated lithiations. *Org. React.* **26**, 1–360 (1976).

P. Hubberstey, Elements of Group I. *Coord. Chem. Rev.* **56**, 1–77 (1984); **49**, 1–75 (1983); **40**, 1–63 (1982); **34**, 1–49 (1981); **30**, 1–51 (1979).

K. Jonas, *Adv. Organomet. Chem.* **19**, 97–122 (1981).

K. Jonas and C. Krüger, *Angew. Chem.* **92**, 513–531 (1980); *Angew. Chem. Int. Ed. Engl.* **19**, 520–537 (1980).

R. G. Jones and H. Gilman, The halogen-metal interconversion reaction with organolithium compounds. *Org. React.* **6**, 339–366 (1951).

E. M. Kaiser, Lithium Annual surveys. *J. Organomet. Chem.* **227**, 1–134 (1982); **203**, 1148 (1980); **183**, 101–105 (1979); **158**, 1–92 (1978); **143**, 1–97 (1977); **130**, 1–131 (1976); **98**, 1–115 (1975); **95**, 1–137 (1974).

E. M. Kaiser, J. D. Petty and P. L. A. Knutson, Di- and polyalkali metal derivatives of heterofunctionally substituted organic molecules. *Synthesis* pp. 509–550 (1977).

J. Klein, Directive effects in allylic and benzylic polymethylations: The question of U-stabilization, Y-aromaticity and cross-conjugation. *Tetrahedron* **39**, 2733–2759 (1983).

A. Krief, Synthetic methods using α-heterosubstituted organometallics. *Tetrahedron* **36**, 2531–2640 (1980).

A. W. Langer, Polyamine-chelated alkali metal compounds. *Adv. Chem. Ser.* No. 130. *Am. Chem. Soc.*, New York, 1974.

J. M. Mallan and R. L. Bebb, Metallations by organolithium compounds. *Chem. Rev.* **69**, 693–755 (1969).

S. F. Martin, Synthesis of aldehydes, ketones, and carboxylic acids from lower carbonyl compounds by C-C coupling reactions. *Synthesis* pp. 633–665 (1979).

N. S. Narasimhan, and R. S. Mali, Syntheses of heterocyclic compounds involving aromatic lithiation reactions in the key step. *Synthesis* pp. 957–986 (1983).

J. P. Oliver, Structures of main group organometallic compounds containing electron-deficient bridge bonds. *Adv. Organomet. Chem.* **15**, 235–271 (1977).

N. Petragnani, M. Yonashiro, The reactions of dianions of carboxylic acids and ester enolates. *Synthesis* pp. 521–578 (1982).

D. Seebach, Methods of reactivity umpolung. *Angew. Chem., Int. Ed. Engl.* **18**, 239–258 (1979).

M. Schlosser, "Polare Organometalle." Springer-Verlag, Berlin and New York, 1973

U. Schöllkopf, Methoden zur Herstellung und Umwandlung von Lithium-Organischen Verbindungen. In "Houben-Weyl's Methoden der organischen Chemie, 11 Vol. 13, Part I, pp. 87–253. Thieme, Stuttgart, 1970.

K. Smith, Lithiation and organic synthesis. *Chem. Bro.* pp. 29–32 (1982).

J. C. Stowell, "Carbanions in Organic Synthesis." Wiley, New York, 1979.

K. Wade, "Electron Deficient Compounds." Nelson, London, 1971.

B. J. Wakefield, "The Chemistry of Organolithium Compounds." Pergamon, Oxford, 1974.

J. L. Wardell, Alkali metals. In "Comprehensive Organometallic Chemistry" (G. Wilkinson, F. G. A. Stone, and F. W. Abel eds.), Vol. 1, Chapter 2. Pergamon Oxford 1982.

Index

Cumulative List of Contributors

Abel, E. W., **5**, 1; **8**, 117
Aguilo, A., **5**, 321
Albano, V. G., **14**, 285
Alper, H., **19**, 183
Anderson, G. K., **20**, 39
Armitage, D. A., **5**, 1
Armor, J. N., **19**, 1
Atwell, W. H., **4**, 1
Behrens, H., **18**, 1
Bennett, M. A., **4**, 353
Birmingham, J., **2**, 365
Blinka, T. A., **23**, 193
Bogdanović, B., **17**, 105
Bradley, J. S., **22**, 1
Brinckman, F. E., **20**, 313
Brook, A. G., **7**, 95
Brown, H. C., **11**, 1
Brown, T. L., **3**, 365
Bruce, M. I., **6**, 273; **10**, 273; **11**, 447; **12**, 379; **22**, 59
Brunner, H., **18**, 151
Cais, M., **8**, 211
Calderon, N., **17**, 449
Callahan, K. P., **14**, 145
Cartledge, F. K., **4**, 1
Chalk, A. J., **6**, 119
Chatt, J., **12**, 1
Chini, P., **14**, 285
Chiusoli, G. P., **17**, 195
Churchill, M. R., **5**, 93
Coates, G. E., **9**, 195
Collman, J. P., **7**, 53
Connelly, N. G., **23**, 1; **24**, 87
Connolly, J. W., **19**, 123
Corey, J. Y., **13**, 139
Corriu, R. J. P., **20**, 265
Courtney, A., **16**, 241
Coutts, R. S. P., **9**, 135
Coyle, T. D., **10**, 237
Craig, P. J., **11**, 331
Cullen, W. R., **4**, 145
Cundy, C. S., **11**, 253
Curtis, M. D., **19**, 213
Darensbourg, D. J., **21**, 113; **22**, 129
de Boer, E., **2**, 115
Dessy, R. E., **4**, 267
Dickson, R. S., **12**, 323
Eisch, J. J., **16**, 67

Emerson, G. F., **1**, 1
Epstein, P. S., **19**, 213
Erker, G. **24**, 1
Ernst, C. R., **10**, 79
Evans, J., **16**, 319
Evans, W. J., **24**, 131
Faller, J. W., **16**, 211
Fehlner, T. P., **21**, 57
Fessenden, J. S., **18**, 275
Fessenden, R. J., **18**, 275
Fischer, E. O., **14**, 1
Forster, D., **17**, 255
Fraser, P. J., **12**, 323
Fritz, H. P., **1**, 239
Furukawa, J., **12**, 83
Fuson, R. C., **1**, 221
Garrou, P. E., **23**, 95
Geiger, W. E., **23**, 1; **24**, 87
Geoffroy, G. L., **18**, 207; **24**, 249
Gilman, H., **1**, 89; **4**, 1; **7**, 1
Gladfelter, W. L., **18**, 207, **24**, 41
Gladysz, J. A., **20**, 1
Green, M. L. H., **2**, 325
Griffith, W. P., **7**, 211
Grovenstein, Jr., E., **16**, 167
Gubin, S. P., **10**, 347
Guerin, C., **20**, 265
Gysling, H., **9**, 361
Haiduc, I., **15**, 113
Halasa, A. F., **18**, 55
Harrod, J. F., **6**, 119
Hart, W. P., **21**, 1
Hartley, F. H., **15**, 189
Hawthorne, M. F., **14**, 145
Heck, R. F., **4**, 243
Heimbach, P., **8**, 29
Helmer, B. J., **23**, 193
Henry, P. M., **13**, 363
Herrmann, W. A., **20**, 159
Hieber, W., **8**, 1
Hill, E. A., **16**, 131
Hoff, C., **19**, 123
Horwitz, C. P., **23**, 219
Housecroft, C. E., **21**, 57
Huang, Yaozeng (Huang, Y. Z.), **20**, 115
Ibers, J. A., **14**, 33
Ishikawa, M., **19**, 51
Ittel, S. D., **14**, 33

464

Cumulative List of Titles